ATOMIC PHYSICS 17

Previous Proceedings in the Series of Conferences on Atomic Physics

Year	Held in	Publisher	ISBN
16	1998 Windsor, Ontario, Canada	AIP Conf. Proceedings vol. 477	1-56396-752-9
15	1996 Amsterdam, The Netherlands	World Scientific	98-102-3186-5
14	1994 Boulder, Colorado	AIP Conf. Proceedings vol. 323	1-56396-348-5
13	1992 Munich, Germany	AIP Conf. Proceedings vol. 275	1-56396-057-5
12	1990 Ann Arbor, Michigan	AIP Conf. Proceedings vol. 233	0-88318-811-2

Other Related Titles from AIP Conference Proceedings

547   Atomic Processes in Plasmas: Twelfth Topical Conference
Edited by Roberto C. Mancini and Ronald A. Phaneuf, December 2000, 1-56396-976-9

543   Atomic and Molecular Data and Their Applications: ICAMDATA—Second International Conference
Edited by Keith A. Berrington and Kenneth L. Bell, November 2000, 1-56396-971-8

535   Fundamental Physics of Ferroelectrics 2000: Aspen Center for Physics Winter Workshop
Edited by Ronald E. Cohen, September 2000, 1-56396-959-9

525   Multiphoton Processes: ICOMP VIII: 8th International Conference
Edited by Louis F. DiMauro, Richard R. Freeman, and Kenneth Kulander, June 2000, 1-56396-946-7

506   X-Ray and Inner Shell Processes: 18th International Conference
Edited by R. W. Dunford, D. S. Gemmell, E. P. Kanter, B. Krässig, S. H. Southworth, and L. Young, February 2000, 1-56396-713-8

500   The Physics of Electronic and Atomic Collisions: XXI International Conference
Edited by Yukikazu Itikawa, Kazuhiko Okuno, Hiroshi Tanaka, Akira Yagishita, and Michio Matsuzawa, February 2000, 1-56396-777-4

467   Spectral Line Shapes: Volume 10, 14th ICSLS
Edited by Roger M. Herman, March 1999, 1-56396-754-5

To learn more about these titles, or the AIP Conference Proceedings Series, please visit the webpage **http://www.aip.org/catalog/aboutconf.html**

ATOMIC PHYSICS 17

XVII International Conference
ICAP 2000

Florence, Italy   4–9 June 2000

EDITORS

Ennio Arimondo
Università di Pisa, Pisa, Italy

Paolo De Natale
INOA, Florence, Italy

Massimo Inguscio
LENS, Università di Firenze, Florence, Italy

Melville, New York, 2001
AIP CONFERENCE PROCEEDINGS ■ VOLUME 551

Editors:

Ennio Arimondo
Dipartimento di Fisica
Università di Fisica
Via Buonarroti 2
I-56126 Pisa
ITALY

E-mail: arimondo@mail.df.unipi.it

Paolo De Natale
Istituto Nazionale di Ottica Applicata
Largo E. Fermi 6
I-50125 Firenze
ITALY

E-mail: denatale@ino.it

Massimo Inguscio
Dipartimento di Fisica
e LENS, Università di Firenze
Largo E. Fermi 2
I-50125 Firenze
ITALY

E-mail: inguscio@lens.unifi.it

The articles on pp. 173–186, 204–217, and 367–381 were authored by U. S. Government employees and are not covered by the below mentioned copyright.

L.C. Catalog Card No. 00-111558
ISBN 1-56396-982-3
ISSN 0094-243X
Printed in the United States of America

CONTENTS

CLOCKS

FUNDAMENTAL CONSTANTS

SINGLE SPECIES

ENTANGLEMENT

DARK RESONANCE

QED WITH IONS, ATOMS, AND MUONIC SYSTEMS

GRAVITATIONAL WAVES AND DETECTORS

LASER COOLING AND DEGENERATE MATTER

QUANTUM CONTROL AND COLLISIONAL DYNAMICS

PREFACE

ICAP 2000 closes a century that began with the introduction of Planck's constant and has seen a revolution in the understanding of the structure of matter. Atomic physics has played a crucial role for the development of quantum mechanics and has also represented the natural ground to test the new theory and to form the basis for a description of more complex systems.

The scientific program of ICAP 2000 demonstrates that, keeping up with the continuous progress in the last century, atomic physics remains a frontier research field, maintaining its central role for understanding the fundamental laws of nature.

The conference program consisted of 40 invited talks, all presented in plenary sessions, and 285 posters divided into 10 sessions. The papers collected in this volume are the written contributions from most of the speakers and represent the cutting edge of research in atomic physics. The 19 sessions of ICAP 2000 covered many topics in very rapid development, such as the explosion of results following the experimental demonstration of Bose-Einstein condensation, including the tuning of the interactions and the evidence of the quantization by "h" of the angular momentum of vortices. Tremendous experimental and theoretical progress has also been registered in the study of entanglement, which is opening new perspectives in quantum computing and cryptography. Frontier experiments in gravitational waves detection were presented, as well as the recent revolution in optical frequency metrology due to the technological advances in femtosecond laser sources and optical fibers, with profound implications for the precise determination of fundamental constants.

ICAP 2000 was organized by conference co-chairs Ennio Arimondo of the University of Pisa and Massimo Inguscio of the European Laboratory for Nonlinear Spectroscopy and University of Firenze in the unique setting of Santa Maria Novella, in Firenze. The conference program, which included contributions from approximately 380 participants coming from 33 countries, was determined by the Program Committee: V. Bagnato, R. Barbieri, R. Blatt, J. Dalibard, P. Hannaford, T. W. Hänsch, E. Hinds, I. Lindgren, P. Meystre, T. Yabuzaki, D. Wineland, and E. Zavattini. Policy guidance as well as numerous suggestions relating to the scientific program were provided by the International Advisory Committee: E. Arimondo, V. I. Balykin, S. Chu, C. Cohen-Tannoudji, G. Drake, N. Fortson, T. W. Hänsch, S. Haroche, V.W. Hughes, M. Inguscio, W. Ketterle, D. Kleppner, K. Kulander, R. R. Lewis, I. Martison, H. Narumi, E. Otten, D. Pritchard, P. G. H. Sandars, F. Shimizu, J. Walraven, C. E. Wieman, and P. Zoller.

In honor of the illustrious citizen Galileo Galilei, who made a fundamental contribution to the measurement of time, a public session on clocks, chaired by Norman Ramsey, was held in the magnificent Salone dei Cinquecento, in the historical building of Palazzo Vecchio.

All poster sessions took place in the fourteenth-century Great Cloister of Santa Maria Novella. All of Santa Maria Novella is nowadays the seat of the Scuola Marescialli e Brigadieri dei Carabinieri, and a special acknowledgment goes to Generale Giuseppe De Gregorio, chief of this military school, who kindly hosted the conference sessions, and to Maresciallo Ennio Robbio for directing the concert played by the Fanfara dei Carabinieri at the conference dinner.

Conference activities were arranged by the other members of the Local Organizing Committee: F. Corsi, M. Davini, P. De Natale, and F. Fuso. Special thanks are also due to the secretary, Donatella Perri, and to Cristina Cipullo, Ilaria Gallotta, and Maila Lanzini of the administrative staff of LENS.
Giovanna Rasario, who is the author of the nice illustrations and of the conference logo, is warmly acknowledged, too.

We gratefully acknowledge the sponsorship of the following institutions: Banca Toscana, Comune di Firenze, Dipartimento di Fisica dell'Università di Firenze, Dipartimento di Fisica dell'Università di Pisa, European Optical Society, European Physical Society, Gruppo Nazionale di Struttura della Materia del CNR, Istituto e Museo di Storia della Scienza, International Union for Pure and Applied Physics, Istituto Nazionale per Fisica della Materia—Sezione Fisica Atomica e Molecolare, Istituto Nazionale di Fisica Nucleare—Sezione di Firenze, Istituto Nazionale di Ottica Applicata, Museo di S. Marco di Firenze, Scuola Marescialli e Brigadieri dei Carabinieri di Firenze, Soprintendenza per i Beni Artistici e Storici di Firenze, Università di Firenze, and Università di Pisa.

Ennio Arimondo
Paolo De Natale
Massimo Inguscio

CLOCKS

The image on the following page reproduces the logo especially designed by Giovanna Rasario for ICAP 2000. In the logo are neatly integrated three crucial documents which allude to the fundamental contribution made by Galileo to perfecting instruments for the measurement of time and to conceiving a new method for solving the quest for longitude. The background image displays one of the many autograph pages in which, as from January 7th, 1610, the Pisan scientist systematically recorded the appearances of Jupiter and of its moons. As Galileo soon perceived, this newly discovered corner of the Universe could be used as a potentially perfect clock. At the centre one observes an elegant curved structure which evokes the refined ebony and ivory frame in which the Medici (to whom the Jupiter moons were dedicated by Galileo) inserted the objective lens of the telescope used in Galileo's early and tremendously successful exploration of the heavens. Finally, on the right, is the reproduction of the sketch of the pendulum-regulated clock which Galileo conceived to improve precision in keeping local time. He rightly believed that this new device could help to solve the quest for longitude.

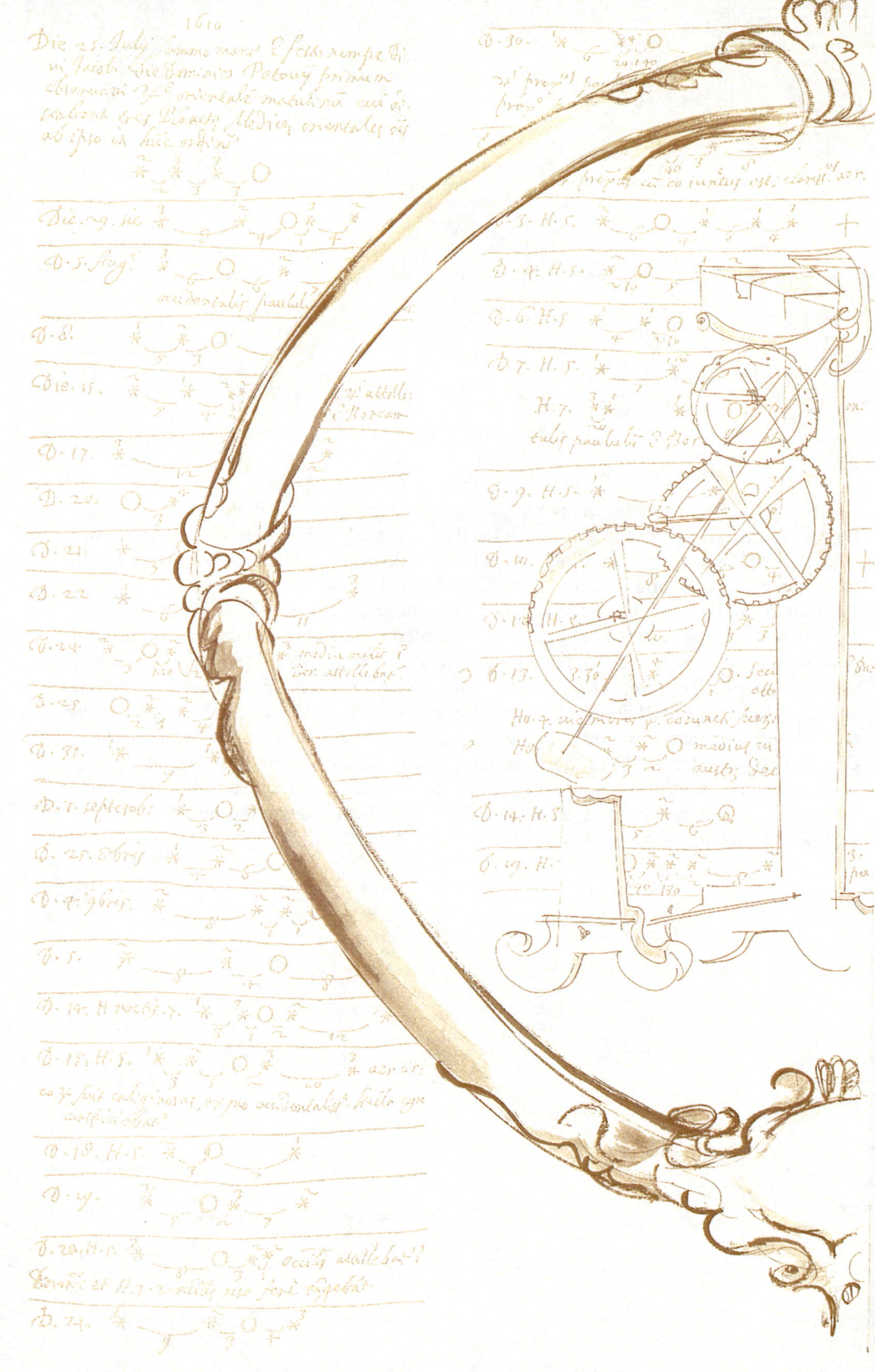

Galileo's Clocks

Paolo Galluzzi

Istituto e Museo di Storia della Scienza, Firenze
Università degli Studi di Firenze

Historians of the Scientific Revolution, the intellectual adventure which produced - from Copernicus to Newton - a radically new conception of the natural world, have rarely given adequate attention to the many crucial consequences of one of Galileo's most sensational discoveries. I refer to the discovery of the four moons around Jupiter, which Galileo, thanks to his telescope [FIG. 1], detected for the first time on the magic night of January 7th [FIG. 2], 1610, in Padua, where he was teaching Mathematics at the local Studium.

As Galileo immediately perceived, the satellites of Jupiter seemed to offer strong evidence in support of the heliocentric system, which, more than 60 years after the publication of Copernicus' *De revolutionibus orbium coelestium* (1543), was still largely considered as an untenable and even absurd cosmological hypothesis. The four companions of Jupiter confirmed to Galileo that Earth centred motions did not constitute a strict rule for all heavenly bodies. Moreover, the campaign of systematic observations of the satellites in which he was engaged, soon convinced Galileo that only once the Sun, and not the Earth, was taken as the Jovial system's point of reference, the satellites' periods and their eclipses displayed perfectly regular rhythms.

Thanks to the contributions by Galileo and by many other gifted scholars, this celestial novelty stimulated, throughout the following decades, a process of radical reformation of traditional knowledge in many fields: from the continuous improvement of the instruments to observe and compute celestial phenomena, to the formulation of revolutionary hypotheses on the physical causes of the motions and on the orbits of the planets; from the discovery of the finite velocity of light, to the scientific foundation of geography and cartography; from the transformation of the traditional organisation of research, with the birth of the first State-founded scientific institutions to which finalised goals were assigned, to the solution of fundamental problems, like the invention of precise clocks, the detection of the cause of variations in gravity with latitude, and - last but not least - the determination of longitude.

It has also to be noted that Galileo's dedication of the Jupiter satellites to the Medici family introduced a new chapter in the history of Renaissance patronage. From now on influential patrons will consider the heavens a promising space for their image-policies. At the same time, scientists will enjoy hitherto unheard of possibilities of social promotion, of high economic reward, and of dignified employment. In a word, the moons of Jupiter contributed a great deal towards stimulating interested public support for the development of scientific research.

After Galileo's discovery, the mysterious and moody satellites of Jupiter attracted the attention of the major protagonists of the Scientific Revolution: Pierre Gassendi and Isaac Newton, Christiaan Huygens and Giovan Domenico Cassini, Ole Roemer and John Flamsteed, Giovanni Alfonso Borelli and Robert Hooke.

CP551, *Atomic Physics 17,* edited by E. Arimondo, P. DeNatale, and M. Inguscio
© 2001 American Institute of Physics 1-56396-982-3/01/$18.00

FIGURE 1. The ebony and ivory trophy encompassing the objective lens of the telescope with which Galileo discovered the Jupiter moons.

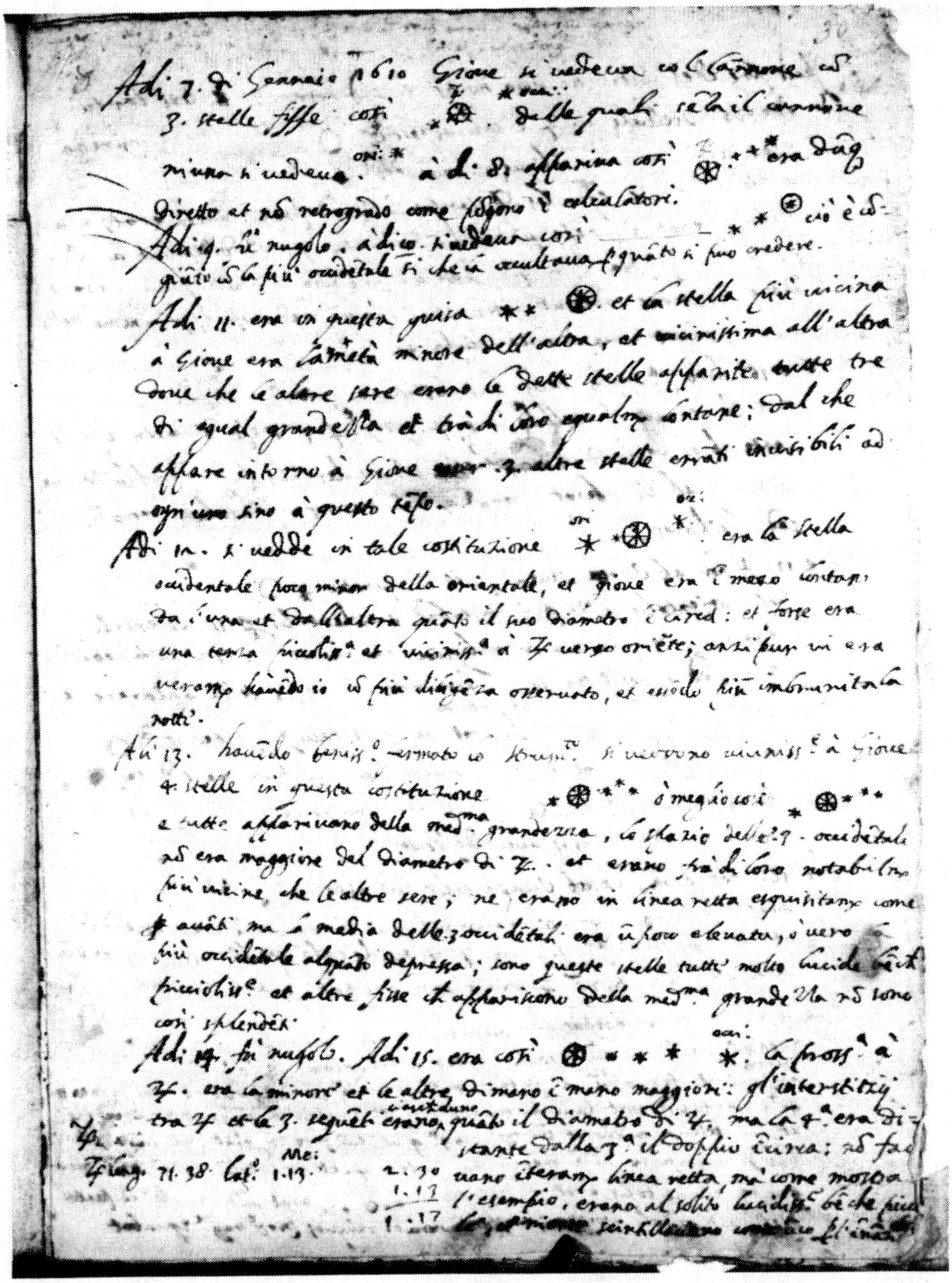

FIGURE 2. Galilean autograph records of his observation of Jupiter's satellites, starting from January 7th, 1610. Biblioteca Nazionale Centrale, Florence, Ms. Gal. 48 (Div. 2a, P. III, t. 3), c.30r.

Among many other exciting possibilities, the discovery of Jupiter's moons opened up new perspectives for the many individuals and public bodies who had been engaged for decades in the search for a viable method of determining longitude.

When Galileo directed his telescope towards Jupiter, the dramatic quest for longitude had an ultra-centenary history and boasted a substantial number of different proposals, none of which had ever proved satisfactory.

The difficulties in determining longitude were known to classic authors. But after the discovery of the New World, the quest for longitude became a fundamental need. Since the first decades of the Sixteenth century, an increasing number of ships had set out to cross the Atlantic Ocean, where no fixed points of reference were available. As time went on, other immense oceans were explored by the fleets of the major European nations in the search for precious goods. In small-scale navigation, like that in the Mediterranean or in the seas of Northern Europe, it was enough, in order to secure a satisfactory degree of orientation, to have a good knowledge of coastal profiles (provided by special maps, called *portolani*), to be able to determine latitudes by measuring the altitude of the Sun at noon with the cross staff and/or with the nautical astrolabe, and to have a magnetic compass on board. These techniques were not sufficient in oceanic navigation, which demanded the precise determination of longitude.

The lack of effective methods for determining longitude produced dramatic consequences both for the seamen and for their vessels, often cargoed with precious goods.

In the best of events, the voyage took many months to complete. As a consequence food and fresh water became scarce, the precarious hygienic conditions resulted in the proliferation of infectious diseases, and the prolonged lack of fruit and fresh vegetables in the seamen's diet led to the disastrous consequences of scurvy. Beyond the enormous economical damage, the price paid in the loss of human life was so enormous that in order to assemble a crew it was necessary to resort to men who had been sentenced to imprisonment or to death for serious crimes. In a word, the exploitation of the Eldorado of the New World was seriously hindered by the incapacity of providing an appropriate solution to the cursed longitude affair. This helps to explain why the European nations invested a great deal of energy and resources into the quest for longitude.

Before a viable solution was found, the methods at hand were based purely on the seamen's intuitions and experience. The conduct of Columbus, Vespucci and early ocean explorers has been labelled as the "dead reckoning" method, depending mainly on a rough evaluation of the distance covered daily by the ship. This was obtained by periodically estimating the velocity of the ship by dropping a log, connected to a regularly knotted line, from the prow of the vessel. The seamen counted how many knots of the line were passed by the ship within a given time (measured by a sand glass), thus making an estimate of the ship's velocity. It is worth remembering that this is the origin of the custom of measuring a ship's velocity and covered distances in "knots"! Periodical single observations of the Sun and the Stars, using the quadrant, the cross-staff and the nautical astrolabe, helped to correct the empirical estimates.

Another method largely employed relied upon the use of the compass by plotting of the declination of the needle of the magnetic compass from true North. But since no simple relationship between magnetic declination and change of longitude was known, this method declined progressively and was abandoned after c. 1635, when it was discovered that magnetic declination at any given location varies with time.

At the beginning of the 16th century, Johann Werner of Nuremberg, and later on the Imperial Astronomer , Peter Apian, recommended using the Moon as a clock [FIG. 3] by measuring its distance from the main stars and/or from the Sun. The idea was excellent, but the impossibility of predicting with any accuracy the position of the Moon made the "lunar distances" method practically useless. Another lunar method depended on the eclipses of the Moon. But apart from the extreme difficulties encountered in timing precisely the eclipses, these phenomena were too rare to offer a solution to a problem which demanded frequent checks and corrections.

The Belgian mathematician and instrument maker Gemma Frisius was the first, in 1530, to advance the much simpler idea of carrying standard time on the ships with a clock. In this way, the difference between local time (established by observing the altitude of the Sun at noon) and standard time would have immediately given longitude. The method, theoretically impeccable and extremely easy to carry out, proved once again impractical: there was no clock - and there would not be one for a long time to come - able to keep standard time with sufficient precision over a long period and under all conditions.

The increasing importance of navigation and a series of maritime disasters as a consequence of the uncertainties of longitude, pushed those nations most seriously involved in world trade, to undertake important initiatives towards solving the problem.

Philip II of Spain, in 1567, offered to the inventor of a practical method of determining longitude a perpetual pension of 6000 ducats. The prize was increased in 1598 by Philip III. In 1600 the Dutch Republic established a prize of 5000 florins. Also the King of France, of Portugal, as well as the Venetian Republic, offered huge rewards. But no practical solution was generated by these incentives.

Such was the frustrating situation when, at the beginning of 1610, Galileo observed for the first time, through the eye-piece of his telescopes, Jupiter, surrounded by 4 satellites. Galileo soon realised that the moons of Jupiter could be used as a celestial clock in the determination of longitude.

This engraving [FIG. 4] from James Ferguson's *Astronomy* effectively illustrates Galileo's idea. He thought of using the eclipses of the moons of Jupiter to determine the difference in longitude between two places on the Earth. In Ferguson's illustration we observe that the eclipse of the innermost satellite K (behind J - Jupiter) is observed from two points on the Earth (R and Q) at precisely the same moment. If the observer standing at Q has tables which allow him to compare the time of the eclipse at point R with his local time , he can easily find the difference in longitude between the two points. If the time at R is 4p.m. and local time at Q is noon, then, because of the difference of 4 hours, point Q is 45° west of R.

It soon became evident that for the success of this operation three major conditions had to be fulfilled. Firstly, it was necessary to have at hand a good telescope to observe the Jupiter system. Secondly, one had to consult tables effectively predicting the immersion and emersion of the moons in and from the shadows of the planet.

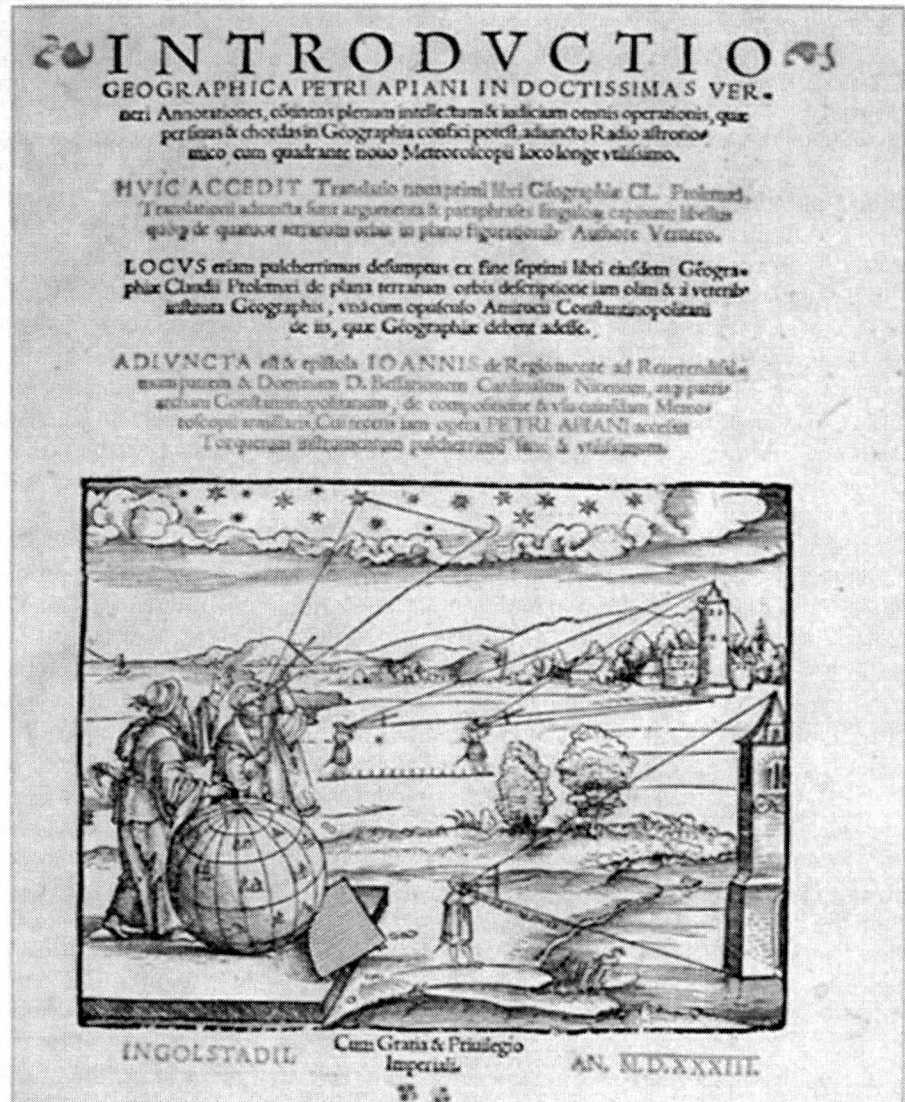

FIGURE 3. The engraving which adorns the front page of Petrus Apianus' *Introductio Geographica* (Ingolstadt, 1533) offers a practical demonstration of his lunar distances method.

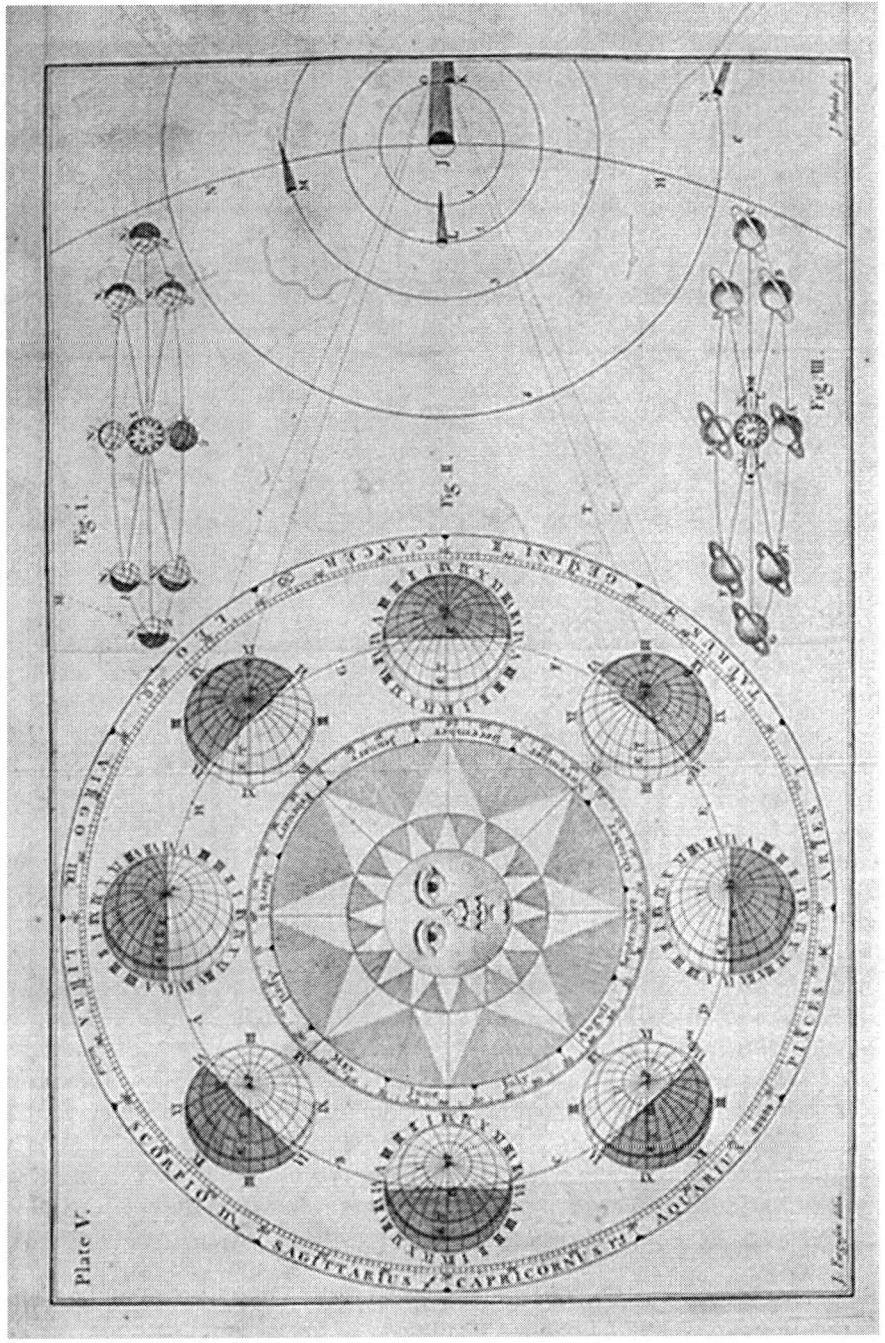

FIGURE 4. Engraving effectively illustrating the effects of prostapheresis in James Ferguson, *Astronomy explained upon Sir Isaac Newton Principles*, London, 1757 [2nd ed.], Plate V.

9

Thirdly, some special system had to be devised to clearly observe the satellites with a long telescope also on the rocking deck of a ship.

Galileo worked intensely during 1610 and 1611 to improve the performances of his telescopes. In this field he achieved excellent results. In fact for at least three decades only the Pisan's telescopes could guarantee a bright and distinct vision of Jupiter and its moons. As to the eclipses of the satellites, Galileo devoted a great deal of time from mid-1611 to their systematic observation with the aim of preparing precise tables. By September 1612 he had determined the periods of the satellites. His autograph records show that, during these months, Galileo had taken into account the prostapheresis, that is the continuously variable angle between the Earth and the Sun (due to the annual motion of the Earth), as seen from Jupiter. Prostapheresis caused the irregular appearance of the periods and eclipses of Jupiter's moons as seen from the Earth. He devised a nomogramm [FIG. 5] called Giovilabio, to make automatic computation of prostapheresis. Later on from Galileo's paper normogramm a neat brass instrument [FIG. 6], a kind of analogic computer, was derived

The quality of the data recorded by Galileo's early moons of Jupiter tables was more than acceptable. He was convinced that the eclipses could be timed with an accuracy of about one minute (corresponding to an error in longitude of only 15'). An excellent outcome, especially when compared with standard errors of other methods at that time, often of more than 4°.

Once he had prepared the tables, Galileo thought of the practical application of his new method. A very important occasion soon presented itself. In 1612, the Great Duke of Tuscany, Cosimo II Medici, was discussing with the Spanish Crown the possibility of Tuscan ships obtaining permission to trade with the West Indies, using as base the harbour of Leghorn. To make the King of Spain's attitude more favourable, the Medici Prince offered to disclose Galileo's method of determining longitude and to train Spanish seamen in it's use. Galileo prepared a formal proposal in which he extolled the superiority of his method when compared to those already in use. Moreover, he offered to construct for the Spanish navy one hundred telescopes that magnified "forty or fifty times" and promised to write out the full instructions for the users of his method.

The discussion with the Spanish authorities lasted a few years and was conducted through the channels of Medici diplomacy. The satellites of Jupiter were considered by the Medici as their personal property (it has not to be forgotten that they were internationally known and quoted as the Medici stars), and the Florentine princes were acting as if they were selling exclusive rights for their practical exploitation. Jupiter's satellites were gaining ever more importance since the modest beginnings of Galileo's private campaigns of celestial observation. The interest in these new celestial bodies on the part of influential princes implied, not only further social promotion for Galileo, but also the emergence of an awareness within the spheres of power of the practical benefits produced by advanced scientific research. During the 17th century, on the wave of the success of the Jupiter satellites, many relevant scientists will obtain a formerly unheard of visibility, reaching an unprecedented level of remuneration through their intellectual work .

Notwithstanding Galileo's full confidence in it, the Jupiter clock proved to be more capricious and moody than had been expected. This was the consequence of many

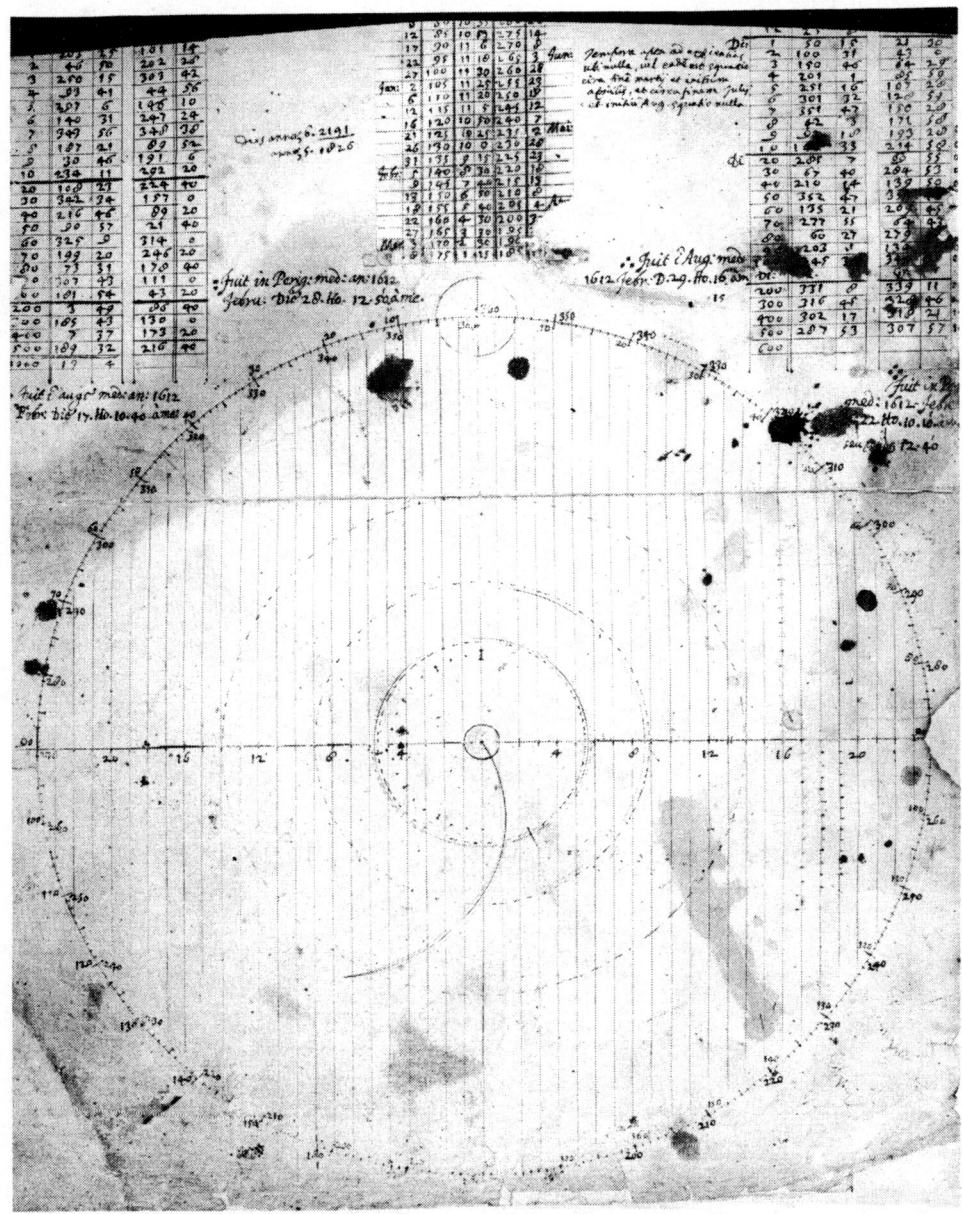

FIGURE 5. Galileo's drawing illustrating his idea for a *giovilabio*. The instrument was conceived to easily compute prostapheresis. Biblioteca Nazionale Centrale, Florence, Ms. Gal. 70 (Div. 2a, P. IV, t. 6), cc. 33-34.

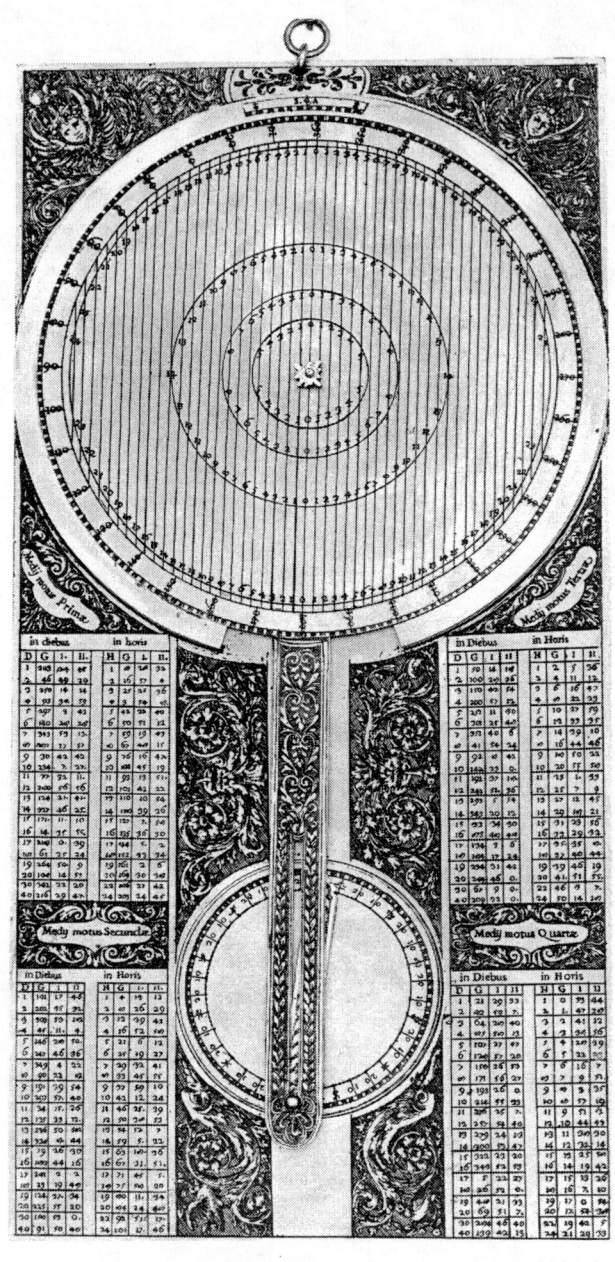

FIGURE 6. The Galilean Giovilabe, Istituto e Museo di Storia della Scienza, Florence, Inv. N. 3178.

different reasons. In fact, in spite of his claims, Galileo's tables could not predict the eclipses of Jupiter's moons with sufficient accuracy. Moreover, the modest quality of the telescopes available at the time made observation of the occultation of satellites very difficult and uncertain. This was mainly due to the fact that the instruments had a very small field of view. This meant that it was difficult to catch the planet and the satellites and almost as difficult to keep them within the field of view. Moreover, Galileo's erroneous conviction that the moon's orbits were circular and traversed with constant velocity did not help the Pisan scientist to accurately time the eclipses. Finally, he was ignorant of the alterations produced by atmospheric refraction on Jupiter appearances, nor he was aware of the perturbing factor of the velocity of light in the exact timing of eclipses, as the consequence of periodical varying of distances between Jupiter and the Earth.

Galileo had also underestimated a few technical problems. When the longitude deal with the Spanish Crown was revived in 1616, he had to confront a major objection that was raised against his method. The Tuscan Ambassador in Madrid, the Count d'Elci, reported this objection directly to Galileo. "In order to put your method into practice it is compulsory and necessary first to see the said stars and their aspects. I do not know how this can be done at sea ...For, leaving aside that the telescope cannot be used in ships because of their motion, even if it could be used, it could serve neither during the day nor during overcast weather at night. But the navigator needs to know hour by hour the degree of longitude at which he is".

Galileo felt the blow. But he did not give up. To make the telescope practically usable on the deck of a rocking ship he designed a special headgear. The headgear, which had to be fixed to the neck and probably to the shoulders, would neutralise, thanks to a kind of Cardanic suspension, the effects of the ship. A short telescope was fitted to the headgear in a way that it remained constantly in line with one of the eyes, while the other eye was left free to easily find Jupiter. This device, of which no graphic description remains, was certainly made and also tested, apparently in a satisfactory way, on Medicean galleys in Leghorn. Evidence remains that Galileo applied to the *celatone* (the Italian name given by Galileo to the tool) a special micrometer, which helped the user to measure, in Jupiter diameters, the distance between the planet and its moons and between the 4 moons. Notwithstanding his claims of having perfected an invention able to provide the final solution to the problem of longitude, the Spanish Crown did not grant the prize to Galileo. Later on, in 1620, and again in 1629 and 1631, negotiations with the Spanish Crown were revived, but always with negative results.

This disappointing conclusion did not discourage the attempts of Galileo and other scientists to disclose the secrets of Jupiter's clock. Galileo's method might have been difficult to put into practice at sea, but was perfectly appropriate to determine with precision longitude on land. And this was not considered a benefit of little importance. The perfect knowledge of the longitude of the coastal profile would have offered tremendous help to navigation.

After the failure of the Spanish deal, Galileo continued periodically to observe the satellites in order to improve the prediction accuracy of his tables. But new problems and interests diverted him from continuous application in this field. He had to face the Church authorities' opposition against his stance in favour of the Copernican system.

He was involved in delicate polemics and he was warned in 1616 against presenting the Copernican cosmology as the true system of the world. He devoted his life to pushing the Roman Church to assume a more open attitude towards the heliocentric hypothesis and he developed fundamental work on the nature and laws of motion, field of research from which he expected the final confirmation of the truth of the Copernican system. These efforts, which produced absolute masterpieces of science and culture, like the *Saggiatore* (*Essayer*), in 1623, and the *Dialogo* (*Dialogue about the two systems of the world*), in 1632, ended in the dramatic trial brought against Galileo in 1633 by the Tribunal of the Inquisition and in his final condemnation for being - as a Copernican - strongly suspected of heresy.

A few months after the condemnation Galileo was authorised to return, "under home arrest", to his Florentine house, the Villa Il Gioiello, at Arcetri. With the help of a young researcher, the Olivetan monk, Vincenzo Renieri, he went back to his old "longitude method". Prevented from celestial observation by the poor state of his eyesight (he will become soon completely blind) he conferred this task to Vincenzo Renieri who perfected Galileo's earlier tables. Unfortunately his work was lost due to the sudden death of the monk in 1647. During those years, Galileo was immersed in the preparation of his last work, the *Discourses and demonstrations on two new sciences,* published in Leyden in 1638, a work destined to radically alter the face of mechanics and of the science of motion. During the preparation of this work, Galileo resumed his early reflections, demonstrations and experiments on pendulums. According to his last disciple, Vincenzo Viviani, Galileo, in 1584, had observed that pendulums of the same length have isochronous oscillations (no matter what the amplitude of their oscillations). In 1602, he produced indirect geometric demonstration of pendulum isochronisms. And a little later, he discovered the proportion between length and periods of pendulums.

When, in 1636, thanks to the mediation of his friend, Elia Diodati, in Paris, Galileo entered new negotiations on the longitude method with the Dutch Republic, he reproposed substantially the same method unluckily submitted to the Spanish court, but with the exception of one fundamental innovation.

Since 1627 Galileo had been aware of the prize of 30,000 scudi established by the States General of the United Provinces of the Netherlands for anyone who could establish a viable method of finding longitude. At the beginning Galileo was reluctant to address the proposal. Apparently he was still convinced that Madrid would have finally accepted it. Encouraged by the fact that the Jury of the Dutch prize was made up of renowned scientists (like the astronomer Abrahams Oretelius, the distinguished cartographer J. Blaew, and the outstanding mathematician and natural philosopher Isaac Beeckman), he decided, in 1636, to compete. It seems that he had also the temptation to leave Italy - where he was living under the conditions of a prisoner - to enjoy freedom in the prosperous and active Amsterdam. The proposal of the longitude method communicated to Amsterdam extolled the high performances of his telescopes and the accuracy of his ephemerides of the appearances of the Jupiter satellites. For making easier observations on the deck of the ship he suggested a special chair, floating in a bath of water or oil, in order to eliminate the perturbing effects of the agitation of the sea.

Galileo's proposal included a totally new element, consisting in the description of a perfectly working time-piece, less subject to external alterations than any other similar instrument. This device was intended by Galileo to keep precisely the local time between two successive astronomical measurements. Galileo's idea of the new time-piece, exploiting the isochronous oscillations of pendulums, was disclosing new promising land. It has to be stressed that the Pisan scientist did not think of the construction of a true clock, capable of precisely transporting standard time across the ocean, as Gemma Frisius had suggested one century before. He intended his time measurer to serve for very short time intervals, only in order to keep local time. It had to work in combination with, not as an alternative to, the astronomical method. The latter would be used to establish the tables thanks to which standard time could be checked and compared with local time. Thus, Galileo's new clock was conceived as an accessory to the true astronomical clock of the Jupiter systems. After so many years of accurate and patient observations, after having promoted world-wide the virtues of Jupiter's moons, he was so involved with these celestial objects that he did not realise that the pendulum offered in itself the potential solution, making the complicated astronomical measurements no longer indispensable. He missed perceiving that a perfectly transportable clock like the one he was proposing to the Dutch could entirely and much more easily solve the problem.

Only a few years later in 1641, one year before his death, Galileo worked out a device, a pin-wheel escapement [FIG. 7], thanks to which the regular oscillations of the pendulum could be recorded by a counter.

Even the Dutch affair became complicated. First, Galileo had to face the suspicious attitude on the part of the Roman Church, which did not consider in a favourable light the fact that Galileo was dealing with representatives of a Reformed nation. Moreover, after their initially promising attitude, the Dutch authorities' interest in the proposal seemed to cool down. Their final negative decision was also affected by the deaths of many of those members of the Jury who had manifested a more open attitude. It seems that longitude was a curse, not only for the poor seamen, but also for those engaged in the search for a solution!

As the final resolution of the States General recites, the main concern was, again, the impracticability of the method at sea. Galileo's proposal required accurate observations and refined computation. On the ships' decks there were not scientists - stressed the members of the Jury - "but sailors, who are rude people, men only superficially acquainted with mathematics and astronomy, who content themselves with those few propositions which are useful to their needs. Men, moreover, who still find insuperable the problem of using this discovery on a moving ship, continuously tossed about".

The failure of the second negotiation of longitude did not dampen either the continual perfecting of the ephemerides of Jupiter's moons, or the attempts to improve mechanical time-measurement - the two directions to which Galileo had made impressive contributions.

Both the celestial and the mechanical clocks witnessed dramatic improvements during the decades following Galileo's death.

FIGURE 7. Drawing made by Vincenzio Galilei, son of Galileo, of his father's conception of the application of pendulum to the clock. Istituto e Museo di Storia della Scienza, Florence, Inv. N. 2433.

The observation work on Jupiter's satellites went on thanks to the efforts, above all, of Giovan Domenico Cassini, who began his brilliant career in Bologna. In 1668 he published improved ephemerides, based on the Bologna meridian. The favourable reception of this work on the part of astronomers throughout Europe earned Cassini the offer of a high salary and a permanent position in Paris at the court of Louis XIV, the Sun King, in 1669. In Paris he continued his work and organised expeditions to observe simultaneously, from different places, celestial phenomena, with the help of competent assistants. Cassini was nominated Director of Paris Observatory, the first permanent astronomical institution, founded by Louis XIV in 1667 with the specific goal of favouring the perfectioning of geographical studies and of finding the final solution for longitude. During his long life Cassini continuously referred to the moons of Jupiter as the perfect clock to determine longitude, at least on land. In fact, during the second half of the 17th century, the continuous improvement of the Galilean method of Jupiter's satellites produced a radical reform of geographical studies. Many expeditions were organised to systematically map the French possessions, both in Europe and in the New World.

Cassini relied on the cooperation of many outstanding scholars. Jean Picard helped him in restructuring the map of the coastal regions of northern France, a project which appears not to have satisfied King Louis XIV who had promoted it. It is reported that the Sovereign, on seeing that the new coast line reduced the surface of the French lands [FIG. 8], lamented that he had lost more territory to his astronomers than to his enemies!

It was while working on the ephemerides of the Jupiter moons by Cassini, that the Danish mathematician Ole Roemer realised that errors in table predictions depended on the finite velocity of light. In a memory read by Roemer in November 1676 in Paris, at the Academie des sciences, he proposed his explanation of the perturbing role of the finite velocity of light in the prediction of eclipses of Jupiter satellites. When Jupiter was at the maximum elongation from the Earth, the eclipses were seen many minutes later than the times predicted by the tables. Working on the estimated distance between Jupiter and the Earth and on the difference in time between ephemerides, predictions of eclipses (based on mean motions of the satellites) and real observation of them, he arrived at the conclusion that the velocity of light is circa 140,000 miles per second (that is about 20% less than its actual value).

It has to be noted that this extraordinary deduction was again a consequence of the discovery of the Jupiter moons. Another major discovery of the workings of Nature had derived from the observation of the Galilean Jupiter clock!

Similar good fortune was encountered by the other great idea related to time measurement envisaged by Galileo: the application of the pendulum to the clock. And its consequences appeared to be of at least equal importance.

As everybody knows, the Dutch scientist Christiaan Huygens put Galileo's invention into practice. He not only actually made and progressively perfectioned the pendulum- regulated clock which Galileo had simply thought of. Huygen's improvements were of such quality and dimension to make untenable the claim of Galileo's priority advanced by the Medici and by the Pisan's disciples in Florence. He also showed with tests and with extraordinary geometrical demonstrations that the oscillations of the Galilean circular pendulum are not really isochronous, the real

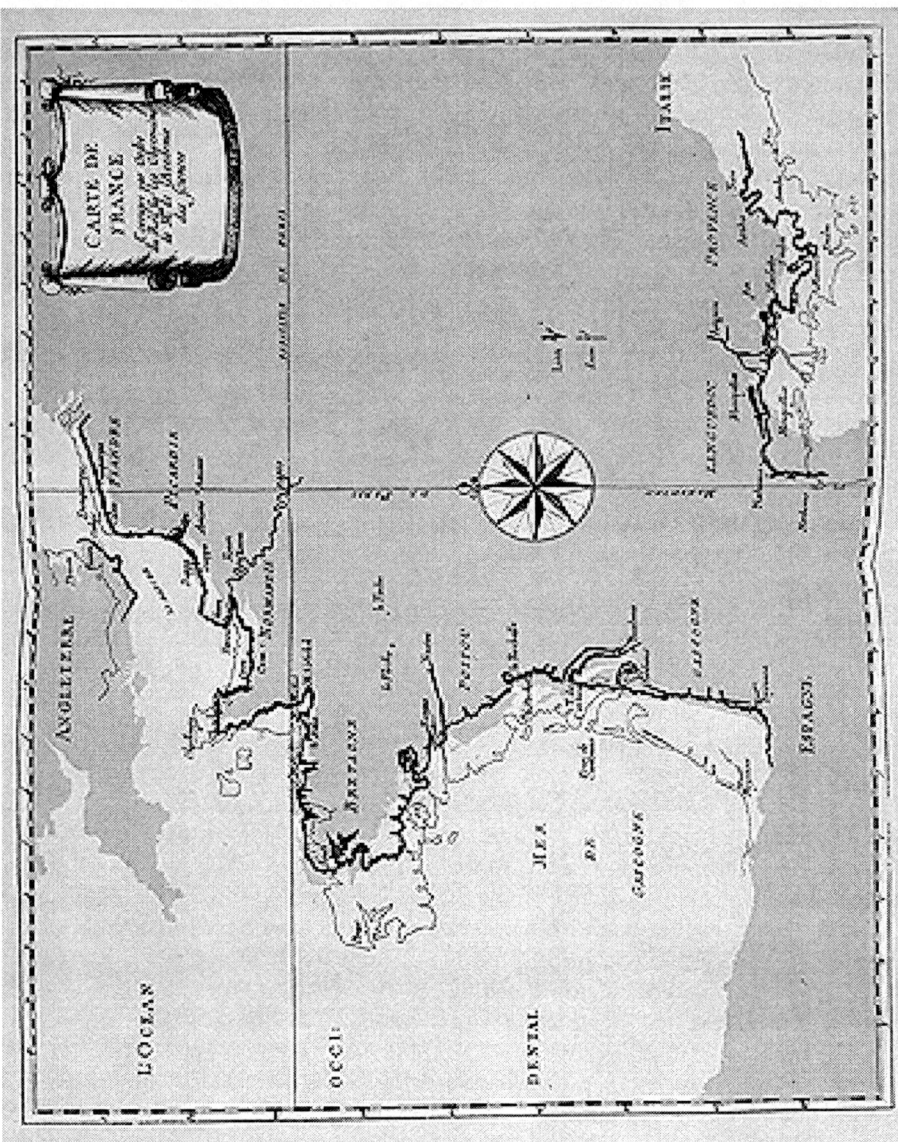

FIGURE 8. The coastline of Northern France retraced (bold line) thanks to accurate surveying by J. Picard and Ph. De La Hire. The map is superimposed upon the earlier map (fine line) by Sanson (1679), showing the dramatic correction introduced thanks to the adoption of the method of Jupiter moons eclipses for determining longitude on land.

isochronous pendulum being the one describing a cycloidal arc. After this important discovery, Huygens used cycloidal pendulums as time regulators [FIG. 9].

Huygens' clock represents an enormous advancement in the history of time measurement. The Dutch scientist was convinced that, thanks to his perfect machines, it was now possible to solve definitively the problems of longitude by purely mechanical means (that is, by keeping standard time on the ship, thanks to a clock). No more observation of lunar distances, of lunar eclipses or of the eclipses of the Jupiter satellites would be necessary.

Huygens' clocks were tested many times in actual navigation. Notwithstanding the care with which they had been designed and built, the results were very disappointing. First of all, Huygens' clock suffered from the alterations in motion, humidity and temperature on the decks of ships. Moreover, the unsatisfactory performances of his clocks contributed to revealing that periods of pendulums (even of the cycloidal pendulums) changed in different regions of the Globe. This evidence convinced Huygens that gravity varies with latitude, another great discovery which we owe to the quest for longitude and of which Newton would demonstrate rigorously the physical cause.

It is common knowledge that the final solution to the dramatic search was provided, not by one of the many outstanding scientists who confronted these problems, but by a modest and almost unlettered craftsman, a British clock-maker, John Harrison. He moved forward along the path traced by Galileo with his research on pendulum isochronisms and by Huygens' proposal of the cycloidal pendulum as time regulator. He worked with tenacity and with great success on making the clock insensible to the physical stress suffered on ships. Harrison, too, carried out a lengthy series of tests, which for a long time yielded disappointing results. He also clashed on many occasions with the famous astronomers of the Royal Observatory at Greenwich, an institution founded by Charles II in 1675 - like the Paris Observatory - with the task of developing observational astronomy for practical purposes and especially to solve the problem of longitude. The Royal Astronomers would not admit that an effective solution could be devised by a craftsman poorly trained in mathematics and ignorant of astronomy. They will strongly oppose Harrison's mechanical proposal, stressing that only Astronomy could offer the final answer.

Harrison had perfectly understood the reason why Huygens' pendulum clocks had failed at sea and he put all of his energy into finding the technical solutions to neutralise the perturbing effects of the regular working of the clock, caused by alterations in temperature and humidity and by the rolling of the vessels. The first Harrison marine time-keeper was completed in 1735 (construction took 5 years). It was tested by the Royal Navy and gave excellent results. But Harrison was not satisfied and went on to a new project. A few years later, the second marine time-keeper was ready. It showed a totally new conception and was more compact. But the outcome of his personal tests was not encouraging and Harrison did not accept that the clock be carried on board ship. He then dedicated himself to the construction of a third chronometer, a project which took 19 years to be completed in 1757.

Many parts of the clock were of a totally new conception and its key organs were made of bi-metallic components in order to compensate automatically for any changes in temperature that could affect the clocks working. Special effort was put into

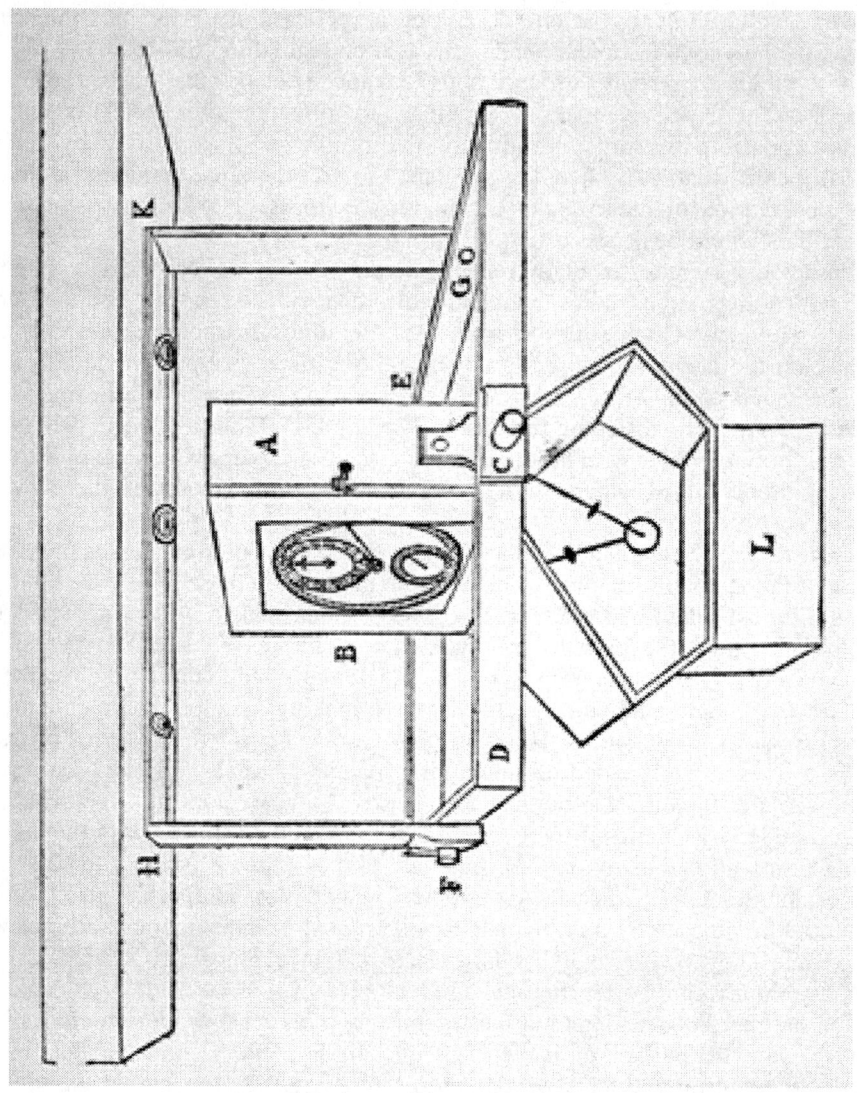

FIGURE 9. *Application of triangular pendulum to clock by C. Huygens (Oeuvres completes, 22 vols., The Hague, 1888-1950, IX, p. 56).*

making the clock an extraordinarily compact instrument (considering the unheard of complexity of its mechanics). This led to the fourth chronograph, completed in 1759. At last perfection had been almost reached!

The clock was tested during a long transoceanic trip by a British ship from Portsmouth to the Antilles and back. William Harrison, son of John, was aboard and took care of the clock. The chronometer surpassed John Harrison's expectations. Back in the harbour of Portsmouth, in little less than three months after departure, it had lost only 5 seconds. The moment had come for the elderly John Harrison to finally cash the first half of the 20.000 pounds prize (an enormous sum!) put aside by the Longitude Act of 1717 for the inventor of an effective method of determining longitude. A few years later he was able to put his hands on the rest of the prize.

Thus, in the long awaited final solution, astronomy took a separate path from that of mechanical clocks. And the latter imposed themselves over astronomical methods. But the challenge would continue and mathematicians and astronomers would soon take their revenge on unlettered clock-makers.

As we have seen, the early history of time-measurement seems to be one of those issues which develops on the borders between theoretical knowledge and practical expertise. The long quest for longitude clearly shows these characteristics. Galileo's clocks proved not to work well – neither the astronomical clock nor the mechanical one. But he had perfectly foreseen the two major paths to be explored and had paved the way both for the astronomer- mathematician and for the clock-maker.

The re-proposal of the crucial role in the history of time measurement of the Jupiter satellites, the planets which Galileo devoted to the Medici family, assumes a particularly evocative meaning in Florence, in the context of an International conference on atomic physics and clocks. This power of evocation seems further emphasized in the room in which this presentation was offered – the Salone dei Cinquecento in Palazzo Vecchio - which embeds the spirit of Cosimo I, the founder of the Great Duchy of Tuscany. Cosimo moved his residence into Palazzo Vecchio, entrusting his distinguished architect Giorgio Vasari with the task of transforming this marvellous Salone into an effective propaganda machine of his enormous ambition. Playing on the derivation of his own name, Cosimo, from the Greek word *Cosmos* [FIG. 10], the Duke intended suggesting that he was destined to become Lord of the Universe. One might say that he foresaw what was to happen a few decades later, when Galileo discovered and dedicated to the Medici the moons of Jupiter. At that point, the Medici would assume the exclusive privilege of being the only dynasty to have a possession in the heavens. A privilege which was to invite the envy of the most powerful monarchs of Europe. Thanks to Galileo's brilliant exploitation of the Jupiter clock, the Medici seemed now to legitimately claim the no less impressive title of Lords of time.

LA SFERA DI
PROCLO LICEO
TRADOTTA DA MAESTRO
Egnatio Danti;
Cofmografo del Sereniffimo Gran
Duca di Tofcana.

Con le Annotazioni, & con l'vfo della Sfera
del medefimo.

IN FIORENZA
Nella Stamperia de' Giunti.
M D L X X I I I.

FIGURE 10. The direct connection between *Cosmos* and *Cosmus* (for Cosimo I Medici) ambiguously proposed on the front page of Egnatio Danti, *La sfera di Proclo Liceo,* Florence, 1573.

Cold Atom Clocks

C. Salomon[*], Y. Sortais[†], S. Bize[†], M. Abgrall[†], S. Zhang[†], C. Nicolas[†], C. Mandache[†], P. Lemonde[†], P. Laurent[†], G. Santarelli[†], A. Clairon[†], N. Dimarcq[††], P. Petit[††], A. Mann[†††], A. Luiten[†††], S. Chang[†††]

[*] *Laboratoire Kastler Brossel, Ecole Normale Supérieure, 24 rue Lhomond, 75231, Paris, France. E-mail: salomon@physique.ens.fr*
[†] *BNM-LPTF, Observatoire de Paris, 61 Av. de l'Observatoire, 75014, Paris, France*
[††] *Laboratoire de l'Horloge Atomique, bat. 221, Université Paris sud, 91405 Orsay, France*
[†††] *Physics Department, University of Western Australia, Nedlands, 6907 WA, Australia*

Abstract.

This paper reviews recent progress on microwave clocks using laser cooled neutral atoms. With an ultra-stable cryogenic sapphire oscillator as interrogation oscillator, a cesium fountain operates at the quantum projection noise limit. With $6 \ 10^5$ detected atoms, the relative frequency stability $\delta\nu/\nu$ is $4 \ 10^{-14} \ \tau^{-1/2}$ where τ is the integration time in seconds. This stability is comparable to that of hydrogen masers. At $\tau = 2 \ 10^4$ s, the measured stability reaches $6 \ 10^{-16}$. Equally important is the accuracy of the frequency standard since ^{133}Cs is the primary reference for the definition of the time unit, the second. The accuracy of our cesium fountain FO1 is presently $1 \ 10^{-15}$, currently the best reported value.

A ^{87}Rb fountain has also been constructed and the ^{87}Rb ground-state hyperfine energy has been compared to the Cs primary standard with a relative accuracy of $2.5 \ 10^{-15}$. Comparing the hyperfine energies of atoms with different atomic numbers Z, one can search for possible variations of the fine structure constant $\alpha = e^2/\hbar c$ with time. Measurements of the ratio $\nu(^{87}Rb)/\nu(^{133}Cs)$ spread over an interval of 24 months indicate no change at a level of $3.1 \ 10^{-15}$/year, placing a new upper limit for $1/\alpha(d\alpha/dt)$. The second attractive feature of ^{87}Rb fountains is the smallness of the frequency shift induced by the mean field interaction between atoms. This shift is found to be at least ~ 50 times below that of cesium.

Finally, the interest of the microgravity of space for cold atom experiments is outlined. A space mission, ACES, carrying ultra-stable clocks, is presented. ACES has been selected by the European Space Agency to fly on the International Space Station in 2004.

CP551, *Atomic Physics 17*, edited by E. Arimondo, P. DeNatale, and M. Inguscio
© 2001 American Institute of Physics 1-56396-982-3/01/$18.00

I INTRODUCTION

A From Galileo's pendulum to atomic fountains

At the beginning of the XVIIth century, Galileo noticed that a pendulum is a stable oscillator. The period of small oscillations is remarkably constant and depends only on simple parameters, the length of the pendulum and the gravity acceleration (fig.1). By counting the number of oscillations one easily measures time intervals, thus realizing a clock. Since 1967, the SI unit of time, the second, is based on a quantum oscillator using the ground-state hyperfine transition in atomic ^{133}Cs. By definition, the frequency of this transition is 9 192 631 770 Hz. The best realization of the second is currently achieved by laser cooled ^{133}Cs atomic fountains which combine both high stability and high accuracy. The accuracy is an estimate of how well the definition of the second is realized by a particular clock device. For instance, in the case of Galileo's pendulum, temperature variations induce variations in the length of the pendulum, or tides induce local variations in the gravity. These are systematic effects which must be be corrected for, and the accuracy of this *basic* oscillator depends on the accuracy of their evaluation. Over several centuries, many mechanical improvements were found to stabilize further the pendulum period [1] and in the 1920 's, Shortt designed a pendulum clock which had a drift of only a few seconds per year. The evolution of the relative accuracy of ^{133}Cs primary frequency standards with time is shown in fig.1. Since the 1960's, it has improved on average by a factor of 10 every 10 years, reaching currently 10^{-15} in cold atom fountains. For comparison the microwave trapped ion frequency standard based on Hg^+ [2] and the recent optical measurements of the 1s-2s transition in atomic hydrogen [3] and of the 657 nm transition in calcium [4] are also shown in fig.1.

In passive atomic clocks, the relative frequency stability $\sigma_y(\tau)$ is given by the Allan variance of the frequency fluctuations of a macroscopic oscillator (a quartz crystal, or a sapphire cryogenic oscillator), slaved to the resonance of a sample of microscopic oscillators (Cs atoms in a primary standard). Here, τ is the measurement time. This Allan variance is always measured using the beat note between two independent oscillators. The accuracy of an atomic clock depends on various parameters such as the temperature of the experiment, the surrounding magnetic field, or the collisions between the atoms.

Cold atom clocks use the method of separated oscillatory fields introduced by Ramsey in 1950, but adapted to a vertical geometry [5,6]. The atoms are first prepared in one of the two levels $|g\rangle$ and $|e\rangle$ of the clock transition, and then launched upwards. On the way up, they pass through a resonant microwave cavity. The microwave frequency is generated from the macroscopic oscillator and tuned to the clock transition. The atoms interact with the microwave field sustained inside the cavity, and undergo a first $\pi/2$ pulse. On the way down, they pass through the

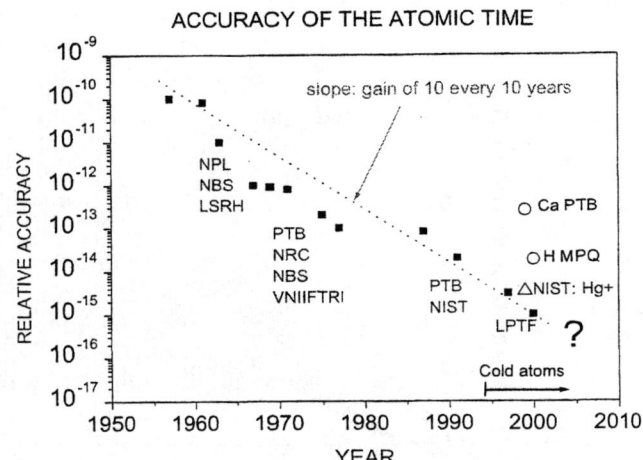

ACCURACY OF THE ATOMIC TIME

FIGURE 1. Left : Galileo by Leoni. Right : accuracy of cesium atomic clocks over the last 40 years (black squares). Open triangle: Hg^+ ion at 40 GHz [2]. Open circles : H [3] and Ca [4] optical lines.

same cavity, where they undergo after a time T a second $\pi/2$ pulse. The longer T, the narrower the Ramsey resonance, and the higher the frequency resolution. After the Ramsey interactions, the populations N_g and N_e of both clock levels are measured. From the transition probability signal, $N_e/(N_g + N_e)$, as a function of the microwave detuning, an error signal is computed to lock the macroscopic oscillator to the atomic transition. In fig.2, the Ramsey resonance as a function of the microwave frequency is shown for the BNM-LPTF Cs fountain FO1. The atoms are cooled down to 0.8 μK and launched upwards at $\sim$ 4 m/s. They reach apogee $\sim$ 30 cm above the microwave cavity, corresponding to $T \sim 0.5$ s. Ramsey fringes have a width $\Delta\nu = 1$ Hz FWHM. Each point is one cycle of duration 1.1 s, and the signal to noise ratio is 2000 for 8 10^5 detected atoms.

B   Scope of the paper

In this paper, we discuss the present performance and limits of ^{133}Cs and ^{87}Rb fountains. We have recently shown that Cs fountains can be operated with a frequency stability limited by the quantum projection noise [7]. The accuracy of BNM-LPTF CS fountains has been improved up to 1.0 10^{-15}. Going much beyond seems a hard task, because of the large frequency shift induced by cold collisions. In contrast, ^{87}Rb fountains exhibit a collisional shift $\sim$ 50 times smaller than in ^{133}Cs [8,9]. ^{87}Rb fountains are thus interesting candidates for future atomic clocks with better performance. We present the measurement of the collisional shift in our ^{87}Rb fountain and we give a new limit for the variation of the fine structure constant $\alpha = e^2/\hbar c$ with time. Finally, micro-gravity environment is capable of

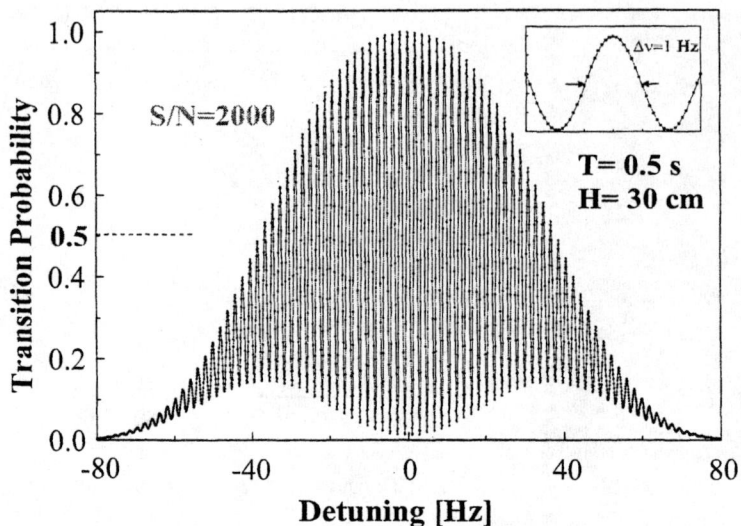

FIGURE 2. Ramsey fringes of the BNM-LPTF Cs fountain FO1. The fringe width is 1 Hz and signal to noise ratio is 2000 per cycle of 1.1 s.

providing even longer interaction times than in Earth fountains [10,11].

II   CS FOUNTAINS

Until recently, the frequency stability of atomic fountains was limited by the phase noise of the quartz oscillator used for the atom interrogation. With state-of-the-art quartz oscillators, this limit is above $10^{-13}\,\tau^{-1/2}$ [12]. With such a stability, a 10^{-16} accuracy goal is challenging since a single measurement with this resolution implies an averaging time of more than one week. By using an ultra-stable cryogenic sapphire oscillator (SCO) developed by the University of Western Australia [13], we have been able to overcome this limitation and to reach the quantum projection noise limit [7,14,15].

Before being detected, the atomic internal state is a linear superposition of the two states $|e\rangle$ and $|g\rangle$ of the clock transition: $|\psi\rangle = \alpha|g\rangle + \beta|e\rangle$, with $|\alpha|^2 + |\beta|^2 = 1$. The measurement process projects atoms in state $|e\rangle$ or $|g\rangle$. The probability of finding an atom in $|e\rangle$ is $|\beta|^2 = \langle\psi|P_e|\psi\rangle = p$, with P_e the projection operator onto state $|e\rangle$. The measurement exhibits quantum fluctuations of standard deviation $\sigma = (\langle\psi|P_e^2|\psi\rangle - (\langle\psi|P_e|\psi\rangle)^2)^{1/2} = \sqrt{p(1-p)}$. For a set of N_{at} uncorrelated atoms, σ scales as $N_{at}^{-1/2}$. Itano *et. al.* [15] named this effect *quantum projection noise* and have observed it both in the case of repeated measurements on a single particle prepared successively under identical conditions, and in the case of an ensemble average with a number N_{at} of identical trapped particles, up to $N_{at} = 380$.

The SCO possesses a mode which is only 1.3 MHz away from the Cs hyperfine splitting frequency ν_0. Oscillating on this mode, its frequency stability is below 10^{-14} from 0.1 to 10 s. The excess noise is then negligible compared to the projection noise in our atomic fountain for $N_{at} \leq 10^6$. The Allan standard deviation of the relative frequency fluctuations $y(t)$ of an atomic fountain can be expressed as:

$$\sigma_y(\tau) = \frac{1}{\pi Q_{at}} \sqrt{\frac{T_c}{\tau}} \left(\frac{1}{N_{at}} + \frac{1}{N_{at} n_{ph}} + \frac{2\sigma_{\delta N}^2}{N_{at}^2} + \gamma \right)^{1/2} \tag{1}$$

In (1) τ is the measurement time in seconds, T_c is the fountain cycle duration (~ 1 s) and $\tau > T_c$. $Q_{at} = \nu_0/\Delta\nu$ is the atomic quality factor. The first term in parentheses is the atomic projection noise $\propto N_{at}^{-1/2}$. The second term is due to the photon shot-noise of the detection fluorescence pulses and is less than 1% of the projection noise. The third term represents the effect of the noise of the detection system. $\sigma_{\delta N}$, the uncorrelated rms fluctuations of the atom number per detection channel, is about 85 atoms per fountain cycle. This noise contribution becomes less than the projection noise when $N_{at} > 2 \cdot 10^4$. γ is the contribution of the frequency noise of the interrogation oscillator [16,12]. With the SCO, this contribution is at most $10^{-14}\tau^{-1/2}$ and can be neglected. As an example, for $N_{at} \sim 6 \cdot 10^5$ detected atoms, $\Delta\nu = 0.8$ Hz, $Q_{at} = 1.15 \cdot 10^{10}$ and $T_c = 1.1$ s, the expected frequency stability is $\sigma_y(\tau) = 4 \cdot 10^{-14}\tau^{-1/2}$.

To observe the quantum projection noise, we vary the number of atoms in the fountain and measure the frequency stability $\sigma_y(\tau)$ by comparison with the free running sapphire oscillator which is used as a very stable reference up to $10 - 20$ seconds. A plot of the normalized Allan standard deviation as a function of atom number is presented in fig.3. Since we explored several values of Q_{at} and of the cycle duration T_c, we plot the quantity $\sigma_y(\tau)\pi Q_{at}\sqrt{\tau/T_c}$ for $\tau = 4$ s. This quantity should simply be equal to $N_{at}^{-1/2}$ when the detection noise is negligible. At low atom numbers, the $1/N_{at}$ slope indicates that the stability is limited by the noise of the detection system (third term in eq. 1). For $N_{at} > 4 \cdot 10^4$, the experimental points are in good agreement with the relation $y = aN_{at}^{-1/2}$ with $a = 0.91(0.1)$, close to the expected value of 1 (fig.3).

With the largest number of detected atoms, $N_{at} = 6 \cdot 10^5$, the stability is $4 \cdot 10^{-14}\tau^{-1/2}$ for $Q_{at} = 1.15 \cdot 10^{10}$, $T_c = 1.1$ s. This is a significant improvement over the fountain operation with a quartz oscillator. It is comparable to the best short-term stability achieved with microwave ion clocks using uncooled [199]Hg[+] and [171]Yb[+] samples [17,18]. In a second experiment, we have locked the SCO to the fountain signal for $N_{at} = 5 \cdot 10^5$ and compared it to a hydrogen maser. The Allan standard deviation of this frequency comparison is $7 \cdot 10^{-14}\tau^{-1/2}$ and reaches $6 \cdot 10^{-16}$ at $\tau \sim 2 \cdot 10^4$ s, a value close to the flicker floor of the H-maser.

Table 1 presents the current accuracy budget of the Cs fountain FO1. The accuracy evaluation was performed by amplifying the systematic effects when possible as described in [19] and with a resolution for each measurement of 10^{-15}. This fountain has the best accuracy ever reported for frequency standards. This accuracy

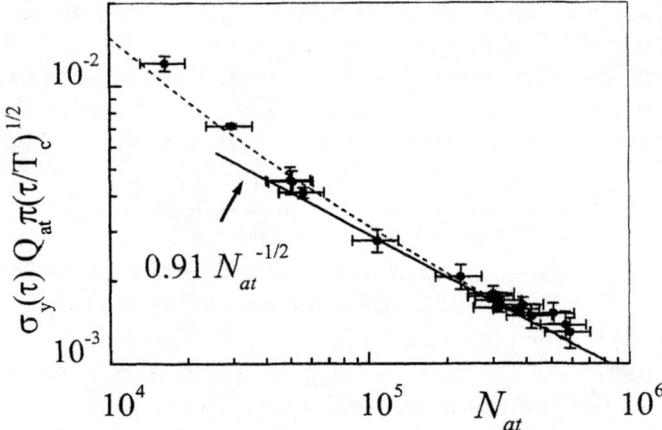

FIGURE 3. Normalized frequency fluctuations as a function of the number of detected atoms N_{at}. The expected quantum projection noise law is $y = N_{at}^{-1/2}$. The thick line $y = 0.91(0.1)N_{at}^{-1/2}$ is a least square fit to the experimental points for $N_{at} > 4\ 10^4$. The dashed line is the quadratic sum of the detection noise and quantum projection noise.

is currently a factor of three better than the trapped Hg^+ ion frequency standard operating at 40 GHz [2].

In the near future, direct comparison of our Cs fountains should lead to a stability on the order of $\sqrt{2} \times 4 \times 10^{-14}\tau^{-1/2}$, reaching 2×10^{-16} for $\tau \simeq 1\,\text{day}$. However under these conditions, the collisional frequency shift of the Cs clock transition is on the order of -5×10^{-15}. Reaching the 10^{-16} level of accuracy requires a stability of the average atomic density and an extrapolation to zero density at the 2% level, a delicate task.

III ^{87}RB : AN INTERESTING ALKALI

In 1997, Kokkelmans *et al.* predicted that the collisional frequency shift for ^{87}Rb was 15 times smaller than for ^{133}Cs [20]. Rubidium fountains could then operate with larger atom numbers, with a better short term stability and better accuracy. ^{87}Rb fountains appear thus as an attractive alternative to Cs frequency standards. Our ^{87}Rb fountain has already been described in detail in ref. [21]. In this section, we present measurements of the two main atom number dependent effects in a ^{87}Rb fountain, the collisional frequency shift and the cavity pulling effect.

As explained above, in a fountain the atoms interact twice with a microwave field (Ramsey method), inducing transitions between the two internal states $|g\rangle$ and $|e\rangle$. The collisional frequency shift is due to the interaction between atoms inside the atomic cloud during the Ramsey interaction. At very low temperatures,

TABLE 1. Present accuracy budget of the ^{133}Cs fountain FO1. The device operates with a stability of $1.5 \, 10^{-13}\tau^{-1/2}$, and a number of detected atoms of 10^5. The MOT is disabled and atoms are captured in optical molasses. The temperature is $0.8 \, \mu K$. The gravitational redshift is not intrinsic to the clock and is not included in the table.

Effect	Correction $[10^{-15}]$	Uncertainty $[10^{-15}]$
2nd order Zeeman shift	-133	≤ 0.1
Blackbody radiation	$+17.6$	≤ 0.3
Microwave spectrum	0	0.2
First Order Doppler	0	≤ 0.5
Microwave leaks	0	0.2
Pulling by other lines	0	0.4
Second order Doppler and gravitation	0	$<< 0.1$
Microwave photon	-0.3	0.3
Cold collisions + Cavity Pulling (Molasses)	1	≤ 0.5
Background gas collisions	0	≤ 0.5
Total 1σ uncertainty		$1.0 \, 10^{-15}$

only s-wave scattering occurs and the collision between two atoms in internal states $|i\rangle$ and $|j\rangle$ is described using the scattering length $a_{i,j}$. Here, $|i\rangle$ and $|j\rangle$ refer to any atomic Zeeman substate $|F, m_F\rangle$, while $|g\rangle$ and $|e\rangle$ are notations for the clock levels only. In the case of ^{87}Rb, the lower and higher clock levels are $|g\rangle = |1,0\rangle$, and $|e\rangle = |2,0\rangle$. In the general case where states other than $|g\rangle$ and $|e\rangle$ are populated, the clock frequency shift is related to the partial densities n_j of states $|j\rangle$ by the following expression [20]:

$$\Delta\nu = -\frac{2\hbar}{m} \times \sum_j n_j(1 + \delta_{g,j})(1 + \delta_{e,j})(a_{g,j} - a_{e,j}) \qquad (2)$$

where the Kronecker symbol $\delta_{i,j}$ accounts for quantum statistics. In a fountain, the partial densities n_j are space- and time-dependent and the measured frequency shifts appear as an average over these variables.

The second effect which depends on atom number is the cavity frequency pulling [22]. In our fountain, the atomic transition is probed by a 6.8 GHz microwave field in a TE_{011} copper resonator with a quality factor $Q \simeq 10^4$. The pulling effect is due to the interference inside the microwave resonator between the field radiated by the input coupler and the field radiated by the atomic magnetic dipoles when the atoms pass through the cavity. This interference induces a time dependent phase shift between the field inside the resonator and the signal delivered by the oscillator. It produces a clock shift that exhibits a dispersive dependence as a function of the cavity detuning with respect to the atomic resonance ($\Delta\nu_{cav} = \nu_{cav} - \nu_{at}$). The amplitude of this dispersion curve is proportional to Q and to the number of atoms N_{cav} crossing the cavity. The width is proportional to $1/Q$. When the atoms enter the cavity in the upper state (resp. lower state) and deposit (resp. extract) energy in the cavity, the clock frequency is *pulled* towards (resp. *pushed* away from) the cavity resonance frequency.

Our treatment of the frequency shift measurements takes into account both collisional frequency shift and cavity pulling simultaneously. By tuning the cavity around the Rb resonance frequency, we display the caracteristic dispersive lineshape of the cavity pulling effect and find good agreement with theory (fig.4). A quantitative model of the pulling effect allows us to make a precise determination of the number of atoms crossing the cavity, N_{cav}. The measurements of the velocity distribution and of the initial size of the atomic cloud, combined with a Monte-Carlo simulation and the measurement of N_{cav}, give the average atomic density during the Ramsey interrogation ($\overline{n}$). After subtraction of the pulling contribution, we thus deduce the collisional frequency shift. In order to test the Monte-Carlo simulation, other methods have been used to measure the number of atoms in the fountain (experimental details are given in ref. [9]). These methods agree with the value deduced from the cavity pulling within 20%.

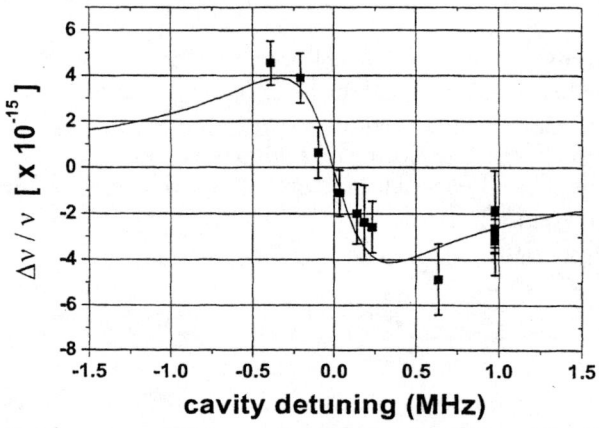

FIGURE 4. Relative frequency shift vs cavity detuning for $\Delta N = 3.5 \times 10^7$ in case (i). Solid line : fit from eq. (2).

Our measurement approach uses a differential method. Every 200 fountain cycles ($\sim 350\,$s), we change the trapping light intensity so that the number of atoms in the fountain alternates between two values N_{high} and $N_{low} \simeq N_{high}/4$. We measure the frequency difference between the two situations. This method efficiently rejects slow frequency fluctuations which are not related to atom number or density, in particular the H-maser drift. In the differential method the frequency resolution is $\sim 3 \times 10^{-13}\tau^{-1/2}$ where τ is the averaging time in seconds. At 350 s the maser stability is $\sim 2.7 \times 10^{-15}$, at least 4 times better than the fountain stability. The relative frequency resolution on effects which depend on atom number reaches $\sim 1 \times 10^{-15}$ per day.

After launch, atoms are spread among the various m_F substates of the $F = 2$ manifold. Data have been taken in three cases. (i) normal clock operation : atoms in the $|2, 0\rangle$ substate are transferred to $|1, 0\rangle$ using a microwave π pulse. Atoms remaining in $|2, m_F \neq 0\rangle$ are pushed away by radiation pressure. Only three scattering lengths a_{gg}, a_{ge} and a_{ee} in eq. (2) are involved. (ii) no selection : all Zeeman sublevels in $F = 2$ are equally populated to within $2 - 3\%$. The frequency shift of the $|2, 0\rangle \rightarrow |1, 0\rangle$ transition is measured in the presence of $|2, m_F \neq 0\rangle$ populations. (iii) Atoms are pumped in $F = 1$ by a laser pulse tuned to the $F = 2 \rightarrow F' = 2$ optical transition. Atoms remaining in $F = 2$ state ($\leq 2\%$) are pushed away by radiation pressure. The launched atoms are equally spread among the $F = 1$ substates to within 1%. The frequency shift of the $|1, 0\rangle \rightarrow |2, 0\rangle$ transition is measured in the presence of $|1, m_F \neq 0\rangle$ populations.

Fig. 4 shows the differential measurement of the clock frequency shift as a function of the cavity detuning $\Delta\nu_{\text{cav}}$ with respect to the atomic resonance, for atoms initially selected in $|1, 0\rangle$, with $N_{\text{high}} \simeq 4.3 \times 10^7$ entering the microwave cavity and $N_{\text{low}} \simeq 0.8 \times 10^7$. The simulation shows that this modulates the effective density by a factor ~ 3.3, because both velocity distribution and initial size of the atomic cloud depend on the regime of operation of the fountain (high or low number of atoms). The frequency and quality factor of the cavity are measured with an uncertainty of $20\,\text{kHz}$ and 4% respectively and the cavity is tuned by temperature [23]. The solid line is a least square fit of the data using the following function :

$$\delta\nu/\nu = f(\Delta\nu_{cav})\Delta N + B\Delta\bar{n} \qquad (3)$$

where B is the free parameter of the fit, and $\Delta N = N_{\text{high}} - N_{\text{low}}$ [24]. $f(\Delta\nu_{\text{cav}})$ is the calculated dispersive line shape of the cavity pulling effect for a single atom in the cavity. $\bar{n}$ is the effective atomic density and is related to the partial densities n_j in eq.(2) by $\bar{n} = \sum_j \bar{n}_j$, where $\bar{n}_j$ are the partial densities $n_j(t)$, averaged over time, over the initial spatial distribution and over the velocity distribution.

We find $B = -7.2(20.0) \times 10^{-24}\,\text{cm}^3$. B, the collisional shift coefficient, is surprisingly consistent with zero even with a frequency standard deviation of 3×10^{-16} given by the fit. Thus in case (i), at a density of $1.5 \times 10^7\,\text{cm}^{-3}$, ^{87}Rb exhibits no detectable shift (fig. 5(i)) whereas ^{133}Cs would exhibit a shift of 3×10^{-14}. Our upper value for the ^{87}Rb collisional shift is consistent with the recent measurement of C. Fertig and K. Gibble giving $-0.38(8)\,\text{mHz}$ for a density of $1.0(6) \times 10^9\,\text{cm}^{-3}$ [8].

Fig. 5 summarizes the collisional shifts measurements versus the effective density $\bar{n}$ in the three cases (i), (ii) and (iii) [25]. By contrast to case (i), data recorded in case (ii) and (iii) show a clear dependence with the density after subtracting the cavity pulling effect.

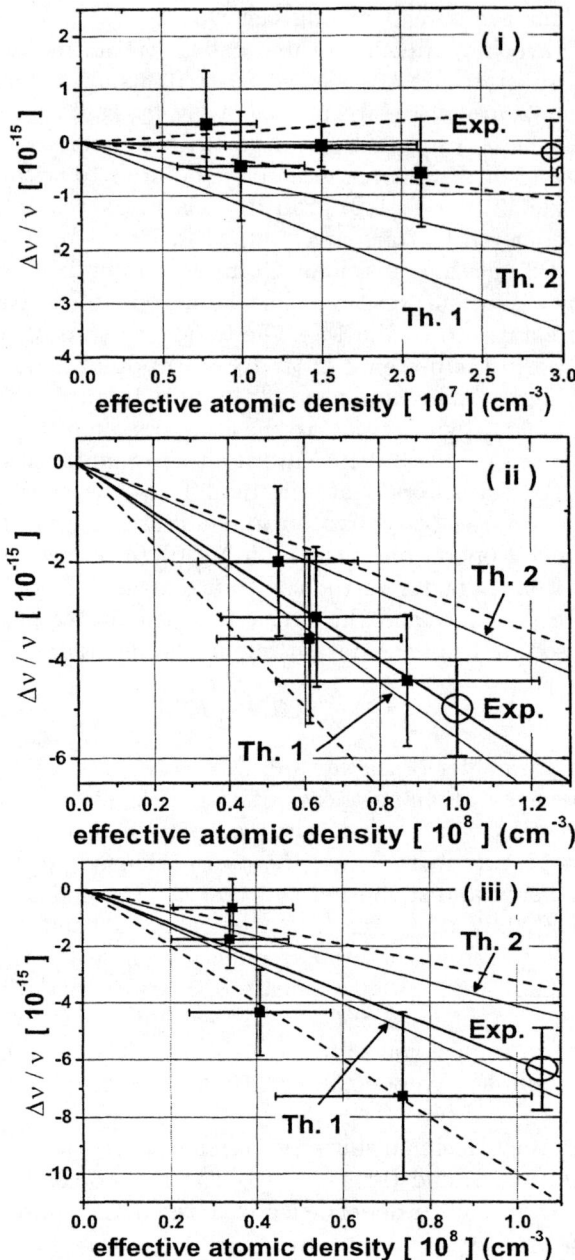

FIGURE 5. Frequency shift of the $|1,0\rangle \rightarrow |2,0\rangle$ transition vs effective density in cases (i),(ii), and (iii), and comparison with theories of ref. [20] (Th.1) and ref. [26] (Th.2). Open circles :linear fits to the experimental data with statistical frequency uncertainty at the given density. Horizontal error bars represent the 40% density uncertainty. Dotted lines : experiment with combined frequency and density uncertainties.

The measured value in case (ii) $-50(10)(^{+22}_{-34}) \times 10^{-24}\,\mathrm{cm^3}$ is in reasonable agreement with the theoretical predictions $-56 \times 10^{-24}\,\mathrm{cm^3}$ [20] and $-33 \times 10^{-24}\,\mathrm{cm^3}$ [26]. The first parenthesis refers to the frequency statistical uncertainty (1 standard deviation); the second to the linear combination of frequency uncertainty and density calibration uncertainty. From variations of the parameters in the simulation and experimental calibrations , we deduce a 35% type B uncertainty on our density. Adding quadratically the 11% statistical uncertainty on atom number, we get a 40% combined uncertainty on the density. This corresponds to a scale factor of 1.4 and 1/1.4 on the density axis of fig. 5, defining the acceptance domain of the measurements (dotted lines). In case (iii), the agreement with theory is similar. We find $-60(16)(^{+29}_{-46}) \times 10^{-24}\,\mathrm{cm^3}$ to be compared to $-68 \times 10^{-24}\,\mathrm{cm^3}$ (ref. [20]) and $-41 \times 10^{-24}\,\mathrm{cm^3}$ (ref. [26]). Finally, our data in case (i), $-7.2(20.0)(^{+25}_{-31}) \times 10^{-24}\,\mathrm{cm^3}$ show a disagreement at $\sim 3\sigma$ with theory of ref. [20] and seem to favor the more recent theory of ref. [26].

IV   A NEW LIMIT FOR THE VARIATION OF α WITH TIME

A   Present tests

Comparing the hyperfine energies of different alkali or alkali-like species provides an interesting test of the Local Position Invariance, which is one of the statements of the Einstein Equivalence Principle (EEP). *"In local freely falling frames, the outcome of any non-gravitational test experiment is independent of where and when it is performed"*. Experimental tests of the EEP thus aim at determining the relative rate of change of non-gravitational fundamental constants, such as the fine structure constant $\alpha = e^2/\hbar c$, either with time or with the gravitational potential. A variation of α would violate the EEP, and give support to recent many-dimensional cosmologies (Kaluza-Klein models, superstring theories) that attempt to unify gravity and other fundamental forces ([27–29]). Two types of tests of the stability of α are being performed : at the cosmological and at the laboratory scale. An example of cosmological measurement is the Oklo test. This test analyses the isotope ratios $^{149}Sm/^{147}Sm$ in the natural uranium fission reaction that took place around 1.8×10^9 years ago (corresponding to a redshift $z \approx 0.1$) at the present day site of the Oklo mine in Gabon, West Africa. Recent analysis of the Oklo data [30] has shown that:

$$-0.9 \times 10^{-7} \leq (\alpha_{Oklo} - \alpha_{now})/\alpha_{now} \leq 1.2 \times 10^{-7} \qquad (4)$$

so that, integrating over 1.8×10^9 years, one obtains:

$$|\dot{\alpha}/\alpha| \leq 5 \times 10^{-17}\,yr^{-1} \qquad (5)$$

In a similar way, astrophysical methods make it possible to estimate α at earlier evolutionary phases of the Universe and determine $|\dot{\alpha}/\alpha|$ as a function of redshift z (i.e. cosmological time), by making spectroscopic observations of gas clouds against background quasars. Extensive analysis has been carried out by various groups, measuring the relativistic fine-structure splitting of alkali-type resonance doublets. This method is based on the fact that the separation between lines within one multiplet is proportional to α^2. Measuring the doublet separation in an absorption system and comparing with laboratory values, one can estimate the difference between α at the epoch z and the present value. D. A. Varshalovich et al. [31] focused on the analysis of $SiIV$ doublets absorbed by 3 quasars located at $z \approx 3$, paying special attention to the non-linearities in the dispersion curves of the echelle spectrograph used for the observations. Their analysis leads to an upper bound, at the 2σ level:

$$|\Delta\alpha/\alpha| \leq 1.6 \times 10^{-4} \tag{6}$$

and, averaged over 10^{10} years,

$$|\dot{\alpha}/\alpha| \leq 1.6 \times 10^{-14}\, yr^{-1} \tag{7}$$

J.K. Webb et al. [32,33] recently analysed $MgII$ and $FeII$ doublets absorption spectra emitted by 25 quasars, thus spanning a wide range of epochs ($0.6 < z < 1.6$). Their measurements display a clear departure from zero at the 3σ level, when considering the sample as a whole:

$$|\Delta\alpha/\alpha| = (-1.09 \pm 0.36) \times 10^{-5} \tag{8}$$

However, the $0.6 < z < 1.0$ points alone show no significant trend:

$$|\Delta\alpha/\alpha| = (-0.17 \pm 0.39) \times 10^{-5} \tag{9}$$

These results make the issue of the fine structure constant temporal variation particularly acute.

As noted in [34,35], laboratory measurements based on clock comparisons contrast with cosmological measurements, because they are repeatable and span over monthly to yearly durations ($z \ll 1$). They thus provide complementary information to the cosmological determinations. These clock comparisons involve ultrastable oscillators of different compositions such as superconducting cavities vs H-maser or Cs hyperfine structure transition (case (i)) [34], or atomic clocks with different species (case (ii)): Mg fine structure transition vs Cs hyperfine transition [36], Hg^+ ion frequency standard vs H-maser [35]. In case (i), tests rely on the fact that the cavity frequency is proportional to α while the H-maser or Cs clock transition is proportional to α^2 (at first order). In case (ii), the method is based on the increasing importance of Casimir relativistic contribution to the hyperfine energy splitting as atomic number Z inceases in the group I alkali elements and alkali-like ions [35,37]. A temporal variation of the ratio of the hyperfine energies of different alkali would thus provide a signature of the temporal variation of α.

Prestage *et al.* [35] derived an upper bound for $|\dot{\alpha}/\alpha|$ from the comparison of Hg^+ vs H-maser:

$$\frac{d}{dt}\ln(\frac{\nu_{Hg^+}}{\nu_H}) = +2.2 \times |\dot{\alpha}/\alpha| \tag{10}$$

and:

$$|\dot{\alpha}/\alpha| \leq 3.7 \times 10^{-14}\, yr^{-1} \tag{11}$$

B Comparisons between ^{87}Rb and ^{133}Cs fountains

With our two Cs fountains and one Rb fountain in operation, we can compare Cs and Rb hyperfine frequencies in time, and search for a possible temporal variation of α. Since 1998, three campaigns of measurements of the ^{87}Rb absolute frequency have been performed with reference to the Cs fountain. During each campaign, several parameters such as the launch height of the atomic cloud in the fountain, the microwave power, and the magnetic field have been varied. The accuracy budget of the Rb fountain was thus re-evaluated each time. Each frequency measurement was corrected for systematic shifts. The obtained values were then binned, giving one frequency data point for each campaign. In fig. 6, we have plotted the relative frequency shift $(\frac{\Delta\nu}{\nu})_{Rb}$, where $(\Delta\nu)_{Rb} = (\nu(t) - \nu(t_0))$, versus time. Here, $\nu(t)$ accounts for the frequency measurement at time t. t_0 corresponds arbitrarily to our second campaign (April 1999). A linear fit to the data points over two years leads to :

$$\frac{d}{dt}\ln(\frac{\nu_{Rb}}{\nu_{Cs}}) = (1.9 \pm 3.1) \times 10^{-15}\, yr^{-1} \tag{12}$$

The 3.1×10^{-15} uncertainty is the quadratic sum of the accuracy of the Rb fountain (1.4×10^{-15}), the accuracy of the Cs fountain (1.1×10^{-15}), and of the statistical noise in the frequency comparison(2.5×10^{-15}). This uncertainty of 3.1×10^{-15} represents a 20-fold improvement in the accuracy of long term clock frequency comparisons over previous work [35].

In the interpretation of these results, we first assume, as in [35], that there is no time dependence of the ^{133}Cs and ^{87}Rb nuclear magnetic moments. Then, all possible drift of the hyperfine energy ratio ν_{Rb}/ν_{Cs} can be attributed to the Casimir relativistic correction $F_{rel}(\alpha Z)$. Under this assumption, we obtain :

$$|\dot{\alpha}/\alpha| = (4.2 \pm 6.9) \times 10^{-15}\, yr^{-1} \tag{13}$$

an improvement by a factor 5 over the previous H-maser-Hg$^+$ comparison [35]. Conversely, as pointed out by S. Karshenboim [37], our measurements can also set an upper limit to the time variation of the proton magnetic moment g_p, if we now

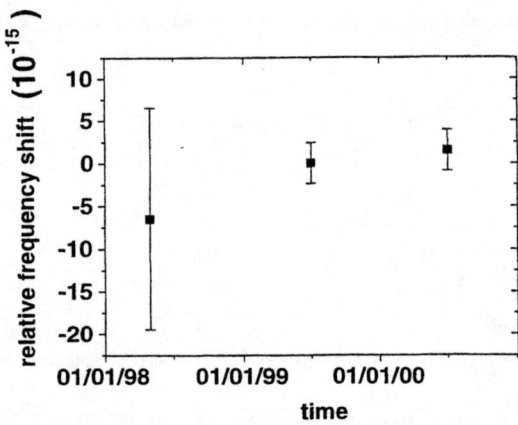

FIGURE 6. Measured $^{8}7$Rb frequencies referenced to the ^{133}Cs fountain FO1. The measurement in April 1999 is 6834682610.904343 Hz. The errors bars represent the combined inaccuracies of ^{87}Rb and ^{133}Cs fountains.

assume no time variation of α. This test is then 4 times more sensitive than the test over α, and we find :

$$|\frac{\dot{g}_p}{g_p}| = (9.5 \pm 16) \times 10^{-16}\, yr^{-1} \tag{14}$$

Of course any combination of these two extreme assumptions is possible and more tests with different alkali and alkali-like atoms will severely constrain these limits in the future.

V   ACES : ATOMIC CLOCK ENSEMBLE IN SPACE

ACES is a space mission which has been selected by the European Space Agency (ESA) to fly on the international space station (ISS) in 2004 [38]. ACES consists of two clocks, a cold atom clock (PHARAO) and a hydrogen maser (SHM, Neuchâtel Observatory), together with microwave and optical links for time and frequency transfer to ground users. These equipments will fit on a nadir oriented express pallet of dimensions $863 \times 1168 \times 1240$ mm. The total mass will not exceed 225 kg and the electrical power 500 Watts (fig.7).

Because micro-gravity allows long interaction times between the atoms and the microwave field and because the atomic velocity is smaller than in earth fountains and constant, we expect an excellent accuracy for a cold atom space clock [10,11,39]. The specified frequency stability of PHARAO for ACES is better than $1\,10^{-13}\tau^{-1/2}$. Averaged over one day, this stability should be $3\,10^{-16}$. Its projected accuracy is $1\,10^{-16}$.

FIGURE 7. Mock-up of the ACES plattform

Scientific objectives of ACES also include measurement of the gravitational redshift with a 25-fold improvement over the Gravity Probe A mission of Vessot *et al.* [40], a better test of the isotropy of speed of light and a search for a possible drift of the fine structure constant α. Finally, with the time transfer equipments, ACES will allow synchronization of time scales of distant ground laboratories with 30 ps accuracy and frequency comparisons with 10^{-16} accuracy. Similar space clock projects are also under development in the USA, PARCS [41] and RACE [42].

VI PROSPECTS

Microwave frequency standards using fountains are now in their maturity; a frequency stability of $6\ 10^{-16}$ has been measured and the present accuracy is $1\ 10^{-15}$. Operating permanently with an SCO as interrogation oscillator, the fountain frequency stability should enter into the low 10^{-16} range in the reasonable averaging time of one day. The two dominant terms in the present accuracy budget are the blackbody radiation shift and the collisional shift. The latter prevents the use of large number of atoms, unless more complicated techniques such as juggling or cw fountains are used. Calculations show that the cavity phase shift is negligible at the 10^{-15} level, but it must be carefully evaluated when asking for a one order of magnitude gain. Rb fountains, because of the strongly reduced collisional shift are likely to push the accuracy further, i.e. in the 10^{-17} range. Fountains now operate at the fundamental quantum limit for uncorrelated particles. An attractive line of research is to apply the recently proposed spin squeezing techniques to improve the frequency stability beyond the projection noise limit proportional to $1/\sqrt{N_{at}}$ [43–45]. A factor of $1/\sqrt{N_{at}}$ can ultimately be gained.

With their relative simplicity, fountain devices compete favorably with laser cooled trapped ion frequency standards. However, practically unlimited interaction times can be achieved with trapped charged particles. This is an advantage for increasing the quality factor. On the other hand, ions interact much more by Coulomb repulsion than neutrals and traps with a single ion (or a small number) must be used [2]. This is then made at the expense of signal to noise ratio hence of short term stability, even if quantum limited detection is implemented using the quantum jump method.

Microwave cold atom standards ultimately will be limited by the quality factor of the atomic line. Very narrow transitions in the optical domain offer attractive possibilities for fountains with neutrals [46]. A prominent example is the $1s - 2s$ transition of hydrogen [47–49], measured recently with a frequency resolution of $1.8 \ 10^{-14}$ by comparison with the PHARAO transportable fountain [3]. Alkaliearth atoms are good candidates since they have already been laser cooled and possess forbidden transitions in the visible domain. Progress with laser cooled Ca, Mg and Sr using intercombination lines has been spectacular [4,50–53]. All these frequency standards, with ions or neutrals, require ultra-stable laser sources to interrogate these narrow lines and a recent breakthrough has been made by the NIST Boulder group measuring sub-Hertz laser linewidth in the visible range [54]. In addition, the recent progress on frequency combs using femtosecond lasers opens new possibilities for comparing microwave and optical frequency standards at 10^{-16} or below [55,49,56].

These atomic frequency standards can already be compared at $1 \ 10^{-15}$ using the transportable PHARAO device. Frequency comparisons of distant clocks using satellite time transfer, GPS-phase or two way techniques, are presently at the level of $\sim 3 \ 10^{-15}$ for an averaging time of $2 \ 10^4$ s [57] and $\sim 2 \ 10^{-15}$ for one day [58]. Global comparisons between fountains will improve the quality of the TAI (Temps Atomique International). In fine, these high performance frequency standards will allow new tests in fundamental physics as well as new applications in navigation, positioning and geodesy.

We acknowledge fruitful discussions with Y. Castin, C. Cohen-Tannoudji, J. Dalibard, G. Shlyapnikov and the ENS group. We are thankful to A. Gérard, M. Lours, M. Dequin, L. Volodimer, P. Petit and to the SERT for technical support. Work supported in part by BNM and CNRS. BNM-LPTF and Laboratoire Kastler Brossel are Unités Associées au CNRS, UMR 8630 and UMR 8552.

REFERENCES

1. Galluzzi P., this volume
2. Berkeland D., Miller J., Bergquist J., Itano W., and Wineland D., *Phys. Rev. Lett.*, **80**, 2089 (1998).
3. Niering M. *et al.*, *Phys. Rev. Lett.*, **84**, 5496 (2000).
4. Riehle F. *et al.*, *IEEE Trans. Inst. Meas.*, **48**, 608 (1999).

5. Kasevich M., Riis E., Chu S. and de Voe R., *Phys. Rev. Lett.*, **63**, 612 (1989).
6. Clairon A., Salomon C., Guellati S., and Phillips W., *Europhys. Lett.*, **16**, 165 (1991).
7. Santarelli G. *et al.*, *Phys. Rev. Lett.*, **82(23)**, 4619 (1999).
8. Fertig C. and Gibble K., *Phys. Rev. Lett.*, **85**, 1622 (2000).
9. Sortais Y. *et al.*, *Phys. Rev. Lett.* in press, scheduled 09 oct. 2000.
10. Clairon A. *et al.*, in *Proc. of the* 6th *European Frequency and Time Forum*, ESA : **SP-340**, pp. 27–33 (1992).
11. Lounis B., Reichel J., and Salomon C., *C.R. Acad. Sci. (Paris)*, **316(2)**, 739 (1993).
12. Santarelli G.*et al.*, *IEEE Trans. Ultr. Ferr. Freq. Contr.*, **45**, 887 (1998).
13. Luiten A., Mann A., Costa M., and Blair D., *IEEE Trans. Instr. Meas.*, **44**, 132 (1995).
14. Wineland D., Itano W., Bergquist J., and Walls F., in *Proc.* 35th *Ann. Freq. Contr. Symp.* (1981).
15. Itano W. *et al.*, *Phys. Rev. A.*, **47**, 3554 (1993).
16. Dick G., in *Proc. of Precise Time and Time Interval*, pp. 133–147, Redendo Beach (1987).
17. Tjolker R., Prestage J., and Maleki L., in J. Bergquist ed., *Proc. of* 5th *Symp. on Freq. Stand. and Metr.*, World Scientific, Singapore (1996).
18. Fisk P., Sellars M., Lawn M., and Coles C., *ibid.*
19. Clairon A. *et al.*, *IEEE Trans. on Inst. and Meas.*, **44(2)**, 128 (1995).
20. Kokkelmans B., Verhaar B., Gibble K., and Heinzen D., *Phys. Rev. A.*, **56**, R4389 (1997).
21. Bize S. *et al.*, *Europhys. Lett.*, **45(5)**,558 1999.
22. Vanier J. and Audoin C., *The Quantum Physics of Atomic Frequency Standards*, Bristol and Philadelphia : Adam Hilger, 1989.
23. Fertig C. and Gibble K., *IEEE. Trans. Instrum. Meas.*, **48**, 520–523 (1999).
24. In fig.(1) we have rescaled the data to a fixed $\Delta N = 3.5 \times 10^7$. During the 6 month data acquisition, ΔN typically fluctuated by 20%.
25. To clarify fig.(5i), we represent, in place of individual measurements, the weighted average of 10 of them performed at very similar densities ($1.5(2)\ 10^7\ cm^{-3}$).
26. Williams C. *et al.*, private communication.
27. Damour T. and Polyakov A., *Nucl. Phys.*, B**423**, 532 (1994).
28. Wu Y. and Wang Z., *Phys. Rev. Lett.*, **57**, 1978 (1986).
29. Marciano W., *Phys. Rev. Lett.*, **52**, 489 (1984).
30. Damour T. and Dyson F., *Nucl. Phys.*, B**480**, 37 (1996).
31. Varshalovich D., Panchuk V., and Ivanchik A., *Astr. Lett.*, **22**, 6 (1996).
32. Webb J., Flambaum V., Churchill C., Drinkwater M., and Barrow J., *Phys. Rev. Lett.*, **82(5)**, 884 (1999).
33. Flambaum V. *et al.*, this volume.
34. Turneaure J., Will C., Farrel B., Mattison E., and Vessot R., *Phys. Rev. D*, **27**, 1705 (1983).
35. Prestage J., Tjoelker R., and Maleki L., *Phys. Rev. Lett.*, **74**, 3511 (1995).
36. Godone A., Novero C., Tavella P., and Rahimullah K., *Phys. Rev. Lett.*, **71**, 2364 (1993).
37. Karshenboim S., this volume and submitted to *Can. Journ. of Physics*.

38. Salomon C. and Veillet C., in *Proc. on "Space Station Utilisation"*, ESA Symposium, **SP-385**, 295 (1996).

39. Laurent P. *et al.*, *Euro. Phys. J. D*, **3**, 201 (1998).

40. Vessot R. *et al.*, *Phys. Rev. Lett.*, **45**, 2081 (1980)

41. Robinson H. *et al.*, in *Proc. of IEEE Int. Freq. Contr. Symp.*, pp. 37–40 (1998)

42. Gibble K., *ibid.*, pp. 41–45.

43. Wineland D., Bollinger J., Itano W., Moore F., and Heinzen D., *Phys. Rev. A*, **46**, 6797 (1992).

44. Kuzmich A., Mölmer K., and Polzik E., *Phys. Rev. Lett.*, **79**, 4782 (1997).

45. Sorensen J., Hald J., and Polzik E., *Phys. Rev. Lett.*, **80**, 3487 (1998).

46. Hall J. and Zhu M., *Journ. Opt. Soc. A. B*, **6**, 2194 (1989).

47. Beausoleil G. and Hänsch T., *Phys. Rev. A*, **33**, 1661 (1986).

48. Udem T. *et al.*, *Phys. Rev. Lett.*, **79**, 2646 (1997).

49. Udem T. *et al.*, in *Proc. of the Opt. Freq. Synth. and Metr. Symp.*, Springer Verlag (2000).

50. Hollberg L. *et al.*, *ibid.*

51. Ruschewitz F. *et al.*, *Phys. Rev. Lett.*, **80**, 3173 (1998).

52. Vogel K. , Dinneen T., Gallagher A., and Hall J., *IEEE Trans. Inst. Meas.*, **48**, 618 (1999).

53. Katori H., Ido T., Isoya Y., and Kuwata-Gonokami M., *Phys. Rev. Lett.*, **82**, 1116 (1999).

54. Young B., Cruz F., Itano W., and Bergquist J., *Phys. Rev. Lett.*, **82**, 3799 (1999).

55. Diddams S. *et al.*, *Phys. Rev. Lett.*, **84(22)**, 5102 (2000).

56. Luiten A. *et al.*, in *Proc. of the Opt. Freq. Synth. and Metr. Symp.*, Springer Verlag (2000).

57. Petit G., Thomas C., Jiang Z., Uhrich P., and Taris F., *IEEE Trans. Ultr. Ferr. Freq. Contr.*, **46**, 941 (1999).

58. Larson K. and Levine J., in *Proc. of IEEE Int. Contr. Symp.*, p. 292 (1998).

FUNDAMENTAL CONSTANTS

Progress towards a Measurement of $\hbar/M_{Cs}$

Joel M. Hensley, Andreas Wicht, Brent C. Young[*] and Steven Chu

Physics Department, Stanford University, Stanford CA 94305-4060, USA
*Present Address, Jet Propulsion Laboratory, California Institute of Technology, Pasadena, CA 91109

Abstract. We review our progress in a measurement of $\hbar/M_{Cs}$. Using an atom interferometer method based on adiabatic transfer between atomic states, we measure the recoil velocity of cesium due to the scattering of a photon. Our current statistical uncertainty is on the order of 3 parts per billion in $\hbar/M_{Cs}$. This paper will summarize our current status in dealing with the systematic effects of this measurement.

INTRODUCTION

Over the past eight years we have been developing an interferometer method for measuring the ratio of Planck's constant to the mass of an atom, h/M. In quantum mechanical equations the mass typically appears in the ratio h/M, so tests of quantum theories require only knowledge of h/M and not of M. As a particularly important example, the fine structure constant, α, can be written as

$$\alpha^2 = (\frac{2R_\infty}{c})(\frac{m_p}{m_e})(\frac{M_{Cs}}{m_p})(\frac{h}{M_{Cs}}) . \tag{1}$$

Of the quantities on the right hand side, c is a defined quantity. The fractional uncertainties of R_∞ and m_p/m_e are 7.6×10^{-12} and 2.1×10^{-9}, respectively [1]. M_{Cs}/m_p has been measured by D. Pritchard and colleagues to better than 1 part per billion (ppb) [2]. An accurate measurement of h/M_{Cs} would yield a more accurate measurement of α.

At present, α is determined by a range of measurements in elementary particle physics, atomic physics, and condensed matter mesoscopic and macroscopic systems. Comparison of various accurate measurements of α constitute one of the most demanding tests of the consistency of physics. For a detailed discussion, the reader is referred to a review article by Kinoshita [3] and the CODATA recommended values [1]. The most accurate determination of α is found by equating the quantum electrodynamic calculation of the magnetic moment of the electron with the measured value. Assuming that both the experiment and theory are correct, a value of α can be deduced with an uncertainty of 3.8 ppb. Other accurate measurements of α (and their relative uncertainties) include measurements based on the muonium hyperfine structure (58 ppb) [4], the quantum Hall effect (20 ppb), neutron diffraction (24 ppb), and the ac Josephson effect (31 ppb) [1]. The first measurement of the recoil

CP551, *Atomic Physics 17*, edited by E. Arimondo, P. DeNatale, and M. Inguscio
© 2001 American Institute of Physics 1-56396-982-3/01/$18.00

frequency shift was made in a heroic experiment by Hall, Bordé and Uehara [5], achieving a resolution $\Delta v/v = 2.3\times10^{-3}$ using a laser with a linewidth of 200 Hz, 32 cm diameter optics and an absorption cell with a 13 meter path length. Systematic effects were responsible for a 6×10^{-3} discrepancy between the known value of h/M and the measured value.

EXPERIMENTAL METHOD

Our approach to this measurement has been described in detail elsewhere and only a summary of the experiment will be presented here [6,7]. Consider an atom initially at rest and in state $|g\rangle$ that absorbs a photon of momentum $p_1 = \hbar k_1$. The photon will cause the atom to make a transition to state $|e\rangle$ and recoil with velocity $v_r = \hbar k_1/M$. Energy conservation demands that the energy of the photon inducing the transition is $\hbar\omega_1 = \hbar\omega_{eg} + Mv_r^2/2$, where $\hbar\omega_{eg}$ is the energy separating the states $|g\rangle$ and $|e\rangle$. Now consider the atom in state $|e\rangle$ and moving with velocity v_r. A photon of frequency ω_2 propagating in the opposite direction from ω_1 will induce the atom to return to state $|g\rangle$ with final velocity $2v_r$. Momentum and energy conservation demands that the two frequencies differ by $\omega_1 - \omega_2 = 2\hbar k_1^2/M \equiv \omega_r$. Thus, $\hbar/M$ can be measured in terms of the frequencies ω_r and $\omega_1 = k_1 c$.

If the atom has some initial velocity $\mathbf{v}_{int}$, there will be an additional Doppler term $2\mathbf{k}_1\cdot\mathbf{v}_{int}$ to the frequency difference ω_r. In order to eliminate this dependence on $\mathbf{v}_{int}$, a $\pi/2$-pulse at frequency ω_1 is used to initially put the atom into a superposition of states $|e\rangle$ and $|g\rangle$. A second π-pulse at frequency ω_2 is used to drive either of the two

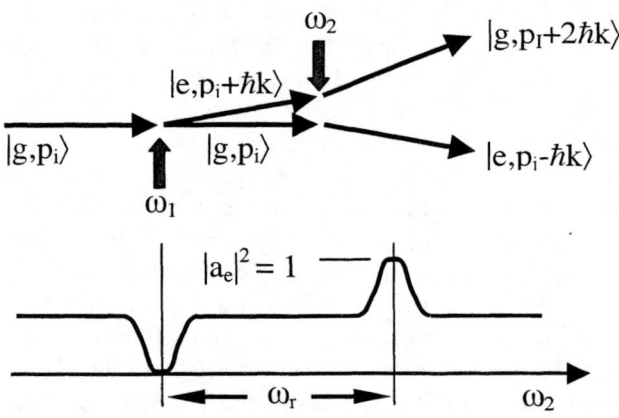

FIGURE 1. (a) An atom in $|g,p_i\rangle$, where p_i is its initial momentum, irradiated with a $\pi/2$-pulse at frequency ω_1 is put into a superposition of states $|e,p_i+\hbar k\rangle$ and $|g,p_i\rangle$. A π-pulse at ω_2 can either put the part of the atom in state $|e,p_i+\hbar k\rangle$ into state $|g,p_i+2\hbar k\rangle$ or put the part of the atom in the state $|g,p_i\rangle$ into $|e,p_i-\hbar k\rangle$. (b) The probability of finding the atom in state $|e\rangle$ as a function of ω_2.

transitions shown in Fig. 1a. The probability of finding the atom in state $|e\rangle$ as a function of ω_2 is shown in Fig 1b.

In order to get higher frequency resolution, the pulse lengths of the $\pi/2$- and π-pulses can be increased. However, this would mean that the pulses would only address atoms with a well-defined velocity relative to the wavevector of the photon. A much larger class of initial atomic velocities can be used if two sets of $\pi/2$-pulses with short pulse lengths are applied, as shown in Fig. 2. Thus, we are led naturally to the experimental configuration that was first used to extend the Ramsey method of separated oscillatory fields into the optical domain [8]. It was later realized by Bordé that this configuration of pulses would cause the wave packets to spatially separate to form a type of atom interferometer [9].

The Ramsey-Bordé interferometer is actually two atom interferometers given by the paths that define the two trapezoids in Fig 2. Each interferometer produces a set of Ramsey fringes where the central fringe of the upward going interferometer is separated from the central fringe of the other interferometer by ω_r. Fig 3 shows a fringe pattern taken with square-shaped $\pi/2$-pulses and is the Ramsey equivalent of the excitation pattern in Fig. 1b.

The precision of this interferometer is greatly increased by sandwiching N π-pulses in between the two sets of $\pi/2$-pulses. These π-pulses increase the spatial separation of the end points of the two interferometers. In the limit of many π-pulses, the photon recoil measurement becomes a measurement of the separation Δx of the

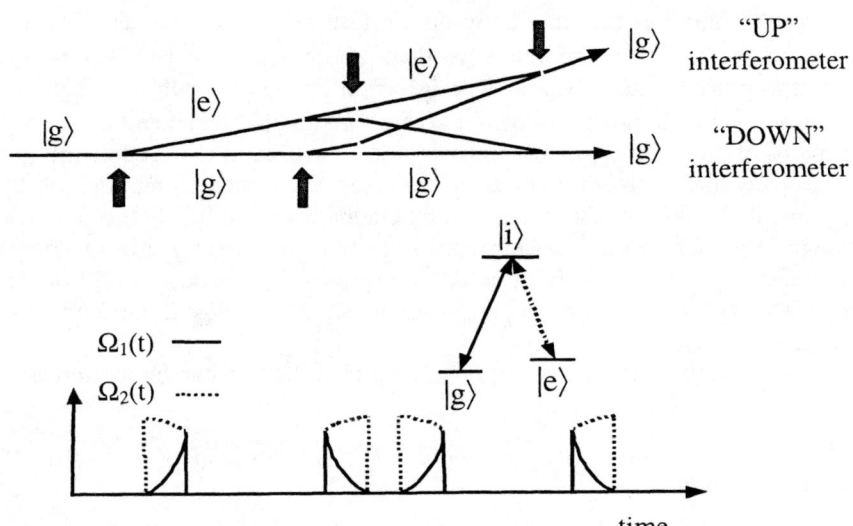

FIGURE 2. The Ramsey-Bordé interferometer. Each arrow indicates the direction of the effective wavevector of the $\pi/2$-pulses. The momentum impulses are given to the atoms via adiabatic passage between two atomic levels $|g\rangle$ and $|e\rangle$. The shapes of the 4 optical $\pi/2$-pulses used to construct the "down" interferometer are shown in the lower half of the figure. The "up" and "down" interferometers are selected by altering the shapes of middle two $\pi/2$-pulses.

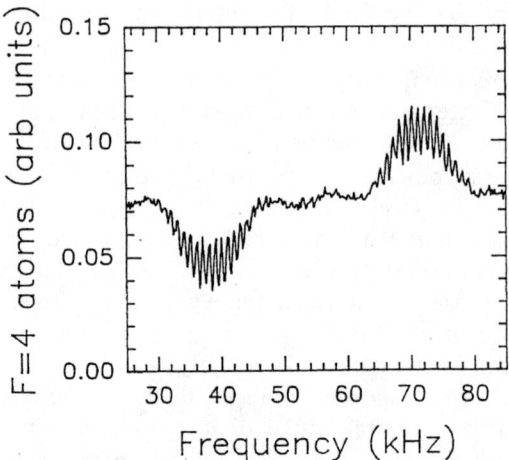

FIGURE 3. Older data from a four $\pi/2$-pulse interferometer using off-resonant Raman pulses at 852 nm. The time separation between each $\pi/2$-pulse pair was 1 ms. Currently, we use 895 nm light and the adiabatic transfer method to take data with 120 ms between $\pi/2$-pulses and the two interferometers separated by ~160 photon momenta instead of 4 photon momenta as in the case above. Thus, our Ramsey linewidths are ~8 Hz and the difference between the central fringes is ~30 kHz × 41.

positions of the two interferometers at the time of the fourth $\pi/2$-pulse, where Δx $=2v_r(N+1)T$, and T is the time between the last two $\pi/2$-pulses. The phase difference between the two interferometers is then $\Delta\phi = k_{eff}\Delta x$, where $k_{eff} = k_1 + k_2$ for counterpropagating laser beams. The insertion of many π-pulses in between the two sets of $\pi/2$-pulses is fairly insensitive to systematic effects, since the pulses act on the two paths of the interferometer, both of which correspond to the same internal state.

Our current atom interferometer uses an adiabatic method of transferring momenta to the atoms first introduced by Gaubatz, et al. [10]. In this form of adiabatic passage, time-delayed and on-resonant light fields efficiently transfer atoms between two states. By generating the time delayed pulses with acousto-optic modulators, we are able to tailor the shape of the pulses as shown in Fig. 2 and construct an atom interferometer.

For the three level atom in Fig. 2, the Hamiltonian can be written as

$$\hat{H} = \frac{\hat{p}^2}{2m} + \hbar\omega_e|e\rangle\langle e| + \hbar(\omega_i - i\Gamma/2)|i\rangle\langle i| + \hbar\omega_g|g\rangle\langle g| - \boldsymbol{d}\cdot\boldsymbol{E}, \quad (2)$$

where the electric field is given by

$$\boldsymbol{E} = \boldsymbol{E}_1\cos(k_1\cdot x - \omega_1 t + \phi_1) + \boldsymbol{E}_2\cos(-k_2\cdot x - \omega_2 t + \phi_2). \quad (3)$$

The interaction Hamiltonian can be shown to be [7]

$$\hat{H}_{\text{int}} = \frac{\hbar}{2} \begin{pmatrix} 0 & \Omega_e^* \, e^{i(\Delta_2 t - \phi_2)} & 0 \\ \Omega_e \, e^{-i(\Delta_2 t - \phi_2)} & -i\Gamma & \Omega_g \, e^{-i(\Delta_1 t - \phi_1)} \\ 0 & \Omega_g^* \, e^{i(\Delta_1 t - \phi_1)} & 0 \end{pmatrix}, \tag{4}$$

where Δ_1 and Δ_2 are the detunings of ω_1 and ω_2 from $|g\rangle$ and $|e\rangle$ to $|i\rangle$.

The atoms adiabatically follow an eigenstate of the atom-field interaction Hamiltonian that is never coupled to the excited state. This so-called "dark state" is given explicitly by

$$|\Psi_D\rangle = \begin{pmatrix} c_{e,p+\hbar k_{\text{eff}}}(t) \\ c_{i,p+\hbar k_1}(t) \\ c_{g,p}(t) \end{pmatrix} = \begin{pmatrix} \sin\theta \, e^{-i\phi} \\ 0 \\ \cos\theta \end{pmatrix}, \quad \tan\theta \, e^{-i\phi} = -\frac{\Omega_g}{\Omega_e} e^{i(\phi_1 - \phi_2)} \tag{5}$$

for ω_1 and ω_2 tuned exactly on resonance. Note that the relative phases of the states $|e\rangle$ and $|g\rangle$ are locked to the phases ϕ_1 and ϕ_2 of the driving fields as long as the states can follow the fields adiabatically.

Also, note that the atoms experience no ac Stark shift for a three level system. The presence of additional energy levels introduces a Stark shift, but if the detuning of these states is sufficient, the ac Stark shift will contribute less than one ppb error in the final interferometer phase. Our numerical calculations of transfer efficiencies and ac Stark shifts for the cesium $P_{1/2}$ intermediate state performed by integrating the time-dependent Schrödinger equation show that adiabatic transfer is superior to stimulated Raman transitions for both transfer efficiency and ac Stark shifts [11].

Experimentally, we found that the adiabatic transfer efficiency is ~94% as compared to off-resonant Raman efficiency ~85% used in our first measurements of h/M [6]. In our adiabatic transfer experiment, we use σ^+ polarized light tuned between the two 6 $S_{1/2}$ ground states $|F = 3, m_F = 0\rangle$ and $|F = 4, m_F = 0\rangle$ and the excited state $6P_{1/2}, |F = 3, m_F = +1\rangle$. Atoms that are *not* coherently transferred are mostly optically pumped to the states $|4,+4\rangle$ and $|3,+3\rangle$ and thus the interferometer contrast is preserved. The small fraction of atoms that fall back into the $|3,0\rangle$ or $|4,0\rangle$ states will not cause a net average phase shift, because their coherence is destroyed by the spontaneous emission process.

The recoil measurement is made in an atomic fountain geometry in which atoms are first collected and cooled in a MOT, further cooled in optical molasses, and then launched upwards by shifting the frequencies of the molasses beams. On the way up the atoms are optically pumped into the $|F = 4, m_F = 0\rangle$ state. The atoms then enter a region with a triple-layer magnetic shielding. A solenoid inside the innermost shield creates a bias field, adjustable between 0-600 mG. The bias field shifts the energy levels of the magnetic-field sensitive states so that the Raman lasers only address the atoms in the $m_F = 0$ states. As shown in Fig. 4, a Ti:sapphire laser generates the light for the two Raman beams locked to the 895 nm D_1-lines.

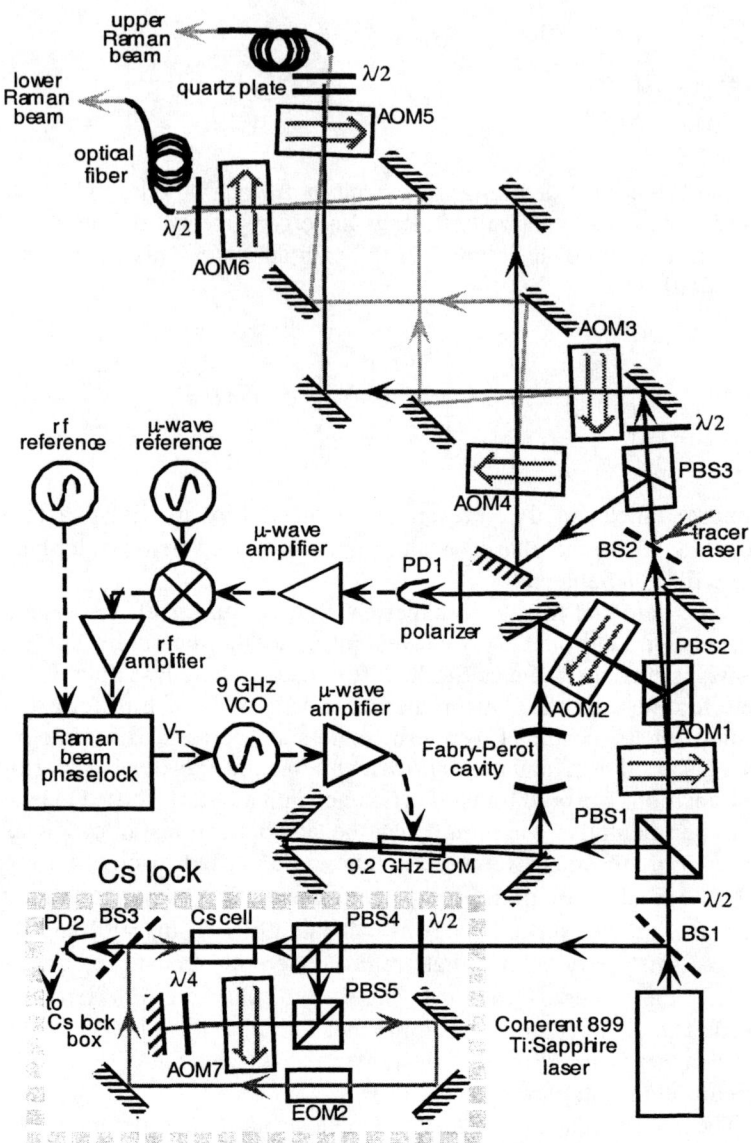

FIGURE 4. Schematic of the method used to generate the laser beams used in the experiment. A Ti:sapphire laser, locked to a Cs line, is used to produce both Raman beams, ω_1 and ω_2. The carrier frequency is at ω_1, while ω_2 is produced by a 9.2 GHz electro-optic modulator. After filtering out the unwanted sideband and carrier frequencies with a Fabry-Perot cavity, ω_1 and ω_2 are recombine at PBS2. The two beams then enter a switchyard consisting of acousto-optic modulators AOM3 - AOM6 that are used to switch the direction of k_{eff}. After the switchyard, the light is directed into optical fibers to spatially filter the beams and stabilize the optical alignment.

SYSTEMATIC EFFECTS

The photon recoil experiment provides many handles for testing potential systematic effects. Within a few seconds, we can completely alter the geometry of the atom interferometer by changing the time between the $\pi/2$-pulses, the number of π-pulses, the positions in the fountain trajectories where the light-atom interactions occur, the intensity and shape of the optical pulses, etc. We can also change the frequency offsets, polarization, alignment and wavefront curvature of the laser beams and vary environmental factors such as the magnetic bias field. We have now placed upper bounds on most of these systematic effects, but have not yet completed our search. Some of this work is described in the Ph.D. thesis of Brent Young [7].

Many of the potential systematic effects are cancelled by the fact that the value of the recoil shift is a difference between two measurements, corresponding to the "up" and "down" interferometer of Fig. 2. Also, we routinely change the direction of k_{eff} with our beam switchyard, thereby interchanging the role of the up and down interferometers. This reversal of k_{eff} is useful in subtracting out any residual non-linear frequency dependent electronic phase shifts, magnetic field phase shifts, and laser frequency dependent phase shifts. (See Fig. 5.)

Note that for our most sensitive interferometers (with 120 ms between $\pi/2$-pulses and 30 π-pulses) a 1 ppb uncertainty in the measurement of the photon recoil (proportional to α^2) requires a 0.7 mrad phase uncertainty.

We briefly discuss the most important systematic effects we have investigated.

Quadratic Zeeman Shifts

Using the Zeeman shift of the atomic resonances, we can map out the magnetic field inside the interferometer region and calculate what the expected shift should be as a function of applied bias field. We have made a preliminary study of the magnetic field dependence of h/M between 20-580 mG and see a very small dependence (-0.012 ±0.033 ppb/mG) in the recoil momentum. The uncertainty to this correction can be made less than 1 ppb.

Frequency Detuning

In order to calculate h/M from our measurement of the recoil shift, we must know the wavelength of the Cs D_1-line. A recent measurement of the D_1-line eliminates this uncertainty [12]. We also have to know the accuracy of our feedback system that locks our laser to this line. A residual uncertainty of 0.0±0.6 ppb is due to errors in our frequency lock (~100 kHz, rms) to a reference cesium cell, which we have calibrated with respect to laser cooled Cs atoms in the vacuum chamber.

Beam Alignment and Wavefront Curvature

We have developed an alignment procedure that can produce counter-propagating beams to within ~10 µrad, corresponding to a 0.03 ppb error. Another systematic effect that has to be considered at this level of precision is the Gouy phase shift of a laser beam in the vicinity of the focal point; $\Phi_G = \arctan(\lambda z / \pi w_0^2)$ [13]. For $w_0 = 1$ cm, this phase shift results in a 0.8 ppb correction, provided the atoms are centered at the waist of the 2 cm diameter beam.

Missed Photon Kicks

A crucial ingredient to the precision of our measurement is the addition of N π-pulses in between the two sets of $\pi/2$-pulses. We demand that all of the atoms receive exactly N additional photon recoils. A phase error will result if some of the atoms miss one or more momentum changing π-pulses. If an atom were to miss a π-pulse, it would be in the bright state at the beginning of the next pulse. This next π-pulse would then induce incoherent, single-photon transitions resulting in no net (averaged) phase shift in the interferometer. Also, since the frequency width of the π-pulses is on the order of the recoil shift, an atom missing two or more π-pulses would be far enough off-resonance that it would not be affected by succeeding π-pulses.

The most serious concern is that some of the atoms may experience a Doppler-free transition induced by two co-propagating beams. To minimize the amount of co-propagating light, we tilt all optics after and including the final polarizers away from normal incidence so that no back reflected light can illuminate the atoms. We also avoid applying π-pulses when the atoms are close to the apogee of their trajectory where they would be most sensitive to Doppler free transitions. An atom that misses one momentum impulse will be drastically shifted in phase by $\Delta\phi = (\hbar k_{eff}^2/M_{Cs})T$. If the fraction of atoms that miss a pulse is small, we can choose the time T so that $\Delta\phi$ is modulo 2π. With this choice of T, the fringe pattern for a single missed recoil will be the same as the fringe pattern for no missed recoils.

As a check on our ability to eliminate back reflections, we looked for this systematic effect by adding 30 π-pulses at the apogee of the atomic trajectory so that the laser was tuned near the Doppler-free transition. We then scanned the time interval T over a range where $\Delta\phi$ changed by 2π. The phase shift we observed had an amplitude of 17 ± 16 ppb, i.e. consistent with no phase shift, indicating that there are no reflected beams that would induce Doppler-free transitions. Under normal operating conditions, with an appropriate choice of T, we expect a suppression of this systematic effect by a factor of at least 100.

Wavefront Distortion (High Spatial Frequency Wavefront Curvature)

High spatial frequency wavefront distortion is difficult to characterize and quantify. We depend on the fibers to filter out all higher order spatial modes, and we insure that all optics after the fibers are optically clean and specified to better than $\lambda/10$. We also refer to the experience gained in our atom interferometer measurement of the acceleration of gravity [14]. In order to test for high frequency wavefront curvature, we added spatial noise by passing the beam through a glass plate with small particles sprinkled onto its surface. Interferometer phase shifts were seen only when the glass plate was severely contaminated by particles.

Electronic Phase Shifts

Our atom interferometer measurement uses light fields as "rulers" for measuring the motion of atoms. The laser field establishes a set of wavefronts with periodicity $\lambda_{eff}=2\pi/(k_1 + k_2)$. Each time the light field causes an atom to make a transition, the atom records (in its phase) its position according to the ruler. Ideally we would want to leave the ruler fixed spatially for the entire measurement. Unfortunately, accelerations of the atoms from gravity and photon recoils causes the velocity to change so much that the Doppler-shift exceeds the frequency width of the transition. Consequently, we need to shift the laser frequency to compensate for the Doppler-shift. In other words, we change the scale of the ruler for each optical pulse. As long as we know exactly how far we have shifted the ruler, this is not a problem. In the process of shifting the laser frequency, however, frequency-dependent phase shifts in the electronics may cause interferometer phase errors. In the thesis of B. Young [7], this systematic effect was on the order of 100 ppb. We are now using a higher bandwidth frequency synthesizer with no rf filters on the output. The electronic frequency dependent phase shift was greatly decreased and the residual shift is 0.0 ± 0.2 ppb.

Coriolis Forces

Phase shifts also arise because of the Sagnac effect. If the momentum impulses are not co-linear with the average atomic trajectory, the interferometer will act as a gyroscope that senses the earth's rotation. We rocked the entire apparatus consisting of the laser table and vacuum chamber in order to induce Coriolis effects. We can establish an upper limit to Coriolis effects by rotating the apparatus much more rapidly than the earth rotation rate. Only with an intentional misalignment, were we able to observe a phase shift. Knowing how well our system is aligned under normal conditions, we estimate this systematic effect to be 0.0±1.9 ppb in h/M_{Cs}.

Index of Refraction Effects [15]

The Cs atoms in the vacuum chamber can be described as a dielectric medium that changes the magnitude of k_{eff}. The effect due to room temperature Cs atoms has been calculated to be less than 0.5 ppb assuming a Cs pressure of 1×10^{-9} torr. The effect due to cold atoms is more complicated. There will be no effects for atoms in the dark state since they do not interact with the excited state $|i\rangle$. For atoms not in the dark state, if the laser is tuned away from the exact atomic resonance, the index of refraction induced by the cold atoms will change the wavevector of the light. The size of this change will depend linearly on the laser detuning (for small detunings) and on the spatial and momentum distributions of the cold atoms. To look for this effect experimentally, we varied both the single and two-photon detunings of the Raman laser beams (Fig. 5a and 5b, respectively). After subtracting out the change in photon momentum $\Delta p_{photon} = \hbar \Delta k$ due to the frequency change, we observe no significant shift.

In addition to modifying the index of refraction, we must also consider the possibility that the momentum impulse delivered to the atom is affected by the presence of a background gas of atoms [16]. Questions regarding the momentum of light in a dielectric medium has generated considerable attention, beginning with Minkowski in 1908 [17]. For our purposes, we can consider our two counter-propagating optical pulses at frequencies ω_1 and ω_2. In vacuum, the momentum of the light fields is well defined. In the medium of dilute gas, the total energy and momentum are shared between the optical field and the energy and momentum stored

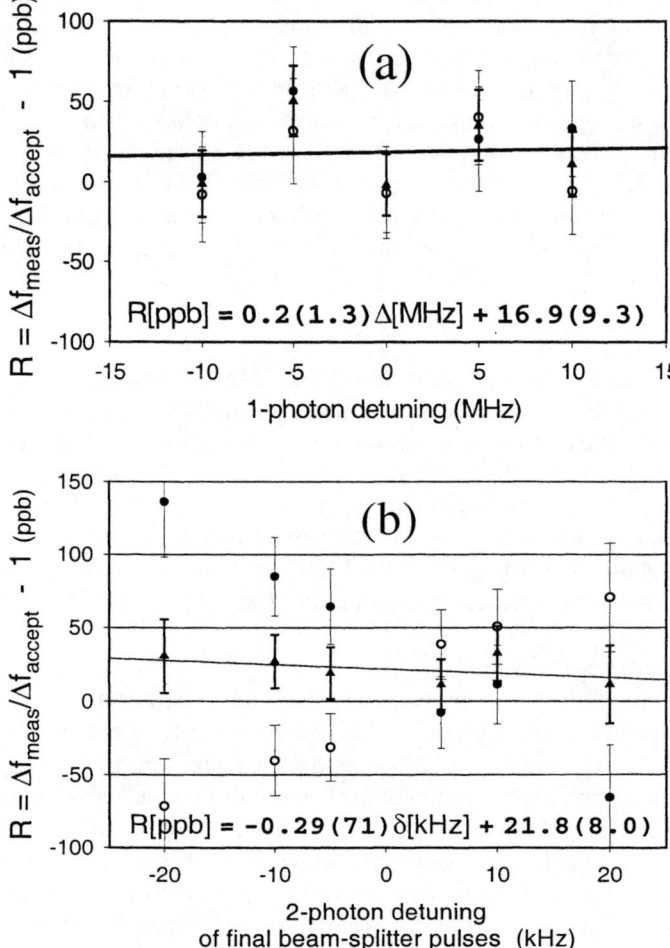

FIGURE 5. Measured deviation of the recoil frequency shift from the accepted value (1998 CODATA) as a function of detuning. (a) The 1-photon detuning Δ is the absolute frequency offset of both laser fields from resonance: $\Delta = \omega_1 - \omega_{ki} = \omega_2 - \omega_{ki}$. (b) The 2-photon detuning δ is the relative detuning of the two laser fields from resonance: $\delta = (\omega_1 - \omega_2) - \omega_{ge}$. The solid circles are the difference between the upper and lower interferometers. The open circles are the same differences taken with the direction of k_{eff} reversed. The triangles are the unweighted average of the data taken with k_{eff} and k_{eff} reversed. The slope and intercept of the straight line fits to these averaged data are given.

with the atoms comprising the dielectric medium [18]. The target atoms (atoms used in the interferometer) will receive impulses by coherently scattering a portion of the electromagnetic field at ω_1 into the field at ω_2. The momentum re-distribution of the two optical pulses can be measured after the two pulses have left the dielectric medium. Since the dielectric medium does not have any net momentum or energy after the pulses have left [19], momentum conservation dictates the atom must receive a momentum impulse equal to the exchange of momentum between the two optical fields. Because the momentum of these two vacuum fields is quantized in units of $\hbar k_1$ and $\hbar k_2$, we conclude that the atoms will receive an impulse $\hbar(k_1 + k_2)|_{\text{vacuum}}$.

If one explicitly calculates the momentum impulse delivered to the atom beginning with the force $F = -\nabla H(r,t) = \nabla(p \cdot E)$, the gradient of the electric field will give a term proportional to $k_{\text{eff}} \times n$. This term suggests that the momentum impulse imparted to the atom is increased by the index of refraction. However, an explicit calculation of the electric and magnetic fields in the dielectric medium shows that the electromagnetic field E acting on the target atom is smaller than the corresponding field in vacuum. This decrease in the E counteracts the increase in k_{eff} so that the momentum imparted to the atom is the same as in vacuum. We will present the details of this calculation in a future publication [20].

AC Stark Shifts

Figs. 5a,b show no unexplained phase shift as we vary the laser frequencies. Thus, we can rule out ac Stark shifts that might perturb the dark state.

The vibration isolation system [21] used in this experiment requires a "tracer beam" that follows the same optical path as the Raman beams and might itself introduce ac Stark shifts. By gating the beam on for different time intervals, we also show that this beam does not affect the measurement.

Phase Errors Due to $\pi/2$-pulses

Erroneous phase shifts caused by the $\pi/2$-pulses are revealed by varying the time T separating the $\pi/2$-pulses. To simplify the analysis, we convert the recoil frequency shift Δf_{rec} into a phase shift using $\Delta\phi = 2\pi(N+1)T \Delta f_{\text{rec}}$. In Fig. 6, we plot this phase shift as a function of T. As T approaches zero, for ideal $\pi/2$-pulses there should be no net phase difference between the interferometers, so any deviation from this result represents a systematic phase error. From the non-zero intercept of the linear fit to the data in Fig. 6, we find a systematic phase error of 23.4±8.0 mrad.

We are in the process of determining the origin of this systematic shift. If the $\pi/2$-pulses induce a phase error, then changes in the shape of the adiabatic $\pi/2$-pulses could result in changes in the interferometer phase shift. For example, in Fig. 7 we plot the shift of the recoil frequency as we change the time $\Delta T_{\pi/2}$ used to turn on and off the two laser fields E_1 and E_2 at the beginning and end of the $\pi/2$-pulses.

We feel that the current systematic error from the $\pi/2$-pulses may result from non-uniformities in the way the light fields are switched on and off, especially when E_1 and E_2 are turned on and off in unison. Currently the intensities of the two light fields are controlled with two independent acousto-optic modulators.

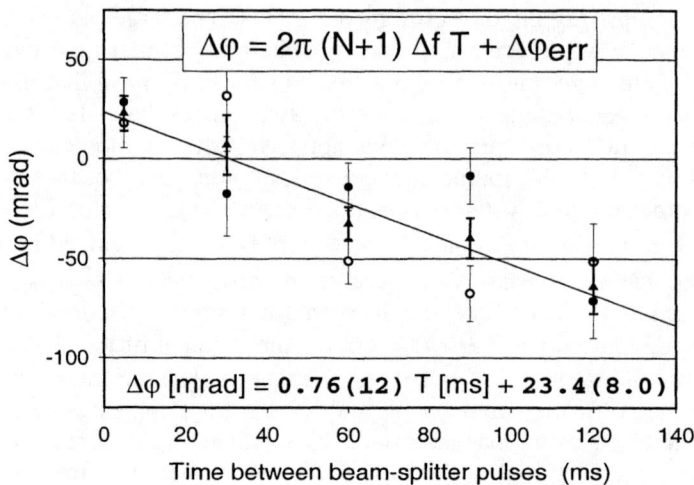

FIGURE 6. The recoil shift Δf_{rec} is measured for different values of T, the time between $\pi/2$-pulses. These frequency shift values (in Hz) are converted to phase (in radians) by $\Delta\phi = 2\pi(N+1)T\,\Delta f_{rec}$, where N is the number of π-pulses. According to this relation the equivalent phase $\Delta\phi$ must vanish as T approaches zero. Any deviation from zero represents a systematic error $\Delta\phi_{err}$. If this error is independent of T, a value for the recoil shift can be extracted from the slope of the linear fit as described in the text.

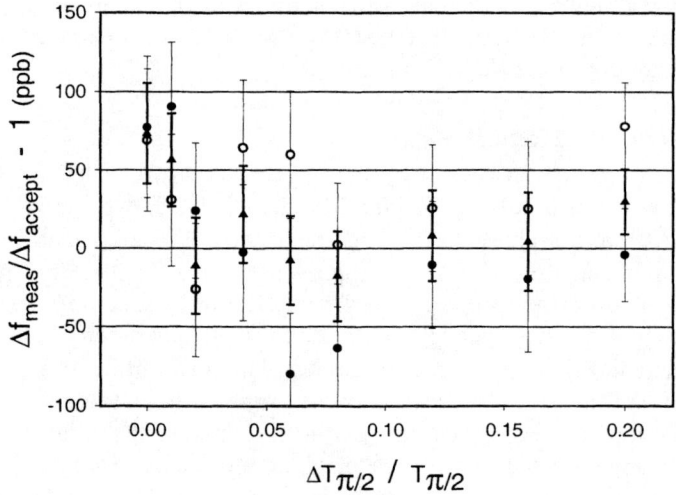

FIGURE 7. Measured deviation of the recoil frequency shift from the accepted value (1998 CODATA) as a function of $\pi/2$-pulse shape. In this case, we vary the time $\Delta T_{\pi/2}$ used to turn on and off the adiabatic transitions at the beginning and end of the $\pi/2$-pulses.

We are particularly concerned that a systematic phase shift could result if these two modulators do not switch identically. To test this idea and improve the overall performance we have installed an additional acousto-optic modulator (between PBS2 and PBS3 in Fig. 4) which will switch both light fields on and off together.

Methods of Dealing with Phase Errors Introduced by the $\pi/2$-pulses

In principle, if we were certain that this shift $\Delta\phi_{err}$ were independent of T, we could correct for this unwanted phase shift by plotting the phase shift $\Delta\phi$ vs. T as in Fig. 6. The slope of the linear fit is a measure of the recoil frequency. Alternatively, the intercept of the linear fit could be used to correct all measurements of $\Delta\phi$ by an amount $-\Delta\phi_{err} / (2\pi(N+1)T)$, where N is the number of π-pulses. In our analysis below, we use the second method to account for the phase shifts introduced by imperfect $\pi/2$-pulses.

CONCLUSION

Based on our search for systematic errors, in Table 1 we list the error budget of our current measurements. With improvements in beam collimation and alignment incorporated into our next set of measurements, we hope to bring most systematic errors below 1 ppb in h/M corresponding to 0.5 ppb in α.

Although we have not completed our search for systematic errors, in Fig. 8 we show the compilation of our data to date, converted to α using Eq. 1 with the 1998 CODATA values [1]. As a result of our search for phase errors from the $\pi/2$-pulses, we have 10 other data sets similar to Fig. 6. Fitting each data set to a line and averaging all values of $\Delta\phi_{err}$ produces a global phase error of 22.7 ± 3.7 mrad (equivalent to a 32.4 ± 5.3 ppb shift in h/M for $T = 120$ ms and $N = 30$ π-pulses). Using this value, we correct all of our data and compute a new average recoil shift, which we also present in Fig. 8 for comparison. One should view the difference

Table 1: Error Budget for α^2 (ppb)

Effect	Current Uncertainty	Expected Uncertainty
Quadratic Zeeman Shift	-0.9 ± 2.4	<1.0
Frequency detuning	0.0 ± 0.6	same
Missed photon kicks	0.0 ± 0.4	same
Wavefront curvature	-2.1 ± 1.2	<0.5
Beam alignment	-2.0 ± 2.0	<0.05
Wavefront distortion	0.0 ± 1.5	<0.5
Electronic phase shifts	0.0 ± 0.2	same
Coriolis effects	0.0 ± 1.9	same
Unresolved shifts:		
in T	±30	?
in pulse shape	±50	?

between the "corrected" and "uncorrected" values as indicative of our unresolved phase errors due to the π/2-pulses. The error bars on both values show only our current statistical uncertainty of less than 3 ppb in h/M. (1.5 ppb in α.)

The absolute value and corresponding uncertainty can not be assigned until we have completed our tests for systematic effects. We are still hopeful we can understand the cause of the phase errors introduced by our π/2-pulses, but even if we can not, we are developing a number of methods that can account for the shifts.

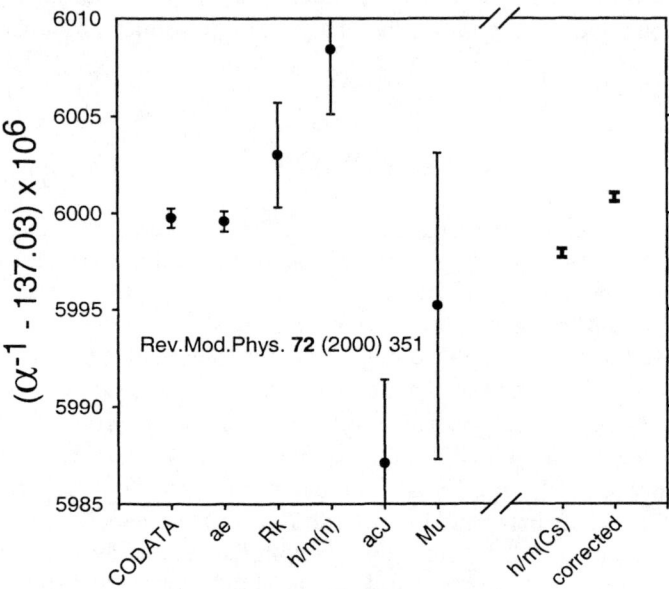

FIGURE 8. Graphical comparison of the five best determinations of fine structure constant α with our preliminary values. The value labeled "h/m(Cs)" is the average of all runs. The "corrected" value assumes that the π/2-pulses add a phase error that is independent of T. We calculate this "corrected value" by using the value of $\Delta\phi_{err}$ obtained from 11 data sets, one of which is shown in Fig. 6. This phase error is then removed from each data set taken for a given pulse separation T and number of π-pulses N. All error bars are one standard deviation uncertainties.

ACKNOWLEDGMENTS

This work was sponsored in part by grants from the AFOSR and the NSF. A.W. acknowledges support form the Alexander von Humboldt Foundation.

REFERENCES

1. These are the uncertainties listed in the CODATA Recommended Values of the Fundamental Constants. 1998. This work is published in P.J. Mohr and B.N. Taylor, *Rev. of Mod. Phys.* **72**, 351-495 (2000). A more accurate determination is discussed in another invited contribution to this conference.
2. The measurement of M_{Cs}/m_p is discussed in another invited contribution to this conference proceedings.
3. T. Kinoshita, *Rep. Prog. Phys.* **59**, 1459-1492 (1996).
4. W. Liu, *et al,. Phys. Rev. Lett.* **82**, 711-713 (1999).
5. J.L. Hall, C.J. Bordé and K. Uehara, *Phys. Rev. Lett.* **37**, 1339-1342 (1976).
6. D.S. Weiss, B.C. Young, and S. Chu, *Phys. Rev. Lett.* **70**, 2706-2709 (1992); *Appl. Phys. B* **59** 217-256 (1994); M. Weitz, B.C. Young, and S. Chu, *Phys. Rev. A* **50**, 2438-2566, (1994); *Phys. Rev. Lett.* **73**, 2563-2566 (1994).
7. B.C. Young, unpublished Ph.D. thesis, (1997); S. Chu, Les Houches Lectures in Physics, June 1999.
8. Y.V. Baklanov, B.Y. Dubetsky and P.P. Chebotayev, *Appl. Phys.* **9**, 171 (1976).
9. C.J. Bordé, *et al.*, *Phys. Rev. A* **30**, 1836 (1984).
10. U. Gaubatz, *et al.*, *Chem. Phys. Lett.* **149**, 463 (1988).
11. M. Weitz, B.C. Young and S. Chu, Phys. Rev. A **50**, 2438 (1994).
12. Th. Udem, J. Reichert, R. Holzwarth, and T. W. Hänsch, Phys. Rev. Lett. **82**, 3568 (1999)
13. H. Kogelnik and T. Li, *Appl. Opt.* **5**, 1550 (1966); A.I. Siegman, *Lasers*, University Science Books, Mill Valley, CA, 1986.
14. A. Peters, K.-Y. Chung and S. Chu, Nature **400**, 849-852 (1999); *op cit.*, *Metrologia*, to be published.
15. S. Chu thanks David Pritchard for asking a provocative question after his talk that led to our consideration of these effects.
16. We wish to acknowledge stimulating discussions with Wolfgang Ketterle and Steve Harris concerning this point.
17. H. Minkowski, *Gesell. Wiss. Göttingen, Nachr., Math.-Phy. Klasse*, 1., pp. 53-111, 1908.
18. See J.D. Jackson, *Classical Electrodynamics,* 2nd ed., John Wiley and Sons, New York, 1975, pp. 240-241.
19. J.P. Gordon, *Phys. Rev. A* **8**, 14-21 (1973).
20. A. Wicht, J. Hensley and S. Chu, to be published.
21. J. Hensley, A. Peters and S. Chu, *Rev. of Sci. Instr.* **70**, 2735-2741 (1999).

Optical frequency metrology and its contribution to the determination of fundamental constants

R. Holzwarth, J. Reichert, Th. Udem and T.W. Hänsch

Max-Planck-Institut für Quantenoptik,
Hans-Kopfermann-Strasse 1, D-85748 Garching, Germany
http://www.mpq.mpg.de/~haensch/chain/chain.html

Abstract. Optical frequency metrology plays an important role for the determination of the Rydberg constant and offers an alternative way to determine the fine structure constant α. We have developed a new technique for measuring optical frequencies using femtosecond light pulses culminating in the single laser optical frequency synthesizer. This new technique greatly simplifies the task of measuring optical frequencies.

Measuring the frequency of light

For many years, optical precision spectroscopy of the simple hydrogen atom has inspired advances in nonlinear laser spectroscopy and optical frequency metrology. This work has now culminated in a simple solution to the long-standing problem of measuring the frequency of light which required heroic efforts in the past. These advances are opening intriguing opportunities for ultra precise measurements and fundamental tests in atomic physics.

Here we describe the new approach to optical frequency metrology which takes advantage of the regularly spaced broad comb of modes of a mode-locked femtosecond (fs) laser, and we report on recent precise measurements of the cesium D_1 line and the hydrogen 1S-2S two-photon resonance which yield new values of the fine structure constant and the Rydberg constant.

Optical frequency differences

A typical situation in frequency metrology arises when a frequency difference between a well known optical reference frequency (calibrated e.g. with a traditional harmonic frequency chain) and an unknown optical frequency tens or hundreds of THz apart needs to be measured. If the frequency gap is on the order of 100 GHz

CP551, *Atomic Physics 17*, edited by E. Arimondo, P. DeNatale, and M. Inguscio
© 2001 American Institute of Physics 1-56396-982-3/01/$18.00

or smaller a beat signal from a fast photo diode can be directly counted. For larger frequency gaps one can use optical frequency comb generators, i. e. electro optical modulators that create side bands very efficiently. Beat signals can then be observed with sidebands on different sides of the carrier and frequency gaps on the order of 8 THz can be bridged [1]. Two powerful tools help us to handle even larger frequency differences, these are fs frequency comb generators discussed in detail below and optical frequency interval divider stages [2].

An optical frequency interval divider (OFID) divides an arbitrarily large frequency difference by a factor of two. It receives two input laser frequencies f_1 and f_2. The sum frequency $f_1 + f_2$ and the second harmonic of a third laser $2f_3$ are created in nonlinear crystals. The radio frequency beat signal between them at $2f_3 - (f_1 + f_2)$ is used to phase-lock the third laser at the midpoint $f_3 = (f_1 + f_2)/2$. Phase-locking of two optical frequencies is achieved electronically by locking the phase of their beat signal to zero or, to reduce $1/f$ noise, to a given offset radio frequency, provided by a local oscillator [3]. Techniques of conventional radio frequency phase-locked loops can be applied.

Femtosecond light pulses

It has been long recognized [4] that the periodic pulse train of a mode locked laser can be described in the frequency domain as a comb of equidistant modes as shown in Fig. 1. To understand the mode structure of a fs frequency comb and the

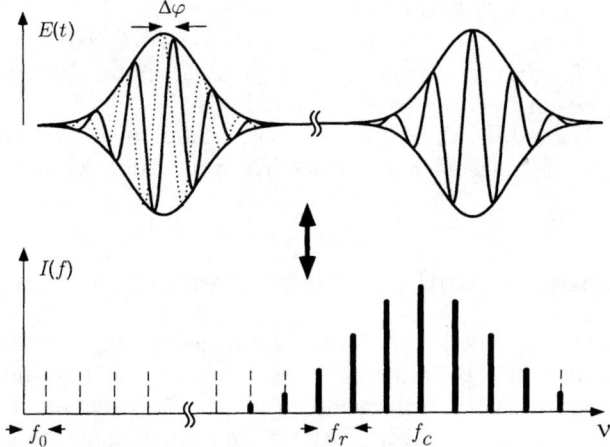

FIGURE 1. Two consecutive pulses of the pulse train emitted by a mode locked laser and the corresponding spectrum. The pulse to pulse phase shift $\Delta\varphi$ results in a offset frequency $f_0 = \Delta\varphi/2\pi T$ because the carrier wave at f_c moves with the phase velocity v_p while the envelope moves with the group velocity v_g.

techniques applied for its stabilization one can look at the idealized case of a pulse circulating in a laser cavity with length L as a carrier wave at f_c that is subject to strong amplitude modulation described by an envelope function $A(t)$. This function defines the pulse repetition time $T = f_r^{-1}$ by demanding $A(t) = A(t - T)$ where T is calculated from the cavity mean group velocity: $T = 2L/v_{gr}$. Because of the periodicity of the envelope function the electric field at a given place (e.g. at the output coupler) can be written as

$$E(t) = Re\left(A(t)e^{-2\pi i f_c t}\right) = Re\left(\sum_q A_q e^{-2\pi i(f_c + q f_r)t}\right) \qquad (1)$$

where A_q are Fourier components of $A(t)$. This equation shows that the resulting spectrum consists of a comb of laser modes that are separated by the pulse repetition frequency. Since f_c is not necessarily an integer multiple of f_r the modes are shifted from being exact harmonics of the pulse repetition frequency by an offset $f_o < f_r$:

$$f_n = n f_r + f_o \qquad n = \text{a large integer} \qquad (2)$$

This equation maps two radio frequencies f_r and f_o onto the optical frequencies f_n. While f_r is readily measurable, f_o is not easy to access unless the frequency comb contains more than an optical octave [5]. In the time domain the frequency offset is obvious because the group velocity differs from the phase velocity inside the cavity and therefore the carrier wave does not repeat itself after one round trip but appears phase shifted by $\Delta\varphi$ as shown in Fig. 1. The offset frequency is then calculated from $f_o = \Delta\varphi/T2\pi$ [6,5].

Note that such a fs frequency comb has two degrees of freedom which are the repetition frequency f_r and the offset frequency $f_0 < f_r$. Depending on the application one or both degrees of freedom have to be stabilized.

As the spectral width of these pulsed lasers scales inversely with the pulse duration the advent of fs lasers has opened the possibility to directly access THz frequency gaps.

Femtosecond combs as frequency rulers

At the high peak intensities of femtosecond laser pulses nonlinear effects due to the $\chi^{(3)}$ nonlinear susceptibility are considerable even in standard silica fibers. The output spectrum of a femtosecond laser can be broadened significantly via self phase modulation in an optical fiber therefore increasing its useful width even further beyond the time-bandwidth limit.

Now the question arises whether or not this broad frequency comb is equally spaced and can therefore be used as a ruler to measure frequency differences. To test this we have compared the fs comb with a divider stage as illustrated in Fig. 2. We have found that the frequency comb is equally spaced even after further spectral broadening in a standard single mode fiber at the level of a few parts in 10^{18} [7]. Note that the coherence between the pulses is obviously preserved. This broadening process can also be understood in the picture of four wave mixing.

Previously we have also shown that the easily accessible repetition rate of such a laser equals the mode spacing within the experimental uncertainty of a few parts in 10^{16} [8]. Following the rapid advances in ultrafast technology recently a 104 THz gap has been bridged with such a broadened comb [9].

To phase-lock the pulse repetition rate to a signal provided by a synthesizer one faces the problem of noise multiplication. It is well known that the total noise intensity grows as N^2 when a radio frequency is multiplied by a factor of N [10]. Fortunately the laser cavity acts as a filter and prevents the high frequency noise components from propagating through the frequency comb. [5].

For most applications it is desirable to fix one of the modes in frequency space and phase-lock the pulse repetition rate simultaneously. For this purpose it is necessary to control the phase velocity (more precisely the round trip phase delay) of that particular mode and the group velocity (more precisely the round trip group delay) independently. A piezo driven folding mirror is changing the cavity length and but leaves $\Delta\varphi$ approximately constant as the additional path in air does have a negligible dispersion. The offset frequency $f_o = \Delta\varphi/2\pi T$ is therefore changed through T. A mode-locked laser that uses two intracavity prisms to produce the negative group velocity dispersion $(\partial^2\omega/\partial k^2 < 0)$ necessary for Kerr-lens mode-

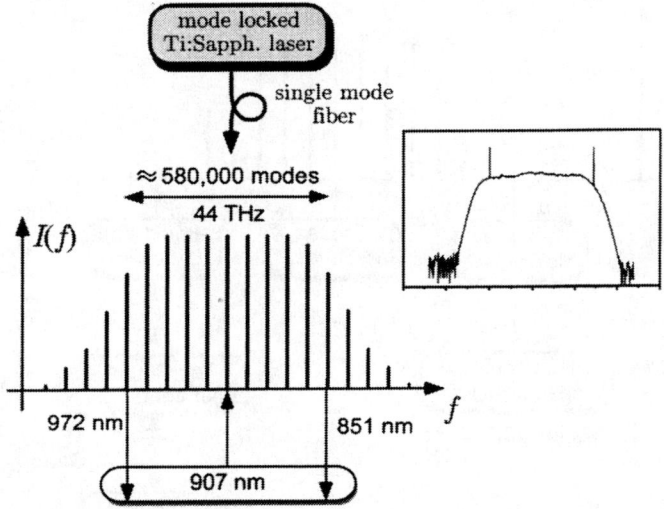

FIGURE 2. Comparison of the broadened fs comb and an optical interval divider (oval symbol). The inset shows on a logarithmic intensity scale the observed spectrum. The peaks mark the position of the 972 nm and 851 nm laser diodes and the comb was generated with a Mira 900 system (Coherent Inc.) delivering 73 fs pulses at a 75 MHz repetition rate.

locking provides us with a means for independently controlling the pulse repetition rate. We use a second piezo-transducer to slightly tilt the mirror at the dispersive end of the cavity about a vertical pivot that ideally corresponds to the mode f_n [5]. We thus introduce an additional phase shift $\Delta\Phi$ proportional to the frequency distance from f_n, which displaces the pulse in time and thus changes the round trip group delay [5]. In the frequency domain one could argue that the length of the cavity stays constant for the mode f_n while higher (lower) frequency modes experience a longer (shorter) cavity (or vice versa, depending on the sign of $\Delta\Phi$). In the case where only dispersion compensation mirrors are used to produce the negative group velocity dispersion one can modulate the pump power or manipulate the Kerr lens by slightly tilting the pump beam [11]. Although the two controls (i.e. cavity length and pump power) are not orthogonal they affect the round trip group delay T and the round trip phase delay differently and this is which allows us to control both, f_0 and f_r.

Cesium D_1 line and fine structure constant α

A first example for an optical frequency measurement using fs frequency comb techniques is the determination of the frequency of the cesium D_1 line. Here we

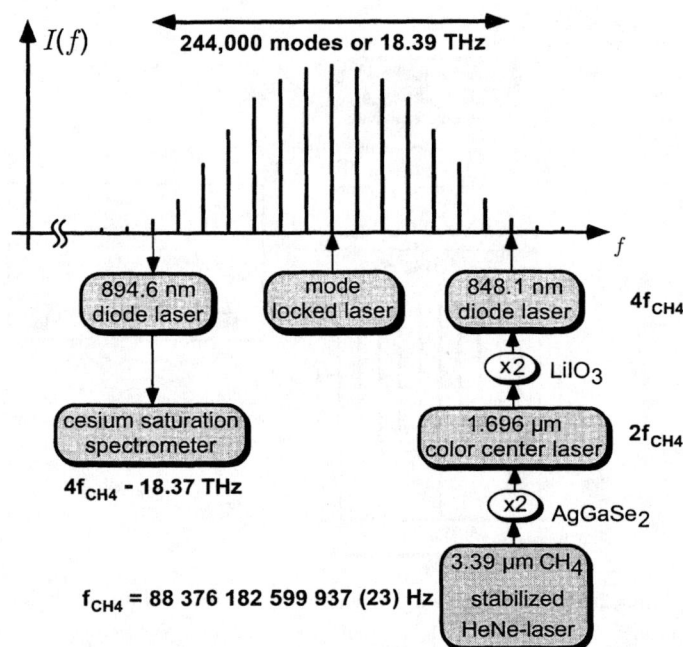

FIGURE 3. Frequency chain used for the determination of the cesium D_1 line.

are facing the situation discussed above of a well known reference frequency, in our case a transportable methane stabilized He-Ne laser, and an unknown frequency.

We compare the frequency of the cesium D_1 line at 895 nm with the 4th harmonic of the methane stabilized He-Ne laser operating at 3.4 μm (88 THz). The frequency of this laser has been calibrated at the Physikalisch Technische Bundesanstalt Braunschweig/Germany (PTB) and in our own laboratory against a microwave cesium atomic clock to within a few parts in 10^{13}. A color center laser at 1.7 μm has been phase-locked to the second harmonic of the He-Ne standard. A laser diode at the second harmonic of the color center laser (848 nm) serves as a link to the femtosecond frequency comb. The femtosecond comb is generated by a Mira 900 system (Coherent Inc.) and bridges the 18.6 THz gap between 848 nm and the cesium D_1 line at 895 nm. To probe the cesium D_1 transition we use a saturation spectrometer with two linearly polarized counter-propagating laser beams of equal intensities (10 μW/cm^2) and a 7.5 cm long cesium cell at room temperature. With this Doppler free method, we observe 4 hyperfine components of the single stable isotope ^{133}Cs for the transitions from the ground states $F_g = 3$ and $F_g = 4$ to the upper states $F_e = 3$ and $F_e = 4$ with a linewidth of about 6 MHz (FWHM). The cross-over resonances were not visible due to the large separation of the hyperfine components. With this technique we have been able to reach an accuracy of 1.2×10^{-10} limited by imperfections in the magnetic shielding of the cesium cell and improve previous measurements by almost 3 orders of magnitude.

One of the experimental problems has been to determine the number of modes between the laser diodes at 848 nm and 895 nm. For this task we employed a cavity that was stabilized to have a free spectral range of exactly 20 times the pulse repetition rate f_r. If the beat notes of the laser diodes with the transmitted light appear at the same frequencies, the mode number was a multiple of 20 (1.5 GHz), and therefore sufficient to be unambiguously identified by our wavemeter [12].

Our measurement of the cesium D 1 line provides an important link for a new determination of the fine structure constant α, one of the most fundamental constants of nature. Because α scales all electromagnetic interactions, it can be determined by a variety of independent physical methods. Different values measured with comparable accuracy disagree with each other by up to 3.5 standard deviations and the currently most accurate value from the electron $g-2$ experiment relies on extensive QED calculations [13]. The 1999 CODATA value [14] $\alpha^{-1} = 137.035\ 999\ 76\ (50)$ (3.7×10^{-9}) follows the $g-2$ results. To resolve this unsatisfactory situation it is most desirable to determine a value for the fine structure constant that is comparable in accuracy with the value from the $g-2$ experiment but does not depend heavily on QED calculations. A promising way is to use the accurately known Rydberg constant R_∞ according to:

$$\alpha^2 = \frac{2R_\infty}{c} \frac{h}{m_e} = 2R_\infty \times \frac{2cf_{rec}}{f_{D_1}^2} \times \frac{m_p}{m_e} \times \frac{m_{Cs}}{m_p} \tag{3}$$

In addition to the Rydberg constant a number of different quantities, all based on intrinsically accurate frequency measurements, are needed. Experiments are underway in Stanford in S. Chu's group to measure the photon recoil shift $f_{rec} = f_{D1}^2 h/2m_{Cs}c^2$ of the cesium D_1 line. Together with the proton-electron mass ratio m_p/m_e, that is known to 2×10^{-9} [15] and even more precise measurements of the cesium to proton mass ratio m_{Cs}/m_p in Penning traps, that have been reported recently [16], our measurement has already yielded a new value of α as discussed by S. Chu in these proceedings.

Absolute optical frequencies

For the absolute measurement of optical frequencies one has to determine frequencies of several 100 THz in terms of the definition of the SI second represented by the cesium ground state hyperfine splitting of 9.2 GHz.

In the past heroic efforts were required to measure optical frequencies with harmonic frequency chains which create successive harmonics from the cesium radio frequency reference. These frequency chains were large, delicate to handle and usually designed to measure only one particular optical frequency. For this reason only a few frequency chains that reach all the way up to the visible or the UV have been built so far [17–20].

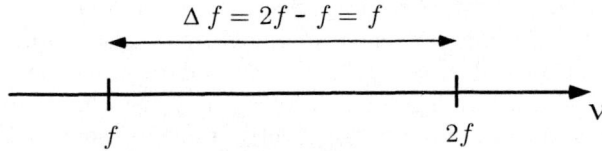

FIGURE 4. The new principle of absolute optical frequency measurements. The interval between Δf between f and $2f$ is just equal to the frequency f itself

Extending our principle of determining frequency differences to the intervals between harmonics or subharmonics of an optical frequency leads naturally to the absolute measurement of optical frequencies. In the most simple case this is the interval between an optical frequency f and its second harmonic $2f$ as illustrated in Fig. 4. But of course other intervals can be used as well.

Hydrogen $1S - 2S$ transition

The first frequency chain following this principle has been used in a recent determination of the hydrogen $1S - 2S$ transition and measured the interval between $3.5f$ and $4f$ where f is the frequency of a HeNe laser at 3.4 μm (88 THz) [21]. Fig. 5 gives a simplified sketch: The He-Ne laser is operated at the 28th subharmonic of the hydrogen transition frequency which is driven by the forth harmonic of a dye laser at $7f$. The femtosecond laser, whose spectrum is broadened in a single mode optical fiber, measures the frequency difference between the output of the optical frequency interval divider at $4f$ and the subharmonic of the dye laser at $3.5f$. This frequency difference of $0.5f \approx 44.2$ THz equals one half of the absolute frequency of the He-Ne standard. Its frequency is therefore determined directly with the mode-locked laser by comparison with a local Cs clock that controls the mode spacing. Here the 10 MHz output from the Cs clock is multiplied in one step to 44.2 THz. Because our 3.39 μm He-Ne laser could not be tuned far enough to reach the 28th subharmonic of the $1S - 2S$ transition we used an additional laser (not shown in Fig. 5) at around $4f$ that was displaced by about 1 THz. The displacement was simultaneously measured with the frequency comb. A more detailed description of the frequency chain is found in Ref. [21].

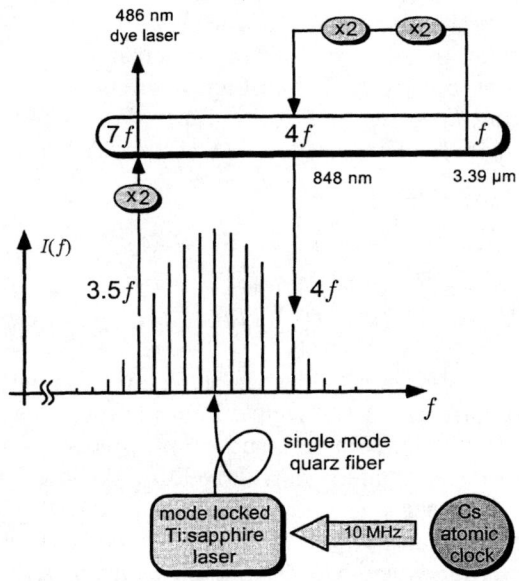

FIGURE 5. Frequency chain used for the frequency determination of the hydrogen $1S-2S$ transition (simplified). The oval symbol represents on optical frequency interval divider as discussed earlier. It receives two input frequencies f and 7 and delivers the mean frequency $4f$.

The hydrogen 1S-2S transition is excited by Doppler-free two-photon spectroscopy with a standing laser wave near 243 nm in a cold atomic beam. The hydrogen spectrometer is operated by the hydrogen team in our group (M. Niering, M. Fischer, M. Weitz) and has been described before in detail [22].

In collaboration with the Paris cesium fountain clock group (P. Lemonde, G. Santarelli, M. Abgrall, P. Laurent, C. Salomon, and A. Clairon) we have used the currently most precise cesium frequency standard in the radio-frequency domain [23] as a reference and reached an accuracy for the $1S - 2S$ transition of 1.8×10^{-14}. This represents by now the most precise measurement of an optical frequency [24]. Our new chain is providing the first phase coherent link from the vacuum UV (121 nm) to the radio frequency domain.

Rydberg constant

Hydrogen is the most simple of all atoms and its properties have been calculated very precisely. For many years now high resolution spectroscopy has been performed on hydrogen to test QED and improve the precision of the Rydberg constant R_∞. The Rydberg constant scales all the energy levels and can be determined from optical frequency measurements on atomic hydrogen with great precision. For highest precision the energy levels within atomic hydrogen are conventionally described as a sum of three contributions: The hyperfine interaction, which is very well known for the states discussed here and the Dirac energy $R_\infty e(nl)$ [25] which include all recoil corrections[1] up to the order $(Z\alpha)^4$. All that is left, e.g. QED contributions, remaining recoils contributions, nuclear size effects etc., is by definition [26] called the Lamb shift L_{nl}. The Lamb shift scales roughly as n^{-3} and is much smaller for P and D states than for S states. We write the frequencies of the most precisely known optical transitions [24,27,28], with the hyperfine structure removed, as follows:

$$f_{1S-2S} = R_\infty \left(e(2S) - e(1S) \right) + L_{2S} - L_{1S}$$
$$f_{2S-8D} = R_\infty \left(e(8D) - e(2S) \right) + L_{8D} - L_{2S}$$
$$f_{2S-12D} = R_\infty \left(e(12D) - e(2S) \right) + L_{12D} - L_{1S} \tag{4}$$

In these equations the left hand sides are determined experimentally to extract some values of the quantities on the right. The measurement of the $1S - 2S$ frequency is more than two orders of magnitude more precise than the other two [24]. Therefore the f_{2S-nD} measurements currently limit the precision of R_∞. Now we have three equations and three unknowns, R_∞, L_{1S} and L_{2S}, if we choose to use theoretical values for L_{8D} and L_{12D}. This is justified as they are comparatively small so that a very crude calculation is enough to avoid any influence on the result. On the other hand precise radio frequency determinations of the $2S_{1/2} - 2P_{1/2}$ and $2S_{1/2} - 2P_{3/2}$ splitting [29,30] allow the determination of R_∞ and L_{2S} from the f_{2S-nD} measurements and R_∞ and L_{1S} from the $1S-2S$ frequency if combined with one of the f_{2S-nD} measurements. Another possibility to replace L_{2S} is to use the $1/n^3$ scaling law of the Lamb shift which allows an accurate theoretical calculation of the small linear combination $L_{1S} - n^3 L_{nS}$ [31]. The differences $7f_{2S-8D} - f_{1S-2S}$ and $7f_{2S-12D} - f_{1S-2S}$ will yield the combination $L_{1S} - 2^3 L_{2S}$. To obtain even more precise results radio frequency beat measurements like $f_{2S-4S} - 1/4 f_{1S-2S}$ [32–34], which should be zero according to the Schrödinger theory, can be used as well as measurements in deuterium.

Most of the combinations to derive the interesting quantities, R_∞, which is needed to fix the values of other constants (e.g. the fine structure constant), and L_{1S}, which allows one of the best tests of QED, yield a comparable accuracy. Therefore

---

[1] Unlike the Schrödinger theory the Dirac theory does not allow to account for a finite mass nucleus simply by replacing m_e by the reduced mass. Instead an expansion in $Z\alpha$ is used.

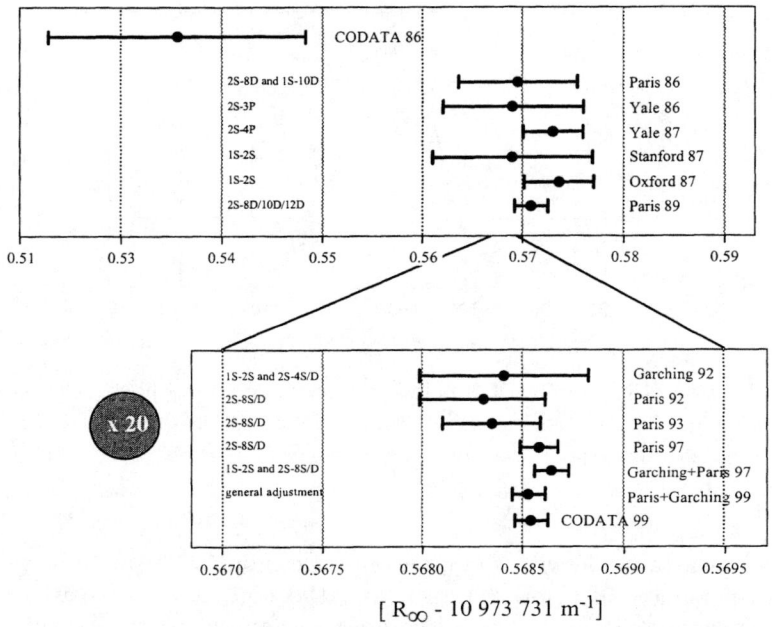

FIGURE 6. A history of measurements of the Rydberg constant

a general adjustment gives the best answers to date [28,14]. In Fig. 6 a history of measurements is visualized. Note the substantially improved accuracy and the quite significant shift from the 1986 CODATA value. The 1999 CODATA value [14] $R_\infty = 10\ 973\ 731.568\ 549\ (83)$ m^{-1} (7.6×10^{-12}) follows closely the 3 most recent measurements from the $1S - 2S$, $2S - 8D$ and $2S - 12D$ transitions made at Garching and Paris [24,27,28]. At this point improvements of the transition frequencies to higher excited states improve the accuracy of the Rydberg constant. To allow improved comparisons of the experimentally determined $1S$ Lamb shift with QED calculations a better value for proton charge radius, that enters the theory, is desperately needed [35].

Frequency combs spanning more than an octave

The first absolute measurement of an optical frequency with a fs frequency comb has inspired further rapid advances in the art of frequency metrology. In collaboration with P. Russell, J. Knight and W. Wadsworth from the University of Bath (UK) we have used novel microstructured photonic crystal fibers (PCF) [36] to achieve further spectral broadening of femtosecond frequency combs. In the fiber used here light is guided in a pure silica core with a diameter of approximately 1.5 μm by surrounding it with an array of air holes. The remarkable dispersion

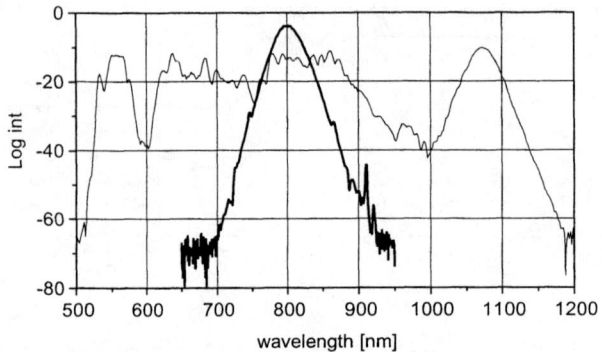

FIGURE 7. Spectral broadening of fs pulses in a photonic crystal fiber. The narrow peaked curve in the middle (bold) denotes the initial pulse directly from the fs laser (25 fs, 170 mW average power, 625 MHz repetition rate). The broadened spectrum stretches from 520 nm to 1100 nm (−10 dB width)

characteristics attainable with the large effective index step (including zero group velocity dispersion well below 800 nm) and the high peak intensities associated with the short pulses and the small core size, enables one to observe a range of unusual nonlinear optical effects [37], including very effective spectral broadening to more than an optical octave.

With such a broad spectrum as shown in Fig. 7 we can directly access the interval between an optical frequency f and its second harmonic $2f$. Such a frequency chain [21,38,39,11] allows the direct comparison of radio and optical frequencies without the need of any divider stages or further nonlinear steps.

Although self phase modulation is likely the dominant mechanism of spectral broadening there are other processes like stimulated Raman and Brillouin scattering or shock wave formation that might spoil the usefulness of these broadened frequency combs. And indeed in an experiment using 8 cm of PCF and 73 fs pulses at 75 MHz repetition rate from a Mira 900 system (Coherent Inc.) we have seen an exceptionally broad spectrum from 450 to 1400 nm but with excessive broadband noise. We did not observe these problems with the 25 fs pulses at 625 MHz repetition rate resulting in the spectrum shown in Fig. 7 and used for the $f : 2f$ frequency chain reported below.

The $f : 2f$ frequency chain

The $f : 2f$ interval frequency chain [21,38,39,11] sketched in figure 4 is based on a Ti:sapphire 25 fs ring laser with a 625 MHz repetition rate (GigaOptics, model GigaJet). While the ring design makes it almost immune to feedback from the fiber, the high repetition rate increases the available power per mode. The highly efficient spectral broadening of the PCF compensates for the decrease of available peak power connected with a high repetition rate. To generate an octave spanning comb we have coupled 190 mW average power through 35 cm PCF. The pump beam at 532 nm (Verdi, Coherent Inc.) is modulated by an EOM (LM 2002, Gsänger). With 7 W of pump power we achieve above 650 mW average power from the fs laser. In our setup a Nd:YAG laser (Prometheus, Innolight) serves to pick and amplify one of the modes by phase locking it to that mode. The Nd:YAG laser is then frequency doubled in a periodically poled KTP crystal and a beat signal in the green with another mode of the comb is observed. This beat signal gives direct access to the offset frequency f_0 introduced earlier as shown in Fig. 8. The offset frequency f_o is phase locked with the help of EOM in the pump beam while the repetition rate f_r is phase locked with a PZT mounted folding mirror. Both are referenced to our cesium atomic clock. By this means the absolute frequency of each of the modes is phase coherently linked to the rf reference and known with the same relative precision.

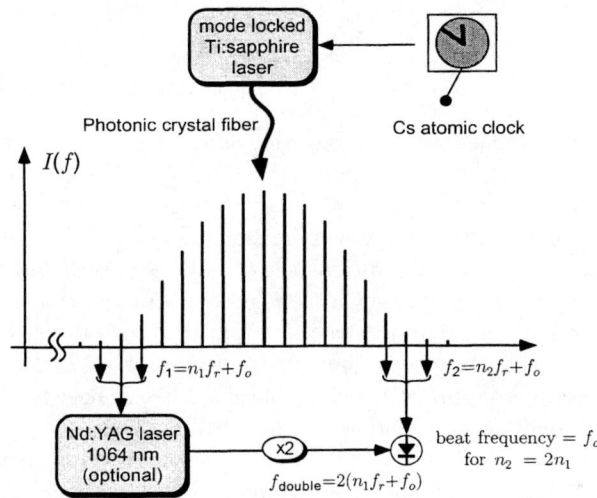

FIGURE 8. Principle of the $f : 2f$ frequency chain with photonic crystal fiber.

Single laser optical frequency synthesizer

The Nd:YAG laser can be omitted, as first demonstrated by J. Hall's group at JILA in Boulder (USA) [38–40,11], if one does not take a single mode of the comb but several modes and generates the sum and second harmonic frequencies directly from the infrared part of the fiber output.

As sketched in Fig. 9 sum and second harmonic frequencies of the modes in the infrared are generated in a 7 mm long KTP crystal. The beat signal between this frequency doubled pulse and the green part of the spectrum directly gives access to the offset frequency. A grating serves as a bandpass filter. Not shown in Fig. 9 are the delay line in one of the arms to overlap the pulses and the EOM for phase locking the offset frequency.

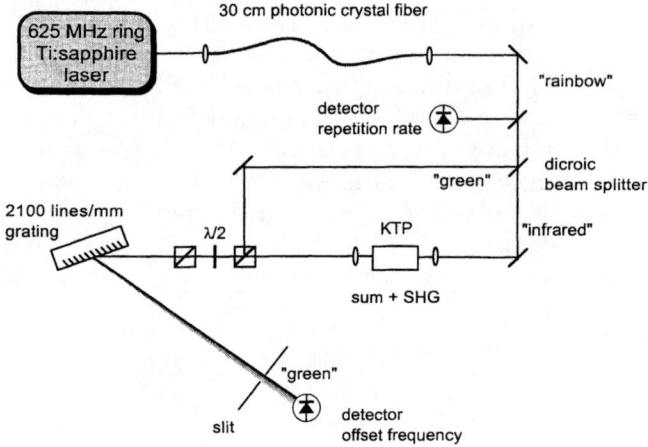

FIGURE 9. Frequency chain consisting of one single laser (and pump).

We have now arrived at a frequency chain that consists of one fs laser (and an optional Nd:YAG laser) only and nevertheless links a 10 MHz rf reference phase coherently in one step with the optical region. It occupies only 1 square meter on an optical table with the potential for further miniaturization. At the same time it supplies us with a reference frequency grid across much of the visible and infrared spectrum with comb lines that are separated by 625 MHz and can easily be distinguished with a commercial wavemeter. This makes it a ideal laboratory tool for precision spectroscopy and a compact solid state system that is ready to serve as a clockwork in future optical clocks. In the reverse direction we expect this clockwork to transfer not only the accuracy but also the superior stability of optical oscillators to the rf domain.

Validation of the $f : 2f$ frequency chain

To check the integrity of the broad frequency comb and evaluate the overall performance of the $f : 2f$ interval frequency chain we have compared it with the $3.5f : 4f$ frequency chain used for the measurement of the hydrogen $1S - 2S$ transition frequency as described above.

The fs comb used here is only 44.2 THz wide and has been thoroughly tested [8]. This chain was modified to replace the dye laser by a frequency doubled diode laser/tapered amplifier combination at 969 nm [41]. The additional frequency gap of 1 THz in the previous setup [21] has been removed by operating the diode laser at exactly $3.5f$.

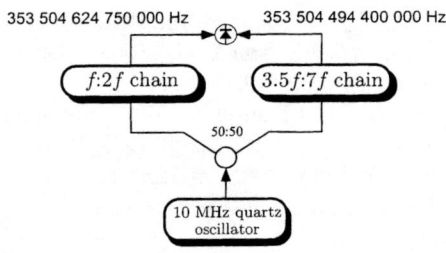

FIGURE 10. Comparision of two frequency chains.

To compare the two frequency chains we use the 848 nm laser diode and a second 848 nm laser diode locked to the frequency comb of the $f : 2f$ chain. The setup is schematically shown in Fig. 10. After averaging all data we obtained a mean deviation from the expected beat frequency of 71 ± 179 mHz at 353 THz. This corresponds to a relative uncertainty of 5.1×10^{-16}. No systematic effect is visible at this accuracy [11].

Conclusion

To summarize we have developed a new concept to measure optical frequencies. This new technique has been applied to the measurement of the Cesium D_1 line which is a cornerstone for a new determination of α and the hydrogen $1S - 2S$ transition needed for the determination of the Rydberg constant and the 1S Lamb shift. This development culminates in the fully phase locked single laser optical frequency synthesizer. It uses a single fs laser and is nevertheless capable of phase coherently linking the rf and the optical domain.

Other important applications of this $f : 2f$ frequency chain arise in the time domain where the carrier offset slippage frequency is an important parameter and needs to be controlled for the next generation of ultrafast experiments. In collaboration with F. Krausz we have applied fs comb techniques to control the phase evolution of ultra-short pulses lasting for only a few optical cycles [40].

Future applications of precise optical frequency measurements also include the search for variations in the fundamental constants and the test of CPT invariance with anti-hydrogen now underway at CERN.

We believe that the development of accurate optical frequency synthesis marks only the beginning of an exciting new period of ultra-precise physics.

Finally we would like to thank our collaborators, with out their help the work presented here would not have been possible.

REFERENCES

1. Kourogi M. et al., *IEEE J. Quantum Electron.* **31**, 2120 (1995).

2. McIntyre D., and Hänsch T. W., *Digest of the Annual Meeting of the Optical Society of America*, paper ThG3, Washington D.C. (1988). Telle H. R., Meschede D., and Hänsch T. W., *Opt. Lett.* **15**, 532 (1990).

3. Prevedelli M., Freegarde T., and Hänsch T. W., *Appl. Phys.* **B 60**, S241 (1995).

4. Eckstein J. N., Ferguson A. I., and Hänsch T. W., *Phys. Rev. Lett.* **40**, 847 (1978).

5. Reichert J., et al., *Opt. Commun.* **172**, 59 (1999).

6. Wineland D. J., et al., Proceedings of *The Hydrogen Atom*, edited by Hänsch T. W., Berlin: Springer 1989.

7. Holzwarth R. et al., to be published.

8. Udem Th. et al., *Opt. Lett.* **24**, 881 (1999).

9. Diddams S.A. et al., *Opt. Lett.* **25**, 186 (2000).

10. Walls F. L. et al., *IEEE Trans. Instrum. Meas.* **24**, 210 (1975).

11. Holzwarth R. et al., accepted for publication at *Phys. Rev. Lett.*

12. Udem Th., et al., *Phys. Rev. Lett.* **82**, 3568 (1999).

13. Kinoshita T., *Rep. Prog. Phys.* **59**, 1459 (1996), and references therein.

14. Mohr P. J. and Taylor B. N., *Rev. Mod. Phys.* **72**, 351 (2000)

15. Farnham D. L., et al., *Phys. Rev. Lett.* **75**, 3598 (1995).

16. Bradley M. P. et al., *Phys. Rev. Lett.* **83**, 4510 (1999).

17. Udem Th. et al., *Phys. Rev. Lett.* **79**, 2646 (1997).

18. Schnatz H. et al., *Phys. Rev. Lett.* **76**, 18, (1996).

19. Schwob C. et al., *Phys. Rev. Lett.* **82**, 4960 (1999).

20. Bernard J. E. et at., *Phys. Rev. Lett.* **82**, 3238 (1999).

21. Reichert R. et al., *Phys. Rev. Lett.* **84**, 3232 (2000).

22. Huber A. et al., *Phys. Rev.* **A 59**, 1844 (1999).

23. Santarelli G., et al., *Phys. Rev. Lett.* **82**, 4619 (1999).

24. Niering M., et al., *Phys. Rev. Lett.* **84**, 5456 (2000).

25. Pachucki K., et al., *J. Phys.* **B 29**, 177 (1996) + erratum p. 1573.

26. Sapirstein J. R., and Yennie D. R., *Quantum Electrodynamics*, edited by Kinoshita T., Singapore: World Scientific, 1990.

27. de Beauvoir B., et al., *Phys. Rev. Lett.* **78**, 440 (1999).

28. Schwob C., et al., *Phys. Rev. Lett.* **82**, 4960 (1999).

29. Lundeen S. R., and Pipkin F. M., *Phys. Rev. Lett.* **46**, 232 (1981).

30. Hagley E. W., and Pipkin F. M., *Phys. Rev. Lett.* **72**, 1172 (1994).

31. Karshenboim S.G., *Z. Phys.* **D 39**, 109 (1997).

32. Weitz M, et al., *Phys. Rev.* **A 52**, 2664 (1995).

33. Berkeland D. J., Hinds E. A., and Boshier M. G., *Phys. Rev. Lett.* **75**, 2470 (1995).

34. Bourzeix S. et al., *Phys. Rev. Lett.* **76**, 384 (1996).

35. Taqqu D. et al., *Hyperfine Interactions* **119**, 311 (1999).

36. Knight J. C. et al., *Opt. Lett.* **21**, 1547 (1996).

37. Wadsworth W. J. et al., *Electron. Lett.* **36**, 53 (2000).

38. Diddams S. A. et al., *Phys. Rev. Lett.* **84**, 5102 (2000).

39. Jones D. et al., *Science* **288**, 635 (2000).

40. Apolonski A. et al., *Phys. Rev. Lett.* **85**, 740 (2000).

41. Zimmermann C., et al., *Appl. Phys. Lett.* **66**, 2318 (1995).

Single Ion Mass Spectrometry and the Fine Structure Constant

David E. Pritchard, Michael P. Bradley, James V. Porto, Simon Rainville, James K. Thompson

Department of Physics and Research Laboratory of Electronics, Massachusetts Institute of Technology.
Cambridge, Massachusetts, USA, 02139
E-mail: dpritch@mit.edu

Abstract. Using a Penning trap single ion mass spectrometer, we have measured the atomic masses of 13 isotopes, many important for fundamental metrology and fundamental constants. The accuracy of the measurements, $\approx 10^{-10}$, is typically two orders of magnitude better than previously accepted values. A wide variety of self consistency checks greatly reduces the possibility of unknown systematic errors.

As part of a program to determine the Molar Planck constant $N_A h$ and the fine structure constant α, we measured the masses of ^{133}Cs, ^{87,85}Rb, and ^{23}Na. Our high accuracy atomic mass measurements can be combined with values of h/m_{atom} from atom interferometry measurements and accurate wavelength measurements for different atoms to give several independent determinations of $N_A h$ and α. This route to α through $N_A h$, the atomic mass of the proton M_p, the electron to proton mass ratio m_e/m_p, and the Rydberg constant R_∞ is based on simple physics. It can potentially achieve the several ppb accuracy needed to test the QED determination of α extracted from measurements of the electron g factor.

I INTRODUCTION

Using a Penning trap single ion mass spectrometer, we have measured the atomic masses of 13 isotopes several of which are important for fundamental metrology and fundamental constants. The accuracy of the measurements, typically 0.1 ppb or 10^{-10}, represents one to three orders of magnitude improvement over previously accepted values. A wide variety of self consistency checks greatly reduces the possibility of unknown systematic errors. Our measurements [1] have contributed to a precise recalibration of the γ-ray spectrum and provided an atomic reference for a realization of an atomic definition of the kilogram to replace the artifact standard currently in use.

Most recently, we measured [2] the masses of ^{133}Cs, ^{87,85}Rb, and ^{23}Na as part of a program to determine the Molar Planck constant $N_A h$ and fine structure constant α

CP551, *Atomic Physics 17*, edited by E. Arimondo, P. DeNatale, and M. Inguscio
© 2001 American Institute of Physics 1-56396-982-3/01/$18.00

TABLE 1. Measured neutral alkali masses.

Species	MIT Mass (u)	ppb	1995 Mass (u) [5]	ppb	$difference/\sigma_{1993}$
^{133}Cs	132.905 451 931 (27)	0.20	132.905 446 800 (3200)	24.0	1.6
^{87}Rb	86.909 180 520 (15)	0.17	86.909 183 500 (2700)	31.0	-1.1
^{85}Rb	84.911 789 732 (14)	0.16	84.911 789 300 (2500)	29.0	0.2
^{23}Na	22.989 769 280 7 (28)	0.12	22.989 769 670 0 (2300)	9.8	-1.7

α from measurements of h/m_{atom}. A further motivation for our measurements is that Cs and Rb are used as reference masses for measurements of heavy radioactive nuclei which are important for modeling astrophysical heavy element formation [3], [4].

A   Molar Planck constant $N_A h$

The Molar Planck constant $N_A h$ is an important quantity in metrology and for fundamental physics. New values of $N_A h$ at the few ppb level in combination with measurements of h (such as a recent 87 ppb measurement [6]) can yield values of N_A with ppb level accuracy. Precise values of $N_A h$ would also provide a way to check QED and test the unity of physics across disciplines by helping to accurately determine the fine structure constant α.

Avogadro's number N_A is the ratio of the SI and atomic units of mass [7]. The unified atomic mass unit is defined by setting the atomic mass of ^{12}C to be exactly 12. N_A is defined as the number of elementary entities in one mole (the amount of substance whose mass in grams equals its atomic mass) and has an approximate value of $N_A \approx 6.022 \times 10^{23}$/mole. Avogadro's number can then be written as the ratio of any atom's mass in atomic mass units denoted by M_{atom} and in SI units denoted by m_{atom}

$$N_A = \frac{M_{atom}}{m_{atom}} \times 10^{-3} \tag{1}$$

where the factor of 10^{-3} arises because of the definition of Avogadro's constant in terms of grams rather than the SI unit kilogram.

Transposing Eq. 1 shows that $1/N_A$ can be regarded as the universal mass quantum (in grams). The mass of any elementary entity is then its atomic mass (i.e. mass quantum number) times this mass quantum. (Unlike most other quantized quantities, the quantum number is not a simple rational number.)

Thus, $N_A h$ is the ratio of h to the mass quantum, a universal h/m. It can be obtained from a particular value of h/m_{atom} by multiplying by M_{atom}

$$N_A h = \frac{h}{m_{atom}} M_{atom} \times 10^{-3} \ . \tag{2}$$

Our technique for measuring M_{atom} therefore allows measurements of h/m_{atom} using different atoms to be compared with $\approx 10^{-10}$ accuracy.

In both Schroedinger's equation for a free particle and the expression for magnetic moments of elementary entities, h and m always occur in the ratio h/m. Thus h/m is often measured in experiments involving simple quantum expressions. By equating the classical ($p = m_x v$) and quantum ($p = h/\lambda_{dB}$) expressions for the momentum of a particle, we see that measurements in SI units of the deBroglie wavelength λ_{dB} and the velocity v of a particle combine to measure h/m_X in SI units,

$$v\lambda_{dB} = \frac{h}{m_X} .$$ (3)

Comparison of the energy and wavelength of a photon would also yield a value of $N_A h$, but at accuracies of ≈ 100ppb [1].

Precision mass spectrometry now allows several independent determinations of $N_A h$ from measurements of h/m_{atom} using different atoms possessing very different experimental systematic errors to be compared with no reduction in accuracy at the 0.1 ppb level.

B   Fine Structure constant α

An accurate value of the Molar Planck constant leads to a new determination of the fine structure constant α. Noting the definitions of $\alpha \equiv e^2/\hbar c$ and the infinite-nuclear-mass Rydberg constant $R_\infty \equiv (2\pi^2 m_e e^4)/(h^3 c) \approx 1.09 \times 10^5$ cm^{-1} (cgs units) makes it easy to see that

$$\alpha^2 = \frac{2R_\infty}{c}\frac{h}{m_e} = \frac{2R_\infty}{c}\frac{m_p}{m_e}\frac{N_A h}{M_p}10^3 .$$ (4)

R_∞ is known with an accuracy of 0.008 ppb [8]. m_p/m_e has been measured to 2 ppb [9]. The mass of the proton in atomic units M_p has been measured by our group to 0.5 ppb [1], and Van Dyck et al. have recently reported a value of M_p accurate to 0.14 ppb [10]. The speed of light c is a defined constant. Thus an independent measurement of $N_A h$ is capable of determining α to 1 ppb.

The possibility of redundancy in the experimental determination of $N_A h$ would greatly enhance the confidence in determinations of α from Eq. 4. The mass ratio m_p/m_e would be the only quantity without more than a single direct measurement at the ppb level (a recent value of m_e/m_{12C} extracted from theory and boundstate electron g factor measurements in hydrogenic ^{12}C has confirmed the value to about 2 ppb [11]). This is not a trivial point since it would take a considerable weight of evidence to believe that disagreement between the QED and $N_A h$ determinations of α signifies some error in QED.

II MEASURING ATOMIC MASSES

A Experimental Technique

We obtain absolute atomic masses M with relative accuracies 0.1 ppb from mass ratios relating the unknown mass to the atomic mass standard ^{12}C. Experimentally, we make a mass comparison by measuring the cyclotron frequency (which is inversely proportional to the mass) of a single molecular or atomic ion in a large and highly-uniform magnetic field (8.5T). The ion is held in a small region of space by the magnetic field which provides radial confinement and by an additional weak dc quadrupole electric field which provides confinement along the axial direction. This combination of confining fields is known as a Penning trap. Trapping the ion allows the long observation time necessary for high precision. Using a single ion is crucial for high accuracy since this avoids the complex frequency perturbations caused by the coulomb interaction between multiple ions.

The combination of magnetic and electric fields in our Penning trap results in three normal modes of motion: trap cyclotron, axial, and magnetron, with frequencies $\omega_c'/2\pi \approx 5$ MHz $>> \omega_z/2\pi \approx 0.2$ MHz $>> \omega_m/2\pi \approx 0.002$ MHz, respectively. The free-space cyclotron frequency ω_c is recovered from the following expression (invariant with respect to trap tilts and ellipticity) [12]:

$$\omega_c = qB/mc = \sqrt{(\omega_c')^2 + (\omega_z)^2 + (\omega_m)^2}. \tag{5}$$

We have developed ultrasensitive superconducting electronics to detect the miniscule currents ($\approx 10^{-14}$ amperes) that the ion's axial motion induces in the trap electrodes. The detector consists of a DC SQUID coupled to a low loss superconducting resonant transformer ($Q \approx 4 \times 10^4$) connected across the endcaps of the Penning trap. Our detection noise is currently dominated by the 4 K Johnson noise present in the resonant transformer. Detection damps the axial motion at a rate $\gamma_z \sim 1$ s^{-1} quickly bringing the axial motion to equilibrium at 4 K.

The trap cyclotron motion is detected phase coherently via an RF coupling to the axial motion (the same coupling is also used to "cool" both the trap cyclotron and magnetron modes). In the spirit of the separated oscillatory fields technique, RF drives and couplings are applied only briefly at the beginning and end of a measurement thus eliminating systematics and uncertainties involved with continuously observing the trap cyclotron motion. Briefly driving the trap cyclotron motion with a fixed phase and then measuring the accumulated phase versus delay time yields the trap cyclotron frequency. A typical phase accumulation time of 1 minute yields a precision of $\approx 2 \times 10^{-10}$. The precision of a single cyclotron frequency measurement is limited by the $\approx 2.5 \times 10^{-10}$ short term fluctuations of the magnetic field. The typical precision with which we can compare the cyclotron frequencies of two ions ($\approx 1 \times 10^{-10}$) is limited mainly by magnetic field drift over time scales between 1 and 15 minutes.

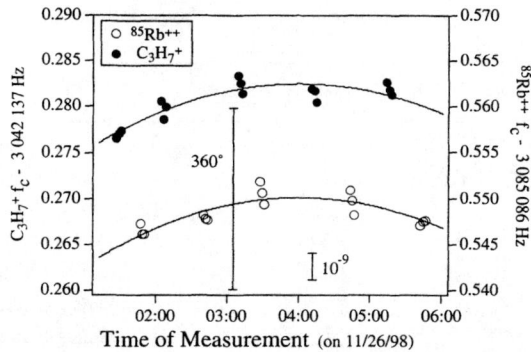

FIGURE 1. Typical night of data. The solid line is a second order polynomial fit to the data. The 360° bar shows the magnitude in Hz of a 360° error in phase unwrapping.

From Eq. 5, it is clear that the magnetron frequency does not need to be known very accurately to recover the free-space cyclotron frequency. It is accurate enough just to use the measured values of ω_c' and ω_z to calculate $\omega_m \approx (\omega_z^2/2\omega_{cyc}')(1 + 9/4\sin^2\theta_m)$ where $\theta_m = 0.16°$ is the measured angle between the B-field and trap axes obtained by actually measuring the magnetron frequency once. The effect of θ_m on a cyclotron frequency ratio is at most 0.002 ppb.

B Recent Alkali Measurements

To determine the masses of the alkali atoms ^{133}Cs, ^{87,85}Rb, and ^{23}Na, we measured the free-space cyclotron frequency ratios $r \equiv \omega_{c2}/\omega_{c1}$ listed in Table 2. The reference ions were selected because of the similar mass to charge ratios (aiding in the reduction of systematic errors) and because we have previously measured the atomic masses of each of the consituent atoms.

A cyclotron frequency ratio r of two different ions was determined by a run measuring a cluster of ω_c values for an ion of type A, then for type B, etc. In a typical 4-hour run period (from 1:30-5:30 am when the nearby electrically-powered subway was not running), we recorded about 5 alternations of ion type (Fig. 1). The measured free-space cyclotron frequencies exhibited a common slow drift. We fit a common polynomial $\Omega(t)$ plus a frequency difference to the data. From this we obtained the frequency ratio r_n and the uncertainty σ_n for a single night. The average order of $\Omega(t)$ was 3 and was chosen using the F-test criterion [13] as a guide.

The distribution of residuals from the polynomial fits had a Gaussian center with a standard deviation $\sigma_{resid} = 0.28$ ppb and a background ($\approx 2\%$ of the points) of non-Gaussian outliers, as in our earlier measurements [1]. As in [1] we chose to handle the non-Gaussian outliers using a robust statistical method to smoothly deweight them [14].

TABLE 2. Measured ion cyclotron frequency ratios, corrected for systematics.

A/B	$\frac{\bar{\omega}_c}{2\pi}$ (MHz)	Nights	$\omega_c[A]/\omega_c[B]$
$^{133}Cs^{+++}/CO_2^+$	2.968	5	0.992 957 580 983 (135)
$^{133}Cs^{++}/C_5H_6^+$	1.977	4	0.993 893 716 487 (427)
$^{87}Rb^{++}/C_3H_8^+$	2.994	2	1.013 992 022 591 (266)
$^{87}Rb^{++}/C_3H_7^+$	3.028	3	0.990 799 127 824 (174)
$^{85}Rb^{++}/C_3H_7^+$	3.064	2	1.014 106 122 230 (164)
$^{85}Rb^{++}/C_3H_6^+$	3.100	2	0.990 367 650 976 (285)
$^{23}Na^+/C_2^+$	5.578	2	1.043 943 669 690 (076)
$^{23}Na^{++}/C^+$	11.155	2	1.043 944 716 614 (098)

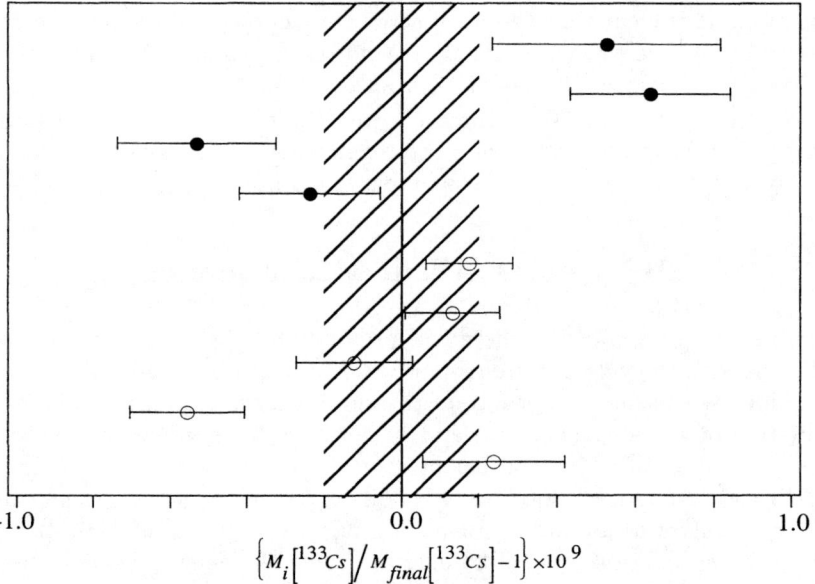

$$\left\{ M_i\left[^{133}Cs\right] \Big/ M_{final}\left[^{133}Cs\right] - 1 \right\} \times 10^9$$

FIGURE 2. Example of Variation of Mass Ratio from Night to Night. A measurement of the neutral mass of ^{133}Cs is extracted from each night's run of cyclotron frequency ratio measurements and plotted in ppb relative to our final published value of the neutral mass of ^{133}Cs. The open and closed circles are from frequency ratio measurements of Cs^{+++}/CO_2^+ and $Cs^{++}/C_5H_6^+$ respectively. The error bars on each night's measurement are extracted from the low order polynomial fit to both ion's cyclotron frequencies and reflects the distribution of the cyclotron frequency measurements during that night. The shaded region represents the one sigma confidence interval arrived at in the final analysis.

As shown in Table 2 and Figure 2, we measured each frequency ratio on more than a single night. For ratios involving Cs and Rb the measured ion mass ratios were distributed from night to night with a scatter larger than the uncertainty predicted from the statistical scatter within a single night ($\chi_\nu^2 \approx 5$) . By contrast $\chi_\nu^2 \approx 0.8$ for ratios involving Na. None of the earlier data taken using this apparatus [1] exhibited these excess night-to-night variations. A search for the source of these fluctuations is discussed elsewhere [2] and was unsuccesful. To account for this excess scatter, the uncertainties in the weighted average of the ion mass ratios involving $Cs^{++}/C_5H_6^+$ and Cs^{+++}/CO_2^+ were increased by factors of 2.6 and 2.2 respectively so that $\chi_\nu^2 = 1$. Since the Rb measurements all had similar m/q, we assumed that the night-to-night fluctuations involving the Rb ratios were drawn from a common statistical distribution. Therefore, we increased the uncertainties for the Rb ion ratios by a factor of 2.2 so that the overall Rb χ_ν^2 was reduced to 1. For Na, $\chi_\nu^2 \approx 0.8$ so the uncertainties were not adjusted.

By correcting for molecular binding and electron ionization energies, we obtained a set of neutral mass difference equations. We added to this the set of mass difference equations used to determine the atomic masses in [1]. Solution of this overdetermined set of linear equations gave the neutral masses of the alkali metals (see Table 1) with uncertainties σ_{od} as well as the previously published neutral masses with $\chi_\nu^2 = 0.83$. The previously published masses were essentially unchanged and so are not reported here. Uncertainties in $M[^{16}O]$ and $M[H]$ (the only atoms other than ^{12}C in the ratios of Table 2) contributed < 0.1 ppb uncertainty to the alkali masses.

The use of two distinct reference ions gave a check on systematics by providing two independent values for each neutral mass. For Rb and Cs χ_ν^2 is less than 1. However, because of the larger uncertainty on $M[Cs]$ from $Cs^{++}/C_5H_6^+$ we quote a final uncertainty of 0.20 ppb $> (\sigma_{od}(Cs) = 0.16$ ppb). For $^{87,85}Rb$ we quote $\sigma_{od}(^{87,85}Rb)$ as the final uncertainties. For the neutral masses from Na^{++}/C^+ and Na^+/C_2^+, the statistical uncertainties are 0.09 and 0.07 ppb respectively. The 0.2 ppb disagreement of the two values may be evidence for a systematic at the 0.1 ppb level. To reflect this we assigned $M[^{23}Na]$ a 0.12 ppb uncertainty $> (\sigma_{od}(Na) = 0.06$ ppb) which spans both independent measurements.

Table 1 quotes the final values for $M[^{133}Cs]$, $M[^{87}Rb]$, $M[^{85}Rb]$ and $M[^{23}Na]$ obtained from the solution of the overdetermined set of mass difference equations with uncertainties from the above discussion. Also included in Table 1 are the alkali masses from the 1995 mass evaluation [5]. Our values differ from the 1995 values by typically $1.5\sigma_{1995}$, which suggests that the uncertainties on the masses from the 1995 evaluation were slightly underestimated. Our value for $M[^{133}Cs]$ lies within the uncertainty of the recent measurement of $M[^{133}Cs]$ reported by the SMILETRAP collaboration [16].

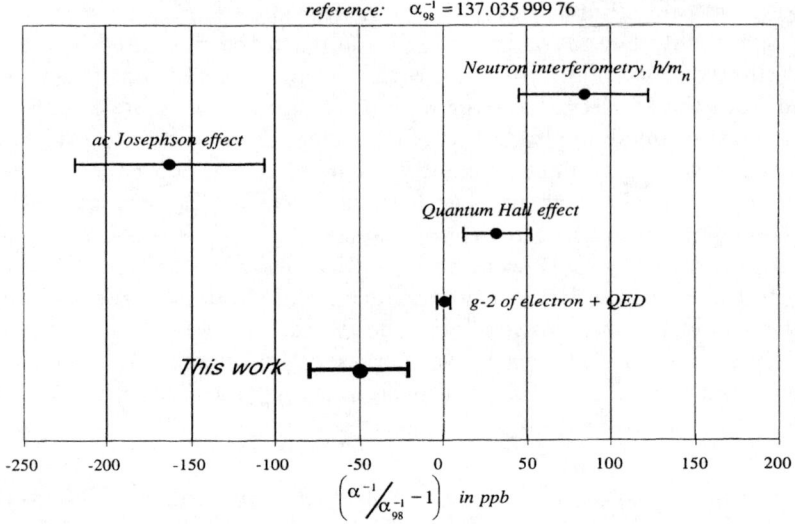

FIGURE 3. Precision Measurements of the Fine Structure Constant plotted with respect to the 1998 CODATA recommended value of α [17]. The relative uncertainties are: ac Josephson effect (56 ppb). Quantum Hall effect (24 ppb). g-2 of electron and QED (4 ppb). Neutron Interferometry, h/m_n (35 ppb). This Work yields a preliminary value with the hope of further improvement in accuracy in the future.

III   PRECISE ROUTES TO α

The fine-structure constant α appears in many contexts in physics and arises in diverse physical systems because of its role as the dimensionless coupling constant describing the interaction of electrons and positrons with the electromagnetic field. This results in numerous experimental routes to measuring α involving several different disciplines. The diversity of these methods allows stringent tests across these disciplines, in principle allowing an outside check on each discipline. Figure 3 gives a compilation of some of the most accurate measurements of α which we will briefly review.

A   AC Josephson Effect

The AC Josephson effect is the oscillation of frequency ν observed when a voltage V is applied across a Josephson junction. Theory predicts that ν and V are related by [18]

$$\nu = (\frac{2e}{h})V \ .$$

$$(6)$$

This relationship has been shown to be independent of the materials used to fabricate the Josephson junction to a level of 2 parts in 10^{16} [19]. Thus measurements of the AC Josephson effect provide a way to measure e/h to high precision.

The highest-precision route relating $2e/h$ to α is

$$\alpha^2 = (\frac{4R_\infty \gamma_p'}{c})(\frac{\mu_B}{\mu_p'})(\frac{h}{2e}) \tag{7}$$

where R_∞ is the infinite-nuclear-mass Rydberg constant, γ_p' is the proton gyromagnetic ratio (the prime refers to the fact that γ_p' is measured using nuclear magnetic resonance on an H_2O sample), μ_p' is the proton magnetic moment, and $\mu_B = e\hbar/2m_e c$ is the Bohr magneton.

R_∞ is known to 0.008 ppb [8], γ_p' is known to 110 ppb [20], (μ_p'/μ_B) is known to 10 ppb [21], and $(2e/h)$ is known to 30 ppb. Thus the overall uncertainty on α via this route is limited by γ_p' to 56 ppb.

B   Quantum Hall Effect

For any effectively two-dimensional electron gas system in a magnetic field and cooled to cryogenic temperatures, the Hall resistance $R_H = V_T/I$ (where V_T is the voltage generated across the sample transverse to the direction in which current I flows) is quantized:

$$R_H = \frac{h}{e^2 n} \qquad n = 1, 2, 3 \ldots \tag{8}$$

Measurement of R_H by comparison to a reference impedance based on a calculable capacitor can provide a high precision measurement of $h/e^2 \approx 25\ 813\ \Omega$; this is essentially a direct measurement of α. Two measurements of α by this route have been made by the same group at NIST. Both measurements have 24 ppb precision, but the values differ by 42 ppb. It is felt that the 1997 measurement is the correct value and is the value shown in Fig. 3 [22], [23], [24].

C   Electron & Positron Anomalous Magnetic Moment

The spin magnetic moment of the electron μ_e may be written as $\mu_e = g(1/2)\mu_B$ where $\mu_B = e\hbar/2mc$ is the Bohr magneton, $1/2$ is the electron spin in units of $\hbar$ and g is the electron "g-factor". Simple classical models of the electron predict $g = 1$ while the Dirac's relativistic electron equation predicts $g = 2$. The development of QED was necessary to explain why careful measurements revealed that in fact g was not exactly equal to 2. Precisely predicting the so-called "anomaly" $a_e \equiv (g - 2)/2 \approx 0.001 \ldots$ is one of the great successes of QED.

The electron anomaly a_e can be measured to high precision by comparing the electron spin-flip frequency $\omega_s = g\mu_B B/\hbar$ to the cyclotron frequency $\omega_c = eB/mc$

(cgs units). Simple algebra gives $\omega_s = (g/2)\omega_c$ so that $a_e = (g/2)-1 = (\omega_s/\omega_c)-1 = (\omega_s - \omega_c)/\omega_c$. Using an electron confined in a Penning trap, $(\omega_s - \omega_c)$ can be measured directly. VanDyck et al. at the University of Washington pioneered the trapping of single electrons for the measurement of $(g-2)$. The most precise results from the UW group are [25]:

$$a_{e^-} = 1\ 159\ 652\ 188.4\ (4.3) \times 10^{-12} \quad \text{electron anomaly} \tag{9}$$

$$a_{e^+} = 1\ 159\ 652\ 187.9\ (4.3) \times 10^{-12} \quad \text{positron anomaly} \tag{10}$$

Complementary to high-precision measurements of a_e are high-order QED calculations of a_e. These calculations have been contributed to by a number of authors since Schwinger and Feynman. The dean of this field is Tochiro Kinoshita of Cornell University. Using the VEGAS Monte-Carlo integration routines [26], [27], [28] running on a supercomputer he has calculated QED corrections up to and including fourth-order in α. The QED contribution to a_e is

$$A_1 = A_1^{(2)}(\alpha/\pi) + A_1^{(4)}(\alpha/\pi)^2 + A_1^{(6)}(\alpha/\pi)^3 + A_1^{(8)}(\alpha/\pi)^4 + A_1^{(10)}(\alpha/\pi)^5 + \ldots \tag{11}$$

This is the dominant contribution to the total expression:

$$a_e = A_1 + A_2(m_e/m_p) + A_2(m_e/m_\tau) + A_3(m_e/m_\mu, m_e/m_\tau) \tag{12}$$

A_2 and A_3 are small mass-dependent QCD terms. The current best values for the QED terms are [28]:

$$
\begin{aligned}
A_1^{(2)} &= \ 0.5 \quad \text{exact} \\
A_1^{(4)} &= -0.328\ 478\ 965\ldots \quad \text{purely analytic} \\
A_1^{(6)} &= \ 1.181\ 259\ (40) \quad \text{numerical} \\
A_1^{(6)} &= \ 1.181\ 241\ 456\ldots \quad \text{analytic} \\
A_1^{(8)} &= -1.409\ 2\ (384) \quad \text{numerical} \\
A_1^{(10)} &= \ \text{unknown}
\end{aligned}
\tag{13}
$$

As discussed above, $A_1^{(2)} = 0.5$ is an exact analytic result. $A_1^{(4)}$ has contributions from 7 Feynman diagrams, all obtained analytically. $A_1^{(6)}$ has contributions from 72 Feynamn diagrams. Until recently this term was calculated as a hybrid of analytical results for 57 diagrams and numerical results for 15 diagrams. More precise numerical work by Kinoshita [26] led to the discovery of an error in one of the analytical integrals which has since been corrected. In addtion, the remaining diagrams have also since been calculated analytically [29], [30], giving a result which agrees well with the precise numerical result. $A_1^{(8)}$ is the current challenge. It consists of 891 Feynman diagrams which can be reduced to 86 integrals using the Ward-Takahashi identity. Each integrand has about 20000 terms. Thus the computations are extensive. A point-by-point cancellation of singularities is used to renormalize the integrals and make them integrable [28].

An extremely precise value of α can be obtained from the average of the experimental measurements of a_{e-} and a_{e+} and the QED calculations described above. This value is:

$$\alpha^{-1} = 137.035\ 999\ 93\ (52) \tag{14}$$

This value has a precision of 3.8 ppb and is the most precise available by a factor of 5. It can be seen from Fig. 3 that the $(g-2)/$QED measurement is in reasonable agreement with the 1997 Quantum Hall effect measurement, and disagrees with the 1998 h/m_n measurement (to be discussed below). The interest in the $N_A h$ route to α using atoms stems from the fact that it is an independent method with an accuracy which is potentially comparable to that of the $(g-2)/$QED measurement, and thus it may help to shed light on the apparent discrepancy in the measured values of α.

D $N_A h$: Neutron Interferometry and Photon Recoil

As discussed earlier, equating the classical $(p = mv)$ and quantum $(p = h/\lambda_{dB})$ expressions for particle momentum we see that measurements of the deBroglie wavelength λ_{dB} and the velocity v of a particle provides a way to measure h/m_X. Thus measurements of λ_{dB}, v, and M_X can provide a (nearly) QED-independent measurement of α, like the AC Josephson and Quantum Hall effect methods. It is true that QED calculations of the $2P_{1/2}$, $2P_{3/2}$ and 8D Lamb shifts are needed to allow R_∞ to be obtained from measurements of the 1S-2S and 2S-8D transitions in hydrogen [8], but these corrections are small and do not need to be known to high accuracy (the largest calculational uncertainty contribution to R_∞ is 0.00026 ppb due to the 8D Lamb shift). Thus even in the event that QED is approximate, this method for measuring α is robust and reliable. If a ppb-level measurement of α can be made by this method QED could be tested for the first time at the ppb-level allowed by the QED/(g-2) measurement.

Kruger et al. have used neutron interferometry to precisely measure λ_{dB} and v for a beam of neutrons, resulting in a measurement of h/m_n with an accuracy of 73 ppb [31]. Combining this with a precise measurement of M_n (from Penning trap measurements of $M[^2\mathrm{H}]$, $M[^1\mathrm{H}]$ and γ-ray measurements of the nuclear binding energy of $^2\mathrm{H}$ [1], [32], [33]) results in a value of α with a precision of 37 ppb (see Fig. 3). This measurement illustrated the promise of the method but was not accurate enough to test QED at the ppb level.

The photon recoil of an atom provides another way to measure h/m. After a photon absorption/emission process an atom recoils with a velocity $v_r = h/m\lambda$ where λ is the wavelength of the photon (which is equal to λ_{dB} of the recoiling atom). The resultant Doppler shift $\Delta\omega = (4\pi^2\hbar)/(m\lambda^2)$ of the atomic absorption and emission frequencies with respect to the laboratory frame provides a way to measure h/m in terms of the resonant wavelength λ. For Cs, the group of S. Chu at Stanford University is using atom interferometry to measure $\Delta\omega$ and the group of T. Hansch at MPI in Garching has developed optical frequency measurement techniques [34] to measure λ. Both of these elegant experiments are described in papers adjacent to this one.

It seems reasonable to expect that atom interferometry experiments exploiting the properties of Bose Einstein Condensation will lead to future measurements of h/m_{atom} with increased accuracy. It is therefore likely that measurements of h/m_{atom} will be performed in ^{23}Na and the two isotopes ^{87,85}Rb [35]. This work lowers the error in M for all of these systems sufficiently that it will not be significant for 1 ppb measurements of $N_A h$.

This work is supported by the National Science Foundation, a NIST Precision Measurements Grant, and the Joint Services Electronics Program.

REFERENCES

1. F. Difilippo, V. Natarajan, K. R. Boyce, and D. E. Pritchard, Phys. Rev. Lett. **73**, 1481 (1994).
2. M. P. Bradley *et al.*, Phys. Rev. Lett. **83**, 4510 (1999).
3. H. Stolzenberg *et al.*, Phys. Rev. Lett. **65**, 3104 (1990).
4. B. Fogelberg *et al.*, Phys. Rev. Lett. **82**, 1823 (1999).
5. G. Audi and A. H. Wapstra, Nucl. Phys. A **595**, 409 (1995).
6. E. R. Williams, R. L. Steiner, D. B. Newell, and P. T. Olsen, Phys. Rev. Lett. **81**, 2404 (1998).
7. B. N. Taylor, Metrologia **31**, 181 (1994).
8. T. Udem *et al.*, Phys. Rev. Lett. **79**, 2646 (1997).
9. D. L. Farnham, R. S. Vandyck, and P. B. Schwinberg, Phys. Rev. Lett. **75**, 3598 (1995).
10. Jr. Van Dyck RS, Farnham DL, Zafonte SL, and Schwinberg PB, in *Trapped Charged Particles and Fundamental Physics.* (AIP, Asilomar, CA, 1998), Vol. 457, pp. 101–110.
11. H. Haffner *et al.*, in *Abstracts of 17th International Conference on Atomic Physics*, edited by F. Fuso and F. Cervelli (AIP, Firenze, Italy, 2000), pp. 27–28.
12. L. S. Brown and G. Gabrielse, Rev. Mod. Phys. **58**, 233 (1986).
13. P.R. Bevington and D.K. Robinson, *Data Reduction and Error Analysis for the Physical Sciences*, 2nd ed. (McGraw-Hill, Boston, 1992).
14. P.J. Huber, *Robust Statistics* (Wiley, New York, 1981).
15. G. Audi and A. H. Wapstra, Nucl. Phys. A **565**, 1 (1993).
16. C. Carlberg, T. Fritioff, and I. Bergstrom, Phys. Rev. Lett. **83**, 4506 (1999).
17. P.J. Mohr and B.N. Taylor, Rev. Mod. Phys. **72**, 351 (2000).
18. B.D. Joseph, Phys. Rev. Lett. **1**, 251 (1962).
19. J. S. Tsai, A. K. Jain, and J. E. Lukens, Phys. Rev. Lett. **51**, 316 (1983).
20. E. R. Williams *et al.*, IEEE Trans. Instrum. Meas. **38**, 233 (1989).
21. E. R. Cohen and B. N. Taylor, Rev. Mod. Phys. **59**, 1121 (1987).

22. M. E. Cage *et al.*, IEEE Trans. Instrum. Meas. **38**, 284 (1989).

23. A. M. Jeffery *et al.*, IEEE Trans. Instrum. Meas. **46**, 264 (1997).

24. B.N. Taylor, personal communication, 1999.

25. R. S. Vandyck, P. B. Schwinberg, and H. G. Dehmelt, Phys. Rev. Lett. **59**, 26 (1987).

26. T. Kinoshita, Phys. Rev. Lett. **75**, 4728 (1995).

27. T. Kinoshita, Rep. Prog. Phys. **59**, 1459 (1996).

28. T. Kinoshita, IEEE Trans. Instrum. Meas. **46**, 108 (1997).

29. S. Laporta, Phys. Lett. B **343**, 421 (1995).

30. S. Laporta and E. Remiddi, Phys. Lett. B **379**, 283 (1996).

31. E. Kruger, W. Nistler, and W. Weirauch, Metrologia **35**, 203 (1998).

32. R. S. Vandyck, D. L. Farnham, and P. B. Schwinberg, Phys. Rev. Lett. **70**, 2888 (1993).

33. G. L. Greene, E. G. Kessler, R. D. Deslattes, and H. Borner, Phys. Rev. Lett. **56**, 819 (1986).

34. T. Udem, J. Reichert, R. Holzwarth, and T. W. Hansch, Phys. Rev. Lett. **82**, 3568 (1999).

35. S. Guellati *et al.*, in *Abstracts of 17th International Conference on Atomic Physics*, edited by F. Fuso and F. Cervelli (AIP, Firenze, Italy, 2000), pp. 210–211.

Does the fine structure constant vary with time and distance?

V.V. Flambaum

School of Physics, University of New South Wales, UNSW Sydney NSW 2052, Australia

Abstract.
   Theories unifying gravity and other interactions predict spatial and temporal variation of physical "constants" in the Universe. Comparison of quasar absorption line spectra with laboratory spectra provide the best probe for variability of the fine structure constant, $\alpha = e^2/\hbar c$, over cosmological time-scales. We have demonstrated [1] that high sensitivity to the variation of α can be obtained from a comparison of the spectra of heavy and light atoms and have obtained an order of magnitude gain in precision over previous methods [2]. Our new constraints [3] on α come from simultaneous fitting of numerous absorption lines of the following species: MgI, MgII, AlII, AlIII, SiII, CrII, FeII, NiII and ZnII. The results are based on an analysis of 49 absorption systems covering cosmological time starting from about 10% of the age of the Universe after Big Bang (the redshift parameters cover $0.5 < z < 3.5$). The data contain *statistical* evidence for a smaller α at earlier epochs at the 4.3σ level. We briefly discuss possible systematic errors and numerous tests done to estimate and reduce these errors. Careful searches have so far not revealed any spurious effect that can explain the observations.

I   INTRODUCTION

   Were the laws of nature the same ten billion light years from us? Theories unifying gravity and other interactions suggest the possibility of spatial and temporal variation of physical "constants" in the Universe (see, e.g. [4]). Current interest is high because in superstring theories – which have additional dimensions compactified on tiny scales – any variation of the size of the extra dimensions results in changes in the 3-dimensional coupling constants. At present no mechanism for keeping the internal spatial scale static has been found (for example, our three "large" spatial dimensions increase in size). Therefore, unified theories applied to cosmology suffer generically from a problem of predicting time-dependent coupling constants. Moreover, there exists a mechanism for making all coupling constants and masses of elementary particles both space and time dependent, and influenced by local circumstances [5]. The variation of coupling constants can be non-monotonic (for example, damped oscillations).

   The strongest terrestial constraint on the time evolution of the fine structure

CP551, *Atomic Physics 17*, edited by E. Arimondo, P. DeNatale, and M. Inguscio
© 2001 American Institute of Physics 1-56396-982-3/01/$18.00

constant, α, comes from a natural uranium nuclear fission reactor in Gabon, West Africa, which was active 1.8 billion years ago. The relative change of α during this time interval does not exceed 1.2×10^{-7} [6]. However, this limit is based on certain assumptions and covers a relatively small fraction of the age of the Universe. Also, it does not exclude oscillatory dependence of α.

Astrophysical measurements enable us to push the probed epoch back to much earlier times. The energy scale of atomic spectra is given by the atomic unit $\frac{me^4}{\hbar^2}$. In the non-relativistic limit, all atomic spectra are proportional to this constant and analyses of quasar spectra cannot detect any change of the fundamental constants. Indeed, any change in the atomic unit will be absorbed in the determination of the redshift parameter z ($1 + z = \frac{\omega}{\omega'}$, ω' is the redshifted frequency of the atomic transition and ω is the laboratory value). However, any change in the fundamental constants can be found by measuring the relative size of relativistic corrections, which are proportional to α^2.

It is natural to search for any changes in α using measurements of the spin-orbit splitting within a specific fine structure multiplet. Indeed, this method has been applied to quasar spectra by several groups. The ratio of fine structure splitting of an alkali-type doublet to the mean transition frequency is proportional to α^2. A comparison of these ratios in cosmic spectra with laboratory values provides powerful constraints on variability. This method was proposed by J. Bachall and M. Schmidt in 1967 [7]. Varshalovich, Panchuk & Ivanchik [8] have obtained very stringent upper limits on any variation at redshifts $z \sim 2.8$–3.1 at the fractional level of $\Delta\alpha/\alpha \equiv (\alpha_z - \alpha_0)/\alpha_0 = 0.2 \pm 0.7 \times 10^{-4}$. Here, α_0 is the present day value of α and α_z is the value at the redshift, z, of the absorbing gas cloud. See [9] for a review of other works.

Recently we developed a new approach which improves the sensitivity to a variation of α by more than an order of magnitude [1,2]. The relative value of any relativistic corrections to atomic transition frequencies is proportional to α^2. These corrections can exceed the fine structure interval between the excited levels by an order of magnitude (for example, an s-wave electron does not have the spin-orbit splitting but it has the maximal relativistic correction to energy). The relativistic corrections vary very strongly from atom to atom and can have opposite signs in different transitions (for example, in s-p and d-p transitions). Thus, any variation of α could be revealed by comparing different transitions in different atoms in cosmic and laboratory spectra.

This method provides an order of magnitude precision gain compared to measurements of the fine structure interval. Relativistic many-body calculations are used to reveal the dependence of atomic frequencies on α for a range of atomic species observed in quasar absorption spectra [1]. It is convenient to present results for the transition frequencies as functions of α^2 in the form

$$\omega = \omega_0 + q_1 x + q_2 y, \qquad (1)$$

where $x = (\frac{\alpha}{\alpha_0})^2 - 1, y = (\frac{\alpha}{\alpha_0})^4 - 1$ and ω_0 is a laboratory frequency of a particular transition. New and accurate laboratory measurements of ω_0 have been carried out specifically for this work by Ulf Griesmann, Sveneric Johansson, Rainer Kling, Richard Learner, Ulf Litzén, Juliet Pickering and Anne Thorne (see also accurate measurements in [10,11,13–20]). We stress that the second and third terms contribute only if α deviates from the laboratory value α_0. The initial observational results [2] for two MgII lines and five FeII lines suggest that α may have been smaller in the past.

This work has been continued in Ref. [3]. A large set of data consists of 49 quasar absorption systems located between 4 and 11 billion light years from us (starting from 10% of the age of the Universe after Big Bang). Many lines of MgI, MgII, AlII, AlIII, SiII, CrII, FeII, NiII and ZnII have been included and a study of both temporal and spatial dependence of α has been performed. For the whole sample, $\Delta\alpha/\alpha = (-7.5 \pm 1.8) \times 10^{-6}$. We should stress that only statistical errors are presented here. This error is now small and the main efforts are directed towards the study of various systematic effects [21].

Note that the data have already passed one crucial test. The relativistic corrections vary very strongly from atom to atom and can have opposite signs in different transitions (for example, in s-p and d-p transitions). It is hard to imagine that the spurious effects "know" about this. Therefore, we measured α variation separately for positive shifters (positive coefficient q_1 in eq. (1)) combined with anchor lines (small q_1) and negative shifters combined with anchor lines. Startlingly, the results for $\Delta\alpha$ are the same in both cases! Spurious shifts of the lines would give the opposite signs for "$\Delta\alpha$" in these two cases.

This cosmic spectroscopy method has been extended to study variation of other fundamental parameters. The ratio of the hydrogen atom hyperfine transition frequency to a molecular (CO, CN, CS, HCO$^+$, HCN etc.) rotational frequency is proportional to $y = \alpha^2 g_p$ where g_p is the proton magnetic g-factor [24]. A new preliminary result here is $\Delta y/y = (-2.4 \pm 1.8) \times 10^{-6}$ about 4 billion light years from us (the average z=0.47). Altogether, we now have 3 independent samples of data: two optical samples (see [2,3]) and one radio sample. All 3 samples hint that $\Delta\alpha$ is negative.

The ratio of rotational and optical frequencies is sensitive to the ratio of the electron and proton masses and a hyperfine/optical comparison constrains $\alpha^2 g_p m_e/m_p$. Note that the proton g-factor and mass are functions of the strong interaction constant α_s and vacuum condensates of the quark and gluon fields.

Another method to search for the time variation of α is to study variation of the ratio of frequencies in the laboratory. The strongest laboratory limit on the α variation was obtained by comparing H-maser vs. HgII microwave atomic clocks over 140 days [22]. This yielded an upper limit $\dot{\alpha}/\alpha \leq 3.7 \times 10^{-14}/\text{yr}$ (see also [23]).

Another possibility is to use optical atomic frequency standards. Any evolution of α in time would lead to a frequency shift. To establish the connection between $\dot{\alpha}$ and $\dot{\omega}$, relativistic calculations of the α dependence of the relevant frequencies for CaI, SrII, BaII, YbII, HgII, InII, TlII and RaII have been performed [1]. The α dependence of the microwave frequency standards (Cs, Hg$^+$) has also been accurately calculated.

Note that we present all results in this paper assuming that the atomic unit of energy $\frac{me^4}{\hbar^2}$ is constant.

II ATOMIC THEORY

A Semi-empirical estimations and advantages of the new method to search for variation of α

To explain the advantages of our proposals let us start from simple analytical estimates of the relativistic effects in transition frequencies. The contribution of the relativistic correction to the energy can be obtained as an expectation value $\langle V \rangle$ of the relativistic perturbation V, which is large in the vicinity of the nucleus only. The wave function of an external electron near the nucleus is presented in, for example, [25]. A simple calculation of the relativistic correction to the energy of external electron gives the following result:

$$\Delta_n = -\frac{me^4 Z_a^2}{2\hbar^2}\frac{(Z\alpha)^2}{\nu^3}\left[\frac{1}{j+1/2} - C(j,l)\right] = \frac{E_n(Z\alpha)^2}{\nu}\left[\frac{1}{j+1/2} - C(j,l)\right]. \quad (2)$$

where Z is the nuclear charge, l and j are the orbital and total electron angular momenta, Z_a is the charge "seen" by the external electron outside the atom, i.e. $Z_a = 1$ for neutral atoms, $Z_a = 2$ for singly charged ions, etc.; ν is the effective principal quantum number, defined by $E_n = -\frac{me^4}{2\hbar^2}\frac{Z_a^2}{\nu^2}$, where E_n is the energy of the electron. For hydrogen-like ions $\nu = n$, $Z_a = Z$, where n is the principal quantum number. To describe the contribution of many-body effects to the relativistic correction, Δ_n, we introduce the parameter $C(j,l)$. Indeed, the single-particle relativistic correction increases the attraction of an electron to the nucleus and makes the radius of the electron cloud smaller. As a result, the atomic potential, which is the nuclear potential screened by the core electrons, becomes weaker. This decreases the binding energy of the external electron. Therefore, the many body effect has the opposite sign to the direct single-particle relativistic effect. Accurate many-body calculations described below give $C(j,l) \sim 0.6$ for s and p orbitals. We see that the relativistic correction is largest for the $s_{1/2}$ and $p_{1/2}$ states, where $j = 1/2$. The fine structure splitting is given by $\Delta_{ls} = E(p_{3/2}) - E(p_{1/2})$.

In quasar absorption spectra, transitions from the ground state are most commonly seen. Therefore, it is important to understand how the frequencies of these transitions are affected by the relativistic effects. The fine splitting in excited states is smaller than the relativistic correction in the ground state since the density of the excited electron near the nucleus is smaller. As a result, the fine splitting of the $E1$-transition from the ground state (e.g., s–p) is substantially smaller than the absolute shift of the frequency of the s–p transition. At $C(j,l) = 0.6$ the relativistic shift of the mean energy of the p-electron ($E(p) = 2/3E(p_{3/2}) + 1/3E(p_{1/2})$)

is small. Therefore, the average relativistic shift of the s-p transition frequency is mostly given by the energy shift of the s-state: $\Delta(p$-$s) \simeq -\Delta(s)$.

The relative size of the relativistic corrections is proportional to Z^2, so they are small in light atoms. Therefore, we can find the change of α by comparing transition frequencies in heavy and light atoms or by comparing s-p and d-p transitions in heavy atoms (like FeII and CrII) where the relativistic frequency shifts have opposite signs.

We stress that the most accurate and effective procedure to search for the change of α must include the analysis of all available lines (rather than the fine splitting in the excited states within one multiplet only). This new method has the following advantages:

- The total relativistic shift of frequencies (e.g. the largest s-electron shift) is included.

- The largest relativistic shift in the ground state is included.

- Very large statistics – all available atomic and ionic lines, different frequency ranges, different redshifts (epoch/distances).

- Many possibilities to search for systematic errors. For example, we can exclude any line or atom to avoid possible effects of unknown line blending, calibration errors, etc. The opposite signs and different values of the relativistic shifts for different lines give us a very efficient method to control the systematic effects.

As a result we can measure the effect which is $\sim$10 times larger than that in the alkali doublet method, have $\sim$100 larger statistics, cover a large range of redshifts/cosmological time and have better control of the systematic errors.

TABLE 1. Dependence on α of the frequencies of the $E1$ atomic transitions of astronomic interest; $Z = 6$ - 20 (units: cm^{-1}). Here $\omega = \omega_0 + q_1 x + q_2 y$ where $x = (\frac{\alpha}{\alpha_l})^2 - 1$, $y = (\frac{\alpha}{\alpha_l})^4 - 1$.

Z	Atom/Ion	Ground state		Upper states		ω_0	q_1	q_2
6	C I	$2s^2 2p^2$	$^3P_0$	$2s^2 2p3s$	$^3P_1$	60352.642 [19]	9	0
				$2s2p^3$	$^3D_1$	64089.861 [19]	143	0
				$2s2p^3$	$^3P_1$	75253.984 [19]	70	0
				$2s^2 2p4s$	$^3P_1$	78116.743 [19]	29	0
				$2s^2 2p3d$	$^3D_1$	78293.490 [19]	52	0
6	C II	$2s^2 2p$	$^2P^o_{1/2}$	$2s2p^2$	$^2D_{3/2}$	74932.617 [19]	177	3
				$2s2p^2$	$^2S_{1/2}$	96493.742 [19]	171	3
				$2s2p^2$	$^2P_{1/2}$	110625.1 [20]	173	-3
				$2s2p^2$	$^2P_{3/2}$	110666.3 [20]	217	3
6	C IV	$1s^2 2s$	$^2S_{1/2}$	$1s^2 2p$	$^2P_{1/2}$	64484.094 [19]	108	8
				$1s^2 2p$	$^2P_{3/2}$	64591.348 [19]	231	-8
7	N V	$2s$	$^2S_{1/2}$	$2p$	$^2P_{1/2}$	80463.211 [19]	196	-4
				$2p$	$^2P_{3/2}$	80721.906 [19]	488	2
8	O I	$2p^4$	$^3P_2$	$2p^3 3s$	$^3S_1$	76794.977 [19]	130	-30
				$2p^3 4s$	$^3S_1$	96225.055 [19]	140	-20
12	Mg I	$3s^2$	$^1S_0$	$3s3p$	$^1P_1$	35051.264(1) [15]	106	-10
12	Mg II	$3s$	$^2S_{1/2}$	$3p$	$^2P_{1/2}$	35669.298(2) [15]	120	0
				$3p$	$^2P_{3/2}$	35760.848(2) [15]	211	0
13	Al II	$3s^2$	$^1S_0$	$3s3p$	$^1P_1$	59851.972(4) [16]	270	0
13	Al III	$3s$	$^2S_{1/2}$	$3p$	$^2P_{1/2}$	53682.880(2) [16]	216	0
				$3p$	$^2P_{3/2}$	53916.540(1) [16]	464	0
14	Si II	$3s^2 3p$	$^2P^o_{1/2}$	$3s3p^2$	$^4P_{1/2}$	42824.297 [19]	437	10
				$3s3p^2$	$^4P_{3/2}$	42932.625 [19]	543	13
				$3s3p^2$	$^2D_{3/2}$	55309.3365(4) [16]	547	-6
				$3s^2 4s$	$^2S_{1/2}$	65500.4492(7) [16]	24	22
				$3s3p^2$	$^2S_{1/2}$	76665.352 [19]	558	-22
				$3s^2 3d$	$^2D_{3/2}$	79338.501 [19]	298	-3
				$3s3p^2$	$^2P_{1/2}$	83801.947 [19]	505	13
				$3s3p^2$	$^2P_{3/2}$	84004.261 [19]	724	3
14	Si IV	$2p^6 3s$	$^2S_{1/2}$	$2p^6 3p$	$^2P_{1/2}$	71287.523 [19]	362	-8
				$2p^6 3p$	$^2P_{3/2}$	71748.625 [19]	766	48
20	Ca I	$4s^2$	$^1S_0$	$4s4p$	$^1P_1$	23652.305 [19]	300	0
20	Ca II	$3p^6 4s$	$^2S_{1/2}$	$3p^6 4p$	$^2P_{1/2}$	25191.512 [19]	192	16
				$3p^6 4p$	$^2P_{3/2}$	25414.427 [19]	420	16

B Relativistic many-body calculations

Accurate calculations of relativistic effects in atoms have been done using many-body theory which includes electron-electron correlations. We used a correlation-potential (self-energy operator) method [26] for atoms with one external electron above closed shells and a combined configuration interaction and many-body perturbation theory method [27] for atoms with several valence electrons. These *ab initio* methods allow us to obtain an accuracy of $\sim 0.1\%$ for energy levels in atoms and ions with one external electron above closed shells and a few per cent in atoms

TABLE 2. Same as Table I; Z = 24 - 32

Z	Atom/Ion	Ground state		Upper states		ω_0	q_1	q_2
24	Cr II	$3d^5$	$^6S_{5/2}$	$3d^44p$	$^6F_{3/2}$	46905.17 [18]	-1624	-25
				$3d^44p$	$^6F_{5/2}$	47040.35 [18]	-1493	-21
				$3d^44p$	$^6F_{7/2}$	47227.24 [18]	-1309	-18
				$3d^44p$	$^6P_{3/2}$	48398.868(2) [17]	-1267	-9
				$3d^44p$	$^6P_{5/2}$	48491.053(2) [17]	-1168	-16
				$3d^44p$	$^6P_{7/2}$	48632.055(2) [17]	-1030	-13
25	Mn II	$3d^54s$	$^7S_3$	$3d^54p$	$^7P_2$	38366.184 [19]	918	34
				$3d^54p$	$^7P_3$	38543.086 [19]	1110	19
				$3d^54p$	$^7P_4$	38806.664 [19]	1366	27
26	Fe II	$3d^64s$	z $^6D_{9/2}$	$3d^64p$	$^6D^o_{9/2}$	38458.9871(2) [11]	1449	2
				$3d^64p$	z $^6D^o_{7/2}$	38660.0494(2) [11]	1687	-36
				$3d^64p$	z $^6F_{11/2}$	41968.0642(2) [11]	1580	29
				$3d^64p$	$z^6F_{9/2}$	42114.8329(2) [11]	1730	26
				$3d^64p$	z $^6F_{7/2}$	42237.0500 [11]	1852	26
				$3d^64p$	z $^6P_{7/2}$	42658.2404(2) [11]	1325	47
				$3d^64p$	z $^4F_{9/2}$	44232.512 [19]	936	278
				$3d^64p$	z $^4D_{7/2}$	44446.878 [19]	1616	3
				$3d^64p$	z $^4F_{7/2}$	44753.799 [19]	1701	141
				$3d^64p$	z $^8P_{7/2}$	54490.2 [20]	1719	-179
				$3d^64p$	z $^4G_{7/2}$	60956.82 [20]	1724	6
				$3d^64p$	z $^4H_{7/2}$	61156.835 [19]	1780	-86
				$3d^64p$	y $^4D_{7/2}$	61726.078 [19]	1342	-51
				$3d^64p$	y $^4F_{7/2}$	62065.528(3) [12]	1110	48
				$3d^64p$	y $^6P_{7/2}$	62171.625(3) [12]	1002	141
28	Ni II	$3d^9$	$^2D_{5/2}$	$3d^84p$	$z^2G_{7/2}$	56371.41 [19]	-134	0
				$3d^84p$	z $^2F_{7/2}$	57080.373(4) [17]	231	0
				$3d^84p$	z $^2D_{5/2}$	57420.013(4) [17]	-1188	0
				$3d^84p$	z $^2F_{5/2}$	58493.071(4) [17]	654	0
				$3d^84p$	z $^2D_{3/2}$	58705.94 [19]	275	0
				$3d^84p$	y $^2F_{5/2}$	67694.63 [19]	-1329	0
				$3d^84p$	y $^2F_{7/2}$	68131.22 [19]	-1158	0
				$3d^84p$	y $^2D_{3/2}$	68154.29 [19]	-585	0
				$3d^84p$	y $^2D_{5/2}$	68735.99 [19]	403	0
				$3d^84p$	z $^2P_{3/2}$	68965.66 [19]	266	0
				$3d^84p$	x $^2D_{5/2}$	71720.82 [19]	-451	0
				$3d^84p$	x $^2D_{3/2}$	72375.40 [19]	-444	0
				$3d^84p$	y $^2P_{3/2}$	72985.67 [19]	-336	0
				$3d^84p$	x $^2F_{7/2}$	75917.64 [19]	-876	0
				$3d^84p$	y $^2G_{7/2}$	79823.05 [19]	-716	0
30	Zn II	$3d^{10}4s$	$^2S_{1/2}$	$3d^{10}4p$	$^2P_{1/2}$	48481.077(2) [17]	1445	66
				$3d^{10}4p$	$^2P_{3/2}$	49355.002(2) [17]	2291	94
32	Ge II	$4s^24p$	$^2P_{1/2}$	$4s^25s$	$^2S_{1/2}$	62403.027 [19]	-575	-16

with several valence electrons. The accuracy was controlled by comparison between the calculated and observed energy levels and fine structure intervals. The values of the relativistic corrections and coefficients (q_1 and q_2) were obtained by repeating the calculations for different values of α (see eq. (1)).

The numerical procedure is the following:

- A relativistic Hartree-Fock (RHF) Hamiltonian was used to generate a complete set of single-electron orbitals, energy levels and Green's functions.

- Many-body perturbation theory in difference between the exact and Hartree-Fock Hamiltonians (perturbation $U = H - H_{HF}$) is used to calculate the effective Hamiltonian for valence electrons. This effective Hamiltonian includes correlations between the valence and core electrons which result in corrections to the valence electron energies and wave functions and screening of the electron-electron interaction by the core electrons.

- Diagonalization of the effective Hamiltonian for the valence electrons (the configuration interaction method).

The results are presented in the tables 1 and 2.

We see that some lines have a large increase in frequency when α increases (e.g. lines with large positive q_1 coefficients – "positive shifters" – like FeII and ZnII), some lines have a large decrease in frequency (e.g. lines with large negative q_1 coefficients – "negative shifters" – like CrII) and there are "anchor" lines which are not sensitive to a variation of α (e.g. lines with small q_1 like MgI, MgII, AlII, AlIII, SiII).

III RESULTS OF OBSERVATIONS

In this section we follow the work in [3].

All our QSO (quasar) spectra used in this work were obtained at the Keck I 10m telescope. The measurements of α-variation are based on two samples of data which loosely separate into two redshift régimes. The low-redshift sample ($\bar{z} = 1$) contains 28 absorption systems in the spectra of 17 QSOs with MgII and FeII lines. Full details of the reduction process are given in [28]. The absorbers in this sample lie in the range $0.5 < z < 1.8$ and so the useful transitions here are the five Iron lines, FeII $\lambda2344$–$\lambda2600$, and the Magnesium transitions, MgII $\lambda2796$ and $\lambda2803$. The MgII lines have small q_1 coefficients and so act as anchors against which the larger FeII shifts can be measured.

The high redshift ($\bar{z} = 2.1$) sample contains 21 systems in the spectra of 13 QSOs. The absorbers in this sample lie in the range $0.9 < z < 3.5$ and contain absorption from some or all of the following species: SiII, NiII, ZnII, FeII and CrII lines. Full details of the reduction procedures can be found in [29].

Our results are presented in Table 3. It shows the weighted mean (including statistical error bars), $\langle \Delta\alpha/\alpha \rangle_w$, the unweighted mean, $\langle \Delta\alpha/\alpha \rangle$ and the significance

TABLE 3. Statistics for the two subsamples and the sample as a whole. We give the average redshift, $\bar{z}$, for each sample and the number of data points, N, contributing to the weighted mean, $\langle \Delta\alpha/\alpha \rangle_w$, and unweighted mean, $\langle \Delta\alpha/\alpha \rangle$ (in units 10^{-5}). We also give the significance of the deviation from zero and the reduced χ^2 when the weighted mean is taken as the model.

Sample	$\bar{z}$	N	$\langle \Delta\alpha/\alpha \rangle_w$	$\langle \Delta\alpha/\alpha \rangle$	Significance	χ^2_{red}
Low z	1.02	28	-0.75 ± 0.23	-0.76 ± 0.32	3.3σ	0.82
High z	2.12	21	-0.74 ± 0.28	-0.62 ± 0.36	2.7σ	0.77
Total	1.49	49	-0.75 ± 0.18	-0.70 ± 0.24	4.3σ	0.78

level of the weighted mean for the low and high redshift samples, together with those statistics for the sample as a whole. We also include the value of the reduced χ^2, χ^2_{red} (i.e. χ^2 per degree of freedom), for each sample where the model is taken to be a constant equal to $\langle \Delta\alpha/\alpha \rangle_w$.

Our results show a 4.3σ variation in α over the redshift range $0.5 < z < 3.5$. We note that the weighted means do not differ significantly from the unweighted means for either sample. This indicates that we have not grossly underestimated the error bars on some small number of points, allowing them to dominate the overall weighted mean. Therefore, our results seem statistically self consistent.

To illustrate the distribution of $\Delta\alpha/\alpha$ over cosmological time, we plot our results in Fig. 1. The upper panel of Fig. 1 shows the raw values of $\Delta\alpha/\alpha$ as a function of fractional look-back time to the absorbing cloud using a flat Λ cosmology ($H_0 = 68$ kms^{-1}Mpc^{-1}, $\Omega_M = 0.3$, $\Omega_\Lambda = 0.7$). The redshift scale is also given for comparison. The lower panel shows an arbitrary binning of the data such that all bins have equal number of points (7 bins $\times$ 7 points per bin = 49 points). We plot the weighted mean for each bin with the associated 1σ error bars. At low redshifts we see that $\Delta\alpha/\alpha$ is consistent with zero – an expected behaviour if indeed cosmological variation of α exists. It is tempting to overinterpret such a diagram but we do note that the results are consistent with a generally smooth evolution of α with redshift. Note that the last point contains large error bars. Therefore, this picture seems to be consistent with both oscillatory and monotonic time dependence of α.

We have also made a more complete analysis of two radio spectra initially treated in [24] to obtain constraints on $y \equiv g_p \alpha^2$. Assuming g_p to be constant, we find $\Delta\alpha/\alpha = (-0.1 \pm 0.1)10^{-5}$ and $\Delta\alpha/\alpha = (-0.2 \pm 0.2)10^{-5}$ at $z = 0.25$ and 0.68 respectively. If we note the low redshift points in the lower panel (binned data) of Fig. 1 then we see that our results are also consistent with the two radio points.

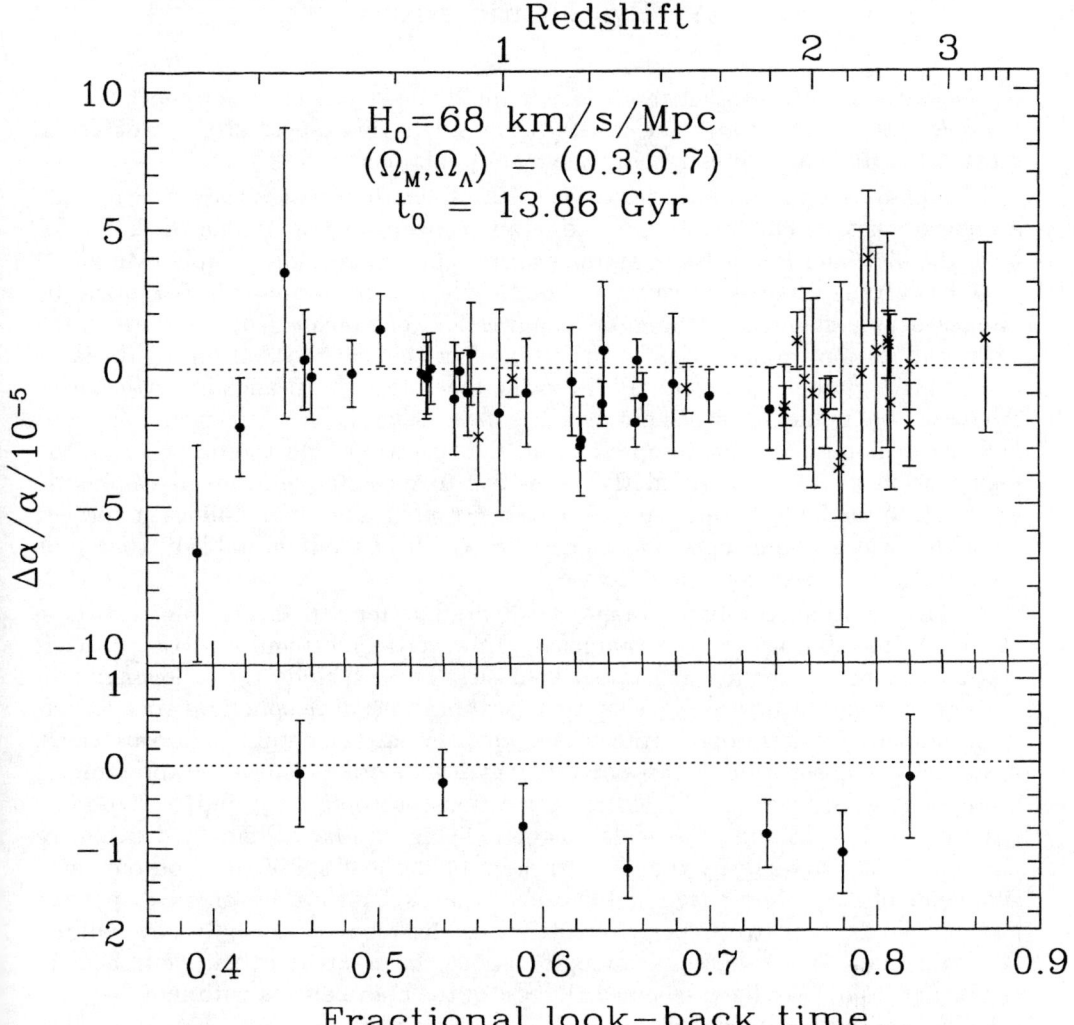

FIGURE 1. $\Delta\alpha/\alpha$ versus fractional look-back time within the current popular cosmology. The upper panel shows our raw results and 1σ error bars: the dots represent the low redshift sample and the crosses mark the high redshift sample. Note that the high redshift sample does contain some lower redshift absorbers. The lower panel shows an arbitrary binning of our results: 7 bins $\times$ 7 points per bin = 49 points. The redshifts of the points are taken as the mean redshift of clouds within that bin and the value of $\Delta\alpha/\alpha$ is the weighted mean with its associated 1σ error bar.

A Systematic errors?

The statistical error in our result is now small and so our attention must turn to possible systematic errors. The work [21] considers this aspect of the problem in great detail and so we present only the main points here.

To explain our results in terms of an effect other than real variation of α, such an effect must be capable of mimicking an non-zero average value of $\Delta\alpha/\alpha$ for both the high and low redshift samples alike. The low redshift sample (MgII/FeII systems only) is most sensitive to systematic effects since a non-zero $\Delta\alpha/\alpha$ can be mimicked by a slight 'stretching' or 'compression' of the spectrum relative to the ThAr calibration frames. This is due to the fact that all the FeII lines have large and positive q_1 coefficients and all lie to the blue of the MgII anchors. A negative $\Delta\alpha/\alpha$ can be mimicked by a slight decrease in the separation between the MgII and FeII lines, that is, a slight compression of the spectrum. On the other hand, the high redshift sample is particularly insensitive to such simple forms of systematic error. Since the high redshift sample contains transitions of many different species, all with various magnitudes and signs of q_1 coefficients, it is unlikely that it is greatly effected by systematic effects.

We have considered a broad range of potential systematic errors. Most of these can be excluded with simple arguments. However, some require further consideration and others require a detailed analysis. These include the following: laboratory wavelength errors, wavelength miscalibration, atmospheric dispersion effects, unidentified interloping transitions, intrinsic instrumental profile variations, positioning of absorption features on different echelle orders, spectrograph temperature variations, heliocentric velocity corrections, isotopic ratio and/or hyperfine structure variations and large scale magnetic fields. We have conducted extensive numerical tests to quantify the effect (if any) of the first six of these on our measurements of $\Delta\alpha/\alpha$ and have reliably excluded all but one, atmospheric dispersion effects. We have yet to properly quantify this effect but it can only have shifted $\Delta\alpha/\alpha$ in the positive sense; removing this effect from our data will yeild a more significant result. We have excluded the rest of the above effects with more general calculations and arguments.

We have also carried out other tests to search for a simple, unknown systematic effect. For example, it is unlikely that such an effect will be able to mimic the very specific q_1 dependence of the various lines: spurious effects do not "know" the magnitude and sign of the relativistic corrections to the transition frequencies. Thus, if we remove the transitions with large positive or negative q_1 coefficients from our fit and find new values for $\Delta\alpha/\alpha$, then we expect the two values to differ in sign if the line shifts are caused by some simple systematic effect. We have conducted such a test on the high redshift sample since it contains a subset of 12 systems in which we observe at least one anchor line, at least one transition with a large positive q_1 co-efficient and at least one with a large negative q_1 co-efficient. The results for $(\Delta\alpha/\alpha)$ are the following:

- No lines removed (actually, we do remove any lines that have "mediocre" shifts as to clearly deliniate the three different types of transitions)
  $(-1.31 \pm 0.39) \times 10^{-5}$

- The anchor lines removed (with absolute value of q_1 less than 300 cm^{-1})
  $(-1.49 \pm 0.44) \times 10^{-5}$

- The positive shifters removed (with $q_1 > 1000$ cm^{-1})
  $(-1.54 \pm 1.03) \times 10^{-5}$

- The negative shifters removed (with $q_1 < -1000$ cm^{-1})
  $(-1.41 \pm 0.65) \times 10^{-5}$

Thus, we find consistent values for the average values of $\Delta\alpha/\alpha$ both before and after the line removal. The interpretation that the line shifts we observe are caused by a varying α seems robust.

We also removed individual lines and all lines of an ion to test for possible effects of unknown line blending, change of isotopic ratio and laboratory wavelength errors.

From the above analysis we can conclude that if our results are due to some effect other than real variation in α, then this will only be revealed with further independent observations. Optical spectra are very useful for probing the range of redshifts we have explored here. However, it is clear that a statistical sample of radio spectra of H I 21 cm and molecular absorption systems has the potential to increase our precision limit by another order of magnitude. This effort is hampered by the fact that very few such systems are known at present and finding new systems is a somewhat serendipitous affair. Thus, a more systematic approach to finding these systems should be made. Also, as the arsenal of 8–10 m class telescopes with high resolution spectrographs grows, more observations should be devoted to obtaining *carefully calibrated* spectra of absorption systems at both high and low redshift. This will provide a new crucial check on our results and will allow us to discover or rule out any further subtle/unknown systematic effects in our analysis.

ACKNOWLEDGMENTS

The works described in this review have been performed with J.D. Barrow , V.A. Dzuba, C.W. Churchill, M.J. Drinkwater, M.T. Murphy, J.X.Prochaska, J.K. Webb, and A.M. Wolfe. We are very grateful to Ulf Griesmann, Sveneric Johansson, Rainer Kling, Richard Learner, Ulf Litzén, Juliet Pickering and Anne Thorne for conducting laboratory wavelength measurements especially for the present work and for communicating their results prior to publication. We would also like to thank Michael Bessell, Tom Bida, Bob Carswell, Rolf Engleman Jr., Alberto Fernández-Soto, John Hearnshaw, Jochen Liske, Geoff Marcy, Gillian Nave and Steve Vogt for very helpful communications.

REFERENCES

1. V.A. Dzuba, V.V. Flambaum, J.K. Webb. Phys. Rev. Lett. **82**, 888 (1999); Phys. Rev. A**59**, 230 (1999); Phys. Rev A, in press.
2. J.K. Webb, V.V. Flambaum, C.W. Churchill, M.J. Drinkwater, and J.D. Barrow, Phys. Rev. Lett. **82**, 884 (1999).
3. J.K. Webb, M.T. Murphy, V.V. Flambaum, V.A. Dzuba C.W. Churchill, J.X.Prochaska, A.M. Wolfe and J.D. Barrow, submitted to Phys. Rev. Lett; M.T. Murphy, J.K. Webb, V.V. Flambaum, C.W. Churchill, J.X. Prochaska, A.M. Wolfe, and J.D. Barrow, submitted to Mon. Not. Roy. Ast. Soc.
4. W. Marciano, Phys. Rev. Lett., **52**, 489 (1984). J. D. Barrow, Phys. Rev. D, **35**, 1805 (1987).
5. T. Damour, and A. M. Polyakov, Nucl. Phys. B, **423**, 596 (1994).
6. A.I. Shlyakhter, Nature **264**, 340 (1976); T. Damour and F. Dyson, Nucl. Phys. B **480** , 37 (1996).
7. J. Bachall, M. Schmidt, Phys. Rev. Lett., 19, 1294(1967).
8. D. A. Varshalovich, V. E. Panchuk, A. V. Ivanchik, Astron. Lett., 22, 6 (1996).
9. D. A. Varshalovich and A. Y. Potekhin, Space Science Review, **74**, 259 (1995).
10. U. Litzén, J. W. Brault, A. P. Thorne, Phys. Scr., 47, 628 (1993).
11. G. Nave , R. C. M. Learner, A. P.Thorne, C. J. Harris , J. Opt. Soc. Am. B, 8, 2028 (1991).
12. S. Johansson, Private communication.
13. R. E. Drullinger, D. J. Wineland, J. C.Bergquist, Appl. Phys., 22, 365 (1980).
14. W. Nagourney , H. G. Dehmelt , Bull. Am. Phys. Soc., 26, 805 (1981).
15. J. C. Pickering , A. P. Thorne, J. K. Webb, Mon. Not. Roy. Ast. Soc., 300, 131 (1998).
16. U. Griesmann, R. Kling, 2000, Submitted to ApJL.
17. J. C. Pickering, A. P. Thorne, J. E. Murray, U. Litzén, J. K. Webb, 2000, submitted to Mon. Not. Roy. Ast. Soc.
18. J. Sugar, C. Corliss, 1985, J. Phys. Chem. Data, **14**, Supplement No 2.
19. D. C. Morton, ApJS, 77, 119 (1991).
20. C. E. Moore, 1958, *Atomic Energy Levels*, Natl. Bur. Stand. (US), Circ. No. 467 (Washington). vol. **1**.
21. M.T. Murphy, J.K. Webb, V.V. Flambaum, C.W. Churchill, J.X. Prochaska, to be submitted to Mon. Not. Roy. Ast. Soc.
22. J. D. Prestage, R. L. Tjoeker, and L. Maleki, Phys. Rev. Lett. **74**, 3511 (1995).
23. Udem T. et al, Phys. Rev. Lett., 79, 2646 (1997); Udem T., Reichert J., Kourogi M., Hänsch, T., Optics Lett., 23, 1387 (1998).
24. M.J. Drinkwater,, J.K. Webb, J.D. Barrow and V.V. Flambaum. Mon. Not. R. Astron. Soc. **295**, 457 (1998).
25. I. I. Sobelman, *Introduction to the Theory of Atomic Spectra*, Nauka, Moscow, (1977) (Russian).
26. V. A. Dzuba, V.V.Flambaum, O.P.Sushkov, J.Phys. B, **16**, 715 (1983); V. A. Dzuba, V. V. Flambaum, P. G. Silvestrov, and O. P. Sushkov, J.Phys. B, **20**, 1399 (1987).

27. V. A. Dzuba, V. V. Flambaum, M. G. Kozlov, JETP Lett., **63**, 882 (1996); Phys. Rev. A **54**, 3948 (1996).

28. C. W. Churchill, 1995, Lick Technical Report #74. C. W. Churchill, 1997, The low ionization content of intermediate redshift galaxies, PhD. thesis, UC Santa Cruz. C. W. Churchill , J. R. Rigby, J. C. Charlton, S. S. Vogt , ApJS, **120**, 51 (1999).

29. J. X. Prochaska, A. M. Wolfe , ApJ, **470**, 403 (1996); ApJ, **474**, 140 (1997); ApJS, **121**, 369 (1999).

SINGLE SPECIES

Real-Time Tracking and Trapping of Single Atoms in Cavity QED

H. J. Kimble, K. Birnbaum, A. C. Doherty, C. J. Hood, T. W. Lynn, H.-C. Nägerl, D. M. Stamper-Kurn, D. W. Vernooy, and J. Ye

Norman Bridge Laboratory of Physics 12-33

California Institute of Technology

Pasadena, California 91125

Cavity quantum electrodynamics (QED) offers powerful possibilities for the deterministic control of atom-photon interactions quantum by quantum [1]. Indeed, modern experiments in cavity QED have achieved the exceptional circumstance of strong coupling, for which single quanta can profoundly impact the dynamics of the atom-cavity system. The diverse accomplishments of this field set the stage for advances into yet broader frontiers in quantum information science for which cavity QED offers unique advantages, such as the realization of quantum networks by way of multiple atom-cavity systems linked by optical interconnects [2,3].

The primary technical challenge on the road toward such scientific goals is the need to trap and localize atoms within a cavity in a setting suitable for strong coupling. Beginning with the work of Mabuchi et al. [4], several groups have been pursuing the integration of the techniques of laser cooling and trapping with those of cavity quantum electrodynamics (QED). [5–12] Two separate experiments in our group have recently achieved significant milestones in this quest, namely the trapping of single atoms in cavity QED [6,9,10]. Note that these experiments with cold atoms localized over time τ achieve $g_0\tau \geq 10^5\pi$, whereas experiments with conventional atomic beams in cavity QED have $g_0T \simeq \pi$, with T as the atomic transit time through the cavity mode.

In our experiments, the arrival of a single atom into the cavity mode can be monitored with high signal-to-noise ratio in *real time* by a near resonant field with mean intracavity photon number $\bar{n} < 1$. We emphasize that interactions in cavity QED bring an *in principle* enhancement in the capability to sense atomic motion beyond that which is otherwise possible in free space. Stated quantitatively, the ability to sense atomic motion within an optical cavity by way of the transmitted field can be characterized by the optical information $I = \alpha\frac{g_0^2 \Delta t}{\kappa} \equiv \alpha R\Delta t$, which roughly speaking is the maximum possible number of photons that can be collected as signal in time Δt with efficiency α as an atom transits between a region of optimal coupling g_0 and one with $g(\vec{r}) \ll g_0$. Here, the coupling parameter g_0 is defined such that $2g_0$ is the single-photon Rabi frequency and κ is the decay rate of the cavity mode itself. A key enabling aspect of our experiments is that $R = \frac{g_0^2}{\kappa} \gg (\kappa, \gamma)$,

CP551, *Atomic Physics 17*, edited by E. Arimondo, P. DeNatale, and M. Inguscio

© 2001 American Institute of Physics 1-56396-982-3/01/$18.00

leading to information about atomic motion at a rate that far exceeds that from either cavity decay at rate κ or spontaneous scattering at rate γ (as in fluorescence imaging).

For our first experiment [13], an atom's arrival within the cavity mode triggers *ON* an auxiliary field that functions as a far-detuned dipole-force trap (FORT) [9], thereby trapping the atom within the cavity mode. When the FORT is turned *OFF* after a variable delay, strong coupling likewise enables detection of the atom. Repetition of such measurements for single atoms yields a trap lifetime $\tau = (28 \pm 6)$ms, which has been limited by fluctuations in the intensity of the FORT arising from the intracavity conversion of FM to AM [14]. Currently implemented improvements in the FM-noise spectrum of the FORT laser should extend this lifetime to well beyond 1s.

In a second experiment depicted in Figure 1 [15], we rely upon light forces at the single-photon level to trap a single atom within the cavity mode [6,10]. As in the preceding experiment, the arrival of a single atom within the resonator is sensed with high signal-to-noise ratio, and triggers a trapping field *ON*. However, in this case, the trapping field is tuned near resonance with the atom-cavity system with $\bar{n} \simeq 1$ intracavity photon. Because the atomic kinetic energy $E_k < \hbar g_0$, even a single quantum is sufficient to profoundly alter the atomic center-of-mass (CM) motion and indeed to trap an atom as it moves through a region of spatially varying coupling coefficient $g(\vec{r}) = g_0\psi(\vec{r})$ (as arises in the Gaussian mode of our Fabry-Perot cavity, $\psi(\vec{r})$).

Furthermore, strong coupling means that the atomic motion will generate large variations in the transmission of a weak probe laser. From the resulting record of the detected photocurrent generated by the transmitted probe, we are able to reconstruct the trajectory of an individual atom by way of a novel inversion algorithm that we have developed. These reconstructions reveal single atoms bound in orbit by the mechanical forces associated with single photons. Our *atom-cavity microscope* yields 2μm spatial resolution in a 10μs time interval. Over the duration of the observation, the sensitivity is near the standard quantum limit for sensing the motion of a Cesium atom [16].

An important component of this research has been to investigate the extent to which light-induced forces in cavity QED are distinct from their free-space antecedents. The perspective has been to seek qualitatively new manifestations of optical forces at the single-photon level within the setting of cavity QED. Note that quantum character for the relevant fields or phenomena is not ensured the statement that the mean photon number $\bar{n} \sim 1$, since this is trivially the case in an equivalent free-space volume for a field of the same intensity as that inside the cavity.

By way of extensive numerical simulations of the relevant forces and their fluctuations and of comparisons to the standard free-space theory of laser cooling and trapping, we have investigated atom trapping inside optical resonators by the mechanical forces associated with single photons [17]. We have focussed on two points of interest, namely (1) whether or not there are qualitatively different effects of optical forces at the single-photon level within the setting of cavity QED, and (2) the features of the resulting atomic motional dynamics and how these dynamics are mapped onto experimentally observable variations of the intracavity field. Not surprisingly, qualitatively distinct atomic dynamics arise as the fundamental coupling and dissipative rates are varied.

For the experiment of Hood et al. [10], our analysis strongly supports the conclusion that atomic motion is largely conservative in nature with only smaller contributions from fluctuating forces. Atomic motion is predominantly in radial orbits transverse to the cavity axis, which in fact enables our reconstruction algorithm. A comparison of the well-known free-space theory and its cavity QED counterpart demonstrates that the usual fluctuations associated with the dipole force along the standing wave are suppressed by an order of magnitude. This suppression in dipole-force heating is based upon the Jaynes-Cummings ladder of eigenstates for the atom-cavity system, which represents qualitatively new physics for optical forces at the single-photon level within the setting of cavity QED [17].

We have also employed our numerical simulations to investigate another experiment that reports atom trapping in the single-photon regime [11]. Somewhat surprisingly, even in a regime of strong coupling, we find that for the parameters of this experiment, there are only small quantitative distinctions between the free-

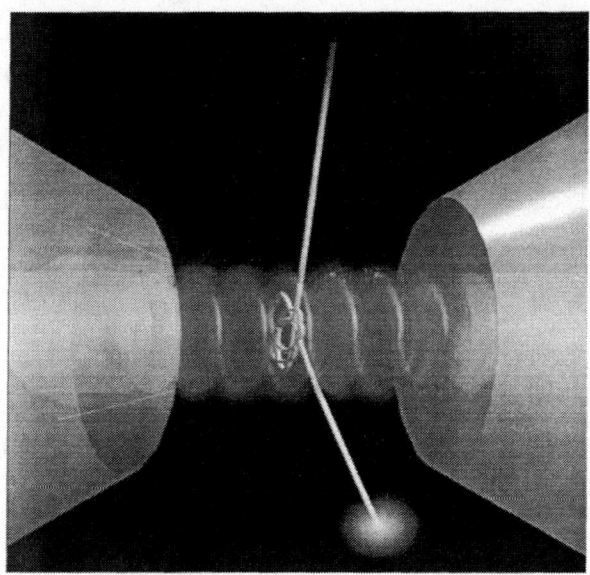

FIGURE 1. Illustration of the tracking and trapping of a single atom with the *atom-cavity microscope* in a regime with $\bar{n} \simeq 1$ photon. Shown is an experimentally reconstructed trajectory for an atom as it enters into the cavity mode, orbits, and finally exits [10,17]. Animated versions of reconstructed atom orbits can be viewed at *www.its.caltech.edu/~qoptics/atomorbits/*.

space theory and the appropriate quantum theory in cavity QED. It is thus not at all clear that description of this experiment as a novel single-quantum trapping effect is necessary [17]. Furthermore, in this setting our simulations demonstrate that atomic motion is dominated by diffusion-driven fluctuations in both the radial and axial dimensions, leading to an average observed localization time comparable to the time for an atom to transit freely through the cavity. The non-conservative character of the dynamics also hampers and often prohibits inference of atomic motion from the record of intracavity photon number, for both radial and axial processes.

To conclude, we emphasize again our goal of utilizing interactions in cavity QED to enable diverse protocols in quantum information science [2,3]. Our initial realizations of trapped atoms in cavity QED are important steps towards these ends. This research has been supported by the NSF, by DARPA via the QUIC (Quantum Information and Computation) program administered by ARO, and by the ONR.

REFERENCES

1. For a recent review, see contributions in the Special Issue of Physica Scripta **T76** (1998).
2. J.-I. Cirac, et al., Physica Scripta **T76**, 223 (1998).
3. A. S. Parkins and H. J. Kimble, Journal Opt. B: Quantum Semiclass. Opt. **1**, 496 (1999).
4. H. Mabuchi, Q. A. Turchette, M. S. Chapman, and H. J. Kimble, Opt. Lett. **21**, 1393 (1996).
5. C. J. Hood, M. S. Chapman, T. W. Lynn, and H. J. Kimble, Phys. Rev. Lett. **80**, 4157 (1998).
6. J. Ye, C. J. Hood, T. Lynn, H. Mabuchi, D. W. Vernooy, and H. J. Kimble, IEEE Trans. Instru. & Meas. **48**, 608 (1999).
7. H. Mabuchi, J. Ye, and H. J. Kimble, Appl. Phys. B **68**, 1095 (1999).
8. P. Münstermann, T. Fischer, P. Maunz, P. W. H. Pinkse, and G. Rempe, Phys. Rev. Lett. **82**, 3791 (1999).
9. J. Ye, D. W. Vernooy, and H. J. Kimble, Phys. Rev. Lett. **83**, 4987 (1999).
10. C. J. Hood, T. W. Lynn, A. C. Doherty, A.S. Parkins, and H. J. Kimble, Science **287**, 1447 (2000).
11. P. W. H. Pinkse, T. Fischer, P. Maunz, and G. Rempe, Nature **404**, 365 (2000).
12. P. Münstermann, T. Fischer, P. Maunz, P. W. H. Pinkse, and G. Rempe, Phys. Rev. Lett. **84**, 4068 (2000).
13. In this experiment, the Fabry-Perot cavity into which the atoms fall is formed from two super-polished spherical mirrors. The cavity length $l = 44.6\mu m$, waist $w_0 = 20\mu m$, and finesse $F = 4.2 \times 10^5$, and hence a cavity field decay rate $\kappa/2\pi = 4\text{MHz}$. The atomic transition employed for cavity QED is the ($g \equiv 6S_{1/2}, F = 4, m_F = 4 \rightarrow e \equiv 6P_{3/2}, F = 5, m_F = 5$) component of the D_2 line of atomic Cesium at $\lambda_{atom} \equiv c/\nu_{atom} = 852.4\text{nm}$. For our cavity geometry and from the atomic transition properties, we have $(g_0, \gamma_\perp)/2\pi = (32, 2.6)\text{MHz}$, with g_0 as the peak atom-field

coupling coefficient and $\gamma_\perp$ as the dipole decay rate for the $e \to g$ transition. These rates lead to critical photon and atom numbers ($m_0 \equiv \gamma_\perp^2/2g_0^2, N_0 \equiv 2\kappa\gamma_\perp/g_0^2$) = (0.003, 0.02).

14. C. W. Gardiner, J. Ye, H.-C. Nägerl and H. J. Kimble, Phys. Rev. **A6104**, 5801 (2000).

15. The optical cavity for this experiment is formed by two 1mm diameter, 10cm radius of curvature mirrors, located on the tapered end of 4mm x 3mm glass substrates. The multi-layer dielectric mirror coatings have a transmission of 4.5×10^{-6} and absorption/scatter losses of 2.0×10^{-6}, giving rise to a cavity finesse $F = 480,000$. For the measured cavity length $l = 10.9\mu m$ and waist $w_0 = 14\mu m$, we have parameters $(g_0, \kappa, \gamma) = 2\pi(110, 14.2, 2.6)$ MHz, where again these rates refer to the transition $\{6S_{1/2}, F = 4, m_F = 4\} \to \{6P_{3/2}, F = 5, m_F = 5\}$ at $\lambda = 852$nm in atomic Cesium. These rates lead to critical photon and atom numbers ($m_0 \equiv \gamma_\perp^2/2g_0^2, N_0 \equiv 2\kappa\gamma_\perp/g_0^2$) = (0.00028, 0.0061).

16. A more extensive gallery of reconstructed orbits can be viewed at http://www.its.caltech.edu/~qoptics/atomorbits/gallery.html.

17. A. C. Doherty, T. W. Lynn, C. J. Hood, and H. J. Kimble, submitted to Phys. Rev. A, available as quant-ph/0006015.

One-Electron Quantum Cyclotron (and Implications for Cold Antihydrogen[1])

G. Gabrielse[2], S. Peil[3], B. Odom and B. D'Urso

Department of Physics
Harvard University[4]
Cambridge, MA 02138

Abstract. Quantum jumps between Fock states of a one-electron oscillator reveal the quantum limit of a cyclotron accelerator. The states live for seconds when spontaneous emission is inhibited by 140 within a cylindrical Penning trap cavity. Averaged over hours the oscillator is in thermal equilibrium with black-body photons in the cavity. At 80 mK, quantum jumps occur only when resonant microwave photons are introduced into the cavity, opening a route to improved measurements of the magnetic moments of the electron and positron. The temperature demonstrated is about 60 times lower than the 4.2 K temperature at which charged elementary particles were previously stored. Implications for the production of cold antihydrogen are discussed.

INTRODUCTION

The harmonic oscillator is one of a few fundamental systems of physics that can be exactly solved. Its Hamiltonian $H_c = h\nu_c(a^\dagger a + 1/2)$ and equally spaced energy levels (Fig. 1a) are familiar to every student of quantum mechanics. The eigenstates of H_c are number states ($|n = 0\rangle$, $|n = 1\rangle$, ...) that are often called Fock states. Surprisingly, the production, observation and use of Fock states for experiments has proven to be difficult. Zero- and one-photon Fock states, $|n = 0\rangle$ and $|n = 1\rangle$, have been observed for a radiation mode of a cavity [1,2], but efforts are underway to observe higher Fock states, $|n \geq 2\rangle$ [3]. Vibrational Fock states of a laser-cooled Be^+ ion in a potential well have also been selectively excited [4]. Their formation was deduced from repeated measurements which transferred the population of identically prepared states to internal energy levels whose monitored

---

[1] Progress towards cold antihydrogen by our ATRAP team at CERN's new AD facility is summarized at http://hussle.harvard.edu/~atrap.
[2] gabrielse@physics.harvard.edu
[3] Current address is NIST, Gaithersburg, MD 20899.
[4] Support comes from the NSF, the AFOSR and the ONR.

CP551, *Atomic Physics 17*, edited by E. Arimondo, P. DeNatale, and M. Inguscio
© 2001 American Institute of Physics 1-56396-982-3/01/$18.00

time evolution revealed the original state. Very recently, the $|n = 0\rangle$ and $|n = 1\rangle$ Fock states of neutral atoms oscillating in a one dimensional harmonic well were also observed [5].

We are now able to display Fock states of a one-electron cyclotron oscillator [6]. The states are readily produced and used for experiments, last for a long time, and are directly observed without destroying them. A quantum nondemolition (QND) measurement [7,8] of the oscillator's energy shows that the electron oscillator remains in a Fock state until it makes an abrupt quantum jump to an adjacent Fock state. Such states can persist for many seconds because spontaneous emission is strongly suppressed by a factor of 140, much larger than the first observation in a trap [9], by detuning the oscillator from the radiation modes of a cylindrical trap cavity. Averaged over a much longer time (hours), the one-electron cyclotron oscillator is shown to be in a thermal state, in equilibrium with the black-body radiation in the cold trap cavity. As the temperature is reduced from 4.2 K to 80 mK, absorption and emission stimulated by black-body radiation essentially stop. The oscillator can then be excited with resonant microwave photons we introduce into the cavity. This demonstration suggests that resonant quantum-jump spectroscopy of the cyclotron oscillator is a promising new approach to an improved measurement of the electron magnetic moment, and a much better determination of the fine structure constant.

Charged elementary particles have not been stored at temperatures below 4.2 K before this work. This demonstration has implications for the positrons and antiprotons that are currently being stored as part of an effort to produce and study cold antihydrogen for the first time, since lower temperature antihydrogen could be more readily confined in a magnetic trap.

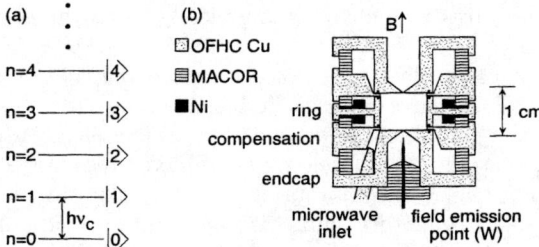

FIGURE 1. (a) Energy levels of the one-electron cyclotron oscillator. (b) Electrodes of the cylindrical Penning trap cavity.

ONE-ELECTRON CYCLOTRON OSCILLATOR

The one-electron cyclotron oscillator is realized in a cylindrical Penning trap (Fig. 1b) [10] that is cooled by a dilution refrigerator. The harmonic potential in the trap is of sufficient quality to allow the storage and nondestructive monitoring of a single trapped electron [11]. At the same time, the trap cavity has been shown to be a good approximation to a cylindrical microwave cavity at frequencies up to 160 GHz [12]. Tiny slits (125 μm) in the walls of the cavity make it possible to apply a trapping potential between the central ring electrode and the two flat endcap electrodes. This potential is made a better approximation to a harmonic potential along the central symmetry axis of the trap by tuning an additional potential applied to the two compensation electrodes. The small slits include quarter wave "choke flanges" to minimize the loss of microwave radiation from the cavity.

Cavity radiation modes that couple to the cyclotron oscillator have been measured and identified [12,13]. They have quality factors as high as $Q = 5 \times 10^4$. Energy in a 150 GHz mode with this Q value would damp exponentially with a 50 ns time constant that is very short compared to all relevant time scales. The radiation modes of the cavity are thus thermal states with the temperature of the trap cavity. Thermal contact to a dilution refrigerator allows us to adjust the trap temperature between 4.2 K and 70 mK (only to 80 mK when our detector is on.) We detune the frequency of the one-electron cyclotron oscillator away from the radiation modes to decrease the spontaneous emission rate.

Two of the three motions of a trapped electron (charge $-e$ and mass m) in a Penning trap [14] are relevant to this work. Our central focus is upon the circular cyclotron motion, perpendicular to our vertical 5.3 Tesla magnetic field, with classical cyclotron frequency $\nu_c = eB/(2\pi m) = 147$ GHz and energy levels separated by $h\nu_c$. The Fock states $|n\rangle$, often called Landau states for the particular case of a charged particle in a magnetic field, decay via spontaneous emission to $|n-1\rangle$ at a rate $n\gamma$, where γ is the classical decay rate of the oscillator. In free space for our field, $\gamma = 16\pi^2\nu_c^2e^2/(3mc^3) = (94 \text{ ms})^{-1}$. (We shall see shortly that the trap cavity greatly inhibits the spontaneous emission.)

The electron is also free to oscillate harmonically along the direction of the vertical magnetic field, $\hat{z}$, at a frequency $\nu_z \approx \nu_c/1000 = 64$ MHz. We drive this axial motion by applying an oscillatory potential between the ring and an endcap electrode, and detect the oscillatory current induced through a resonant tuned circuit attached between the ring and the other endcap. The electron damps as energy dissipates in the detection circuit, yielding an observed resonance width of 5 Hz for the driven axial motion. With appropriate amplification and narrow bandwidth detection we are able to measure shifts $\Delta\nu_z$ as small as 1 Hz. A heterostructure field effect transistor (HFET), constructed with Harvard collaborators just for these experiments [15], provides the radiofrequency gain that is needed while dissipating only 4.5 μW. The dilution refrigerator had difficulty with the nearly 700 times greater power dissipation (3 mW) of the MESFETs used in the past.

The cyclotron and axial motions of the electron would be uncoupled except that we incorporate two small nickel rings into the ring electrode of the trap (Fig. 1b) to deliberately distort the otherwise homogeneous magnetic field. This "magnetic bottle", $\Delta\vec{B} = B_2[(z^2 - \frac{1}{2}\rho^2)\hat{z} - z\rho\hat{\rho}]$ with $\vec{\rho} = x\hat{x} + y\hat{y}$, is similar to but much bigger than what was used to determine the spin state of an electron [16]. It gives rise to a coupling term in the Hamiltonian $V \sim z^2 a^\dagger a$, and shifts the monitored axial frequency in proportion to the energy in the cyclotron motion, $\Delta\nu_z = \delta(n + \frac{1}{2})$. A one quantum excitation of the cyclotron oscillator shifts the monitored axial frequency by $\delta = 13$ Hz, substantially more than the 5 Hz linewidth for the driven axial motion.

QND OBSERVATION OF QUANTUM JUMPS
BETWEEN FOCK STATES

The measurement of the cyclotron energy is an example of a QND measurement [7,8] in that V and H_c commute, $[V, H_c] = 0$. The desirable consequence is that a second measurement of the cyclotron energy at a later time will give the same answer as the first (unless a change is caused by another source). In this sense a QND measurement will not in and of itself change the state of the system. This is not generally true for measurements with a quantum system. For example, measuring the position of a free particle would make its momentum completely uncertain. After additional time evolution a second measurement of the particle's position would thus give a different outcome.

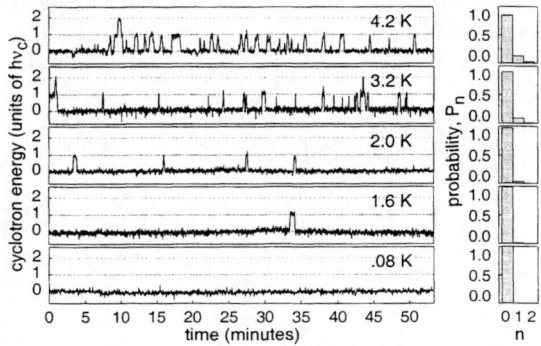

FIGURE 2. Quantum jumps between the lowest states of the one-electron cyclotron oscillator decrease in frequency as the cavity temperature is lowered.

Five one-hour sequences of QND measurements of the one-electron oscillator's energy are shown in Fig. 2. Each is for a different cavity temperature T, as measured with a ruthenium oxide sensor attached to the ring electrode. Greatly expanded views of several quantum jumps are shown in Fig. 3. Energy quantization is clearly visible, as are the abrupt quantum jumps between Fock states. The upward quantum jumps are absorptions stimulated by the black-body photons in the trap cavity. The downward transitions are spontaneous or stimulated emissions. Mostly we see the oscillator in its ground state $|n = 0\rangle$, with occasional quantum jumps to excited Fock states. Fig. 3b shows a rare event in which 4.2 K black-body photons sequentially excite the one-electron cyclotron oscillator to the Fock state $|n = 4\rangle$.

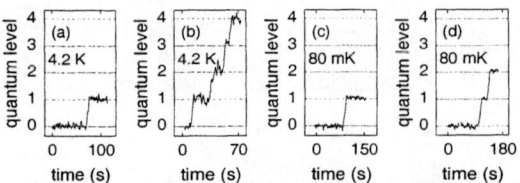

FIGURE 3. Excitations to excited Fock states which are stimulated by 4.2 K black-body photons in (a) and (b), and by an externally applied microwave field in (c) and (d).

QUANTUM BOLTZMANN THERMOMETER

Averaged over hours the oscillator is in a thermal state. We verify this and determine the equilibrium temperature of the the oscillator, T_c, from the measured probabilities P_n for occupying Fock states $|n\rangle$. The observed P_n, measured over many hours, are shown to the right in Fig. 2 for each cavity temperature. The probabilities fit well to the Boltzmann factors $P_n = Ae^{-nh\nu_c/kT_c}$ which pertain for thermal equilibrium, allowing us to deduce T_c from our "quantum Boltzmann thermometer". The solid points in Fig. 4a show that the cyclotron temperature T_c is equal to the cavity temperature T, as would be expected if the cyclotron oscillator is in thermal equilibrium with the black-body photons in the cavity. The solid points in Fig. 4b show the measured average quantum number superimposed upon the curve $\bar{n} = [e^{h\nu_c/kT} - 1]^{-1}$ which pertains for an oscillator in thermal equilibrium at the measured cavity temperature T. For temperatures of 4.2 K, 1 K and 80 mK, $\bar{n}$ varies dramatically from 0.23, to 2×10^{-3}, to 4×10^{-31}.

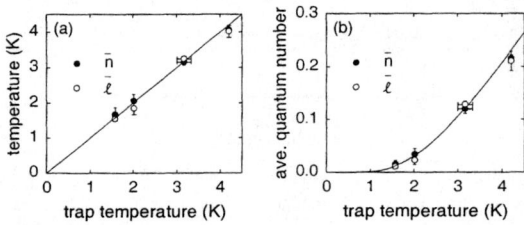

FIGURE 4. (a) The oscillator temperatures deduced from the measured occupation times in each number state (solid points) and deduced from the transition rates (open points) are compared to the temperature of a ruthenium oxide thermometer attached to a trap electrode. (b) Measured average values $\bar{n}$ and $\bar{\ell}$ as a function of cavity temperature.

Below 1 K the oscillator resides in its ground state for so long (we estimate 10^{32} years for 80 mK) that it is difficult to directly measure the oscillator temperature T_c. The best we can do is to establish that at some confidence level C, this temperature is below a limit given by $kT_c \leq h\nu_c/\ln[1 - \gamma t/\ln(1 - C)]$ if we observe no excitation for time t. When no excitation is observed for $t = 5$ hours, for example, we establish that $T_c < 1.0$ K at the $C = 68\%$ confidence level. For temperatures below 1 K, black-body photons have been essentially eliminated, and the one-electron cyclotron oscillator is virtually isolated from its environment.

INHIBITED SPONTANEOUS EMISSION

We can separately measure the rate Γ_{abs} for the upward jumps (corresponding to stimulated absorption), and the rate Γ_{em} for downward jumps (corresponding to stimulated and spontaneous emission together). For $T = 1.6$ K, Fig. 5 shows a histogram of the dwell times in $|n = 0\rangle$ in (a) and for $|n = 1\rangle$ in (b). Both histograms decrease exponentially, indicating random processes, so the fitted lifetimes $(\Gamma_{abs})^{-1}$ and $(\Gamma_{em})^{-1}$ are just the average values of the dwell times. The rates for stimulated emission from $|n\rangle$ to $|n - 1\rangle$ and for stimulated absorption from $|n - 1\rangle$ to $|n\rangle$ are expected to be equal by the principle of detailed balance. Thus the spontaneous emission rate is simply the difference between the observed emission rate and the observed absorption rate, $\gamma = \Gamma_{em} - \Gamma_{abs}$. At $T = 1.6$ K (Fig. 5) the measured stimulated absorption rate is negligibly smaller so that $\gamma^{-1} \approx \Gamma_{em}^{-1} = 13$ s.

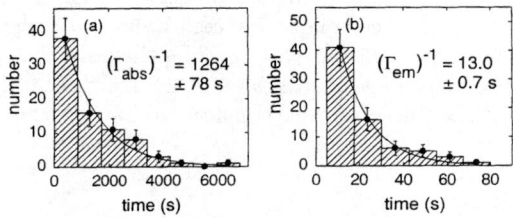

FIGURE 5. Histograms of the dwell times preceeding stimulated absorptions from $n = 0$ to $n = 1$ in (a), and for spontaneous and stimulated emissions from $n = 1$ to $n = 0$ in (b), both for $T = 1.6$ K. Dwell times less than 5 s are excluded since short dwell times are obscured by detection time constants.

Comparing the spontaneous lifetimes that are measured (13 s) and expected for free space (94 ms) shows that spontaneous emission is suppressed by a factor of 140. This large inhibition, much larger than the initial observation [9], is due to the copper trap cavity that encloses the electron oscillator. By tuning the magnetic field, the frequency of the cyclotron oscillator is tuned away from resonance with the radiation modes of the trap cavity. The electron oscillator then couples only very weakly to the modes of the radiation field, and spontaneous emission is suppressed. We would not otherwise be able to signal average sufficiently to observe the quantum jumps so distinctly, nor would the excited Fock states persist so long.

As a final step in our quantitative analysis we can measure the average number $\bar{\ell}$ of resonant black-body photons within the cavity. Quantum electrodynamics indicates that the stimulated emission from $|n\rangle$, and the stimulated absorption into $|n\rangle$, both have rates equal to $\bar{\ell}n\gamma$. Applied to the two lowest states, this means that

$\Gamma_{abs} = \bar{\ell}\gamma$ and $\Gamma_{em} = (1+\bar{\ell})\gamma$. The average number of black-body photons in terms of measurable quantities is thus given by $\bar{\ell} = \Gamma_{abs}/(\Gamma_{em} - \Gamma_{abs})$. The measured open points in Fig. 4b agree well with the expected curve $\bar{\ell} = [e^{h\nu_c/kT} - 1]^{-1}$, and $\bar{n} = \bar{\ell}$ as predicted. Fitting to the measured $\bar{\ell}$ gives an independent measurement of the temperature of the cavity (open points in Fig. 4a). These agree well with the directly measured cavity temperature.

Once black-body radiation is essentially eliminated, by lowering the temperature below 1 K, we are free to make resonant excitations of the one-electron cyclotron oscillator with microwave photons introduced into the cavity from the outside. Fig. 3c-d illustrate such excitations. Quantum jumps take place only when the frequency of the microwave photons is tuned near resonance with the cyclotron frequency. Such quantum jump spectroscopy should make it possible to measure the frequency between the lowest Fock states with the exquisite precision required to improve on the very accurate measurement of the electron magnetic moment reported twelve years ago [16]. Moreover, the large inhibition of spontaneous emission indicates a weak coupling to the cavity modes, weak enough that the cavity shifts would be smaller than those that limited the accuracy of the early measurement. Spectacular theoretical advances made in recent quantum electrodynamics calculations [17] provide the important motivation. The theoretical accuracy now exceeds the accuracy of the experiment, and reducing the experimental uncertainties by a factor of 7 will allow the fine structure constant to be deduced more accurately by this factor.

One challenge that faces measurements using a magnetic bottle is solved by the extremely low temperature demonstrated here. The magnetic bottle (with strength given by δ) not only couples the cyclotron energy to ν_z as is desired for good detection sensitivity. It also shifts the cyclotron frequency in proportion to the axial energy E_z with proportionality constant δ, by $\Delta\nu_c \sim \delta E_z$. Our very low temperature makes this shift small even when we use a large magnetic bottle (i.e. large δ) to enhance the detection sensitivity. If the axial detectors are turned off during the crucial parts of a measurement, the axial motion cools to the 70 mK temperature of the trap cavity, and $\Delta\nu_c < 10^{-9}\nu_c$.

Throughout this report we refer to repeated measurements of the cyclotron energy, rather than to continuous measurement, because it takes 2 s of signal averaging for us to ascertain the quantum state of the cyclotron oscillator. The true measurement time is less, being the time required to establish the quantum state in principle, but is a bit difficult to estimate. The estimate that has been made [18] uses assumptions that do not correspond very well to our experimental conditions. In the future we hope to vary and observe the effect of the measurement time.

FIRST EXAMPLES OF QUANTUM JUMP
SPECTROSCOPY AND SPIN FLIPS

One objective of this work is to measure the magnetic moment of the electron more accurately than ever before, by perhaps as much as a factor of ten. To do so requires that we measure the cyclotron frequency and that we be able to observe spin flips. Fig. 6 shows our first example of using quantum jump spectroscopy to measure the cyclotron frequency. The number of quantum jumps from n = 0 to n = 1 is plotted as a function of the cyclotron drive frequency. The lineshape is well understood [14] and will not be discussed here.

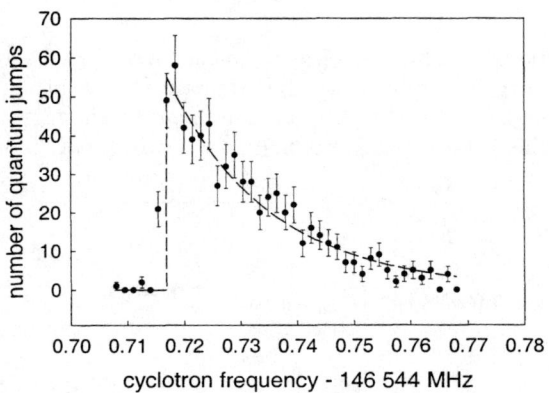

FIGURE 6. Early electron cyclotron resonance using quantum jump spectroscopy for transitions between the ground state and the first excited state shows an axial temperature of 1 K.

Fig. 7 shows the cyclotron and spin energy levels, along with our first observation of a spin flip. The base line axial frequency simply changes by an amount that cannot be distinguished from the shift due to a cyclotron excitation, since the cyclotron and spin frequencies are nearly equal. The difference, of course, is that a cyclotron excitation is soon followed by an emission process that takes the axial frequency back to its base value, while the spin is so decoupled from the environment that no spin relaxation would ever be observed.

A measurement of the electron magnetic moment can either be used to test quantum electrodynamics, or to determine the fine structure constant, α. Quantum electrodynamics gives a relationship between the magnetic moment and α. Right now, the measurement of the electron magnetic moment is already much more accurate than any measurement of the fine structure constant, though there is some hope that this situation will soon change. When is does, then quantum electrodynamics will be tested at an improved accuracy. Already, the accuracy of the comparison between theory and experiment is unprecedented. Meanwhile, with the assumption that the quantum electrodynamics calculations are accurate, the

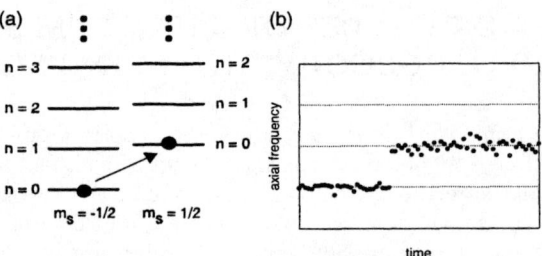

FIGURE 7. (a) Electron cyclotron and spin energy levels shows the spin flip transition. (b) Experimental observation of this transition.

measured magnetic moment of the electron is used with the theory to deduce what is by far the most accurate value of the fine structure constant (Fig. 8). The theory is sufficiently more accurate than the measurements right now that an improved measurement of the electron magnetic moment is warranted.

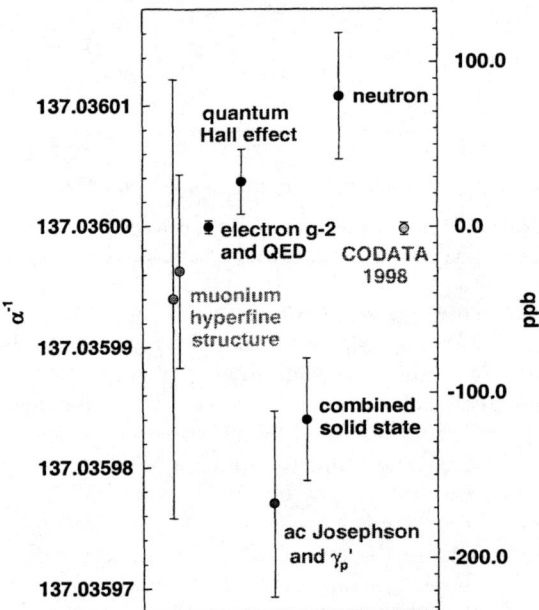

FIGURE 8. Already the fine structure constant is determined mostly by the 1981 measurement of the electron magnetic moment together with QED calculations which relate the measurement to the fine structure constant. We aim to substantially improve upon the measurement.

IMPLICATIONS FOR COLD ANTIHYDROGEN

For about thirteen years our TRAP Collaboration, working at the LEAR facility of CERN, developed and demonstrated all the crucial techniques needed to accumulate cold antiprotons in a particle trap. Antiprotons from LEAR were slowed in matter, captured in a trap, cooled via collisions with cold trapped electrons, kept cold by a resonant circuit, lived for months or more in a vacuum better than 5×10^{-17} Torr, and were monitored nondestructively. The TRAP antiprotons were accumulated at an energy 10^{10} times lower than the energy of the antiprotons received from LEAR. A recent review summarizes these developments [19].

One goal was to compare the charge-to-mass ratios of the antiproton and proton to high accuracy. In a series of three measurements this comparison was improved by almost a factor of a million, to an accuracy of 90 parts per trillion. This is the most accurate test of CPT invariance made with a baryon system. Any difference between antimatter and matter would be extremely interesting since we do not yet understand why we have a universe made of matter. We would expect that the big bang that originated our universe would create equal amounts of antimatter and matter, which would then annihilate, leaving nothing. The great mystery is why enough matter was left over that we and our matter universe could exist.

Another goal was to make antihydrogen atoms that are cold enough that they can be trapped in a magnetic trap, a goal that was enunciated shortly after antiprotons were first trapped [20].

> "For me, the most attractive way ... would be to capture the antihydrogen in a neutral particle trap ... The objective would be to then study the properties of a small number of [antihydrogen] atoms confined in the neutral trap for a long time."

Once trapped, we would probe the cold antihydrogen with our lasers to look for any tiny difference between the structure of antihydrogen and hydrogen; we hope to achieve an accuracy that is substantially better than what we achieved in comparing q/m for the antiproton and proton. The closest approach to cold antihydrogen came with the simultaneous confinement of cold antiprotons and cold positrons in the same trap structure [21].

CERN's unique new antimatter factory, the Antiproton Decelerator (or AD for short) now gives us the opportunity to pursue this goal in a dedicated way. It is about to begin delivering useful numbers of antiprotons to experiments. Continuing in the CERN tradition of supporting important and fundamental antimatter experiments large and small, this new facility is dedicated entirely to the lowest energy antimatter particles. The AD is optimized to slow antiprotons rather than to let them coast or speed up in the manner of the other storage rings in the world. Based upon our demonstration that antiprotons could be accumulated in a tiny ion trap much more inexpensively than in a large storage ring, CERN was able to open the way to the new, low energy frontier, and at the same time save resources by replacing three storage rings with the single AD.

Our ATRAP experiment (the offspring of TRAP) is a quest to make antihydrogen atoms that are cold enough to be trapped, whereupon lasers directed at them can probe for tiny differences between antihydrogen and hydrogen. We have also picked up competitors (ATHENA) who are pursuing similar goals using the antiproton techniques that we developed. A third experiment at the AD, ASACUSA, creates and studies exotic helium atoms. Besides some collision studies, they hope to improve upon laser spectroscopy measurements that, together with our much more accurate comparison of q/m for the antiproton and proton, demonstrate that the q and m of the antiproton and proton are separately the same to about 1 ppm.

At ATRAP we hope to receive pulses of antiprotons shortly after ICAP that will allow us to rapidly trap, cool and accumulate cold antiprotons, along with copious amounts of cold positrons from a radioactive source [5].

Although the relationship between the quantum cyclotron and cold antihydrogen is not immediately clear, there is an important link. The quantum cyclotron demonstrates that charged elementary particles in a trap can be made to have temperatures at least 60 times lower than the 4.2 K which was previously the lowest temperature at which such particles were stored (Fig. 9). The techniques that were used for the quantum cyclotron should also work with positrons, and with antiprotons. With colder ingredients, the hope is that colder antihydrogen can be produced. The colder the antihydrogen atoms, the more of them that can be trapped in the shallow magnetic traps that can be made.

CONCLUSION

In conclusion, a one-electron cyclotron oscillator is realized within a cylindrical Penning trap cavity whose temperature can be varied between 70 mK and 4.2 K. Nondestructive QND measurements of the oscillator's energy show occasional quantum jumps between Fock states that can persist for many seconds because the cavity inhibits spontaneous emission by a factor of 140. Absorption and emission stimulated by black-body photons in the trap cavity are essentially gone below 1 K. At lower temperatures, quantum jump spectroscopy with microwave photons introduced to the cavity offers the possibility of a significantly improved measurement of the electron magnetic moment and the fine structure constant. There is also the interesting prospect of investigating the time required for a quantum measurement. The great reduction in the temperature of trapped elementary particles has important implications for the production of lower temperature antihydrogen atoms.

---

[5] Note added: ATRAP trapped, cooled and stacked the first 4.2 K antiprotons at CERN's new AD in mid July. In late August ATRAP was the first to add 4.2 K positrons in the same trap structure. We used them to cool antiprotons. More than 2 million 4.2 K positrons were accumulated in a bit more than an hour.

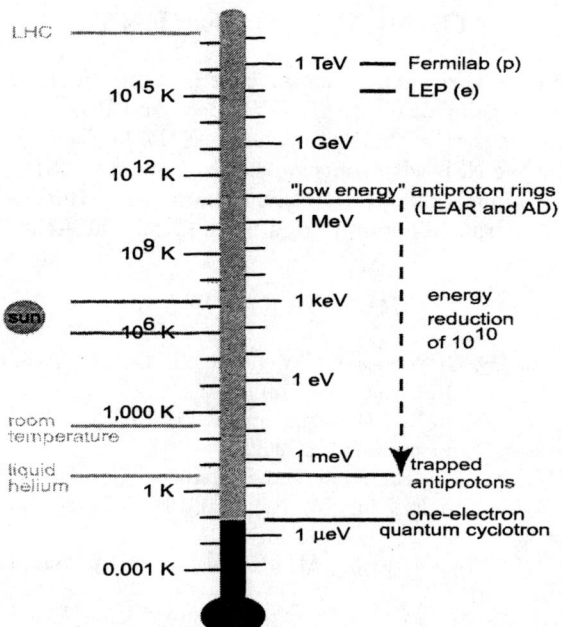

FIGURE 9. The quantum cyclotron is the lowest energy and temperature at which a charged particle has been stored.

ACKNOWLEDGEMENTS

We are grateful for early experimental contributions by K. Abdullah and D. Enzer, for the HFET collaboration with R.B. Beck and R.M. Westervelt. Some of this work was done as part of the Ph.D. work of S. Peil. The electron experiments were supported by the NSF with some assistance from the ONR. B. D'Urso is also supported by the Fannie and John Hertz Foundation. The Harvard contribution to the antihydrogen experiments is supported by the NSF, the AFOSR and the ONR.

REFERENCES

1. X. Maître, E. Hagley, G. Nogues, C. Wunderlich, P. Goy, M. Brune, J. M. Raimond, and S. Haroche, Phys. Rev. Lett. **79**, 769 (1997).
2. M. Weidinger, B. Varcoe, R. Heerlein, and H. Walther, in *Abstracts of ICAP 16* (Univ. of Windsor, Windsor, 1998), p. 362.
3. P. Domokos, M. Brune, J. Raimond, and S. Haroche, Eur. Phys. J. D **1**, 1 (1998).
4. D. Meekhof, C. Monroe, B. King, W. Itano, and D. J. Wineland, Phys. Rev. Lett. **76**, 1796 (1996).
5. I. Bouchoule, H. Perrin, A. Kuhn, M. Morinaga, , and C. Salomon, Phys. Rev. A **59**, R8 (1999).
6. S. Peil and G. Gabrielse, Phys. Rev. Lett. **83**, 1287 (1999).
7. K. Thorne, R. Drever, and C. Caves, Phys. Rev. Lett. **40**, 667 (1978).
8. V. Braginsky and F. Y. Khalili, Rev. Mod. Phys. **68**, 1 (1996).
9. G. Gabrielse and H. Dehmelt, Phys. Rev. Lett. **55**, 67 (1985).
10. G. Gabrielse and F. C. MacKintosh, Intl. J. of Mass Spec. and Ion Phys. **57**, 1 (1984).
11. J.N. Tan and G. Gabrielse, Appl. Phys. Lett. **55**, 2144 (1989).
12. J.N. Tan and G. Gabrielse, Phys. Rev. Lett. **67**, 3090 (1991).
13. G. Gabrielse and J.N. Tan, in *Cavity Quantum Electrodynamics*, edited by P. Berman (Academic Press, New York, 1994), p. 267.
14. L.S. Brown and G. Gabrielse, Rev. Mod. Phys. **58**, 233 (1986).
15. S. Peil, R.G. Beck, R.M. Westervelt, and G. Gabrielse, (to be published).
16. R.S. Van Dyck, Jr., P. Schwinberg, and H. Dehmelt, Phys. Rev. Lett. **59**, 26 (1987).
17. T. Kinoshita, Rep. Prog. Phys. **59**, 1459 (1996).
18. M. Rigo, G. Alber, F. Mota-Furtado, and P. O'Mahony, Phys. Rev. A **55**, 1665 (1997).
19. G. Gabrielse, Adv. At. and Mol. Phys. **45**, 1 (2000).
20. G. Gabrielse, in *Fundamental Symmetries*, edited by P.Bloch, P. Paulopoulos, and R. Klapisch (Plenum, New York, 1987), p. 59.
21. G. Gabrielse, D. Hall, T. Roach, P. Yesley, A. Khabbaz, J. Estrada, C. Heimann, and H. Kalinowsky, Phys. Lett. B **455**, 311 (1999).

Triggered Emission of Single Photons by a Single Molecule

Christian Brunel, Philippe Tamarat, Brahim Lounis, and Michel Orrit

Centre de Physique Moléculaire Optique et Hertzienne, CNRS et Université Bordeaux I,
351 cours de la Libération, 33405 Talence, France

Abstract. The linear Stark effect on a single molecule allows us to sweep its frequency through resonance with a fixed laser frequency by applying a sinusoidal RF voltage on two electrodes. If the conditions of adiabatic following are fulfilled, we can prepare the excited molecular state with near certainty. Spontaneous emission from this state gives rise to a single photon. With our current experimental conditions, up to 75 % of the sweeps lead to the emission of a single photon. Since the adiabatic passage is done on command, the molecule performs as a high rate source of triggered photons. The experimental results are in quantitative agreement with quantum Monte Carlo simulations.

INTRODUCTION

In addition to their fundamental interest [1], squeezed or controlled light states have many potential applications, such as communications, quantum computing, and quantum cryptography [2]. Let us consider a very regular stream of photons, i. e. individual photons periodically coming out of a source. The corresponding light beam would have no amplitude- or photon-noise, and would be very attractive for quantum cryptography. Eavesdropping on a message involves a measurement, and therefore perturbs the quantum state of the light carrying the message. As a stream of single photons is a well-defined quantum state, the perturbation would be easy to detect in the received photon stream.

The emission of photons one by one has already been demonstrated via the antibunching of the fluorescence light [3], but this emission is random. Several other schemes have been used to generate single photons, or twin photons. For example, a cascade process in a calcium atom [4] can produce a pair of correlated photons. Another scheme is parametric generation of twin photons by a laser pulse in the nonlinear crystal of an OPO [5]. In these two schemes, the efficiency of the generation has to be kept low to avoid pile-up, and the emission process is still random. To obtain a single photon on command, one has to prepare a single emitting state with certainty. In a recent work with semi-conductor quantum well structures, Yamamoto and colleagues have used Coulomb blockade effects to let a single electron and a single

CP551, *Atomic Physics 17*, edited by E. Arimondo, P. DeNatale, and M. Inguscio
© 2001 American Institute of Physics 1-56396-982-3/01/$18.00

hole recombine, thereby producing a single photon [6]. The problem of this method is the low efficiency of the photon generation: about one recombination in three leads to the emission of a single photon. Moreover, the distribution of the emitted photons was not measured, and the source was not fully characterised.

In the solution we propose here, we want to prepare a single molecule in its excited state. By spontaneous emission, the molecule will return to its ground state and emit a single photon within a few excited state lifetimes T_1, assuming that the fluorescence quantum yield is close to unity, which is the case of many aromatic molecules. To prepare the molecule in its excited state, we used the method of rapid adiabatic passage [7]. We chose to rapidly sweep the frequency difference between molecule and laser through resonance. To do this, we decided to sweep the molecular frequency by Stark effect around the fixed laser frequency. The sweep has to be slow enough to remain adiabatic, so that the Bloch vector accompanies the effective magnetic field as the molecule is swept through resonance with the laser. On the other hand, the sweep cannot be too slow, because the adiabatic passage is destroyed by relaxation. The conditions for ideal rapid adiabatic passage (RAP) are:

$$T_2 >> T_{pass} >> T_{Rabi} = 1/\Omega \qquad (1)$$

where Ω is the Rabi frequency, and T_2 is the coherence lifetime. T_{pass} can be defined naively as "the time it takes the laser" to cross the saturated resonance:

$$T_{pass} = \Gamma_S \, dt/d\omega \, , \quad \text{with } \Gamma_S = \Gamma \sqrt{1 + I/I_S} \, .$$

Γ_S is the saturated width, Γ the homogeneous width, I the laser intensity and I_S the saturation intensity. $d\omega/dt$ is the rate of angular frequency scanning, whose physical meaning is considered in more detail hereafter. The RAP requires that the Rabi frequency be much larger than the homogeneous linewidth for a convenient passage time to exist. In our experiments, Ω was limited to 5 Γ by the maximum admissible background.

The notion of a passage time for a frequency is self-contradictory in principle. If a frequency is to be followed as a function of time, the sweep must be slow enough, otherwise the Fourier frequency uncertainty T_{pass}^{-1} due to the finite passage time T_{pass} would be larger than the frequency swept during the passage time, $T_{pass} \, d\omega/dt$. Therefore, the naive definition of the passage time is correct only if:

$$T_{pass} >> \sqrt{dt/d\omega}$$

This condition is equivalent to the second inequality in (1), which states that the passage time, defined as the time to pass through the saturated resonance (with a breadth of about Ω) should be longer than the inverse of the Rabi frequency. If this condition is not fulfilled, broadening of the molecular frequency by its swift sweeping must be taken into account. Then, the effective passage time will be $(dt/d\omega)^{1/2} = T_C$.

This effective sweep time T_C will be used in the following instead of $dt/d\omega$, to characterize the sweep rate. The RAP conditions (1) may also be rewritten as:

$$T_2 \gg T_C \gg T_{Rabi} \quad (2)$$

The article is set up as follows: section 2 gives experimental details. The theory and simulations are briefly discussed in section 3. The results [8] are given and discussed in section 4.

EXPERIMENTAL

The system used for the present study is the highly fluorescent dibenzanthanthrene (DBATT) molecule in a n-hexadecane matrix, at 1.8 K. Most of the molecules in the wings of the inhomogeneous band (centered at 589 nm) present a linear Stark effect, with shifts up to 1 GHz for an electric field of 1 MV/m [9]. The optical design for the excitation of single molecules and the collection of fluorescence is of the lens-paraboloid-type with a full silica parabolic mirror (see Figure 1). The sample, a small drop of DBATT in hexadecane, is placed on a thin glass plate carrying two evaporated aluminium electrodes, with a gap of 20 µm. The RF electric field with a frequency of a few MHz is applied to the electrodes and shifts the single molecule's transition frequency sinusoidally with time.

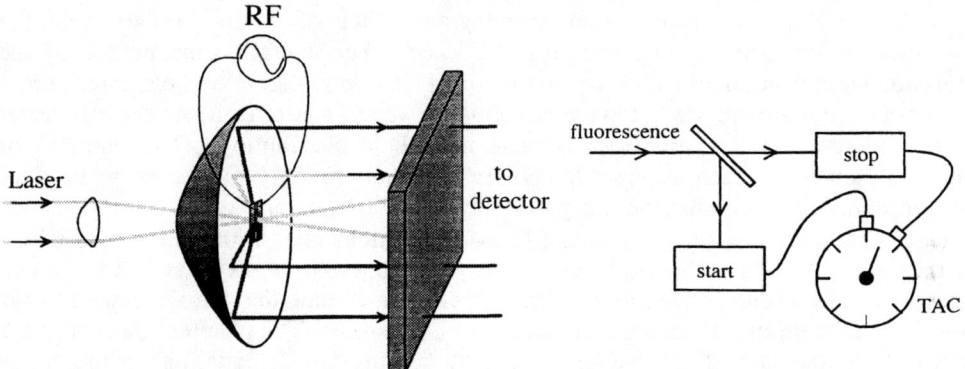

Figure 1: Left part: Schematic view of the optical setup, with the aspheric lens focusing the exciting laser, the full silica paraboloid collecting fluorescence emitted by the sample, and the filter blocking the laser light. The sample is placed in the 20 microns gap between two aluminum electrodes. The right part shows a scheme of the start-stop experiment.The start detector is the avalanche photodiode, the stop detector is a photomultiplier tube. The delay between start and stop is measured by a time-to-amplitude converter (TAC).

The temporal profile of the photon bursts is measured by triggering a multichannel analyser on a rising edge synchronous with the RF voltage, and by accumulating many photons.

To perform intensity correlation measurements, we have built a Hanbury-Brown and Twiss coincidence setup with a beamsplitter and two detectors (Figure 1): a photomultiplier and an avalanche photodiode. The "stop" detector is delayed to investigate negative time intervals. The overall detection yields are a few 10^{-3} in both arms.

THEORY AND SIMULATIONS

We simulated the dynamics of the molecule in the laser field with optical Bloch equations, after we had determined the various parameters by experimental measurements.

We extracted the homogeneous linewidth and the saturation intensity from a study of the optical saturation for each molecule we investigated. For DBATT in n-hexadecane at 1.8 K, dephasing by acoustical phonons is negligible, and the only cause of coherence relaxation is the decay of the excited state ($T_2 = 2\ T_1$), mostly by spontaneous emission: the fluorescence quantum yield is close to unity.

The amplitude of the Stark modulation of each single molecule line was measured by applying a static voltage or a low-frequancy ac-voltage to the electrodes.

The time-dependent differential system of optical Bloch equations was solved by a Runge-Kutta algorithm, with an integration step of $2.\ 10^{-3}\ T_1$, to yield the evolution of the density matrix during the sweep. The average time profile of the emission burst was calculated from the time-dependent population of the excited state.

To simulate our start-stop experiments, however, and to estimate the efficiency of our source, it is important to evaluate the whole distribution of the number of photons emitted in each sweep. Optical Bloch equations only give average rates of photon emission. This distribution was obtained by introducing discrete spontaneous emission processes in quantum Monte-Carlo simulations [10]. Since the period of the sweeps (i.e. half the RF period, typically a few hundreds of ns) was always much longer than the excited state lifetime (about 9 ns), we assume that the molecule always starts from the ground state at the beginning of each sweep. The coherent motion of the molecule in the laser field was simulated by solving Bloch equations without any relaxation, and was suddenly interrupted when a random spontaneous emission process occured. No additional dephasing processes were introduced. In order to obtain reliable statistics of the distribution of emitted photons, a large number of passages were simulated. The pairs of photons were simply counted for each delay τ, and plotted as a histogram to represent $g^{(2)}(\tau)$, the second order correlation function of the fluorescence intensity.

From the simulated emission times of all photons, we could determine the probabilities p(0), p(1), p(2),... for the emission of 0, 1, 2,... photons per sweep. These probabilities in turn gave us the expected probability of detecting a photon pair. The average number of coincidences detected during the same m-photon sweep is $\eta_1 \eta_2 m(m-1)$, η_1 and η_2 being the overall detection yields in the start and stop arms. Therefore, the average number q_0 of photon pairs detected within one sweep is:

$$q_0 = \eta_1 \eta_2 \sum_{m=2}^{\infty} m(m-1)p(m), \quad (3)$$

whereas the average number q_1 of photon pairs with the photons emitted in two different sweeps is simply the square of the average number of photons detected in one sweep, i.e.

$$q_1 = \eta_1 \eta_2 \left(\sum_{m=1}^{\infty} m\, p(m) \right)^2. \quad (4)$$

The ratio q_0/q_1 deduced from simulations will be compared to the ratio of the experimental structures shown in the next section.

RESULTS AND DISCUSSION

In a first experiment, we recorded the average time profile of the emission bursts synchronized with the applied RF voltage, which drives the molecule through its resonance with the fixed laser. The frequency of the laser was adjusted approximately in the middle of the swept interval, corresponding to zero applied voltage. The inset of Figure 2 shows an averaged time trace of the fluorescence signal for many sweeps. The fluorescence bursts are emitted periodically, each half RF period. Figure 2 shows the time profile of one burst. It presents a short rise time (a few ns), and an exponential decay with a characteristic time of about 8-9 ns. This decay corresponds to the measured lifetime of the DBATT molecule [11], and is compatible with the homogeneous width of single DBATT molecules, about 20 MHz. Oscillations are superimposed on the decay. They should be absent for an ideal rapid adiabatic passage, but appear as soon as the RAP conditions are not perfectly fulfilled, in particular when the very strong inequality $\Omega \gg \Gamma$ does not apply: The data of Figure 2 were obtained with $\Omega = 3\Gamma$.

The average time profile of the photon bursts shows that emission indeed takes place just after the molecule has crossed resonance. However, the average number of photons delivered in each burst is difficult to evaluate precisely, because we did not measure the absolute quantum yield of our detection accurately. Moreover, measuring an average number gives no hint about the statistical distribution of emitted photons. Determining which bursts produced no photon would require an ideal detector. But by using a pair of detectors, it is possible to obtain information about the number of bursts giving 2 or more photons. This amounts to measuring the second order correlation function of the intensity, $g^{(2)}(\tau)$. Before explaining the measurements and their results, we examine theoretical simulations of this correlation function.

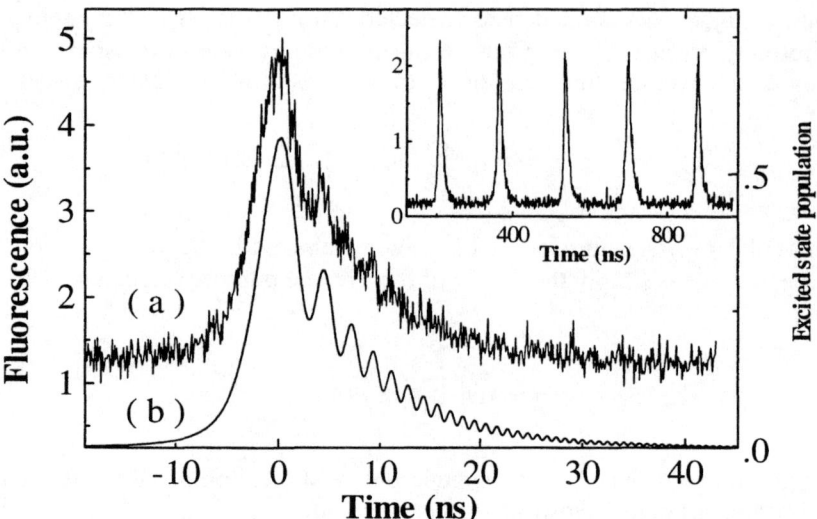

Figure 2: The inset shows the average fluorescence signal triggered on the RF voltage and accumulated for a large number of sweeps. The emission occurs after the molecule has crossed resonance with the laser. The main plot shows the time profile of one of the bursts, with a fast rise and a nearly exponential decay, upon which oscillations are superimposed. The smooth line is a simulation with all parameters determined from experimental measurements.

Quantum Monte-Carlo simulations enabled us to calculate the probabilities for the emission of 0, 1, 2 and more photons in each sweep. Obviously, the faster the sweep, the lower the probability of excitation and of fluorescence. Figure 3 shows plots of p(n), for n = 0 to 4, as functions of $T_c = (dt/d\omega)^{1/2}$, the effective scanning time defined in the introduction, for a given Rabi frequency (here 5 Γ). For a very fast passage (T_c very short), the probability of excitation is very low, and zero-photon bursts are most probable. Note that the theoretical value expected for p(0), given by the Landau-Zener formula [12]

$$P_{LZ}(0) = \exp\left(-\pi \, \Omega^2 T_c^2 / 2\right)$$

(curve in Figure 3) is in full agreement with the simulation. When the scan rate is lowered, the probability of one-photon bursts reaches a maximum (here for $T_c = 0.25$ T_1), then decreases for slower sweeps. The probability of emission of two photons then reaches its maximum, before decreasing for still slower sweeps, etc. For very slow sweeps, we would obtain a Poisson distribution, with an average number determined by the ratio of the passage time to the lifetime of the excited state. Around the maximum of p(1), the distribution is very far from Poissonian. The point of the adiabatic passage is to prepare the molecule in its excited state with certainty, i.e. to create a non-Poissonian distribution, ideally with p(1) = 1.

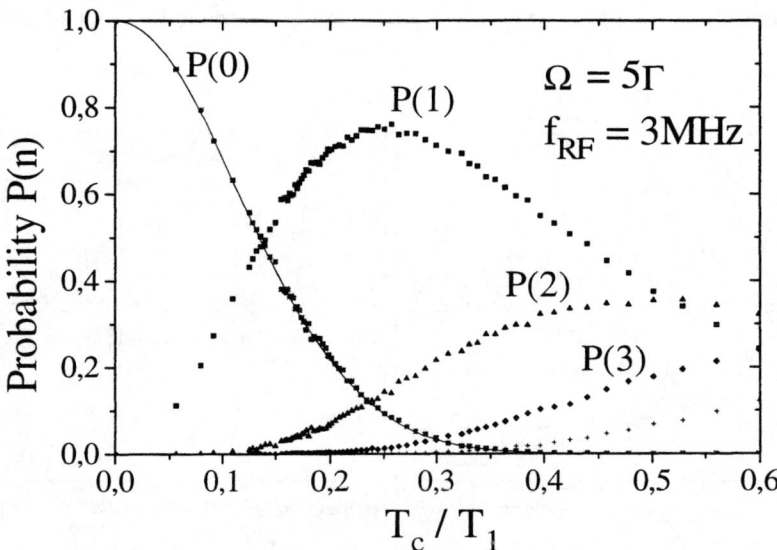

Figure 3: Variations of the probabilities of emission of n photons as functions of effective sweep time, calculated from quantum Monte-Carlo simulations and for the largest Rabi frequency we could achieve experimentally. A maximum probability of 75% for a one-photon burst is reached for $T_c/T_1 = 0.25$. The solid line corresponds to the Landau-Zener formula for p(0).

Figure 4 shows the results of some start-stop measurements for different experimental conditions. As expected for the periodic emission of Figure 2, the correlation function is nearly periodic, with structures separated by half the RF period. In all three histograms the intensity of the central structure, around zero delay, is clearly reduced as compared to that of the lateral structures. The ratio of their areas roughly gives twice the probability of two-photon sweeps, compared to that of one-photon sweeps. If the passage is too fast (upper part of Figure 4), the ratio q_0/q_1 is very low, but there are much more zero-photon sweeps than one-photon sweeps, as Figure 3 shows. On the other hand, if the passage is too slow (lower part of Figure 4), the intensity of the central structure and the ratio q_0/q_1 grow, because the probability of emission of two or more photons increases. The highest probability of 1-photon passage obtained in our experiments was 68 % for $\Omega = 3\Gamma$ and 74 % for $\Omega = 5\Gamma$. For the latter case, this corresponds to the maximum of p(1) in Figure 3.

The central structures of the histograms often show a clear dip for zero delays (see Figure 4). This dip, due to photon antibunching [3], is a direct consequence of our exciting only one molecule. The antibunching phenomenon has been discussed in detail in the literature, and much better data have been obtained with single molecules at cryogenic or at room temperatures [13]. The aim of the present measurement was not to demonstrate or measure antibunching, but to show that the intensity of the central structure as a whole is dramatically reduced as compared to that of the lateral ones.

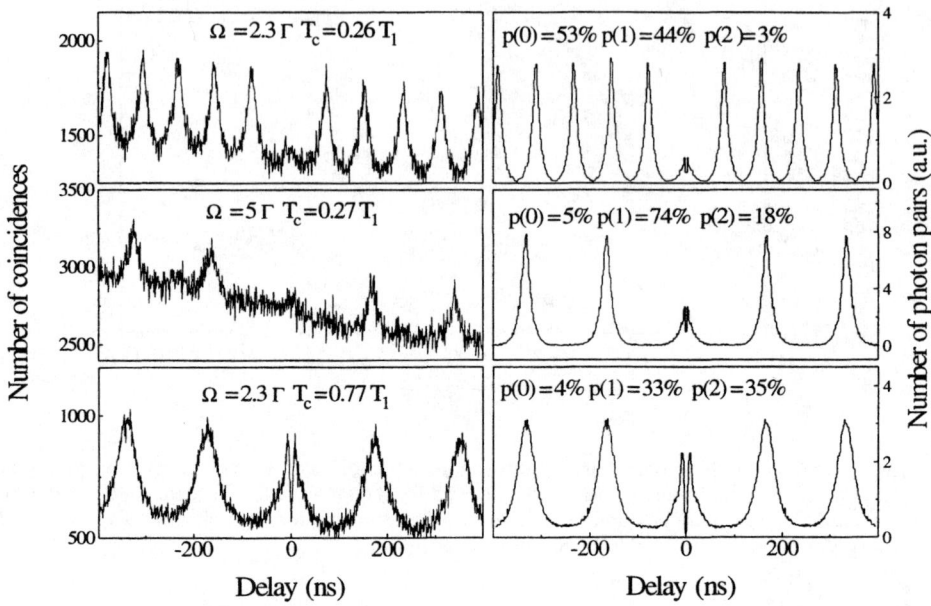

Figure 4: Experimental histograms of consecutive photon pairs measured in the start-stop experiment (left curves), and simulated correlation functions (or distribution of photon pairs) for the same parameters, with the calculated probabilities of emission of 0, 1, and 2 photons. The upper curves are for a fast sweep: the number of two-photon sweeps is very low, but so is the number of one-photon sweeps as compared to zero-photon sweeps. The middle data correspond to the best conditions we achieved (Ω = 5 Γ), but at the price of a very high background. These curves correspond to the maximum of $p(1)$ in Figure 3. The lower curves are for a slow sweep: The intensity of the central structure is comparable to that of the lateral ones.

Finally, some of the experimental spectra of Figure 4 show a clear slant in the background level when going from negative to positive times. This slant is due to the difference between the pair distribution as measured in the start-stop experiment and the true correlation function, where all photon pairs would be measured. Since the stop signal tends to occur more often at the beginning than at the end of the acquisition, especially for strong background, a positive bias appears for short delays. The background not only limits the signal-to-noise ratio, it also prevents measurement of the pair distribution at long times. For practical uses of our single photon source, it would be necessary to reduce the background dramatically, for example by reducing the laser intensity between the passages through the resonance, when the laser light has no effect on the molecule.

CONCLUSION

This works demonstrates how a single molecule can be made to emit single photons on command.The experiment is not only feasible, but even relatively easy. Here, we have prepared the molecule in its excited state by a rapid adiabatic passage, i.e. by simply sweeping the molecular frequency through resonance with a fixed laser. In the best passage we could achieve, limited by the intensity of background fluorescence, we achieved 75% of single photon generation, with 25 % of either zero-photon or two-photon passages. This figure could be improved easily with a much higher Rabi frequency, using an optical shutter or a pulsed excitation laser to reduce the background.

The rate of spontaneous emission could be enhanced if the molecule was placed in a cavity with high Purcell factor [14], or in a photonic band gap structure. This could improve the spectral purity of the emission. The emission spectrum of a single molecule at low temperature consists of several sharp lines (about 1 cm^{-1} broad) spread over several hundreds of cm^{-1} [15]. Selecting one of these lines with a resonant cavity would dramatically narrow the fluorescence spectrum.

Other small quantum systems with well separated electronic levels, such as nanocrystals or quantum dots could also be the core of a single photon source at room temperature. The main problem to solve for practical applications will be to achieve an overall detection yield close to unity. Once efficient collection optics and high-yield photodetectors are available, single quantum systems will provide compact, cheap and reliable sources of single photons.

REFERENCES

[1]. D. F. Walls and G. J. Millburn, Quantum Optics, Springer (Berlin), 1994.
[2]. W. Tittel, G. Ribordy, and N. Gisin, Physics World 11 (1998) 41.
[3]. H. K. Kimble, M. Dagenais, and L. Mandel, Phys. Rev. A 18 (1978) 201.
[4]. P. Grangier and A. Aspect, Phys. Rev. Lett. 54 (1985) 418.
[5]. A. Heidmann, R. J. Horowicz, S. Reynaud, E. Giacobino, C. Fabre, Phys. Rev. Lett. 59 (1987) 2555.
[6]. J. Kim, O. Benson, H. Kan, and Y. Yamamoto, Nature 397 (1999) 500.
[7]. Y. R. Shen, The Principles of Nonlinear Optics, Wiley (New York) 1984, p.399.
[8]. Ch. Brunel, B. Lounis, Ph. Tamarat, and M. Orrit, Phys. Rev. Lett. 83 (1999) 2722.
[9]. Ch. Brunel, Ph. Tamarat, B. Lounis, J. C. Woehl, and M. Orrit, J. Phys. Chem. 103 (1999) 2429.
[10]. K .Moelmer, Y. Castin, and J. Dalibard, J. Opt. Soc. Am. B10 (1993) 524.
[11]. A.-M. Boiron, B. Lounis, and M. Orrit, J. Chem. Phys. 105 (1996) 3969.
[12] L. Landau, Phys. Z. Sowjetunion, 2 (1932) 46; C. Zener, Proc. R. Soc. London A 137 (1932) 46.
[13] Th. Basché, W. E. Moerner, M. Orrit, and H. Talon, Phys. Rev. Lett. 69 (1992) 1516.
[14]. P. Goy, J.-M. Raimond, M. Gross, and S. Haroche, Phys. Rev. Lett. 50 (1983) 1903.

Experiments towards quantum information with trapped Calcium ions

D. Leibfried, C. Roos[†], P. Barton, H. Rohde, S. Gulde,
A. B. Mundt, G. Reymond[*], M. Lederbauer, F. Schmidt-Kaler,
J. Eschner, and R. Blatt

Institut für Experimentalphysik, Universität Innsbruck, A-6020 Innsbruck, Austria
[†]*present address: Ecole Normale Superieure, Paris, France*
[*] *present address: Institute d'Optique, Orsay, France*

Abstract. Ground state cooling and coherent manipulation of ions in an rf-(Paul) trap is the prerequisite for quantum information experiments with trapped ions. With resolved sideband cooling on the optical $S_{1/2}$ -$D_{5/2}$ quadrupole transition we have cooled one and two $^{40}Ca^+$ ions to the ground state of vibration with up to 99.9% probability. With a novel cooling scheme utilizing electromagnetically induced transparency on the $S_{1/2}$ -$P_{1/2}$ manifold we have achieved simultaneous ground state cooling of two motional sidebands 1.7 MHz apart. Starting from the motional ground state we have demonstrated coherent quantum state manipulation on the $S_{1/2}$ -$D_{5/2}$ quadrupole transition at 729 nm. Up to 30 Rabi oscillations within 1.4 ms have been observed in the motional ground state and in the $n = 1$ Fock state. In the linear quadrupole rf-trap with 700 kHz trap frequency along the symmetry axis (2 MHz in radial direction) the minimum ion spacing is more than 5 μm for up to 4 ions. We are able to cool two ions to the ground state in the trap and individually address the ions with laser pulses through a special optical addressing channel.

INTRODUCTION

Even elementary quantum information processing operations put severe demands on the experimental techniques. A single quantum gate, for instance, requires two strongly interacting quantum systems, highly isolated from environmental disturbances. In their proposal for a quantum logic gate, Cirac and Zoller showed that single ions, trapped in a linear radio-frequency (Paul) trap and cooled to the motional ground state have the potential to offer such a system [1] thus initiating experimental work towards quantum logic with ion traps in several groups around the globe.

The work done in Innsbruck is based on $^{40}Ca^+$ ions which offer the advantage that light sources for all transitions involved are derived from diode and solid state lasers in a relatively easy way. So far we have investigated several ways to cool

CP551, *Atomic Physics 17*, edited by E. Arimondo, P. DeNatale, and M. Inguscio
© 2001 American Institute of Physics 1-56396-982-3/01/$18.00

one or two ions to the ground state of motion in two different traps. In the first trap, a mm-sized spherical Paul trap loaded with a single ion, we have developed and refined our cooling techniques and the coherent manipulations necessary for quantum information processing with ions. The second trap, with a linear geometry, was used to manipulate two and more ions, namely for ground state cooling and individual addressing.

EXPERIMENTAL SETUP

Relevant Transitions in ^{40}Ca$^+$

There are two basic prerequisites for an ion species to be suitable for quantum information processing. First they should allow for cooling to the ground state, typically with a Doppler precooling step, and second their electronic levels should contain at least two long lived states that can form an effective two level system to encode the quantum bit (qubit).

Calcium ions can accommodate both these prerequisites with the additional benefit that all necessary light sources can be generated from diode and solid state lasers. The levels relevant to our experiments are shown schematically in Fig. 1. Doppler cooling is achieved by driving the dipole transition from the $S_{1/2}$ to the $P_{1/2}$ level at 397 nm with a frequency-doubled Ti:Sapph laser that is red detuned from the transition line center by about 20 MHz, the natural decay width of the $P_{1/2}$ level. To prevent pumping to the $D_{3/2}$ level (branching ratio 16:1), the ions are simultaneously irradiated with a diode laser at 866 nm that drives the $D_{3/2}$-$P_{1/2}$ transition (Fig. 1 (a)). To provide a well defined direction of magnetic field and to split the Zeeman sublevels we produce a field of about 4 Gauss at the position of the ion. With an additional beam at 397 nm that is σ^+ polarized we can prepare the electronic state of the ion(s) to be $S_{1/2}$, $m = 1/2$.

For the two qubit levels we chose the $S_{1/2}$ ground state and the metastable $D_{5/2}$ state with a natural lifetime of about 1 s (Fig. 1(b)). Quadrupole transitions between these levels are driven with a Ti:Sapphire laser at 729 nm, stabilized with the Pound-Drever-Hall method to a high finesse cavity. To maintain the coherence necessary for qubit manipulations this laser has to be highly stable. We have determined an upper bound of 76(5) Hz (FWHM) for the effective linewidth of our laser system by observing the fringe contrast in high resolution Ramsey spectroscopy on the $S_{1/2}$, $m = -1/2$-$D_{5/2}$, $m = -5/2$ transition as a function of the time delay between the two excitation pulses [2].

To initialize the qubit in the $S_{1/2}$ ground state we use an additional diode laser at 854 nm that repumps the ion(s) from the $D_{5/2}$ level via the $P_{3/2}$ level. All our laser sources can be switched with a timing accuracy of better than 1 μs and frequency tuned with acoustooptic modulators (AOMs).

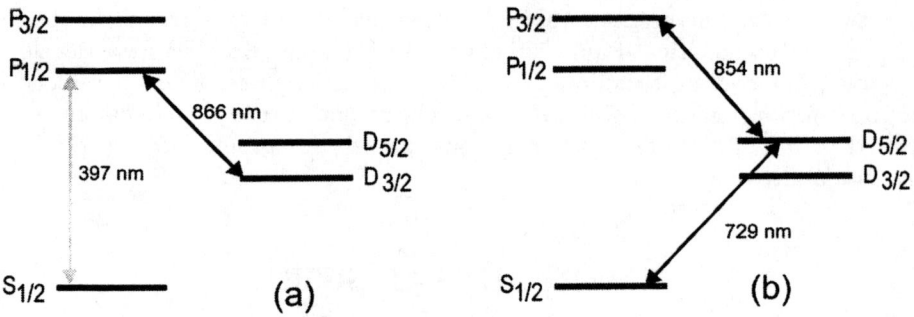

FIGURE 1. Relevant levels in $^{40}\text{Ca}^+$ for (a) Doppler cooling and state detection (b) resolved sideband cooling and coherent manipulations are performed on the qubit effective two level system formed by $S_{1/2}$ and $D_{5/2}$ levels.

Qubit State Detection

The laser beams at 397 nm and 866 nm used for Doppler cooling also provide highly efficient state detection with the quantum jump technique [3]. The two light fields couple the $S_{1/2}$ ground state to a cycling transition so that many photons are scattered if the ion is in the ground state. On the other hand if the ion is in $D_{5/2}$, this level is decoupled from the excitation and no fluorescence photons will be emitted. Although only a small fraction (ca. 10^{-2}) of these fluorescence photons is collected by a lens and imaged onto a photomultiplier with a quantum efficiency of about 10%, one can distinguish the two qubit states within 2 ms of detection time. This allows us to measure the state of our qubit with practically 100% efficiency.

Ion Traps

In our experiments we use two different ion traps. The first trap is a regular spherical Paul trap, with ring and endcaps made of 0.2 mm diameter molybdenum wire. The ring diameter is 1.4 mm and the endcap distance 1.2 mm. With an rf-drive voltage of about 1 kV at 20 MHz we obtain motional frequencies of up to 4.5 MHz and 2 MHz along the axis of symmetry and in the ring plane respectively. This trap is mainly used for experiments with one ion. Due to its high secular frequencies, Doppler cooling leads to relatively low average occupation numbers ($\bar{n}=2$ (4) for the axial (radial) harmonic oscillator) and thus makes it a good test bed for cooling and coherent control techniques with just one ion.

Our second trap is a linear quadrupole trap with four 0.6 mm diameter quadrupole rods and the endcaps made of stainless steel. The trap is held together by Macor spacers. The diagonal distance between quadrupole electrodes is

2.36 mm and the distance between the cylindrical 6 mm diameter endcap rings is 10 mm. In this trap we reach secular frequencies of 2 MHz in the radial direction and up to 700 kHz in the axial direction. For a string of up to 4 ions this leads to inter-ion distances ≥ 5 μm, well above the diffraction limit of our laser beams. With this we are able to individually address such a string (see below), at the expense of relatively low axial secular frequencies and a higher average occupation number ($\bar{n} \simeq 25$ in the axial direction) after Doppler cooling.

INDIVIDUAL ADDRESSING

One of the key features in the Cirac-Zoller [1] gate is that the internal states of ions in a string have to be manipulated individually. Although a number of technically different proposals have been made, the original approach of Cirac and Zoller, to focus a laser beam sufficiently so it will only interact with a single ion in the string, still seems to be the most straightforward way. One obvious limitation of this approach is that the size of the focus is limited by diffraction to roughly a micron so the minimum distance between ions has to be larger than that number. A given minimum spacing of ions restricts the maximum center of mass (COM) frequency for a given number of ions along the axis of symmetry [4]. If four ions of Ca^+ should not be closer together than 5 μm, the maximum COM frequency is about 700 kHz.

To image the fluorescence of ion strings, we use a Nikon MNH-23150-ED-Plan-1,5x macroscope lens with a working distance of 65 mm and an intensified CCD camera (see figure 2) [5]. The spatial resolution of our imaging system is about 2 μm, sufficient to resolve single ions in not too densely packed strings. For individual addressing we use the imaging lens in reverse. Since the addressing beam is at 729 nm we can use a dichroic mirror to superimpose it with the imaging channel (see figure 2). A telescope is used to sufficiently expand the beam diameter [6]. The beam is steered over the ions with an electrooptic deflector. The deflection efficiency of 5 mrad/kV is translated into a displacement on the ions of 23 μm/kV. A high voltage amplifier stage for addressing allows us to switch from one ion to the other in a few μs. We have checked the beam diameter and pointing stability of our system by mapping the Rabi frequency on the $S_{1/2}$-$D_{5/2}$ transition versus the beam displacement and found a 1/e width of 3.7(0.3) μm for this excitation. If we apply a π-pulse to the ion addressed, the probability of exciting a neighboring ion in the ground state and 5 μm away would be about 1%.

GROUND STATE COOLING

Resolved Sideband Cooling

Ground state cooling has been achieved so far with a single $^{199}Hg^+$ ion [7], and with $^9Be^+$ [8], using resolved sideband cooling on either a quadrupole or a Raman

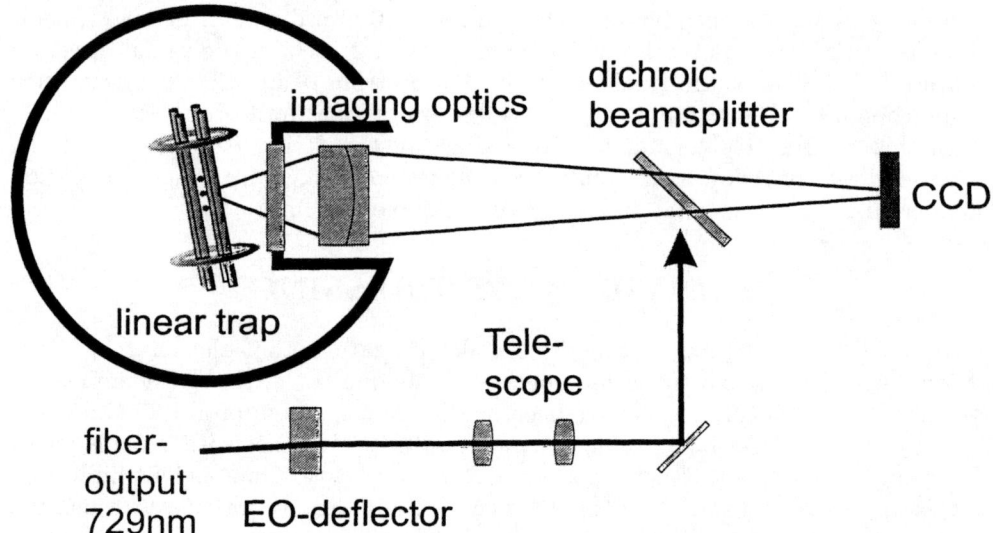

FIGURE 2. Setup for individual addressing of a string of ions. The addressing beam is displaced by an electrooptic (EO) deflector and superimposed onto the CCD imaging channel with a dichroic beamsplitter. By sending it through the imaging lens system in reverse we utilize the high spatial resolution of this system for addressing.

transition. We use a cooling method on the $S_{1/2}$-$D_{5/2}$ quadrupole transition, similar to the Hg experiment [9]. The weak coupling between light and atom on a bare quadrupole transition would necessitate long cooling times. However, the cooling rate is greatly enhanced by (i) strongly saturating the transition and (ii) shortening the lifetime of the excited state via coupling it to a dipole-allowed transition. The $S_{1/2}(m = 1/2) - D_{5/2}(m = 5/2)$ transition, well resolved in frequency from all other possible transitions by the applied magnetic field, is excited with about 1 mW of light focused to a waist size of 30 μm at the position of the ion. At the same time, the decay rate back to the ground state is increased by exciting the $D_{5/2}(m = 5/2) - P_{3/2}(m = 3/2)$ transition. The intensity of this quenching laser is adjusted for optimum cooling during the experiment. Optical pumping to the $S_{1/2}(m = -1/2)$ level is prevented by occasional short laser pulses of σ^+ polarized light at 397 nm. The duration of those pulses is kept at a minimum to prevent unwanted heating. The ground state occupation is found by comparison of the on resonance excitation probability for red and blue sideband transitions [9]. In the spherical Paul-trap we reach up to 99.9% of motional ground state occupation within 6 ms (see figure 3). All earlier successful ground state cooling experiments were plagued by an unexpectedly high motional heating (see [10] and references therein). In our setup we find a motional heating rate of one phonon in 190 ms for a trap frequency of 4 MHz, two orders of magnitude smaller than in the trap used at NIST for the $^9Be^+$ experiment. While this is still a much higher heating rate

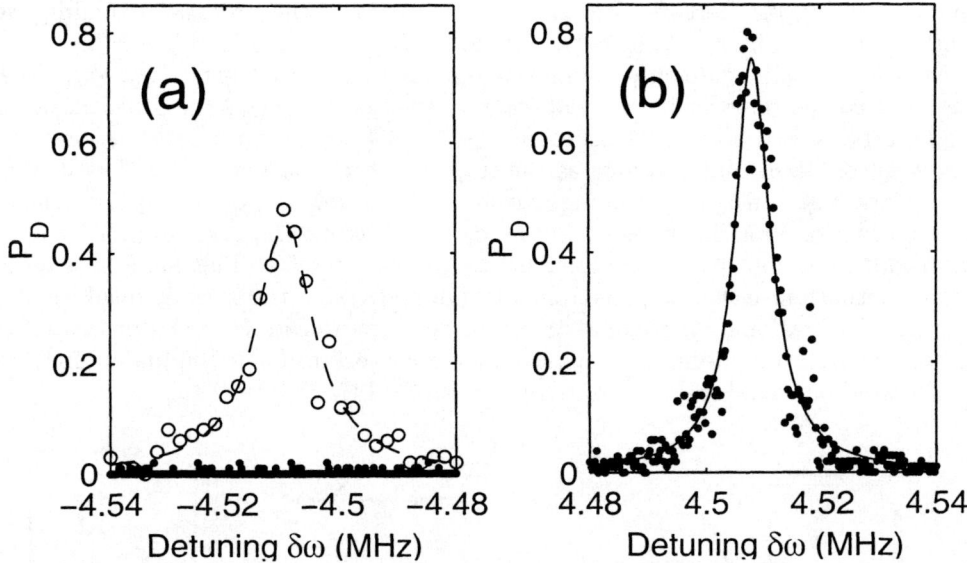

FIGURE 3. Red and blue sidebands of the axial mode (at 4.51 MHZ) of one ion after Doppler cooling (dashed) and after resolved sideband cooling (solid). From the ratio of sideband strengths we determine 99.9% ground state occupation.

than expected from black body radiation it happens on a timescale much longer than the time typically needed for quantum logic gates (we estimate an upper limit of 200 μs for one CNOT gate).

In separate experiments we also cooled all motional modes of two ions in our linear trap to the ground state [11]. Cooling only one of the 6 motional modes at a time we achieved at least 95% ground state occupation for all modes. In these experiments we used the addressing channel to illuminate only one ion with the cooling radiation. The second ion is cooled sympathetically due to the strong inter-ion coupling by the Coulomb-force.

Cooling with Electromagnetically Induced Transparency

Resolved sideband cooling only leads to very low temperatures if the red sideband is excited with a narrow excitation bandwidth. Otherwise nearby nonresonant transitions (e.g. carrier transitions) will lead to unwanted excess heating and severely increase the final temperature of the ion(s). Unless sideband frequencies are degenreate this limits resolved sideband cooling to one motional sideband at a time. For more degrees of freedom, for example the 6 motional modes of two ions, resolved sideband cooling needs involved cooling schemes where all degrees of freedom have to be cooled sequentially with laser detuning and laser power appropriate for their coupling strength. Moreover the phonons scattered in the process of cooling one motional mode will reheat the other modes.

For a Cirac-Zoller gate only the motional mode that is used as the 'quantum-bus' has to be cooled to very high ground state occupation as long as the other motional modes are inside the Lamb-Dicke regime [12]. This regime can be reached in principle by Doppler cooling, as long as the trap frequencies are not too different from the natural linewidth of the cooling transition. Therefore typically trap frequencies on the order of 10 MHz or higher are needed to reach the Lamb-Dicke regime, but will result in an ion spacing that is hard to optically resolve. This leads to a problematic situation: either one has difficulties in addressing the ions or in sufficiently cooling all vibrational modes of a string. In our experiments with two or more ions in the linear trap we decided to maintain good conditions for individual addressing and limited our axial trap COM frequency to 700 kHz or lower.

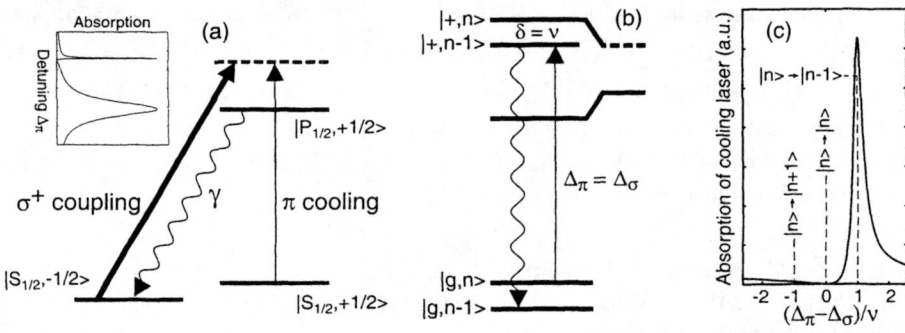

FIGURE 4. (a) Level scheme of the $S_{1/2}$-$P_{1/2}$ manifold used for EIT cooling. The bare levels are dressed with a strong σ^+ polarized beam. This causes a distinct asymmetry in the absorption profile of a weak π-polarized beam. (b) Dressed state picture: with light shift $\delta =$ trap frequency ν, the cooling laser is resonant with the transition $|g,n\rangle \rightarrow |+,n-1\rangle$ where $|g\rangle$ is the S-state and $|+\rangle$ is the dressed state corresponding to the virtual level [dashed line in (a)]. (c) By detuning the π-polarized beam to the carrier transition frequency and choosing the AC-Stark shift caused by the dressing beam to be equal to the trap frequency, the asymmetry of the absorption profile is superimposed over the carrier and sideband transitions in such a way that the cooling red sideband transitions are much more probable as compared to the heating blue sideband transitions, and the carrier is completely suppressed by a dark resonance.

Under these conditions it was desirable to find a cooling technique that is not as narrow-band as resolved sideband cooling but has a lower cooling limit than Doppler cooling. Ideally one would want to cool the ion deeply into the Lamb-Dicke regime for all motional degrees of freedom simultaneously. A very recent proposal to use electromagnetically induced transparency (EIT) for the cooling of trapped particles [13] held this promise.

We adapted this cooling scheme, originally proposed for a three-level system, for the case of the $[S_{1/2}, P_{1/2}]$ four level system in Ca^+ that we also use for Doppler cooling. The manifold is dressed with a σ^+ polarized beam at 397 nm, blue detuned by Δ_σ=60 MHz (3 linewidths of the S-P transition) that connects the $S_{1/2}$, m=-1/2 with the $P_{1/2}$, m=1/2 level. Under these circumstances a second low intensity π polarized beam will experience an absorption (Fano-) profile as depicted in Fig. 4(a). In addition to the usual line profile around $\Delta_\pi = 0$ a dark resonance is created at the point where the detuning of the π-plarized beam Δ_π is equal to Δ_σ and a bright resonance appears at $\Delta_\pi = \Delta_\sigma + \delta$ where δ is the AC Stark shift due to the σ^+ polarized beam. This creates an asymmetry in absorption for carrier and sidebands. The carrier is almost completely suppressed due to the dark resonance, the blue sideband is in the shallow wing of the profile, but the red sideband is greatly enhanced by the bright resonance created by the dressing beam.

For cooling the π-polarized beam is tuned to the *carrier*. Absorption on the carrier is then suppressed by the dark resonance while absorption on the red sideband is enhanced by the bright resonance. When we tuned the Stark-shift δ to be equal to one of the motional modes at 3.34 MHz we were able to cool this mode to 90% ground state occupation or $\bar{n} = 0.1$.

As sketched in figure 4 (b) the bright resonance can have a substantial width and the red sideband has not necessarily to coincide exactly with the maximum of the bright resonance to get a cooling effect. This opens the possibility to cool several motional modes *simultaneously*, as long as they are not too far apart in frequency. To demonstrate simultaneous cooling of two vibrational modes with

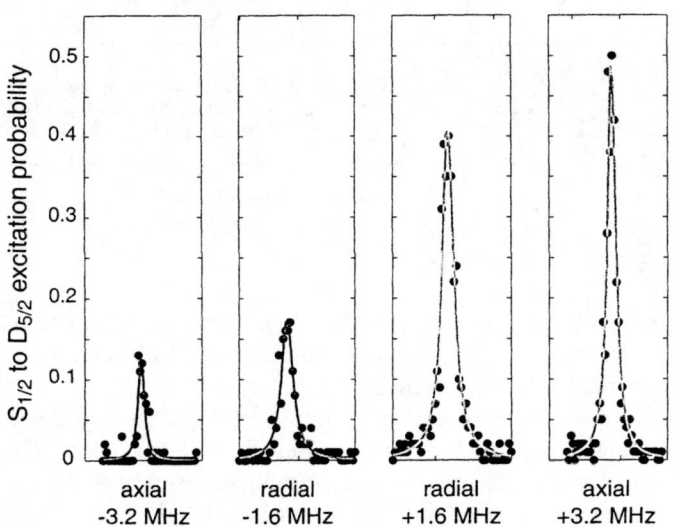

FIGURE 5. Simultaneous cooling of two vibrational modes with an oscillation frequency difference of 1.73 MHz. The sideband asymmetry corresponds to 58% ground state occupation ($\bar{n} = 0.85$) in the mode at 1.61 MHz and 74% ($\bar{n} = 0.35$) at 3.34 MHz.

this method we chose two vibrational modes at 1.61 MHz and 3.34 MHz with an oscillation frequency difference of 1.73 MHz. The AC-Stark shift δ of the σ^+- beam was adjusted to be about 2.5 MHz, halfway in between the two mode frequencies. With this settings we achieved 58% ground state occupation ($\bar{n} = 0.85$) in the mode at 1.61 MHz and 74% ($\bar{n} = 0.35$) at 3.34 MHz. From this result we estimate that we can sufficiently cool all axial degrees of freedom of a string of up to 5 ions with a COM mode frequency of 700 kHz.

COHERENT MANIPULATIONS

For quantum information processing, it is important to know for how long coherent interaction with the ion(s) is possible. To this end we cooled one ion to the ground state and then irradiated the ion with light at the blue sideband frequency (This interaction is used in a quantum gate to transfer the internal state of a qubit into the motion) [9]. We then monitored the occupation probability of the D-state versus the pulse length on the blue sideband. The same interaction was also used after preparing the ion in the $n = 1$ motional Fock state by a π-pulse on the blue sideband followed by a repumping pulse on the $D_{5/2}$-$P_{3/2}$ transition. As figure 6

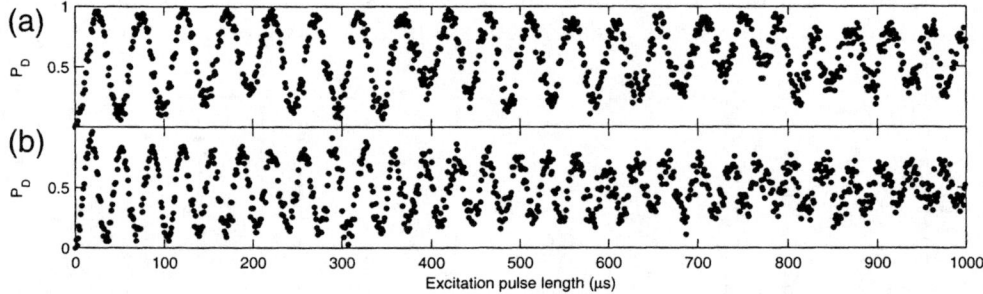

FIGURE 6. (a) Rabi-oscillations on the blue sideband for the initial motional state $|n = 0\rangle$. (b) Rabi-oscillations as in (a), but for $|n = 1\rangle$.

shows, we were able to observe Rabi-flops for both initial motional states with a contrast of better than 50% for 1 ms. The ratio of Rabi-frequencies is $\sqrt{2}$ as expected for this kind of interaction in the Lamb-Dicke regime. These results make us confident that we should be able to apply gate pulses equivalent to at least 40 π-pulses before the fidelity of the total operation drops below 0.5. In our system motional heating is too slow to be the prime source of the observed decoherence, we rather attribute it to time dependent magnetic field fluctuations that shift the levels and to slow vibrations in our setup that introduce a fluctuating Doppler-shift for our optical beam at 729 nm. The linewidth of our laser system could also contribute, but from the results of the Ramsey-experiment described above we would expect this source to contribute with a characteristic time of about 15 ms.

CONCLUSIONS AND OUTLOOK

In conclusion we have demonstrated that we have all necessary ingredients to perform a two bit quantum logic gate with trapped ions. The time scales in our system are well resolved (see figure 7). The fastest characteristic time is given

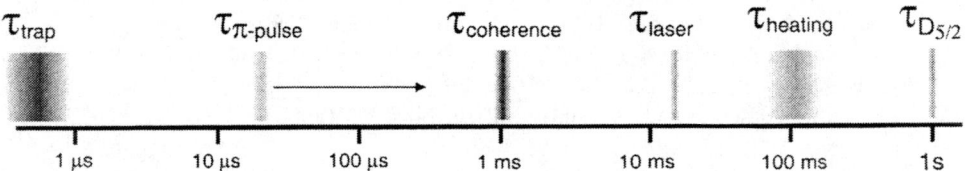

FIGURE 7. Characteristic time scales for our experimental setup. For more details see text.

by the harmonic motion of the ion(s) and on the order of 1 μs. A π-pulse on a motional sideband takes about 20 μs and, as stated above, the fidelity of coherent manipulations remains above 0.5 for times smaller than 1 ms. Our laser system would allow for coherent manipulations for at least 15 ms. Motional heating begins to play a role for times around 100 ms. The ultimate source of decoherence in our experiment is the 1 s lifetime of the $D_{5/2}$ state.

In the near future we plan to use our ability to individually address ions to demonstrate a CNOT quantum logic gate with two ions and to create maximally entangled states with 2 and more ions. We also envision to perform small quantum algorithms and first experiments on error corrections with up to 5 ions.

Acknowledgements This work was supported by the Fonds zur Förderung wissenschaftlicher Forschung (FWF) within the special research grant SFB 15, by the European Commission within the TMR networks 'Quantum Information' (ERB-FMRX-CT96-0087) and 'Quantum Structures' (ERB-FMRX-CT96-0077) and by the Institut für Quanteninformation GmBH.

REFERENCES

1. J. I. Cirac, P. Zoller, Phys. Rev. Lett. **74** 4091 (1995).
2. H. Rohde, S. Gulde, P. Barton, F. Schmidt-Kaler, R. Blatt, unpublished.
3. H. Dehmelt, Bull. Am. Phys. Soc. **20** 60 (1975).
4. A. Steane, Appl. Phys. B **64**, 623 (1998).
5. H. C. Nägerl, W. Bechter, J. Eschner, F. Schmidt-Kaler, R. Blatt, Appl. Phys. B **66**, 603 (1998).
6. H. C. Nägerl, D. Leibfried, H. Rohde, G. Thalhammer, J. Eschner, F. Schmidt-Kaler, R. Blatt, Phys. Rev. A **60**, 145 (1999).
7. F. Diedrich, J. C. Bergquist, W. M. Itano, D. J. Wineland, Phys. Rev. Lett. **62**, 403 (1989).
8. C. Monroe, D. M. Meekhof, B. E. King, S. R. Jeffers, W. M. Itano, D. J. Wineland, P. Gould, Phys. Rev. Lett. **74**, 4011 (1995).

9. C. Roos, T. Zeiger, H. Rohde, H. C. Nägerl, J. Eschner, D. Leibfried, F. Schmidt-Kaler, R. Blatt, Phys. Rev. Lett. **83**, 4713 (1999).

10. Q. A. Turchette, D. Kielpinski, B. E. King, D. Leibfried, D. M. Meekhof, C. J. Myatt, M. A. Rowe, C. A. Sackett, C. S. Wood, W. M. Itano, C. Monroe, D. J. Wineland, Phys. Rev. A **61**, 063418-1 (2000).

11. F. Schmidt-Kaler, Ch. Roos, H. C. Ngerl, H. Rohde, S. Gulde, A. Mundt, M. Lederbauer, G. Thalhammer, Th. Zeiger, P. Barton, L. Hornekaer, G. Reymond, D. Leibfried, J. Eschner, R. Blatt, quant-ph/0003096 (2000).

12. D. J. Wineland, C. Monroe, W. M. Itano, D. Leibfried, B. King, and D. M. Meekhof, Jou. Res. Nat. Inst. Stand. Tech., 103, 259 (1998).

13. G. Morigi, J. Eschner, C. Keitel, quant-ph/0005009 (2000).

ENTANGLEMENT

Step by Step Engineered Entanglement with Atoms and Photons in a Cavity

Serge Haroche, Gilles Nogues, Arno Rauschenbeutel, Stefano Osnaghi,
Patrice Bertet, Michel Brune and Jean-Michel Raimond

*Laboratoire Kastler Brossel, Département de Physique de l'Ecole Normale Supérieure,
24 rue Lhomond, 75231, Paris cedex 05, France*

Abstract. We have performed multiparticle entanglement experiments with circular Rydberg atoms crossing one at a time a high Q superconducting microwave cavity. Two-level atoms and a zero or one photon field stored in the cavity act as qubits carrying quantum information. Controlled qubit entanglement is produced by the quantum Rabi oscillation coupling the atom to the cavity field. Qubit state superpositions are produced and analyzed by classical microwave pulses before and after the atom cross the high Q cavity, using a Ramsey interferometer arrangement. We have demonstrated the coherent operation of a quantum phase gate and used it to perform for the first time a quantum non-destructive measurement of a single photon. Combining this gate with quantum Rabi oscillations of various durations, we have entangled step by step three subsystems – two atom and one field mode – by a controlled succession of one and two qubit operations. Once some limitations of our experiment are overcome, it will be generalized to larger number of particles, opening the way to the study of even more complex entangled states.

INTRODUCTION

Entanglement is a central concept in quantum theory. When particles have interacted and fly apart from each other, they remain usually described by a non-separable entangled multiparticle wave function involving quantum mechanical correlations at a distance between the particles. In the simplest case, the particles can be idealized as two-level systems carrying bits of information, which can be prepared in state superposition of two basic values (0 and 1). Entanglement of these *qubits* makes it possible to process information in ways impossible to do classically [1]. The teleportation [2] of quantum states and the operation of simple quantum logic gates [3] rely on entanglement. The distribution of secure cryptographic keys between a sender and a receiver can also be performed using entanglement [4].

Quantum optics offers various solutions for entanglement production and manipulation. It was first realized with pairs of photons spontaneously emitted in a single atom cascade, a system which has been used to demonstrate the inherent non-local character of quantum physics [5]. Twin photon beams generated by parametric

CP551, *Atomic Physics 17*, edited by E. Arimondo, P. DeNatale, and M. Inguscio
© 2001 American Institute of Physics 1-56396-982-3/01/$18.00

downconversion in a non-linear crystal have replaced more recently the atomic cascade source, leading to improved Bell's inequality tests [6], cryptographic key distribution experiments [7], teleportation studies [8] and the generation of a three particle entangled photon state [9], of the type proposed by Greenberger, Horne and Zeilinger (GHZ state [10]). In these all-photon experiments, the qubits are basically non-interacting and the entanglement is produced by an irreversible spontaneous process.

For quantum information processing, it is often more convenient to store the information in massive particles, which interact with each other and can be manipulated for relatively long times. Qubit manipulation of nuclear spins in molecules has been realized by magnetic resonance [11], but these experiments involve macroscopic samples without clear-cut entanglement [12]. Ion traps [13] and cavity QED set-ups [14] have been recently used to manipulate isolated massive qubits and entangle them. In the first case, the qubits are carried by internal or external degree of freedom of ions. Quantum gates [15], two- [16] and four- ion entanglement [17] have been realized.

In cavity QED, qubits are carried either by two level Rydberg atoms, or by a zero- or one-photon field stored in the cavity. Atom-field and atom-atom entanglement have already been reported [18]. We describe here the recent realization of a quantum phase gate [19] and its use for the quantum non-demolition (QND) detection of a single photon [20]. We also describe an experiment in which we have *engineered* the entanglement of three sub-systems – two atoms and one field mode[21]. Contrary to photon beam experiments, the entanglement is fully prepared by a controlled and reversible unitary evolution. In contrast with present ion experiments in which the entanglement is realized in one single process involving all particles at once, the atom-cavity entanglement is produced *step by step*, by a succession of one and two qubit operations. This gives flexibility to the process and makes it – in principle – straightforward to generalize to larger systems.

THE CAVITY QED ENTANGLING MACHINE

Experimental Set-up

Our set-up is shown in Fig.1. It has already been described in [19-21] and we recall only briefly here its main features. The atoms, effusing from oven O, are velocity selected by laser optical pumping (velocity v = 503 ± 2 m/s) and prepared in the circular Rydberg state with principal quantum number 51 (level e in the following) or 50 (level g) by a succession of laser and radiofrequency induced transitions in zone B. Levels e and g are shown in Fig.1 inset. A third level (the circular state with principal quantum number 49, called i in the following) plays also a role in the experiments.

The atoms cross one at a time the cavity C sustaining a Gaussian field mode (waist = 6 mm), resonant with the e ⇒ g transition at 51.1 GHz. The cavity is made of two superconducting niobium mirrors enclosed by an aluminum ring with small holes for atom access (the ring is cut open in the figure to show the inside of C). The atom and field relaxation times, 30 and 1 ms respectively, are much longer than the atom cavity crossing time (of the order of 20 μs). This insures that quantum coherences survive much longer than the qubit coupling time.

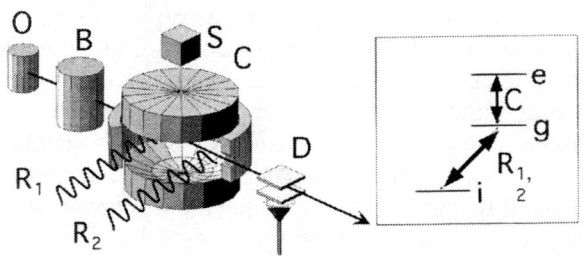

FIGURE 1. Scheme of the cavity-QED entangling experimental set-up. The relevant atomic energy levels are shown in the inset. The cavity mode C is resonant with the e $\Rightarrow$ g transition. Classical Ramsey pulses R_1 and R_2 can be applied on the g $\Rightarrow$ i transition (as shown in inset) or on the g $\Rightarrow$ e transition.

The atomic preparation is pulsed so that the position of each atom is known with ± 1 mm. The atoms are detected, after C, by a state selective field-ionization detector D, which discriminates between e, g and i (detection efficiency 40%). Sequences of atoms can be sent across the cavity with a delay between atoms of the order of 100 μs exceeding the atom-cavity crossing time This insures that one atom at most interacts with the cavity field at any time. In order to avoid the presence of more than one atom in each pulse of each sequence, the lasers preparing the Rydberg states have a low intensity, exciting on average 0.1 to 0.2 atom per pulse. The preparation process being Poissonian, each pulse contains randomly either zero or one atom. The probability to prepare a sequence of two or three atom is accordingly low and we must use the detection to post-select the useful events out of a large number of runs in which at least one of the atoms of the sequence is missing.

A field can be prepared in C either by the emission of one of the atoms of the sequence, called *source* atom, or by an external coherent solid state microwave source S, coupled into C by a waveguide. In this latter case a coherent field with a Poissonian photon number distribution is prepared in C. The whole set-up is cooled to 1.3 K. The mean number of thermal photons in C, dominated by microwave leaks, is 0.7. Radiation cooling is provided, before each experimental sequence, by pulses of atoms crossing C in level g. They absorb the residual thermal field and reduce the field background to 0.12 photon on average. The whole experimental sequence then occurs before the thermal background has had time to relax back to its equilibrium value.

Ramsey Interferometer

The atoms can be prepared in a linear superposition of energy states, before the interaction with C, and mixed again after this interaction, making it possible to study qubit coherences. For this purpose we use two pulses of classical microwave, R_1 and R_2, generated by a second microwave source S' and injected in the cavity structure through a small hole in the ring (S' and this hole are not shown in Fig.1.) This microwave field produces inside the ring-mirror structure a standing wave pattern, which we have mapped by performing auxiliary Rydberg atom spectroscopy experiments. Antinodes of this pattern, sandwiching the central cavity mode, are used to perform the R_1 and R_2 pulses at the time the atom is crossing them. Different sequences of pulses can be applied on successive atoms by commuting the field source S'. Applying on an atom two successive $\pi/2$ R_1 and R_2 pulses, resonant on the g $\Rightarrow$ i transition, amounts to operating a Ramsey separated field interferometer. The pulses mixing Rydberg states act as atomic internal state beam splitters. The probability to detect (after R_2) in level i the atom initially prepared in level g (before R_1) is the squared sum of two amplitudes, corresponding to the atom crossing the R_1-R_2 interval in level g or i. The relative phase of these two amplitudes is swept by tuning the frequency of S'. By repeating the experiment many time, we reconstruct the transition probability, which exhibits Ramsey fringes. Typically, they have a 72% contrast instead of the ideal 100%. The interaction of the atom with the quantum field in C between R_1 and R_2 modifies the phase and the amplitude of the fringes, revealing useful information about the atom-quantum field coupling. Microwave R_1 and R_2 pulses mixing levels e and g can also be applied on the atoms. In this case, the classical transition occurs at the frequency of the quantum mode C. As described in [21], special caution has then to be exercised to prevent the R_1 and R_2 pulses to leak photons in C.

The Quantum Rabi « Stitches »

The joint photon-atom state manipulations rely on the resonant quantum Rabi rotation experienced by each atom in C [22]. When C is exactly resonant with the e $\Rightarrow$ g transition, the atom-cavity system undergoes oscillations between the states $|e,0>$ and $|g,1>$ (atom in e or g with either 0 or 1 photon in C). The frequency of this oscillation at cavity center is $\Omega/2\pi = 47$ kHz. The Rabi frequency varies while the atom is crossing the Gaussian field mode and the Rabi rotation integrates this variation. It is convenient to define the effective atom-C interaction time $t_{i, which}$ would produce, if the coupling Ω were constant, the same final Rabi angle on the system. After the atom-cavity interaction, the combined system ends up in the state :

$$|\psi> = \cos(\Omega t_i/2) |e,0> + \sin(\Omega t_i/2) |g,1> \qquad (1)$$

which is in general an entangled atom-field state. It is essential, in order to adjust the atom-field entanglement to be able to control t_i. This is achieved in two steps. First we fix the atomic velocity to the value v =503 m/s such that $t_i = \sqrt{\pi}\,w/v$ corresponds to $\Omega t_i = 2\pi$ (2π Rabi pulse). Then, we apply, when each atom is inside C, a voltage across the mirrors which produces an electric field F on the atom, Stark switching in this way the atomic transition out of resonance with C. The atom-field evolution is then frozen at a well defined time, thus reducing the effective interaction time. Any time between $\sqrt{\pi}\,w/v$ and 0 can be realized in this way. Three Rabi pulses are useful :

146

The π/2 Rabi pulse

When the condition $\Omega\, t_i = \pi/2$ is realized, the atom and the field end up in the maximally entangled state :

$$|\psi> = \;(1/\sqrt{2})\;(|e,0> +\; |g,1>) \tag{2}$$

This entanglement survives after the atom has left the cavity. A measurement on the atom has then an immediate effect on the cavity field, which is prepared either in a Fock state (0 or 1) by detecting the atom in level e or g, or in a superposition of Fock states if the atomic levels are mixed by R_2 prior to detection. We can perform in this way Eintein-Podolsky-Rosen (EPR, [23]) experiments with a mixed pair, half matter, half radiation.

The π Rabi pulse

When the $\Omega t_i = \pi$ condition is satisfied, we realize an exchange of excitation between field and matter :

$$|e,0> \Rightarrow\; |g,1> \;\; ; \;\; |g,1> \Rightarrow\; -|e,0> \tag{3}$$

This pulse can be used to transfer information between the atomic and field qubits, to copy an atomic coherence $C_e\,|e> + C_g\,|g>$ into a field one and back. The storage of the information into the cavity constitutes a quantum memory for the atomic qubit [24]. Combining a π/2 pulse on a first atom with a π pulse on a second one, we also prepare an EPR pair of maximally entangled atoms [18].

The 2π Rabi pulse

Finally, the $\Omega\, t_i = 2\pi$ condition produces, when there is 1 photon in C, a global dephasing of the combined system's wave function, without changing the photon number. No change is obtained if C is empty because $|g,0>$ is not coupled to any other state by the atom-field interaction :

$$|g,0> \Rightarrow\; |g,0> ; |g,1> \Rightarrow\; -|g,1> \tag{4}$$

This effect is analogous to the well-known phase shift experienced by a spin 1/2 undergoing a 2π rotation. The fact that it is conditioned to the value of the photon qubit (0 or 1) is essential for the operation of quantum logic gates. Similar phase shift effects are exploited in the ion trap gates [15]. Note also that multiple of 2π Rabi pulses leaving the photon number unchanged are essential to generate «trapping states» of the micromaser [25].

By combining Rabi pulses of various duration on successive atoms, we engineer entanglement on string of particles crossing the cavity one at a time. The train of atom can be viewed as a thread with which we *knit* entanglement according to a well defined blue print. The quantum Rabi pulses are the *stitches* used to perform this knitting.

QUANTUM GATE AND QND MEASUREMENT OF A SINGLE PHOTON

Phase Gate

We can prepare in C a field containing 0 or 1 photon, or a linear combination of 0 and 1 photon states, by sending an atom in a superposition of e and g levels and having it undergo a π pulse in C (initially in the vacuum). We can also inject a small coherent field in C, produced by S, with an average photon number much smaller than 1. This field is then a superposition of 0 and 1 states, with a negligible probability to contain more than 1 photon. Such a 0/1 photon field will be called the *field qubit*. We can also send across C an atom prepared in level g or i, or in any linear superposition of g and i. We just have to excite the atom in g, then subject it to a pulse R_1 resonant on the g $\Rightarrow$ i transition. We will call it the *atom qubit*. Assume now that this qubit undergoes a 2π Rabi pulse in C. In the 2 x 2 basis of the field and atom qubits, the evolution of the system is described by the following truth table, which completes Eq.(4) :

$$|i,0> \Rightarrow |i,0>; |i,1> \Rightarrow |i,1>; |g,0> \Rightarrow |g,0>; |g,1> \Rightarrow -|g,1> \quad (5)$$

The first two relations express the fact that the photon mode is not resonant with transitions originating from level i, which plays the role of a *reference* level. Equation (5) defines the operation of a quantum phase gate (QPG). When the initial field and atom qubits are sent through C in state superpositions, the operation of the gate results in an entanglement of the two qubits. For example:

$$(1/\sqrt{2})(|i> + |g>)(C_0|0> + C_1|1>) \Rightarrow \{C_0|0>(|i> + |g>) + C_1|1>(|i> - |g>)\}/\sqrt{2} \quad (6)$$

$$= \{|i>(C_0|0> + C_1|1>) + |g>(C_0|0> - C_1|1>)\}/\sqrt{2} \quad (7)$$

where C_0 and C_1 are the probability amplitudes corresponding to 0 or 1 photon in C. Eqs (6) and (7) exhibit the symmetry of the QPG. Either the field in state 1 is the *control* qubit which produces a phase shift on the atom in g (Eq.6), or conversely it is the atom in g which is the *control* qubit dephasing the field in state 1 (Eq.7).

The correlations implied by Eqs(6) and (7) and the coherent operation of the QPG have been checked by experiments which have shown that the gate does operates as an entangling machine for the field and atom qubits [19]. We have also shown that, combined with classical microwave pulses operating on the atom qubit, this gate can be turned into a XOR gate realizing a QND measurement of a single photon.

QND Measurement of a Single Photon

The transformation of the QPG into a XOR (or controlled-not gate) is realized by the procedure schematized in Fig.2. The atomic qubit is subjected to two classical $\pi/2$ microwave pulses R_1 and R_2 sandwiching the QPG. These pulses present a controllable phase difference ϕ. This phase is adjusted so that, if there is 0 photon in C, the second pulse substracts its effect from that of the first. The atom then comes back to its initial state. If there is 1 photon in C instead, the atomic coherence produced by R_1 has its phase shifted by π when the atom crosses C. The second pulse in R_2 now completes the transition between the two atomic qubit states, exchanging their energies. In short, the atomic qubit flips if and only if there is one photon in C. The field qubit remains in any case unaltered. This defines a XOR gate.

Figure 2a,b show the operation of this gate when the atomic qubit is injected in level g, in the cases when the cavity contains 0 or 1 photon respectively. In the first case, the atom exits the gate in g, in the second case in i. The final atomic state is an indicator of the photon number, which remains unaltered at the end of the process. Detection of the atomic state in D after R_2 amounts to an ideal QND measurement [26] of a single photon field.

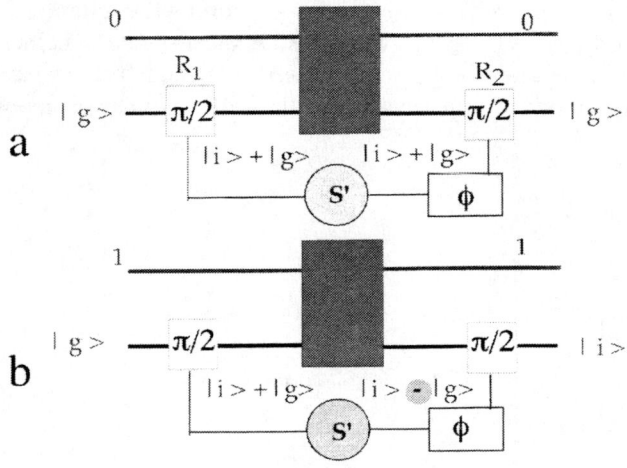

Figure 2. Schematic operation of the cavity QED XOR gate. The quantum phase gate, symbolized by the black rectangle, is sandwiched between two classical microwave pulses R_1 and R_2 acting on the atom qubit, initially in g, with an adjustable phase difference ϕ. If there is 0 photon in C, the second pulse undoes the action of the first and the atom exits in level g (Fig.2a). If there is 1 photon in C, the two pulses add their effects and the atom ends up in i (Fig 2b).

To demonstrate the QND operation, we first generate in C a 0 or 1 photon field. This is achieved by sending first a *source* atom undergoing a π/2 Rabi pulse. According to Equ.2, the detection of this atom in e or g prepares with equal probabilities either 0 or 1 photon. We then record the Ramsey fringes produced on a second *meter* atom undergoing a 2π Rabi pulse. We detect separately the fringes corresponding to the e and g outcomes of the *source* atom measurement. We obtain (see Fig.3) two fringe patterns with opposite phases. This demonstrates that, as expected, a single photon in C dephases by π the atomic coherence. Choosing the right phase difference φ between R_1 and R_2 amounts to setting the interferometer at a fringe extremum (vertical line on Fig.3). We then see that the *meter* atom has a large probability to exit in level g if there is 0 photon in C and in level i if there is 1 photon.

In order to check the non-destructive character of this procedure, we have also performed the repeated measurement of a small thermal field. We have left a blackbody field with an average number of 0.2 photon build up in C. We have then sent a first *meter* atom to perform a QND measurement of this field, then a second *probe* atom, initially in g and undergoing a π Rabi pulse in C, thus absorbing with unit probability the field left in C after the *meter* atom. By analyzing the correlations between the final states of the *meter* and *probe*, we have demonstrated the QND nature of the *meter* atom measuring process. We have found that the *probe* has a much larger probability to be excited if the *meter* has indicated the presence of a photon.

We have also performed a double QND experiment, sending two successive *meter* atoms across C and checking that their indications are correlated. With these experiments, we have analyzed the limitations of our QND scheme. This Ramsey interferometric method allows us to tell, with an 80% success rate, whether there is 0 or 1 photon in C immediately after the passage of the *meter* atom. The probability that this atom absorbs the photon, due to imperfections of the 2π Rabi pulse, is about 10%.

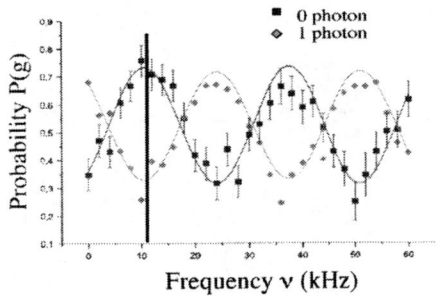

FIGURE 3. Ramsey fringes used for QND detection of a single photon: the probability P(g) to detect the meter atom in g is displayed as a function of the frequency ν of the field applied in R_1 and R_2 (tuning ν amounts to sweeping φ; origin of frequency arbitrary). Squares correspond to the 0 photon case (source atom detected in e) and diamonds to the 1 photon case (source in g). Lines are sinusoidal fits. Setting the interferometer at a fringe extremum (vertical line) correlates the final atomic state to photon number Reprinted by permission from ref. [20], Nature **400**, 239-242, copyright 1999, Macmillan Magazines Ltd.

This QND measurement must be contrasted with usual photodetection procedures based on the destruction of the photon. Here, for the first time, a single photon can be detected and kept stored in a box for further measurements. This opens, as we show in the next section, interesting perspectives for processing quantum information.

It is also interesting to compare our scheme with previous QND measurements in quantum optics, based on the interferometric measurement of the dispersive phase shift induced by a *signal* light field on a *meter* beam [27]. This shift is sensitive to the intensity of the *signal*, via its coupling with the *meter* beam in a non-linear medium. The *signal* field must be intense enough to produce an observable non-linear effect in this medium so that QND detection is generally sensitive only to macroscopic photon fluxes. Formally, our QND scheme looks quite similar. The 0/1 photon signal induces a phase shift on an atom *meter*, which is also detected by an interferometric method. Here however, the non-linearity of the Cavity QED system is extremely large, making it possible to detect non-destructively single light quanta.

Our QND method does not apply to fields containing more than one photon. The Rabi frequency in a field of n photons is $\Omega\sqrt{n}$, so that an interaction time corresponding to a 2π Rabi pulse for n = 1 performs with n photons a $2\pi\sqrt{n}$ rotation. The probability that the photon number changes after the interaction is $\sin^2(\pi\sqrt{n}) \neq 0$ for n >1. This violates the QND requirement as soon as the field state has a component outside the 0/1 photon subspace. We have proposed a dispersive phase shift method that overcomes this limitation [28]. If the cavity mode and the e $\Rightarrow$ g transition are detuned, the *meter* atom can no longer absorb the field, whatever the photon number is. The atomic levels undergo an energy light shift, linear function of n, which can be measured by Ramsey interferometry [29]. We have described detection procedures allowing us to pin-down the photon number [30]. The detection of several atoms is then required to read out n, because we need then at least $Log_2(n)$ bits of information. Experimental realization of this dispersive QND method is in progress.

STEP BY STEP ENGINEERING OF THREE PARTICLE ENTANGLEMENT

Preparation of Entanglement

The atom-cavity QPG can be combined with the $\pi/2$ atom-photon Rabi pulse to prepare a three-particle-entangled state, according to a procedure proposed independently in [31] and [32]. We describe now this experiment [21] which illustrates how our apparatus can be used to *knit* entanglement in a complex system by a *step-by-step* succession of operations. The procedure is schematized in Fig.4a. We have represented, in a space-time diagram, the evolution of the cavity field and the atoms in an elementary sequence. The time axis is horizontal and the space one vertical. The C mode space-time line is thus a horizontal, whereas the atoms, crossing the cavity at a fixed speed, are represented by parallel oblique lines. A first atom A_1 undergoes a $\pi/2$ Rabi pulse in the initially empty cavity mode, resulting in the entangled state of Eq.2. We then use the 0/1 photon field produced in C as a qubit for the operation of a QPG entangling it with a

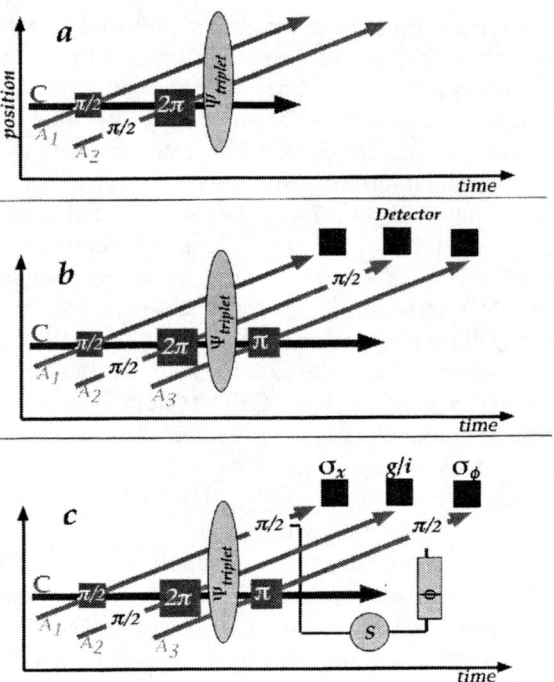

FIGURE 4. Schematics of the three-particle entanglement procedure. The atom and cavity space lines during the experiments are represented qualitatively. The diamonds represent the atom-cavity interactions and the circles the classical pulses acting on atoms. Detection events are indicated by black squares. (a): Preparation of the A_1-A_2-C entangled state. (b) and (c): « longitudinal » and « transverse » pseudospin state detections, respectively.

second atom A_2. This atom, first prepared by R_1 into the linear superposition $(1/\sqrt{2})(|i> + |g>)$, undergoes a 2π Rabi pulse in C. When it exits C, the A_1-A_2-C system is prepared in the entangled state :

$$|\Psi_{triplet}> = (1/2) [|e_1> (| i_2 > + | g_2>) |0> + |g_1> (| i_2 > - |g_2 >) |1>] \quad (8)$$

where the indices correspond to the atom numbers. This state can be rewritten as :

$$|\Psi_{triplet}> = (1/2) [|i_2> (|e_1,0> + | g_1, 1>) + |g_2> (|e_1,0> - |g_1, 1>)] \quad (9)$$

describing an A_1-C entangled pair whose phase is conditioned to the state of A_2.

Because $|\Psi_{triplet}>$ involves two levels for each subsystem, it is equivalent to an entangled state of three spins 1/2. Let us define the states $|+_j> (|-_j>)$ (with j = 1,2) as $|+_1> = |e_1> (|-_1> = |g_1>)$, $|\pm_2> = (|g_2> \pm i_2>)/\sqrt{2}$ and $|+_C> = |0> (|-_C> = |1>)$. With these notations, $|\Psi_{triplet}>$ takes the form of the GHZ three-spin state [10] :

$$|\Psi_{triplet}> = (1/\sqrt{2}) (|+_1,+_2,+_C> - |-_1,-_2,-_C>)$$ (10)

In order to analyze the entanglement, we must detect the two atoms and the field in C. We cannot however measure this field directly. We must first copy it on a third atom A_3 crossing C after A_2 [24]. The A_3-C interaction is set so that A_3, initially in g, is not affected if C is empty but undergoes a π Rabi pulse in a single photon field (Eq.2). Within a phase, A_3 maps the state of C. According to Eq.(3), the cavity state $|+_C>$ is mapped into $|g_3>$ and $|-_C>$ into $|e_3>$. Thus by detecting A_1, A_2 and A_3, we measure a set of observables belonging to the three parts of the entangled triplet. Checking fully the A_1-A_2-C entanglement involves measurements of the *pseudospins* in two different orthogonal bases.

Reading out the Entanglement in the Longitudinal « Spin » Basis

In a first experiment (timing sketched Fig 4b), we checked «longitudinal correlations» by detecting the «spins» along what we define as the z-axis (eigenstates $|\pm_j>$ for j = 1,2 and $|\pm_C>$ for C). For A_1 and C (i.e. A_3) this is a direct energy detection. For A_2, a $\pi/2$ analysis R_2 pulse on the i ⇒ g transition transforms $|+_2>$ and $|-_2>$ respectively into $|i_2>$ and $|g_2>$. The three atoms should then be detected in $\{e_1, i_2, g_3\}$ or $\{g_1, g_2, e_3\}$ with equal probabilities. We show (Figure 5) the histograms giving the probabilities for detecting the atom triplets in the $2^3 = 8$ possible channels. As expected, the $\{e_1, i_2, g_3\}$ and $\{g_1, g_2, e_3\}$ channels largely dominate. The unbalance between these channels and the appearance of the six theoretically forbidden ones is due to experimental imperfections (finite cavity damping time, residual thermal fields, non-ideal Ramsey and Rabi pulses). This experiment demonstrates the existence of three particle correlations. Taken alone, however, it is not enough to prove entanglement because these correlations could be produced by a classical statistical mixture of states.

Reading out the Entanglement in a Transverse « Spin » Basis

In a second experiment, we detect transverse components of the three pseudospins. The sequence of operations is sketched Fig.4c. It differs from the sequence of Fig.4b. by unitary operations, equivalent to $\pi/2$ rotations of the corresponding Bloch vectors, applied to the three atoms prior to detection. For A_1 and A_3, these rotations are performed by two R_2 pulses having a phase difference ϕ. Details about these pulses are given in [21]. No pulse is applied on A_2 after it has interacted with C. The direct detection of this atom in level g or i does correspond to a detection of the transverse spin components $(|+_2> \pm |-_2>)/\sqrt{2}$.

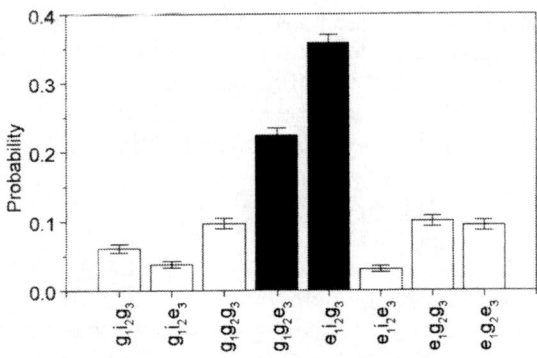

FIGURE 5. Histograms displaying the «longitudinal three-particle correlations». The detections probabilities for the eight relevant detection channels are shown. The two expected channels are in black and the six « spurious » ones in white Reprinted with permission from ref. [21]. Copyright 2000 American Association for the Advancement of Science.

In a control experiment, we do not send A_2. Detecting A_1 and A_3 amounts to measuring transverse «spin» components of the entangled A_1-C state described by Equ.2. By repeating the experiment many times, we reconstruct the average value $<\sigma_{1x}.\sigma_{3\phi}>$ representing the product of «spin 1» component along the « x-axis » and «spin 3» along a direction in the xOy plane making an angle ϕ with Ox. By varying ϕ, we get a modulated signal demonstrating the A_1 - C entanglement. Detecting one «spin» along Ox projects the other along the opposite direction and the probability to find it subsequently along a direction at an angle ϕ is a sinusoidal function of this angle. The signal is shown in Fig.6 (square experimental points). The dashed line is a fit.

We then perform the complete three-atom experiment. We measure the same A_1-A_3 correlations versus ϕ, now conditioned to the final A_2 state. According to Eq.9, we expect that the A_1-A_3 signal is not changed if A_2 is detected in i, whereas the sign of the A_1 - A_3 correlation should be inverted if A_2 is found in g. We have plotted in Fig.6 the experimental results. The curves with experimental circles (diamonds) correspond to A_2 detected in i (g) respectively. The expected sign change of the correlation is observed. This experiment can be considered as the detection of a maximally entangled particle pair (A_1-C i.e. A_1-A_3 correlations), the phase of the entangled superposition being controlled by the operation of a quantum phase gate involving a third particle (A_2).

A tight timing is required to entangle A_1-A_2 and C and to perform the R_2 «rotations» before the first two atoms exit the ring surrounding the cavity. Experimental tricks, described in [21], are required to make it possible. The third atom crosses C however after the first one has exited the hole in the ring, loosing at this time the coherence of its «transverse spin» in the stray electric fields of this hole. In other words, the present timing of the experiment allows us to prepare an A_1-A_2-C entangled triplet, but not an A_1-A_2-A_3 one. The third atom is merely used as a «read out» for the cavity field.

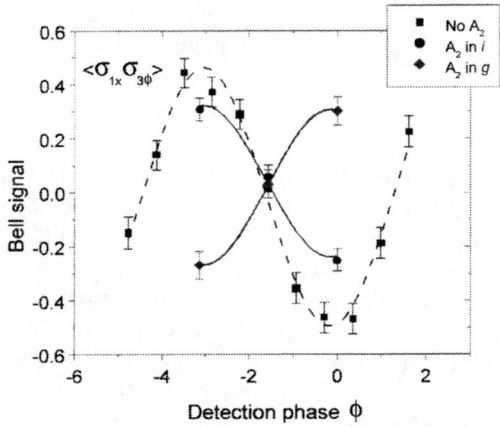

FIGURE 6. Detection of «transverse» three-particle correlations. The squares represent, versus ϕ ,the expectation value $< \sigma_{1X} \; \sigma_{3\phi} >$ Bell signal for the pair A_1- A_3, when atom A_2 is not sent). The dashed line is a sinusoidal fit. The circles (diamonds) show the same signal when A_2 is sent and detected in i (g). The phase shift of the A_1-A_3 (i.e. A_1-C) EPR correlation conditioned to the state of A_2 is clearly observed Reprinted with permission from ref. [21]. Copyright 2000 American Association for the Advancement of Science.

The combination of these «longitudinal» and «transverse» spin component measurements, demonstrate that we have prepared a three particle entangled state. A discussion of the fidelity of our entangled triplet preparation is made in [21]. Due to experimental imperfections, we prepare in fact a mixed state described by a density matrix ρ which projects unto the ideal $|\Psi_{triplet}>$ state with a fidelity F = $<\Psi_{triplet} \; | \; \rho \; | \; \Psi_{triplet}> = 0.54$ (±0.03). The fact that F > 0.5 ensures that genuine three particle entanglement is generated here.

CONCLUSIONS AND PERSPECTIVES

This cavity QED set-up allows us to engineer in a controlled way an entangled multiparticle state. Individual qubit addressing is an important feature of our system. The particles being at centimeter-scale distances, there is no difficulty to manipulate them one at a time. A simple iteration of the three-particle entanglement scheme will prepare a Mermin state [33], generalization of the GHZ one. After having prepared the A_1-C pair in the state described by Equ.1, we will send a stream of atoms A_2, A_3....A_n, all prepared in the state $(| i > + | g >) /\sqrt{2}$ and undergoing, if in g, a 2π Rabi rotation in a single photon field. Because this rotation does not change the photon number, the zero- and one-photon parts of the A_1-C pair will become correlated to an A_2, A_3....A_n

state with all n-1 atoms in $(|i> + |g>)/\sqrt{2}$ for the zero-photon an $(|i> - |g>)/\sqrt{2}$ for the one-photon part. We thus could prepare the entangled state :

$$|\Psi> = (1/\sqrt{2}) (|+_1, +_2, +_{3.........}, +_C> - |-_1, -_2, -_{3,........}, -_C>)$$ (11)

For such experiments, we need an improved version of our apparatus. We have good prospects for realizing a better cavity, without ring, relaxing the tight timing constraints and increasing the entanglement fidelity. The main present limitation is that circular atoms are prepared with Poisson statistics, corresponding to large fluctuations between 0 and 1 of the atom number in each pulse. In consequence, long acquisition times are required to study correlations implying several atoms (the data of Fig.6 have required 8 hours of averaging). However, it seems possible to implement preparation techniques for circular atoms that would generate exactly one atom on demand. For instance, fluorescence from an atomic beam could be used to image a single atom and to excite it deterministically to the circular state in a fully adiabatic process. With such an «atom pistol», the techniques described in this chapter could be extended to more complex situations, such as the one described by Eq.11.

Higher levels of complexity will be reached by realizing entanglement involving several fields modes of one cavity, or field modes belonging to two or more spatially separated cavities crossed successively by a stream of atoms. The study of mesoscopic superpositions involving many atoms, photons, or cavity modes constitutes a challenging goal for the next generation of these cavity QED experiments.

ACKNOWLEDGEMENTS

This work has been supported in part by the European Community and by Japan Science and Technology Corporation (International Cooperative Research Project, Quantum Entanglement Project).

REFERENCES

1. Bennett, C.H. and DiVicenzo, D.P. , *Nature*, **404**, 247-255 (2000).
2. Bennett, C.H., Brassard, G., Crepeau, C., Jozsa, R., Peres, A., Wooters, W.K., *Phys.Rev.Lett.* **70**, 1895-1898 (1993).
3. Ekert, A, and Josza, R., *Rev. Mod. Phys.*, **68**, 733-753 (1996).
4. Bennett, C.H., Brassard, G., and Ekert, A. , *Scientific American*, October, 26-33, (1992).
5. Freedman, S.J. and Clauser, J.F., *Phys.Rev.Lett.* **28**, 938-941 (1972) ; Aspect, A., Grangier, P. and Roger, G. *Phys.Rev.lett.* **47**, 460-463 (1981) ; **49**, 91-95 (1982).
6. Ou, Z.Y., and Mandel, L., *Phys Rev.Lett*, **61**, 50-53 (1988); Kwiat, P.G., Mattle, K., Weinfurter, H., and Zeilinger, A., *ibid*, **75**, 4337-4340 (1995).
7. Rarity, J.G., Owens, P.C.M., and Tapster, P.R., J.Mod.Opt. 41, 2435-2444 (1994).
8. Bouwmeester, D., Pan, J.W., Mattle, K., Eibl, M., Weinfurter, H., and Zeilinger, A., *Nature*, **390**, 575-579 (1997) ; Boschi, D., Branca, S., De Martini, F., Hardy, L., Popescu, S., *Phys.Rev.Lett.*, **80**, 1121-1124 (1998) ; Furusawa, A., Sorensen, J.L., Braunstein, S.L., Fuchs, C.A., Kimble, H.J., Polzik, E.S., *Science*, **282**, 706-709 (1998).
9. Pan, J.W., Bouwmeester, D., Daniell, M., Weinfurter, H., and Zeilinger, A., *Nature*, **403**, 515-519 (2000).

10. Greenberger, D.M., Horne, M.A., and Zeilinger, A., *Am.J.Phys.* **58**, 1131-1143 (1990).

11. Gershenfeld, N.A., and Chuang, I .L., *Science,* **275**, 350-356 (1997).

12. Braustein, S.L., Caves, C.M., Josza, R. , Linden, N., Popesu, S., and Schack, R., *Phys.Rev.Lett.,* **83**, 1054-1057 (1999).

13. Meekhof, D.M., Monroe, C., King, B.E., Itano, W.M., and Wineland, D.J., *Phys.Rev.Lett,* **76**, 1796-1799 (1996).

14. Haroche, S. , « Cavity Quantum Electrodynamics » in *Fundamental Systems in Quantum Optics,* edited by Dalibard, J. et al, North Holland, Amsterdam, 1992, pp. 767- 940.

15. Monroe, C., Meekhof, D.M., King, B.E., Itano, W.M., and Wineland, D.J., *Phys.Rev.Lett.* , **75**, 4714-4717 (1995).

16. Turchette, Q.A., Wood, C.S., King, B.E., Myatt, C.J., Leibfried, D., Itano, W.M., Monroe, C., and Wineland, D.J., *Phys.Rev.Lett.* , **81**, 3631-3634 (1998).

17. Sackett, C.A., Kielpinski, D., King, B.E., Langer, C. , Meyer, V., Myatt, C.J., Rowe, M ., Turchette.Q.A., Itano, W.M., and Wineland, D.J. *Nature,* **404**, 256-259 (2000).

18. Hagley, E., Maître, X., Nogues, G. , Wunderlich, C., Brune, M. , Raimond, J.M., and Haroche, S., *Phys.Rev.Lett.,* **79**, 1-5 (1997).

19. Rauschenbeutel, A., Nogues, G., Osnaghi, S., Bertet, P. , Brune, M. , Raimond, J.M., and Haroche, S., *Phys.Rev.Lett.,* **83**, 5166-51669 (1999).

20. Nogues, G., Rauschenbeutel, A., Osnaghi, S., Brune, M., Raimond, J.M., and Haroche, S. , *Nature,* **400**, 239-242 (1999).

21. Rauschenbeutel, A. , Nogues, G., Osnaghi, S., Bertet, P., Brune, M., Raimond, J.M., and Haroche, S., *Science,* **288**, 2024- 2028 (2000).

22. Brune, M., Schmidt-Kaler, F., Maali, A., Dreyer, J. , Hagley, H., Raimond, J.M., and Haroche, S., *Phys.Rev.Lett* ., **76**, 1800-1803 (1996).

23. Einstein, A., Podolsky, B., and Rosen, N., *Phys.Rev.,* **47**, 777-780 (1935).

24. Maître, X., Hagley, E., Nogues, G., Wunderlich, C., Goy, P., Brune, M., Raimond, J.M., and Haroche, S., *Phys.Rev.Lett.* **76**, 1800-1803 (1997).

25. Weidinger, M., Varcoe,B., Heerlein, R., and Walther, H., Phys.Rev.Lett., 82, 3795-3798 (1999).

26. Braginski, V.B., and Khalili, F.Y., *Rev.Mod.Phys.,* **68**, 1-11 (1996).

27. Grangier, P. Levenson, A.L., and Poizat, J.P., *Nature,* **396**, 537-542 (1998).

28. Brune, M., Haroche, S., Lefevre, V., Raimond, J.M. and Zagury, N., *Phys.Rev.Lett,* **65**, 976-979 (1990).

29. Brune, M., Nussenzveig, P., Schmidt-Kaler, F., Bernardot, F., Maali, A., Raimond, J.M., and Haroche, S., *Phys.Rev.Lett.,* **72**, 3339-3342 (1994).

30. Haroche, S., Brune, M., and Raimond, J.M., *J.Phys.II France,* **2**, 659-670 (1992).

31. Haroche, S., Nogues, G., Rauschenbeutel, A., Osnaghi, S., Brune, M., and Raimond, J.M., « Quantum Knitting in Cavity QED » in *Laser Spectroscopy, XIV International Conference,* edited by Blatt, R., et al, World Scientific, Singapore, 1999, pp 140-149.

32. Zheng, S.B., *J.Opt.B.,* **1**, 534-538 (1999).

33. Mermin, N.D., Phys.Rev.Lett, 65, 1838-1841 (1990).

Quantum computing with trapped ions, atoms and light

Andrew M. Steane

Centre for Quantum Computation
Department of Atomic and Laser Physics
Oxford University, England. OX1 3PU

Abstract. We consider experimental issues relevant to quantum computing, and discuss the best way to achieve the essential requirements of reliable quantum memory and gate operations. Nuclear spins in trapped ions or atoms are a very promising candidate for the qubits. We estimate the parameters required to couple atoms using light via cavity QED in order to achieve quantum gates. We briefly comment on recent improvements to the Cirac-Zoller method for coupling trapped ions via their vibrational degree of freedom. Error processes result in a trade-off between quantum gate speed and failure probability. A useful quantum computer does appear to be feasible using a combination of ion trap and optical methods. The best understood method to stabilise a large computer relies on quantum error correction. The essential ideas of this are discussed, and recent estimates of the noise requirements in a quantum computing device are given.

INTRODUCTION

Quantum computation, considered as the study of the type of information manipulation which quantum mechanics allows, is now a well-established subject, and has been surveyed in several review articles [1–5]. Even text books have begun to be published [6,7]. This is not to say that the subject can yet be regarded as mature, since there remain basic theoretical and experimental issues which are not understood or not explored. On the theoretical side, these include the thorough understanding of channel capacities and of entanglement involving several systems, the hoped-for possibility of further efficient quantum algorithms in addition to Shor's algorithm and its close cousins, and a more complete study of fault tolerant methods in quantum information processing. On the experimental side, only the first steps of manipulating systems of a few quantum bits have been possible so far; the great majority of quantum information processing concepts have yet to be experimentally demonstrated.

The first talk on quantum computing in this conference series was by Ekert in 1994 [8]. In the intervening period, there has been much progress in both theoret-

CP551, *Atomic Physics 17*, edited by E. Arimondo, P. DeNatale, and M. Inguscio
© 2001 American Institute of Physics 1-56396-982-3/01/$18.00

ical and experimental aspects, and now there exist large research programmes in several parts of world whose aim is to realise quantum computing in the laboratory. The intention is to achieve complete coherent control, and repeatable measurement, of a set of a few quantum bits, in a physical system which has one of two properties (or preferably both). The first property is that the structure of the physical system should be amenable to being extended to many hundreds or thousands of quantum bits, with general quantum gate operations between them, hence realising a quantum computer. Even a small processing device (a few qubits) would be very useful if it allowed quantum information to be swapped controllably between its internal qubits and 'flying qubits', e.g. photons travelling over long distances in a specific mode of the electromagnetic field. This is the second desirable property.

My aim here is not to survey the beautiful experiments which have in recent years taken the first steps of manipulating small groups of qubits. Rather, I will discuss the experimental issues which arise when we address the larger goal of building a useful quantum computer. A more thorough account can be found in a special issue of the *Fortschritte der Physik* [9], to be published this year, which is dedicated to a survey of experimental proposals for quantum computers. In particular, results quoted here for ion traps and optical methods are derived or more completely discussed there [10].

It is important to realise that, while the ideas of information processing provide a powerful driving force to the research in this area, the probable impact of the experiments is not restricted to computing. Rather, quantum information ideas yield useful tools which apply to any experiments in which entangled states are manipulated. In addition to computing, such experiments could realise useful physical instruments such as sensors, frequency standards, etc., and explore the physics of decoherence and mesoscopic quantum effects.

WHAT IS QUANTUM COMPUTING?

The essence of quantum computing is *the controlled manipulation of entanglement.* It is significant that entanglement, not merely quantum superposition, is central. This is because quantum superposition is not, in its essential properties, any different from the superposition which occurs in classical vectors and fields, and in particular it cannot be used to realise computational efficiency gains of the type offered by Shor's algorithm, nor can it allow quantum communication protocols such as entanglement-assisted communication and teleportation.

Entanglement, as a mere property, may be present in the low-lying energy eigenstates of many physical systems, owing to quantum statistics, but this is not in itself sufficient for computing, where we require the controlled manipulation of entanglement. Such manipulation is in fact an area of experimental physics which has never previously been explored. It is this aspect of launching into qualitatively new territory which motivates my own interest.

THE "CONTRADICTORY" DEMANDS OF QUANTUM COMPUTING

The physical realisation of quantum computing appears to make contradictory demands on the physical system to be used. One requirement is that the quantum information must be very stable, and this implies the physical degrees of freedom in which it is stored must have only a very weak coupling to one another and to the rest of the universe. However, quantum processing relies on the ability to have fast, controllable quantum gates, and this requires a strong coupling between qubits (which also opens the possibility of undesired coupling between the qubits and the rest of the universe, and among the qubits).

The conflicting demands can be met if we can identify very stable physical variables q, and a means of rapidly introducing and removing a coupling G between them. Candidates for q include electron positions, either in atoms or in solid state quantum dots, electron spins, flux or charge states of superconducting circuits, and nuclear spins. Of these, the nuclear spin is especially interesting because it has both high stability (against *all* changes, including precession as well as spin-flips) and a natural avenue for gate operations. The high stability of nuclear spin states is due to basic features of the natural world. For typical conditions of electric and magnetic field stability, undesired electric field coupling to charges is much stronger than magnetic field coupling to magnetic dipoles, and nuclear magnetic moments are particularly small. However, if the nucleus were, like a neutrino, merely very well decoupled from everything else, it would not be a good candidate for quantum computing. It is significant that the nucleus comes ready-packaged in an electron cloud with highly useful properties for logic gates. The atomic electrons provide the means to take hold of the atom and place it where we wish, and they also provide a ready-made very strong and very stable magnetic field on the nucleus. This results in the hyperfine splitting. The presence of this interaction implies that nuclear spin states can be rendered less stable by fluctuations in the electron orbitals, but the stability of the hyperfine splitting for isolated atoms or ions is well documented, and is in fact used to provide our standards of time and frequency. Thus the stability of free atoms, and the smallness of the coupling between a nuclear spin and anything outside its atom, combine to produce a candidate qubit which is both stable and readily manipulated. For these reasons, we shall be primarily concerned with nuclear spin-based computing.

The coupling G can be brought about in several ways. Experiments in bulk nuclear magnetic resonance [11–16] rely on the very weak spin-spin coupling which is permanently present between adjacent nuclei in a molecule. This coupling is of the order of tens of Hertz and probably is insufficient to allow the large processors which are our goal (but see [17]). All the other proposed experiments make use of the hyperfine splitting, so that during a quantum logic gate the nuclear spin interacts with the atomic electrons (e.g. Fermi contact interaction), which in turn interact with further degrees of freedom, in order to couple qubits.

Several possibilities for a controlled coupling between electrons in different atoms have been proposed. A useful way to characterize what we mean by a controlled coupling is to show how a specific quantum gate operation such as controlled-not ($^{C}X \equiv |00\rangle \langle 00| + |01\rangle \langle 01| + |11\rangle \langle 10| + |10\rangle \langle 11|$) or controlled-phase ($^{C}Z \equiv |00\rangle \langle 00| + |01\rangle \langle 01| + |10\rangle \langle 10| - |11\rangle \langle 11|$) or the square root of swap ($W^{1/2} \equiv (|00\rangle \langle 00| + |10\rangle \langle 01| + |01\rangle \langle 10| + |11\rangle \langle 11|)^{1/2}$) can be implemented. Note that for quantum computing we require not merely a gate between one pair of atoms, but a recipe for producing gates "at will" among all pairs of atoms in the computer. In terms of producing general transformations of the joint state of all the qubits, it is sufficient if only gates between nearest-neighbours in a linear geometry are available, but then the requirements for precision in the gates are more severe (see below).

Ion trap experiments up till now have used a light-induced coupling between the electronic state and the vibrational motion (phonon) of relatively heavy charged particles (ions) in a trap in vacuum [18–22]. Trapped ions or neutral atoms can also be coupled by using light alone to couple the electronic state of one atom and another in a high-quality cavity [23–25]. This concept has been realised (in an experiment not based on nuclear spin or hyperfine interaction) using a beam of neutral atoms [26,27]. A controlled coupling between neutral atoms can also be realised through electric dipole-dipole interactions, in which one atom is moved past another using a movable trap such as a dipole force trap [28,29].

In these examples the atoms and ions are well separated from each other and from other atoms. Their electronic wavefunctions are, to very good approximation, those of a free atom. This means the hyperfine splitting remains very stable, and rivals that of the best frequency standards.

Nuclear spins have also been proposed as qubits in a solid state context by Kane [30]. Here, the nucleus is that of a dopant in a semiconductor. The coupling between qubits uses the hyperfine (Fermi contact) interaction between a dopant nucleus and the electron orbitals centred on it, and the exchange interaction between one such electron and another attached to a neighbouring dopant nucleus situated about 10 nm away. The potential energy in this exchange interaction is given by the Coulomb repulsion between the electron charges; it is adjusted by moving the electrons using d.c. voltages applied to metal electrodes on the surface of the semiconductor. The stability of the hyperfine splitting here will not be as great as that for a quasi-free atom.

Solid state proposals such as that of Kane have excited interest on the grounds that they are "scalable", that is, the method can be extended to many hundreds or thousands of qubits, by laying down larger arrays of electrodes and dopants on the substrate by lithography. It has further been stated that ion trap quantum computers, and in general any methods not based on solid state physics, are "not scalable". This is certainly a premature judgement, since there is plenty of room for ingenuity in fabricating multiple traps, either by lithography or otherwise, and combining atomic and optical techniques (see below). The real issue for experimental quantum computing is not "is it scalable?" but "is it workable?", and to be workable a system must be both scalable and reliable.

RELIABILITY AND SCALABILITY

At this early stage it is not yet clear which idea, or set of ideas, will prove most useful for large-scale quantum computing. However, it is interesting to note that the physical bit-level characteristics of good classical computing devices do not give a good intuition into what are the significant issues for implementing quantum computing.

In the classical case, decoherence is not only present but is an essential ingredient of the reliability of the computing machinery. In an electronic classical computer, transistor amplification, combined with resistive dissipation, provides continuous feedback to the voltage levels in the device, thus stabilizing them. In the quantum case, by contrast, decoherence must be reduced to a minimum throughout the computer, and only introduced in a specific, controlled way to chosen qubits (during syndrome extraction, for example, discussed below).

There is a further difference between quantum and classical in the number of computing elements. A quantum computer of just one hundred reliable logical quantum bits would be a useful machine, in the sense that it could probably address some tasks which are inaccessible even to the largest classical supercomputers. A few thousand reliable logical quantum bits would certainly be very powerful, in contrast to the classical case where a kilobyte computer is almost useless.

Overall, therefore, the quantum computer is a precisely coherent system of comparatively few (100 to 10^4) logical qubits, while the classical computer is a rapidly decohering system of comparatively many (10^8 to 10^{11}) classical bits. In the quantum case, the logical qubits might be contained within a larger number of controllable physical qubits, but the necessary coherence and useful system size remain very great distinctions between what is required of the physical machinery in the quantum and classical cases.

Reliability requirements are sometimes referred to very roughly in terms of a "decoherence time", which is compared to the time to complete a one- or two-qubit logic gate. This is, however, too incomplete a measure to be useful. A sufficient starting point for a consideration of reliability consists of two figures of merit and three design considerations. The figures are the memory noise and the gate noise, defined to be the probability for a failure per qubit per gate time, and the probability of gate failure, respectively. The design considerations are parallelism, i.e. the degree to which many 2-qubit gates can be operated simultaneously; communication, i.e. the ease of transporting quantum information between well-separated qubits (e.g. by repeated gate operations or by a quantum data bus); and measurement, i.e. the speed and reliability with which qubits can be measured.

When all these considerations are not properly combined, misconceptions have arisen. For example, it is sometimes asserted that a precision of 10^{-4} in the gates between physical qubits is, approximately, sufficient to allow large-scale computing using fault-tolerant methods, but this is not true. The figure quoted arises from estimates of fault tolerance which assume unlimited parallelism and no constraints on which qubits can be directly coupled by gates, but no architecture currently proposed has these features.

The "probability of failure" of a gate or memory element is a somewhat subtle concept in the quantum regime. It corresponds roughly to the classical concept of the probability of a bit flip error. In both the classical and the quantum case, the physical parameters of the system (e.g. voltages, quantum state amplitudes) are in fact continuous, not discrete, but the methods used to suppress noise force the system towards a discrete set of states. In the classical computer these are the familiar 'on/off' states, while in the quantum computer they are a discrete set of quantum states which can approximate a continuous set sufficiently well to allow quantum algorithms to be implemented with high probability of success. The concept of 'failure probability' is appropriate when quantum error correction methods are used. The following examples give a rough working definition; a more precise definition will be given in the discussion of quantum error correction below. If during a short time t a qubit is subject to an undesired relaxation process characterized by a term Γt in the master equation for its density matrix, then the failure probability is $1 - \exp(-\Gamma t)$ within a numerical factor of order 1. If a process intended to implement the quantum gate G actually implements the physical operation P, then the failure probability is

$$1 - \left| \langle \psi | \, G^\dagger P \, | \psi \rangle \right|^2,$$

for a computer in the state $| \psi \rangle$, and the average failure probability is this quantity averaged over the possible states. We will be concerned with gate operations whose failure probability is $\leq 10^{-3}$.

Theoretical proposals for quantum computers have, of course, involved estimates of the failure probabilities, but their reliability is hard to assess. The estimates are mostly lower bounds on the failure rates, based on phenomena which are known to be present in the system, but in practice further issues will almost certainly become apparent once experiments are performed. This calls for caution in comparing statements about what is expected in a system which has never been built, and statements about what is observed in a real apparatus. For example, the most severe problem in several of the ion trap experiments has been a wholly unexpected noise source which heats the ions; it is very likely due to fluctuations in the electric field much greater than the expected thermal fluctuations, and probably associated with fluctuating patch potentials on the electrode surfaces [31]. There is now evidence that this noise source can be greatly reduced by a well-chosen construction of the trap [22,31]. However, it illustrates the fact that electric field noise can arise by unexpected routes, and emphasizes the difficulty which can be expected in realizing gates via the controlled motion of charged particles. Such difficulties may be more severe in the solid state proposals where the moving particles (electrons) are lighter and are much closer to the electrode surfaces.

The main disadvantage of a nuclear-spin based approach is that it is slower than other possibilities. The next most obvious candidate for a qubit is the electron spin, e.g. fine structure in atoms with coupling by optical methods, or electrons in quantum dots coupled either by optical methods [25] or via control of the electron position, making use of the exchange interaction between overlapping electrons in adjacent quantum dots [32]. The relative speed of electron- and nuclear-spin based methods is indicated approximately by the ratio of electron to nuclear magnetic moments, i.e. 2000. An electron-spin based method could therefore achieve similar memory noise when the decoherence rate is of order 10^3–10^4 times higher than that needed for nuclear spins. Currently the ratio of observed decoherence rates of electron spins in a semiconductor [33], compared to nuclear spins of trapped ions [34], is of order $(1 \text{ MHz})/(1 \text{ mHz}) = 10^9$.

OPTICAL METHODS

Of the methods outlined above, there are two reasons to pick light as the most promising way to couple electrons in different atoms: it offers both a fast communication across the computer and a reliable operation. The former is implied by the speed of light, and the latter by the fact that light does not couple to electromagnetic fields, which would otherwise be the main noise source; optical fibres and waveguides can pass very close to electronic circuits without ill effect.

We are thus lead to examine the requirements for coupling atoms via photons [23–27]. The most natural concept is to drive a stimulated Raman transition, in which one photon comes from (or is emitted into) a laser beam incident on an atom A, and the other photon is emitted into (or obtained from) a mode of a high-finesse optical cavity. This couples atom A with the cavity mode. A subsequent (or simultaneous) laser pulse on another atom B in the same cavity couples the photon mode to B, and thus A and B are coupled indirectly. In order to permit a fast Raman transition, the laser and cavity-resonance frequencies are chosen similar to the frequency of allowed single-photon transitions in the atom, detuned from single-photon resonance by $\Delta \gg \Gamma$, where Γ is the natural lifetime of the atomic excited state.

Let the Rabi frequencies for atom-laser and atom-cavity couplings be Ω, g respectively, and κ the decay rate for the photon to leave the cavity. A straightforward calculation shows [10] that the best compromise between the undesired processes of atomic relaxation and cavity photon relaxation is obtained when $\Omega/\Delta = (2\kappa/\Gamma)^{1/2}$, leading to a failure probability $p = 2\pi(2\kappa\Gamma/g^2)^{1/2}$ during a π-pulse between zero-photon and one-photon cavity states. The gate rate (inverse of the time required for the π pulse) is

$$\frac{1}{T} = \frac{g^2}{\Gamma} \frac{p}{8\pi^2}. \tag{1}$$

The best high-finesse cavities available today yield $\kappa \Gamma \simeq 0.01\, g^2$ for transitions at optical frequencies [10]. This gives $p \simeq 0.9$, an unworkable system. Considering that p scales as the square root of the cavity finesse, it is clear that very great improvements would be needed before this method could be used in the optical domain. In the microwave domain it can be used successfully [26,27], but the scaling to many qubits is more difficult.

In order to circumvent the problem of cavity relaxation, a dark state method has been proposed by Pellizzari et al. [23], in which two atoms are coupled (via the cavity photon) during an adiabatic passage in which the occupation of the cavity mode can be made arbitrarily small. The failure probability is now dominated by the non-adiabaticity introduced if the process is driven too fast, and by the cavity decay. An approximate estimate of these contributions [10] yields a gate rate

$$\frac{1}{T} \simeq \frac{1}{9} p^2 \frac{g^2}{\kappa},\tag{2}$$

where now the failure probability can be made arbitrarily low, at the cost of slower gate rate. Comparing equations (1) and (2), we see that the adiabatic passage gate rate scales as a higher power of the failure probability, compared to the direct Raman transition. This is an example of a very broad principle that a more precise operation is only obtained at a cost in speed. For this reason, gate rate and gate precision in quantum computers should always be discussed together.

Putting $p = 10^{-3}$ into (2), with $g/(2\pi) = 100$ MHz, $\kappa/(2\pi) = 10$ MHz, which are reasonable numbers for currently available optical cavities and resonance lines in alkali-like atoms or ions, we obtain a gate rate around a kilohertz, i.e. disappointingly slow, considering that the light-atom coupling strength is 100 MHz.

MOTIONAL GATES IN ION TRAPS; HYBRID COMPUTER

Theoretical studies have lead to useful improvements on the original proposal of Cirac and Zoller [18] for implementing gates in an ion trap, which adopts the motional degree of freedom of a trapped ion string as a shared qubit, or quantum data bus. In particular, Molmer and Sorensen [35,36] showed how to make better use of the motional degree of freedom, limiting the degree to which the quantum information is sensitive to relaxation of the motion. In the case that the motional relaxation is small, the gate rate is limited by the need to avoid driving unwanted transitions in the ions by off-resonant excitation [37]. Jonathan et al. [38] proposed a method which allows the gates to be faster by around an order of magnitude at given failure probability. This method exploits the light-shift induced resonance which occurs when the internal transition of an ion is driven with a Rabi frequency which equals the vibrational frequency ν_z of a normal mode of the ion string in the trap. The gate rate is $\eta \nu_z \equiv (R\nu_z)^{1/2}$ where η is the Lamb-Dicke parameter, R is the recoil frequency, and for a range of η the failure probability is $p \simeq \eta^2/2$.

Typical recoil frequencies for optical transitions in a pair of trapped ions are in the range 10 to 500 kHz, implying a gate rate 200 kHz to 10 MHz at $p = 10^{-3}$. In practice, more than two ions will be in the trap, which lowers the recoil frequency, and a method must be used to allow separate ions to be addressed. This means either the trap frequency is reduced, to increase the separation between ions, or ions are moved around by switching d.c. electric fields. In either case the gate rate is further reduced, and one may expect typical values in the region $(400–20000)/N$ kHz for a trap containing N ions.

Experiments in ion trap quantum information processing have recently achieved significant milestones, such as the controlled production of a 4-ion entangled state by Sackett *et al.* [21], and observation of very low temperatures, and low heating rates, for one or more trapped ions [22,31].

We conclude that although an all-optical approach remains promising, considerable technological improvements in optical cavities are required before it could rival the motional method in an ion trap. However, this does not mean our original argument for the benefits of light-based coupling was wholly misleading. A very promising concept for a large computer combines the optical and motional methods. One can envisage a set of traps, with an optical cavity around each one. The cavity is used to implement quantum communication between the traps, and gates within any trap are achieved by the motional method. This allows the ion trap concept to be scaled up to very large numbers of qubits, while maintaining rapid quantum communication across the computer. This hybrid construction has been envisaged by several authors [39–41,24] and a detailed theory of how to implement the communication has been put forward, including error correction protocols [42]. When used for quantum communication, rather than gates, the optical cavity does not require such a high finesse, making the technological requirements less severe. Taken together, these ingredients make a quantum computer based on multiple atom or ion traps, linked by cavities and fibres, a serious proposition.

QUANTUM ERROR CORRECTION

A really useful quantum computer, which could produce $Q = 10^6$ to 10^{12} operations on $K = 100$ to 1000 logical bits, requires a logical gate failure probability below $1/Q$ and quantum memory in the machine with failure probability $1/(KQ)$ per logical qubit per time step. These failure probabilities might just be achievable for the moderate computer size $K = 100$, $Q = 10^6$ with standard methods such as adiabatic passage and very careful control of timing, pulse strengths, etc. However, for a large computer size $K = 1000$, $Q = 10^{12}$, sufficiently stable operation by standard methods is unrealistic.

We can nevertheless realise large scale quantum computation by using the power of the computer itself as a way to combat noise. There are, broadly, two approaches. One is to exploit ideas such as the geometric phase, which depends only on topological properties of the paths traced out by the computer states in some parameter space [43]. The parameter space need not be one directly connected to the physics of the computing machinery, since the versatility of the quantum computer allows it to simulate some other desired (but physically allowable) dynamics. The other approach is that of quantum error correction (QEC). These two approaches can be linked mathematically by interpreting QEC using the tools of quantum field theory [44].

QEC is now a fairly mature subject, so I will not survey it here. Introductory material may be found in [6,7,1,2,45]. I will limit myself to addressing some commonly held misconceptions. In order to do so I will need to give a brief sketch of the essential ideas.

Noise, in all its manifestations, including those called "quantum noise", "decoherence", "relaxation", and so on, is exactly and precisely *loss of information.* Its partner, imprecision, which includes the effects of stray fields, unwanted cross-coupling, and wrong timing, is *lack of information.* In either case, the essential point is that some information about the system dynamics is unavailable. In general, an error process is any process which carries the computer away from the state which the algorithm it is running supposes it to be in (therefore a coupling between qubits which is known to be present, but which the algorithm can't be designed to take account of, constitutes error, as well as undesired coupling to the environment [46]). Powerful ways to circumvent these problems are available if we make use of information processing. This is the essence of why QEC can work.

It is useful to consider QEC in terms of three nested Hilbert spaces. First we have the Hilbert space of the whole quantum computing apparatus and any other systems it interacts with. Call this the 'universal' Hilbert space. Within the computing apparatus is a set of n physical qubits which we can manipulate; they define the 'physical' Hilbert space. For example the physical space could be defined by the $M_z = 0$ (Zeeman field-insensitive) ground state hyperfine levels of a set of atoms, while the universal space contains these and all the other atomic energy levels as well as the surrounding electromagnetic field and controlling machinery. In order to operate quantum computations on a set of K qubits with QEC, we use a physical apparatus of $n > K$ physical qubits. By definition, K qubits of quantum information live in a Hilbert space of 2^K dimensions, and this *logical Hilbert space* is a sub-space of the 2^n-dimensional physical Hilbert space.

The logical Hilbert space is spanned by a set of 2^K orthonormal states called quantum codewords. The set of codewords is called a quantum error correcting code. We choose the codewords to have the special property that when the real physical computer evolves in time, subject to all the noise and imprecision which it has in practice, then the codeword states are either unaffected, or they evolve in a particular useful way, to be described. If the codeword states are unaffected by a process, they are said to form a "decoherence free subspace" with respect to that process. Every (linear) quantum error correcting code is a decoherence free subspace with respect to a set of operators called its stabilizer. In other words, the codewords $|u_L\rangle$ are simultaneous eigenstates, with eigenvalue 1, of all the operators S_i in the stabilizer, $S_i |u_L\rangle = |u_L\rangle$. There are 2^{n-K} such operators.

QEC consists of three steps. First, we suppress "leakage", which is any tendency of the computer state to move out of the physical Hilbert space, for example through weak excitation of transitions to $M_z \neq 0$. Plumbing of leaks can be done by a controlled relaxation process such as optical pumping or cooling, which is tailored to relax the computer into the n qubit physical space, but which does not cause relaxation within that space (or only does so to a degree which can be corrected by the next step). If a suitable relaxation process is not readily available, then one can be constructed from a set of quantum gates and measurements acting at the physical qubit level [47].

The next step is syndrome extraction. This consists of making a set of appropriately-chosen measurements on the system. These are typically joint measurements of several physical qubits, which are carried out by first applying a network of gates, and then making single-qubit measurements on physical qubits reserved for the purpose. The observables thus measured are the operators S in the code stabilizer, hence, their measurement will not disturb the computer if its state is in the logical Hilbert space.

More generally, the computer's state will be projected by this measurement either into the logical Hilbert space, or into one of $2^{n-K} - 1$ other eigenspaces which are mutually orthogonal and orthogonal to the logical Hilbert space. The set of eigenvalues which are the measurement outcomes is called the error syndrome; it indicates which of these sub-spaces the computer is now in. The error correction is completed by applying the rotation which will map the indicated space onto the logical space. This typically requires simple rotation of one or more physical qubits by Pauli spin operators σ_x, σ_y or σ_z.

Note that we have not said anything yet about the error process we are trying to recover from. This is to underline the point that any kind of noise or imprecision can be tackled by QEC [46,44]. The success of the method relies on matching the stabilizer (and hence the measurements made) as well as possible to the error process which is to be resisted. The overall chance of success is measured by the fidelity $\langle \psi_L | \rho' | \psi_L \rangle$ between the computer's density matrix ρ' after QEC, and the logical state $|\psi_L\rangle$ which the computer is intended to be in, according to the algorithm it is running.

In order to understand this situation, it is useful to consider the combination of error process followed by leakage suppression as a single process, which causes the transformation $\rho \equiv |\psi_L\rangle \langle \psi_L| \rightarrow \rho'''$ in the physical state of the computer. The syndrome extraction produces a further transformation $\rho''' \rightarrow \rho'' = E\mathcal{L}(\rho)E^\dagger = E\rho'E^\dagger$ where $\mathcal{L}$ is a transformation within the logical Hilbert space, which is uncorrectable, and E is a correctable error. Here, ρ'' is the state produced for one particular outcome of the syndrome measurement, i.e. it is an eigenstate of the stabilizer. The set of correctable errors is directly connected to the stabilizer by the basic theory of QEC. Typically, stabilizer operators, and correctable errors, are tensor products of Pauli operators acting on the physical bits.

For example, consider the case that the error process affects different physical qubits independently, and its effect is small, but where we make no other assumptions about the error process. In this case we choose the stabilizer so that all tensor products of Pauli operators of weight up to t are correctable[1]. For such a correctable set, we have the striking result that *any change* $\rho \rightarrow \rho'''$ *whatever*, if it is restricted to a set of $\leq t$ physical qubits, is completely correctable ($\rho' = \rho$), including non-unitary changes. We deduce that the only uncorrectable terms in the erroneous state ρ'' are those affecting more than t qubits at once. These terms are very small for independent noise, being of order $C(n, t+1)\epsilon^{t+1}$ where ϵ is the size of the changes produced by the error process in the density matrix of any single physical qubit, and $C(n, t+1)$ is the binomial coefficient $n!/(t+1)!(n-t-1)!$. The quantity ϵ is the 'failure probability' introduced above to quantify the quality of physical qubit gate operations and quantum memory.

The explicit construction of QEC codes which can correct arbitrary errors of weight up to t is a central part of QEC theory, and so is the fact that the coding rate K/n can be efficient even for large t. As an example, consider a set of 127 atoms with two hyperfine levels in the ground state used as one physical qubit per atom. We suppose optical pumping can prevent leakage without affecting the physical qubit space, but collisions and uncontrolled fields such as thermal radiation cause the nuclear spins to relax and decohere independently with a lifetime $\tau = 100$ s. We use an error-correcting code which can correct errors of weight up to 7, and which stores 29 logical qubits in 127 physical qubits [48]. The network to perform syndrome extraction requires about 2000 controlled-not operations and 98 single-bit measurements; we assume these can be done with very high precision in 10 ms (the precision assumption will be reconsidered in a moment). QEC is carried out once every $t = 300$ ms, therefore the error process occurs mostly between corrections. The failure probability is then $\epsilon \simeq t/\tau = 0.003$, and the fidelity of each correction is approximately $1 - 10^{-8}$. We repeat the QEC a million times, once every 300 ms. After this the net fidelity from beginning to end has fallen to $1 - 0.01 \simeq 0.99$. This means the state of 29 qubits has been preserved with high fidelity for 83 hours, i.e. three thousand times the relaxation lifetime of the underlying nuclear spins. Indeed, because the relaxation of the spins occurred independently of anything else, every one of them must have spontaneously decayed and was subsequently re-excited about 1500 times during the whole process, and yet the quantum information survived because it was encoded in multi-spin correlations, not directly in the spins themselves.

The use of QEC in a real quantum computer requires the further concepts of *fault tolerance*, in which the network which performs syndrome extraction is constructed so that it is reliable even though each physical gate and measurement is noisy. This is possible by careful use of repetition, and through further ideas for which there is not space for a description here [6,7,47]. These ideas lead to the possibility of an error threshold, that is, failure probabilities for gates and memory, below which fault-tolerant methods allow arbitrarily long computations to be successful,

---

[1] The weight $|E|$ of a tensor product E of Pauli operators is the number of terms in it, e.g. $|\sigma_x^{(2)} \otimes \sigma_z^{(3)} \otimes \sigma_x^{(7)}| = 3$ where $\sigma_i^{(j)}$ acts on qubit j.

at the cost of an overhead of more and more physical qubits required for longer computations. The exact trade-off between the noise tolerance and the system size scale-up is a subject of active research, and it depends significantly on the other considerations mentioned above, namely the available parallelism, communication, and measurement in the computer. For a system with all the desirable qualities at once, preliminary calculations indicate that the threshold is around 10^{-3} to 10^{-4} for gate failure and 10^{-4} to 10^{-5} for memory failure. For systems with only nearest neighbour interactions, the thresholds are unknown but will be considerably more stringent. The overhead involved for these results is high, the number of physical qubits exceeding the number of logical qubits by a factor 1000 to 10^5. A more efficient type of coding allows this scale-up to be reduced to a factor 20 to 100, but then the constraint on memory noise is more severe by a factor 10 to 100. There is clearly a great advantage in having a very stable quantum memory.

CONCLUSION

It is now possible to envisage a fairly large number of possible designs for a quantum computer, and much work is being done to evaluate them. We also have a preliminary grasp of the noise requirements which are sufficient and necessary to allow large computation to be stabilized by quantum error correction. I have argued for the attractiveness of nuclear spins for the qubits of a computer, especially when they are in atoms or ions which are well isolated and therefore whose electronic wavefunctions are stable, and the use of light for coupling the atomic electrons. An examination of the atom–light coupling reveals, however, that quantum gates using light alone are far from being achievable except at a gate rate slower than the one already achieved using the quantum motion of trapped ions. However, the use of light to communicate quantum information between well-separated qubits in a computer is a very attractive and more immediately achievable possibility. This could be used, for example, to couple a set of ion traps together to make a large quantum computer.

In this discussion I have concentrated on the goal of realising moderate to large scale quantum computing, rather than describing the details of currently achievable experiments. Historically, much interest has focussed on experiments involving the simplest possible quantum interference of single particles, such as the Young's slits experiment or simple interferometers, and Rabi flopping between energy eigenstates. More recently, interference experiments involving two particles, such as a pair of photons, or a photon and an atom, have been studied in minute detail. However, quantum computing provides a cogent language which enables us boldly to discuss much richer types of quantum behaviour, a language which experiments to date have hardly begun to speak. We can already envisage that the current generation of experiments, impressive though they are, will be viewed in retrospect as primitive when moderate scale quantum computers are experimentally realised, since few-particle interference effects, GHZ states, entanglement swapping, teleportation and so on will merely be routine behaviour which we take for granted. Owing to the high degree of precision and coherent control provided by the steady development of atomic physics techniques over many years, atomic physics is now on the verge of entering the richer territory of quantum computing.

I thank D. Stacey and D. Lucas for helpful comments. This work was supported by EPSRC and the EC QUBITS IST-1999-13021.

REFERENCES

1. Steane, A. *Rep. Prog. Phys.* **61**, 117–173 (1998).
2. Bennett, C. H. and Shor, P. W. *IEEE Transactions on Information Theory* **44**, 2724–2742 (1998).
3. Reiffel, E. and Polak, W. *A.C.M. Computing Surveys*, to be published (2000); quant-ph/9809016.
4. Preskill, J. and Kitaev, A. Course notes, California Institute of Technology, http://www.theory.caltech.edu/people/preskill/ph229/index.html.
5. Bennett, C. H. and DiVincenzo, D. P. *Nature* **404**, 247–255 (2000).
6. Lo, H.-K., Popescu, S., and Spiller, T., editors. *Introduction to quantum computation and information.* World Scientific, Singapore, (1998).
7. Bouwmeester, D., Ekert, A., and Zeilinger, A., editors. *The physics of quantum information.* Springer-Verlag, Berlin, (2000).
8. Ekert, A. In *Atomic Physics 14*, 450–466 (AIP Press, New York, 1995).
9. *Fortschritte der Physik* (2000). S. Braunstein and H.-K. Lo, eds.
10. Steane, A. M. and Lucas, D. M. *Fortschritte der Physik* (2000).
11. Cory, D. G., Fahmy, A. F., and Havel, T. F. In *Proc. 4th Workshop on Physics and Computation* (Complex Systems Institute, Boston, MA, 1996).
12. Gershenfeld, N. A. and Chuang, I. L. *Science* **275**, 350–356 (1997).
13. Jones, J. A. and Mosca, M. *Journal of chemical physics* **109**, 1648–1653 (1998).
14. Chuang, I. L., Gershenfeld, N., and Kubinec, M. *Phys. Rev. Lett.* **80**, 3408–3411 (1998).
15. Cory, D. G., Price, M. D., and Havel, T. F. *Phyica D* **120**, 82–101 (1998).
16. Knill, E., Laflamme, R., Martinez, R., and Tseng, C. H. *Nature* **404**, 368–370 (2000).
17. Lloyd, S. *Science* **261**, 1569 (1993).
18. Cirac, J. I. and Zoller, P. *Phys. Rev. Lett.* **74**(20), 4091–4094 (1995).
19. Monroe, C., Meekhof, D. M., King, B. E., Itano, W. M., and Wineland, D. J. *Phys. Rev. Lett.* **75**(25), 4714–4717 (1995).
20. Turchette, Q. A., Wood, C. S., King, B. E., Myatt, C. J., Leibfried, D., Itano, W. M., Monroe, C., and Wineland, D. J. *Phys. Rev. Lett.* **81**(17), 3631–3634 (1998).
21. Sackett, C. A., Kielpinski, D., King, B. E., Langer, C., Meyer, V., Myatt, C. J., Rowe, M., amd W. M. Itano, Q. A. T., Wineland, D. J., and Monroe, C. *Nature* **404**, 256 (2000).
22. Roos, C., Zeiger, T., Rohde, H., Nägerl, H. C., Eschner, J., Leibfried, D., Schmidt-Kaler, F., and Blatt, R. *Phys. Rev. Lett.* **83**, 4713–4716 (1999).

23. Pellizzari, T., Gardiner, S. A., Cirac, J. I., and Zoller, P. *Phys. Rev. Lett.* **75**(21), 3788–3791 (1995).

24. Cirac, J. I., van Enk, S. J., Zoller, P., Kimble, H. J., and Mabuchi, H. *Physica Scripta* **T76**, 223–232 (1998).

25. Imamoḡlu, A., Awschalom, D. D., Burkard, G., DiVincenzo, D. P., Loss, D., Sherwin, M., and Small, A. *Phys. Rev. Lett.* **83**(20), 4204–4207 (1999).

26. Hagley, E., Maitre, X., Nogues, G., Wunderlich, C., Brune, M., Raimond, J., and Haroche, S. *Phys. Rev. Lett.* **79**, 1 (1997).

27. Rauschenbeutel, A., Nogues, G., Osnaghi, S., Bertet, P., Brune, M., Raimond, J. M., and Haroche, S. *Science* **288**, 2024–2028 (2000).

28. Jaksch, D., Briegel, H.-J., Cirac, J. I., Gardiner, C. W., and Zoller, P. *Phys. Rev. Lett.* **82**(9), 1975 (1999).

29. Cirac, J. I. and Zoller, P. *Nature* **404**, 579–581 (2000).

30. Kane, B. E. *Nature* **393**, 133 (1998).

31. Turchette, Q. A., Kielpinski, D., King, B. E., Leibfried, D., Meekhof, D. M., Myatt, C. J., Rowe, M. A., Sackett, C. A., Wood, C. S., Itano, W. M., Monroe, C., and Wineland, D. J. *Phys. Rev. A*, to be published (2000); quant-ph/0002040.

32. Loss, D. and DiVincenzo, D. P. *Physical Review A* **57**, 120 (1998).

33. Kikkawa, J. M. and Awschalom, D. D. *Phys. Rev. Lett.* **80**, 4313 (1998).

34. Wineland, D. J., Monroe, C., Itano, W. M., Leibfried, D., King, B. E., and Meekhof, D. M. *J. Res. Natl. Inst. Stand. Technol.* **103**, 259–328 (1998).

35. Molmer, K. and Sorensen, A. *Phys. Rev. Lett.* **82**, 1835–1838 (1999).

36. Sorensen, A. and Molmer, K. *Phys. Rev. Lett.* **82**, 1971–1974 (1999).

37. Steane, A. M., Roos, C. F., Stevens, D., Mundt, A., Leibfried, D., Schmidt-Kaler, F., and Blatt, R. *Phys. Rev. A*, to be published (2000).

38. Jonathan, D., Plenio, M. B., and Knight, P. L. *quant-ph/0002092* (2000).

39. Cirac, J. I., Zoller, P., Kimble, H. J., and Mabuchi, H. *Phys. Rev. Lett.* **78**, 3221 (1997).

40. Pellizzari, T. *quant-ph/9707001* (1997).

41. Sorensen, A. and Molmer, K. *Physical Review A* **58**, 2745–2749 (1998).

42. van Enk, S. J., Cirac, J. I., and Zoller, P. *Phys. Rev. Lett.* **78**, 4293–4296 (1997).

43. Kitaev, A. Y. *quant-ph/9707021* (1997).

44. Knill, E., Laflamme, R., and Viola, L. *Phys. Rev. Lett.* **84**, 2525–2528 (2000); quant-ph/9908066.

45. Steane, A. M. *Phil. Trans. Roy. Soc. Lond. A* **356**, 1739–1758 (1998).

46. Steane, A. M. *Proc. Roy. Soc. Lond. A* **452**, 2551–2577 (1996).

47. Preskill, J. *Proc. R. Soc. Lond. A* **454**, 385–410 (1998).

48. Steane, A. M. *Nature* **399**, 124–126 (1999); quant-ph/9809054.

Scalable Entanglement
of Trapped Ions

C. Monroe, C.A. Sackett, D. Kielpinski, B.E. King,
C. Langer, V. Meyer, C.J. Myatt, M. Rowe,
Q.A. Turchette, W.M. Itano, and D.J. Wineland

*Time and Frequency Division, National Institute of Standards and Technology[1]
Boulder Colorado 80303*

Abstract. Entangled states are a crucial component in quantum computers, and are of great interest in their own right, highlighting the inherent nonlocality of quantum mechanics. As part of the drive toward larger entangled states for quantum computing, we have engineered the most complex entangled state so far in a collection of four trapped atomic ions. Notably, we employ a technique which is readily scalable to much larger numbers of atoms. Limits to the current experiment and plans to circumvent these limitations are presented.

INTRODUCTION

At the heart of quantum mechanics lies the principle of superposition, where physical properties of a system can exist in two or more states simultaneously. When a system is composed of more than one degree of freedom, superpositions can be prepared where distinct degrees of freedom are perfectly correlated, yet the state of each degree of freedom is by itself in superposition. The prototypical example is Bohm's version [1] of the Einstein-Podolsky-Rosen paradox [2], where a spin-zero particle decays into a pair of spin-1/2 daughters, resulting in the singlet state

$$|\Psi_{EPR}\rangle = \frac{|\uparrow\rangle_1 |\downarrow\rangle_2 - |\downarrow\rangle_1 |\uparrow\rangle_2}{\sqrt{2}}. \tag{1}$$

This state is *entangled*, since it cannot be expressed as a direct product of states representing each particle. When one of the subsystems in such a state is measured,

[1] This work was supported in part by the U.S. National Security Agency and the Advanced Research and Development Activity under contract MOD7037.00, the U.S. Army Research Office, and the U.S. Office of Naval Research.

CP551, *Atomic Physics 17*, edited by E. Arimondo, P. DeNatale, and M. Inguscio
2001 American Institute of Physics 1-56396-982-3

the other subsystem is also determined, even when the particles are not in physical contact or outside each other's lightcones. In general, entangled states such as $|\Psi_{EPR}\rangle$ highlight the nonlocal character of quantum mechanics. Quantitatively, this is usually expressed in terms of Bell's inequality violations [3], where measured correlations between the entangled subsystems can be shown to be incompatible with what would be expected under conditions of local realism.

Although the correlation in the above state cannot be used for superluminal communication, it can be harnessed for enhanced communication rates over what can be obtained classically [4,5]. Furthermore, such states are useful in a variety of quantum communication schemes such as quantum cryptography [6] and quantum "teleportation" [7–10].

Entangled states of larger systems are a defining feature of a quantum computer. Here, for example, a collection of N spin-1/2 particles are prepared in an arbitrary entangled state of the form

$$|\Psi_{QC}\rangle = a_0|000\ldots0\rangle + a_1|000\ldots1\rangle + \ldots + a_{2^N-1}|111\ldots1\rangle, \tag{2}$$

where $|0\rangle$ and $|1\rangle$ refer to the two spin states of each particle, and the a_k are the amplitudes of the number k being stored by the register of particles. By choosing appropriate entangled states and making appropriate state measurements of the particles, quantum computers can solve certain problems much faster than any classical computer [11,12]. The reason quantum computers are mere speculation at this point is that $|\Psi_{QC}\rangle$ is very difficult to produce in the laboratory.

SCALABLE ENTANGLEMENT WITH TRAPPED IONS

Nearly every demonstration of entanglement to date has relied upon a random or selection process which prohibits scaling to large numbers of particles. This can be quantified in terms of the *entanglement efficiency* parameter ϵ, or the probability per unit time that a perfect entangled pair is created [13]. The probability of realizing a perfect N-particle entangled state typically scales as ϵ^{cN}, where c is of order unity and depends on the particular experiment.

The first measured Bell's inequality violations were seen in atomic cascade experiments involving the entanglement of a pair of spontaneously-emitted photons [14,15]. Spontaneous parametric downconversion is now a popular source of entangled photons, where typically ultraviolet photons traverse a nonlinear crystal and downconvert into a pair of polarization-entangled infrared beams [16,17]. Unfortunately, the probability of each input photon being converted leads to an efficiency $\epsilon \simeq 10^{-4}$, so the probability of entangling larger numbers of photons becomes very small. (Nevertheless, by waiting long enough, three-photon entangled states were recently observed from simultaneous downconversion into two pairs [18]). Experiments in cavity-QED have recently shown entanglement of two atoms [19] and two atoms with a photon [20], where a thermal (random) source of atoms traverse a common microwave cavity. In these experiments, $\epsilon \simeq 0.005$. Experiments with

174

optical parametric oscillators can also entangle the continuous quadratures of two optical field modes [21]. Although this source has near-unit entanglement efficiency, scaling to larger numbers of degrees of freedom appears difficult.

The Cirac-Zoller Scheme

In 1995, Cirac and Zoller showed that a collection of trapped ions may be suitable for storing large-scale entangled states such as $|\Psi_{QC}\rangle$ [22]. In their proposal, each atomic ion stores a quantum bit (qubit) of information in a pair of electronic energy levels, and a collective mode of harmonic vibration is used to entangle any pair of ion qubits. By applying laser beams to an individual ion in the collection, its internal qubit state can be mapped onto the collective ion motion, and subsequent quantum logic gates can be applied between the motion and a second ion, effectively entangling the two ions. The entanglement can be extended to any number of ions by repeating these steps on other pairs of ions. When accompanied by single-ion rotations, the Cirac-Zoller scheme allows the creation of an arbitrary entangled state [Eq. (2)], and therefore forms a set of universal quantum logic gates.

The basic elements of the Cirac-Zoller scheme were demonstrated on a single trapped ion in 1995 [23]. A variation of this scheme was later used to entangle a pair of trapped ions [24] with entanglement efficiency $\epsilon \simeq 0.8$, representing the first scalable entanglement source with near-unit efficiency.

The Mølmer-Sørensen Scheme

Instead of entangling the ions sequentially, Mølmer and Sørensen showed how to create the N-ion entangled state

$$|\Psi_N\rangle = \frac{|\uparrow\rangle_1|\uparrow\rangle_2\cdots|\uparrow\rangle_N + e^{i\phi_N}|\downarrow\rangle_1|\downarrow\rangle_2\cdots|\downarrow\rangle_N}{\sqrt{2}} \tag{3}$$

with a single pulse of laser radiation [25]. The Mølmer-Sørensen operation applied to any pair of qubits in a collection of ions (accompanied by single ion rotations) allows the creation of any entangled state [Eq. (2)], and thus forms a set of universal quantum logic gates alternative to the Cirac-Zoller scheme [26]. We have employed the Mølmer-Sørensen scheme to create the entangled state of Eq. (3) for $N = 2$ and $N = 4$ trapped ions [27]. In both cases, the entanglement efficiency was $\epsilon \simeq 0.8$, as discussed below.

The Mølmer-Sørensen entanglement technique can be understood by considering a pair of identical spin-1/2 charged particles confined together in a harmonic potential [28]. The energy levels of this system are illustrated in Fig. 1, where $\hbar\omega_0$ is the internal energy splitting of each qubit, and ν is the oscillation frequency of a particular collective mode of the particles in the trap.

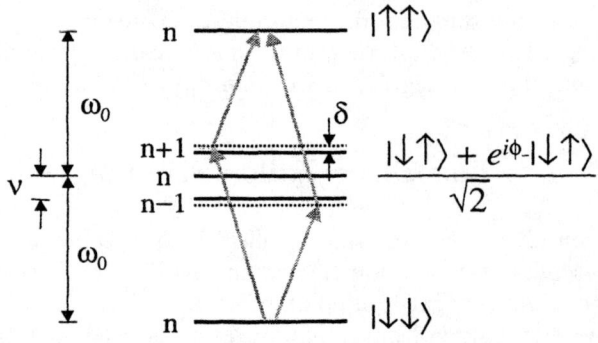

FIGURE 1. Entanglement scheme for two particles. Each ion is initially prepared in the $|\downarrow\downarrow\rangle$ internal state, and the collective motion of the pair initially contains exactly n quanta. Laser fields equally illuminating the two ions and oscillating near $\omega_0 + \nu + \delta$ and $\omega_0 - \nu - \delta$ couple the $|\downarrow\downarrow\rangle$ and $|\uparrow\uparrow\rangle$ states as shown. For sufficient detuning δ, the populations of the middle states are kept small. By driving the double transition for the appropriate time, the entangled state $(|\uparrow\uparrow\rangle + e^{i\phi_2}|\downarrow\downarrow\rangle)/\sqrt{2}$ is generated. For four ions, the same procedure generates the state $(|\uparrow\uparrow\uparrow\uparrow\rangle + e^{i\phi_4}|\downarrow\downarrow\downarrow\downarrow\rangle)/\sqrt{2}$.

The ions are initially prepared in the $|\downarrow\downarrow\rangle$ internal state, and we assume the ions are in a collective motional eigenstate $|n\rangle$. By simultaneously applying optical fields near the first upper and lower motional sidebands (oscillating at $\omega_0 + \nu + \delta$ and $\omega_0 - \nu - \delta$ respectively) with equal illumination on the two ions, the two-step transition from $|\downarrow\downarrow\rangle|n\rangle$ to $|\uparrow\uparrow\rangle|n\rangle$ is driven through the intermediate states

$$|\Psi_{int}\rangle_{\pm} = \frac{|\downarrow\rangle|\uparrow\rangle|n\pm1\rangle + e^{i\phi_-}|\uparrow\rangle|\downarrow\rangle|n\pm1\rangle}{\sqrt{2}}, \tag{4}$$

where ϕ_- is the phase difference of the field at the two ion positions. For sufficiently large detuning δ from the sidebands, these intermediate states are negligibly occupied, so that the motional state is not altered. We also assume $\delta \ll \nu$ so that intermediate states involving other motional modes are not involved in the coupling. As shown in Fig. 1, there are two paths from $|\downarrow\downarrow\rangle$ to $|\uparrow\uparrow\rangle$, and their respective couplings are given by the product of the two resonant sideband Rabi frequencies divided by the detuning $\pm\delta$ from the relevant virtual intermediate level. For the upper- then lower-sideband path (arrows on left side of Fig. 1), this coupling is $(\eta g\sqrt{n+1})^2/\delta$, and for the lower- then upper-sideband path (arrows on right side of Fig. 1), it is $-(\eta g\sqrt{n})^2/\delta$, where g is the single ion resonant carrier Rabi frequency and η is the Lamb-Dicke parameter of the motional mode involved. (These expressions are valid only in the Lamb-Dicke limit $\eta^2(n+1) \ll 1$.) Adding the couplings from these two paths results in a net Rabi frequency from $|\downarrow\downarrow\rangle$ to $|\uparrow\uparrow\rangle$ of $\Omega = \eta^2 g^2/\delta$, independent of the motional state $|n\rangle$ within the Lamb-Dicke regime. The net interaction Hamiltonian is proportional to $\sigma_x^{(1)}\sigma_x^{(2)}$, where $\sigma_x^{(i)}$ is the transverse Pauli spin-1/2 operator of ion i. Entanglement is achieved by simply applying these beams for a time $\tau = \pi/2\Omega$, creating the desired spin state

$$|\Psi_2\rangle = \frac{|\uparrow\uparrow\rangle + e^{i\phi+}|\downarrow\downarrow\rangle}{\sqrt{2}}, \tag{5}$$

where ϕ_- is the sum of the field phases at the two ion positions.

1   Fast Entanglement

In order for the intermediate states $|\Psi_{int}\rangle_\pm$ to be negligibly occupied, the detuning δ must be large compared to both single-spin sideband Rabi frequencies $\eta g\sqrt{n+1}$ and $\eta g\sqrt{n}$, meaning the entangling operation must be much slower than the resonant sideband operations. (This is the characteristic slowdown of driving higher-order transitions through virtual levels.) However, it is possible to violate this condition and still generate the state $|\Psi_2\rangle$ [29,30]. In this case, the intermediate states $|\Psi_{int}\rangle_\pm$ are occupied during the operation (and the motional state becomes entangled with the spins), but this occupation can vanish at exactly the moment the desired entangled spin state $|\Psi_2\rangle$ is created. Without regard to the the spin states, we find that for arbitrarily small δ (and within the Lamb-Dicke limit), the motion evolves during the operation as a coherent superposition of its original state $\rho_m(0)$ and an oscillating displaced state [29]

$$\rho_m^{dis}(t) = \mathcal{D}\left[\sqrt{\frac{\Omega}{\delta}}(e^{i\delta t} - 1)\right]\rho_m(0)\mathcal{D}\left[\sqrt{\frac{\Omega}{\delta}}(e^{i\delta t} - 1)\right]^\dagger, \tag{6}$$

where $\mathcal{D}(\alpha)$ is the displacement operator with phase space argument α [31]. The overall motion is thus in a "Schrödinger Cat"-type superposition state [32], with maximum separation in phase space $2\sqrt{\Omega/\delta}$. The phase space trajectory of the displaced component $\rho_m^{dis}(t)$ follows a circle from its original state with radius $\sqrt{\Omega/\delta}$, returning to the initial motional state $\rho_m(0)$ at times $t_m = 2\pi m/\delta$, where the positive integer m is the number of complete circular cycles of the displacement [29]. Setting the entanglement pulse time τ defined above equal to t_m, we find that the condition for a return to the initial motional state following the entanglement step is $\Omega/\delta = 1/(4m)$. The entangling time can thus be rewritten as $\tau = \pi\sqrt{m}/(\eta g)$, which is only a factor of $2\sqrt{m}$ slower than an analogous resonant sideband transition. To maximize the speed of the Mølmer-Sørensen operation in the experiment, we operate with $m = 1$.

2   Scalable Entanglement

Surprisingly, the Mølmer-Sørensen entangling scheme is scalable in the sense that precisely the same operation can be used to generate the N-particle entangled state of Eq. (3) for any even number of ions N. (For N odd, $|\Psi_N\rangle$ can be generated using one entanglement pulse accompanied by a separate independent rotation of each particle's spin.) The Mølmer-Sørensen interaction is proportional to J_x^2, where J_x is the transverse spin operator for the effective spin-$N/2$ particle. Physically, this interaction simultaneously flips all pairs of ions in the collection. Through the properties of angular momentum rotations [25], this results in the desired entangled state $|\Psi_N\rangle$. In scaling to larger numbers of ions, the only difference (for a given motional mode frequency) is that the operation is $\sqrt{N}$ times slower, since the Lamb-Dicke parameter is proportional to $1/\sqrt{N}$. In addition, the phase which appears in Eq. (3) is the sum of the field phases at each ion position.

If the ions are uniformly illuminated, the Mølmer-Sørensen scheme requires that they all participate equally in the intermediate motional excitation, which implies that the only suitable mode for arbitrary N is the center-of-mass mode. However, this mode has a practical disadvantage that fluctuating ambient electric fields cause it to heat at a significant rate [33]. For large δ, the entanglement operation is independent of the motion, so that heating is unimportant, so long as the ions remain in the Lamb-Dicke regime [29]. In the small-δ case however, motional decoherence of the Schrödinger-Cat state discussed above must be avoided. Modes involving only relative ion motion couple to higher moments of the field, so heating of them is negligible [34]. For $N = 2$ and $N = 4$ ions, such modes do exist in which each particle participates with equal amplitude [35]. In both cases, they are uniform "stretch" modes, in which alternating ions oscillate out of phase; we use these modes here. Excitation of the center-of-mass mode does still affect the experiment, as the ion eventually can get heated out of the Lamb-Dicke regime. For this reason, we initially sideband cool both the center-of-mass and uniform stretch modes to near their ground state. We note that other modes of motion can also be used for entanglement, as long as the laser intensity on each ion is adjusted to compensate for the difference in mode amplitude of that ion, resulting in equal sideband couplings for all ions.

EXPERIMENT

The experiment was performed using $^9\mathrm{Be}^+$ ions confined in a miniature linear RF trap [33], with the N ions lying in a line along the trap's weak axis. Two spectrally resolved ground-state hyperfine levels compose the effective spin-1/2 system, with $|\downarrow\rangle \equiv |F = 2, m_F = -2\rangle$, $|\uparrow\rangle \equiv |F = 1, m_F = -1\rangle$. The hyperfine splitting between these states is $\omega_0/2\pi \simeq 1.25$ GHz.

Coherent coupling between $|\downarrow\rangle$ and $|\uparrow\rangle$ is provided via stimulated Raman transitions. The two Raman laser beams have a wavelength of $\lambda \simeq 313$ nm with a difference frequency near ω_0. Their wave-vectors are perpendicular with their difference wave-vector lying along the line of ions with magnitude $\delta k = 2\pi\sqrt{2}/\lambda$. They are detuned ~ 80 GHz blue of the $2P_{1/2}$ excited state, with intensities giving $g/2\pi \simeq 500$ kHz.

178

The Raman beam frequencies can also be tuned to coherently flip the spins while simultaneously affecting the collective motional state of the ions. For modes considered here (having equal amplitudes of motion for all the ions), the spin-motional coupling is determined by the Lamb-Dicke parameter $\eta = \delta k(\hbar/2Nm_1\nu)^{1/2}$ of the mode with frequency ν, where m_1 is the mass of a single particle in the collection.

Fig. 2 displays a stimulated-Raman absorption spectrum of four trapped ions in a linear array, with the four axial modes as well as higher-order features clearly visible. For both the two- and four-ion experiments, the desired stretch-mode frequency was set to $\nu/2\pi \simeq 8.8$ MHz, giving a Lamb-Dicke parameter of $\eta_{STR} = 0.23/N^{1/2}$. For the Mølmer-Sørensen operation, the two driving frequencies required to generate a coupling near the first blue- and first red-sidebands are generated by frequency modulating one of the Raman beams using an electro-optic modulator. The spectral positions of the relevant difference frequencies of the Raman beam pairs is indicated by the two arrows and dashed vertical lines in Fig. 2.

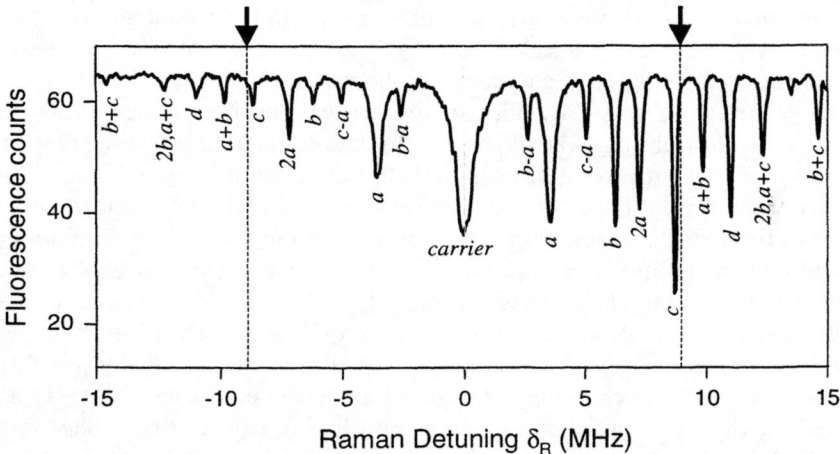

FIGURE 2. Raman absorption spectrum of four ions confined in a linear crystal and Doppler laser-cooled. The ordinate is the detuning δ_R of the Raman beams' difference frequency from the carrier and the abscissa shows the average counts of ion fluorescence per experiment ($200\mu s$ integration time), proportional to the number of ions in the state $|\downarrow\rangle$ (the ions are initially prepared in state $|\downarrow\downarrow\downarrow\downarrow\rangle$). The carrier appears at $\delta_R = 0$, and the first sidebands of the four axial normal modes of motion (labelled by letters a-d) appear at $\delta_R = \pm 3.62, \pm 6.23, \pm 8.67$ MHz, and ± 11.02 MHz in agreement with the theoretical frequency ratios $1 : \sqrt{3} : 2.410 : 3.051$. Several higher order sidebands also appear at sums and differences of harmonics of the normal mode frequencies, as indicated. The sideband asymmetry (upper sidebands are always stronger) indicates cooling to the quantum regime with not-too-many thermal phonons. The two arrows and the dashed lines, just outside the first upper and lower stretch sidebands, indicate the frequencies used for the four ion Mølmer-Sørensen scheme.

After an interaction with the stimulated Raman beams, the ion internal states

TABLE 1. Characterization of two-ion and four-ion states. P_j denotes the probability that j ions were measured to be in $|\downarrow\rangle$, and $|\rho_{\uparrow...\uparrow,\downarrow...\downarrow}|$ denotes the coherence between $|\uparrow...\uparrow\rangle$ and $|\downarrow...\downarrow\rangle$. Uncertainties in the $N = 2$ measurements are ±0.01; uncertainties in the $N = 4$ populations are ±0.02.

N	P_0	P_1	P_2	P_3	P_4	$|\rho_{\uparrow...\uparrow,\downarrow...\downarrow}|$
2	0.43	0.11	0.46	-	-	0.385
4	0.35	0.10	0.10	0.10	0.35	0.215

are measured by illuminating them with a circularly-polarized laser beam tuned to the $2S_{1/2}(F = 2, m_F = -2) \leftrightarrow 2P_{3/2}(F = 3, m_F = -3)$ cycling transition. Each ion in $|\downarrow\rangle$ fluoresces brightly, leading to the detection of ∼15 photons/ion on a photomultiplier tube during a 200 μs detection period. In contrast, an ion in $|\uparrow\rangle$ remains nearly dark. For a single ion, we are able to discriminate between $|\uparrow\rangle$ and $|\downarrow\rangle$ with approximately 99% accuracy, as shown in the histograms of Figs. 3a and 3b. This accuracy is limited by off-resonant optical pumping which causes the dark state $|\downarrow\rangle$ to eventually partake in the cycling transition and fluoresce [24]. This 1% error rate could be improved considerably by appropriately weighting the photon counts by their arrival time, as this optical pumping will contaminate later counts more so than earlier counts. Fig. 3c shows a histogram of four ions prepared in an initial state with incoherent populations in all five possible states of excitation without distinguishing the individual ions. Here, the number of ions in state $|\downarrow\rangle$ can be determined with an accuracy of about 80% on any a given experiment, although this number could be improved to better than 95% by weighting the counts as discussed above. These statistical detection errors can be averaged away by repeating the experiment many times and fitting the resulting photon-number distribution to a sum of Poissonians to determine the probability distribution P_j of having exactly j ions in the state $|\downarrow\rangle$ [24].

N-particle entanglement results

Following the Mølmer-Sørensen entangling procedure, the probability distribution P_j is measured. The results are given in Table 1, and show that in both cases, the probabilities for all N ions to be in the same state are large compared to the probabilities for the other cases. This is characteristic of the state $|\Psi_N\rangle$ [Eq. (3)], although the fact that the middle probabilities are nonzero indicates that we do not generate the entangled states with perfect accuracy.

In order to prove that we are generating a reasonable approximation to $|\Psi_N\rangle$, it is necessary to prove that the populations of $|\uparrow\uparrow\uparrow\uparrow\rangle$ and $|\downarrow\downarrow\downarrow\downarrow\rangle$ are coherent. In terms of the N-spin density matrix ρ_N, we must measure the far off-diagonal

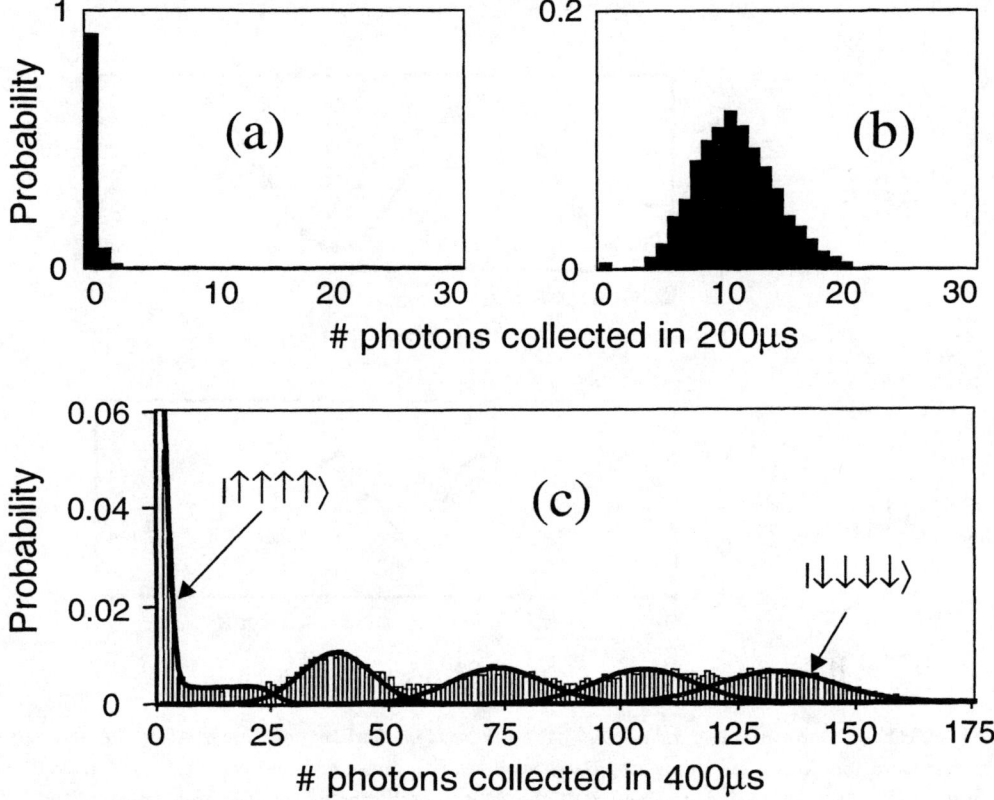

FIGURE 3. Measured probability distribution of detected fluorescence counts of a single trapped ion in (a) state $| \uparrow \rangle$ and (b) state $| \downarrow \rangle$ after 200 μs of integration (1000 measurements). (c) Measured probability distribution of detected fluorescence counts of four trapped ions after 400 μs of integration (1000 measurements). The lines are least-squares fits to reference distributions for having anywhere from 0 ions (leftmost curve) to 4 ions (rightmost curve) in state $| \downarrow \rangle$, providing relative probabilities P_j of j ions in state $| \downarrow \rangle$.

element $\rho_{\uparrow \ldots \uparrow, \downarrow \ldots \downarrow}$. This can be achieved by viewing the first entanglement pulse as the first pulse in a Ramsey experiment [36], and applying a second (non-entangling) $\pi/2$ pulse to the ions before observing them, closing the Ramsey interferometer. The relevant observable after this modified Ramsey experiment is the parity of the number of ions in state $| \downarrow \rangle$ [37]

$$\Pi(\phi) \equiv \sum_{j=0}^{N} (-1)^j P_j(\phi). \tag{7}$$

As the parity is measured while ϕ is varied, the resulting Ramsey fringes oscillate as

181

$\cos N\phi$ for N ions, as seen in Fig. 4. The amplitude of the fringes is just twice the

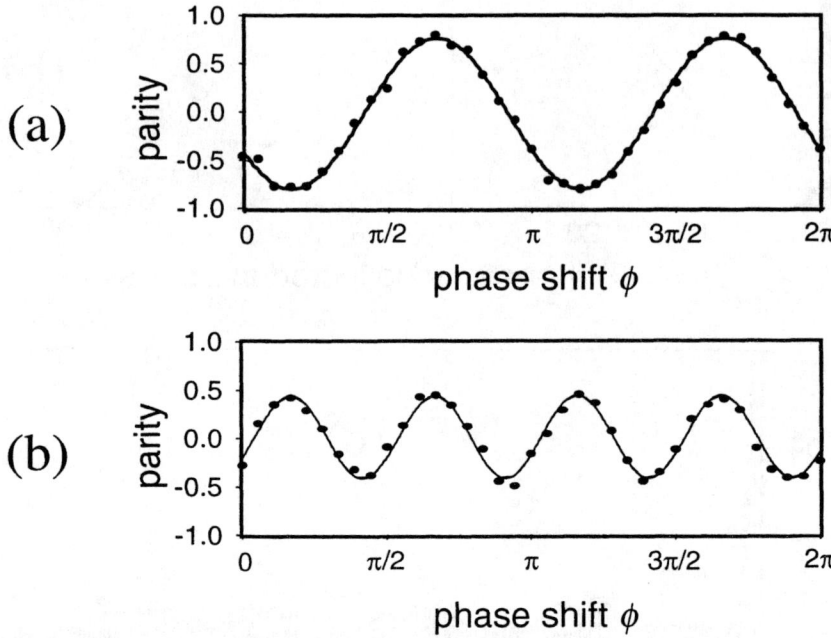

FIGURE 4. Determination of $\rho_{\uparrow...\uparrow,\downarrow...\downarrow}$ for (a) two ions and (b) four ions. After the entanglement operation, a non-entanling $\pi/2$ pulse with relative phase ϕ drives the $|\downarrow\rangle \leftrightarrow |\uparrow\rangle$ transition in each ion. As ϕ is varied, the parity of the N ions oscillates as $\cos N\phi$, and the amplitude of the oscillation is twice the magnitude of the density-matrix element $\rho_{\uparrow...\uparrow,\downarrow...\downarrow}$. Each data point represents an average of 1000 experiments, corresponding to a total integration time of roughly 10 s for each graph.

desired coherence $2|\rho_{\uparrow...\uparrow,\downarrow...\downarrow}|$. This compression of the Ramsey fringes by a factor of N is the basis for extracting Heisenberg-limited signal-to-noise in spectroscopy of entangled states, where the frequency uncertainty $\Delta\omega$ is limited by the N-particle Heisenberg uncertainty relation $\Delta\omega\Delta t \geq 1/N$ for observation time Δt [37]. Fig. 5 shows an analog of this effect in a Mach-Zender interferomer.

The measurements of $|\rho_{\uparrow...\uparrow,\downarrow...\downarrow}|$ are listed in the last column of Table 1 for both 2- and 4-ion cases. The fidelity of our state generation, or the overlap between the idealized state $|\Psi_N\rangle$ in Eq. (3) and the observed density matrix, is

$$\mathcal{F}_N \equiv \langle\Psi_N|\rho_N|\Psi_N\rangle = \frac{P_0 + P_N}{2} + |\rho_{\uparrow...\uparrow,\downarrow...\downarrow}|. \tag{8}$$

For $N = 2$ we achieve $\mathcal{F}_2 = 0.83\pm0.01$, while for $N = 4$, $\mathcal{F}_4 = 0.57\pm0.02$. In both cases the fidelity is above 0.5, indicating N-particle entanglement [27].

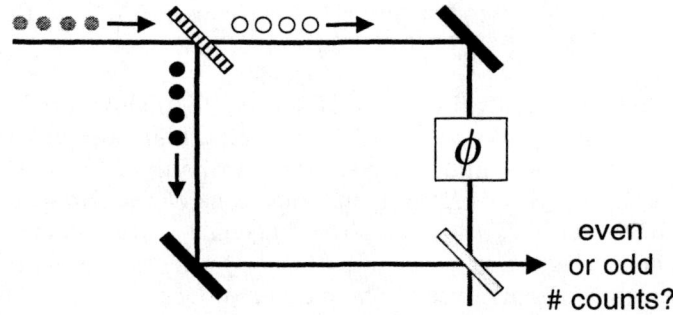

FIGURE 5. Mach-Zender interferometer analog for four-particle entanglement observation. Four photons propagate through a "super-beamsplitter," which sends the photons in a superposition of all going through and all reflecting. One arm contains a phase shifter, and the two paths are recombined on a normal beamsplitter. The parity of the number of photons received in one of the output ports is measured as the interferometer phase is scanned. Because the photons all take the same path, the measured phase shift is amplified by a factor of four (fringe period $= \pi/2$), providing enhanced interferometric sensitivity.

Quantifying the *amount* of entanglement is a more difficult question. A variety of measures of entanglement have been proposed, but most are difficult to calculate even numerically [38,39]. For $N = 2$, Wootters has given an explicit formula for the *entanglement of formation*, $\mathcal{E}(\rho_2)$ [40]. Although we have not reconstructed the entire two-particle density matrix, the populations measured place sufficient bounds on the unmeasured elements to determine that $\mathcal{E}(\rho_2) \approx 0.5$. This indicates that roughly two pairs of our ions would be required to carry the same quantum information as a single perfectly entangled pair.

In the four-ion case, no explicit formula for entanglement is known. The data does indicate that our density matrix can be expressed as

$$\rho_4 \simeq 0.43|\Psi_4\rangle\langle\Psi_4| + 0.57\rho_4^{incoh}, \tag{9}$$

where $|\Psi_4\rangle$ is the desired state of Eq. (3) and ρ_4^{incoh} is completely incoherent (ie., diagonal). The coefficients of Eq. (9) are determined directly from the value of $\rho_{\uparrow\uparrow\uparrow\uparrow,\downarrow\downarrow\downarrow\downarrow}$ in Table I, together with the fact that no evidence for other off-diagonal matrix elements was observed. (Other coherences involving less than four ions would have given fringes varying as $cos\phi$, $cos2\phi$, or $cos3\phi$ in the measured populations $P_j(\phi)$ and parity $\Pi(\phi)$).

A measurement of $|\rho_{\uparrow...\uparrow,\downarrow...\downarrow}| > 0$ does not by itself guarantee N-particle entanglement. For instance, consider the four-particle states

$$|\Psi_A\rangle = \left(\frac{|\downarrow\downarrow\downarrow\rangle + |\uparrow\uparrow\uparrow\rangle}{\sqrt{2}}\right) \otimes \left(\frac{|\downarrow\rangle + |\uparrow\rangle}{\sqrt{2}}\right) \tag{10}$$

and

183

$$|\Psi_B\rangle = \left(\frac{|\downarrow\downarrow\downarrow\downarrow\rangle + i|\uparrow\uparrow\uparrow\rangle}{\sqrt{2}}\right) \otimes \left(\frac{|\downarrow\rangle - i|\uparrow\rangle}{\sqrt{2}}\right). \tag{11}$$

An equally weighted statistical mixture of Ψ_A and Ψ_B exhibits only three-particle entanglement, yet has $|\rho_{\uparrow\ldots\uparrow,\downarrow\ldots\downarrow}| = 0.25$ (larger than our observed value) without any other coherences. A similar mixed state with pairs of two-particle entangled states also has $|\rho_{\uparrow\ldots\uparrow,\downarrow\ldots\downarrow}| = 0.25$ without other coherences. However, these states significantly differ from $|\Psi_4\rangle\langle\Psi_4|$ along the diagonals, so the observed populations P_j following the entanglement procedure (Table 1) can set an upper bound on how much these states can contribute to the measured density matrix. We decompose ρ_4 as a sum of the desired state $|\Psi_4\rangle\langle\Psi_4|$ and a "worst-case" factorizable density matrix ρ_4^F which includes mixed states such as above. We find that an upper bound on the amount of four-particle coherence in ρ_4^F is

$$\rho_{\uparrow\uparrow\uparrow\uparrow,\downarrow\downarrow\downarrow\downarrow}^F(max) = Min\left[P_0, P_4, \frac{P_2}{2} + Min(P_1, P_3)\right]. \tag{12}$$

From the data in Table (1), we find that $\rho_{\uparrow\uparrow\uparrow\uparrow,\downarrow\downarrow\downarrow\downarrow}^F(max) = 0.15$, leaving the remainder of the observed four-particle coherence (0.065) to be unambiguously associated with the four-particle entangled state $|\Psi_4\rangle$. This gives the worst-case decomposition

$$\rho_4 = 0.13|\psi_4\rangle\langle\psi_4| + 0.87\rho_4^F, \tag{13}$$

where ρ_4^F contains mixtures of particular two- and three-particle entangled states (such as Eqs. (10) and (11)) which are very unlikely to occur in the experiment.

OUTLOOK

The data on two-ion and four-ion entanglement are consistent with an entanglement efficiency of $\epsilon \approx 0.8$. Although this represents the only demonstrated source of 4-particle entanglement and uses a scalable method, the imperfect contrast of (Fig. 4) indicates that even this efficiency will limit how many particles can be entangled in this experiment. It may be required to achieve entanglement efficiencies $\epsilon > 0.9999$ in order to implement fault-tolerant error correction schemes which may allow entanglement of arbitrarily large numbers of particles [41].

Several technical noise sources degrade the observed efficiency, including laser intensity and beam-pointing noise, nonuniform illumination of the ions during the Mølmer-Sørensen operation, and magnetic field noise. The chief limitation in the current experiment appears to be stochastic heating of the ions to outside the Lamb-Dicke regime. The center-of-mass (CM) motion of the ions is observed to heat at a rate of $\langle \dot{n}_{CM} \rangle \approx 0.02N \ \mu s^{-1}$ [33], so after a 10 μs four-ion entangling operation, $\langle n_{CM} \rangle$ approaches ≈ 10 thermal quanta. This invalidates the Lamb-Dicke criterion $\eta_{CM}^2 \langle n_{CM} \rangle \ll 1$, and severely limits the fidelity of the operation. Mølmer and Sørensen have shown [29] that the expected fidelity of the entangled state $|\Psi_N\rangle$ of N ions is

$$\mathcal{F} \approx 1 - N(N-1)\eta_{CM}^4 \langle n_{CM} \rangle^2, \tag{14}$$

to lowest order in the center-of-mass Lamb-Dicke parameter η_{CM} with $N \gg 1$ and $\langle n_{CM} \rangle \gg 1$. The factor $N(N-1)/2$ comes from the number of pairs of N ions which are simultaneously flipped during the Mølmer-Sørensen entangling operation. We find for the $N = 4$ experiment, the above expression is consistent with the observations.

The source of ion heating has not been pinpointed, but it appears to be related to fluctuating microscopic potentials on the electrodes. The observed heating is not a fundamental limitation, as it has been observed to be orders of magnitude smaller under some conditions [33]. Moreover, by trapping multiple ion species and continuously laser-cooling one, the other qubit ions can be sympathetically cooled to remain in the Lamb-Dicke regime while not disturbing the qubit coherence [42].

Producing entangled states of very large numbers of ions (tens or hundreds) for relevance to large-scale quantum computing will require a different approach. This is because a trap confining more than several ions will have low oscillation frequencies, and mode cross-coupling from the complicated mode structure will be unavoidable. A promising path to large numbers is to use a multiplexed ion trap structure of many separated ion traps [43]. Here, entangling operations are done only in traps holding a few $(2 - 5)$ ions, and the ions are be shuttled between traps to extend the entanglement to larger numbers. Because the quantum bits are stored in magnetic dipole (hyperfine) internal states and the ions are moved around with electric fields acting on their charge, the coherence of the qubits should not be disturbed. Peeling away an ion from or introducing an ion to other ions in a trap will obviously introduce a significant amount of motional energy, but this energy can be removed again by trapping multiple species and relying on sympatheic cooling to return the motion to well inside the Lamb-Dicke regime for subsequent entangling operations.

REFERENCES

1. D. Bohm, *Quantum theory* (Prentice Hall, Engelwood Cliffs, NJ, 1951).
2. A. Einstein, B. Podolsky, and N. Rosen, Phys. Rev. **47**, 777 (1935).
3. J. S. Bell, Physics **1**, 195 (1964).
4. C. H. Bennett and S. J. Weisner, Phys. Rev. Lett. **69**, 2881 (1996).
5. K. Mattle, H. Weinfurter, P. G. Kwiat, and A. Zeilinger, Phys. Rev. Lett. **76**, 4656 (1996).
6. A. K. Ekert, Phys. Rev. Lett. **67**, 661 (1991).
7. C. H. Bennett *et al.*, Phys. Rev. Lett. **70**, 1895 (1993).
8. D. Bouwmeester *et al.*, Nature **390**, 575 (1997).
9. D. Boschi *et al.*, Phys. Rev. Lett. **80**, 1121 (1998).
10. A. Furusawa *et al.*, Science **282**, 706 (1998).
11. P. W. Shor, *Proceedings of the 35th Annual Symposium on the Foundations of Computer Science* (IEEE Computer Society Press, New York, 1994), p. 124.

12. L. K. Grover, Phys. Rev. Lett. **79**, 325 (1997).
13. A. M. Steane and D. M. Lucas, quant-ph/0004053 (2000).
14. S. J. Freedman and J. F. Clauser, Phys. Rev. Lett. **28**, 938 (1972).
15. A. Aspect, P. Grangier, and G. Roger, Phys. Rev. Lett. **49**, 91 (1982).
16. Z. Y. Ou and L. Mandel, Phys. Rev. Lett. **61**, 50 (1988).
17. Y. H. Shih and C. O. Alley, Phys. Rev. Lett. **61**, 2921 (1988).
18. D. Bouwmeester *et al.*, Phys. Rev. Lett. **82**, 1345 (1999).
19. E. Hagley *et al.*, Phys. Rev. Lett. **79**, 1 (1997).
20. A. Rauschenbeutel *et al.*, Science **288**, 2024 (2000).
21. Z. Y. Ou, S. F. Pereira, H. J. Kimble, and K. C. Peng, Phys. Rev. Lett. **68**, 3663 (1992).
22. J. I. Cirac and P. Zoller, Phys. Rev. Lett. **74**, 4091 (1995).
23. C. Monroe *et al.*, Phys. Rev. Lett. **75**, 4714 (1995).
24. Q. A. Turchette *et al.*, Phys. Rev. Lett. **81**, 1525 (1998).
25. K. Mølmer and A. Sørensen, Phys. Rev. Lett. **82**, 1835 (1999).
26. K. Mølmer and A. Sørensen, Phys. Rev. Lett. **82**, 1971 (1999).
27. C. A. Sackett *et al.*, Nature **404**, 256 (2000).
28. E. Solano, R. L. de Matos Filho, and N. Zagury, Phys. Rev. A **59**, 2539 (1999).
29. A. Sørensen and K. Mølmer, quant-ph/0002024 (2000).
30. G. J. Milburn, quant-ph/9908037 (1999).
31. D. F. Walls and G. J. Milburn, *Quantum Optics* (Springer Verlag, Berlin, 1994).
32. C. Monroe, D. M. Meekhof, B. E. King, and D. J. Wineland, Science **272**, 1131 (1996).
33. Q. A. Turchette *et al.*, Phys. Rev. A **61**, 063418 (2000).
34. B. E. King *et al.*, Phys. Rev. Lett. **81**, 1525 (1998).
35. D. James, Appl. Phys. B **66**, 181 (1998).
36. N. F. Ramsey, *Molecular Beams* (Oxford University Press, London, 1963).
37. J. J. Bollinger, W. M. Itano, D. J. Wineland, and D. J. Heinzen, Phys. Rev. A **54**, R4649 (1996).
38. V. Vedral, M. Plenio, M. Rippin, and P. Knight, Phys. Rev. Lett. **78**, 2275 (1997).
39. M. Lewenstein and A. Sanpera, Phys. Rev. Lett. **80**, 2261 (1998).
40. W. K. Wootters, Phys. Rev. Lett. **80**, 2245 (1998).
41. W. H. Zurek, Physics Today **52**, 24 (1999).
42. D. Kielpinski *et al.*, Phys. Rev. A **61**, 032310 (2000).
43. D. J. Wineland *et al.*, J. Res. Nat. Inst. Stand. Tech. **103**, 259 (1998).

DARK RESONANCE

Collinear Light Scattering Using Electromagnetically Induced Transparency

S. E. Harris, A. V. Sokolov, D. R. Walker,
D. D. Yavuz, and G. Y. Yin

Edward L. Ginzton Laboratory,
Stanford University, Stanford, California 94305

Abstract. The paper describes two types of nonlinear optical processes which are based on electromagnetically induced transparency. These are: (1) Collinear generation of FM-like Raman sidebands and (2) a type of pondermotive light scattering which is inherent to the interaction of slow light with cold atoms. Connections to other areas of EIT-based nonlinear optics are also described.

INTRODUCTION

The essential features of electromagnetically induced transparency (EIT) may be understood with reference to atomic hydrogen. Figure 1(a) shows an energy schematic of atomic hydrogen and the absorption profile which is obtained as a probe frequency ω_p is tuned through the resonance transition. Figure 1(b) shows the energy level diagram and spectrum in the presence of a dc field. The effect of the dc field is to mix and split the bare $|2s\rangle$ and $|2p\rangle$ states. Although one might expect that the absorption spectrum would consist of a pair of separated Lorentzian lines, that is not the case; instead, one finds a quantum interference and a perfect zero in absorption.

In the more general situation, an ac field with Rabi frequency Ω_c is used to establish EIT. Figure 2 shows an energy level diagram for a prototype generation process [1]. In this figure the detuning from resonance is normalized to one-half the Einstein A transition width denoted by γ_{13}. The absorption has the perfect zero described above. The refractive index is equal to unity on line center and has a steep dispersive slope. This slope is the cause of the slow group velocity to be discussed subsequently. Two additional frequencies are applied so as to generate a sum frequency on the resonance line. One notes from the bottom part of this figure that the quantum interference which was destructive and led to the zero in absorption is, for the case of generation, constructive and produces a maximum

CP551, *Atomic Physics 17*, edited by E. Arimondo, P. DeNatale, and M. Inguscio
© 2001 American Institute of Physics 1-56396-982-3/01/$18.00

in the nonlinear susceptibility at line center. A first experiment demonstrating nonlinear optics with EIT was done by Hakuta, Marmet, and Stoicheff [2]. Reviews and discussions of population trapping, EIT, and EIT-based nonlinear optics are given in Ref. [3].

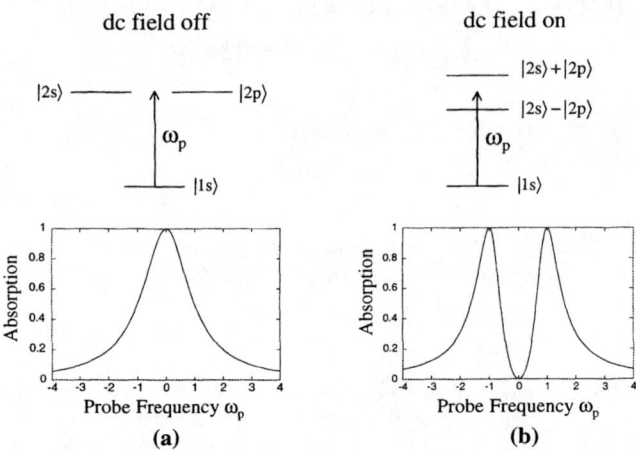

FIGURE 1. Absorption spectrum of atomic hydrogen. (a) Without a dc field the absorption profile is a Lorentzian line centered on the resonance transition. (b) With a dc field present there is a perfect zero in absorption at the position of the bare $|2s\rangle$ state.

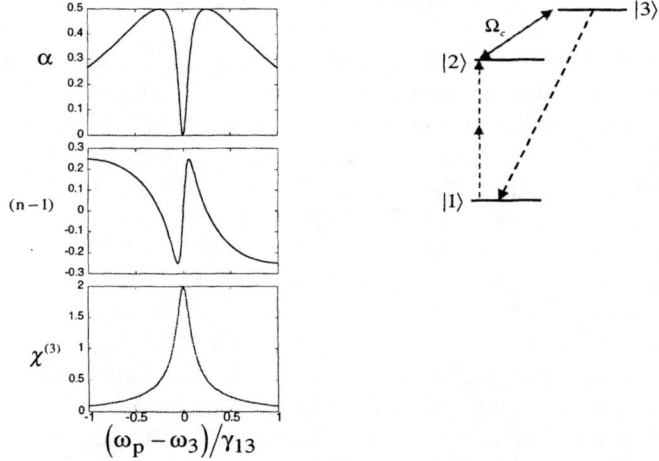

FIGURE 2. Nonlinear optics with EIT. Prototype system for nonlinear optics with EIT. A coupling laser with Rabi frequency Ω_c is used to establish transparency. Additional frequencies drive the sum frequency process shown by the dashed lines. At line center, the absorption is zero, the refractive index is unity, and the nonlinear susceptibility $\chi^{(3)}$ is maximum at line center. In this figure the Rabi frequency is equal to $\Omega_c = \gamma_{13}/2$.

We next ask what happens when the probe frequency is detuned from the resonance transition [4]. This topic was studied experimentally by Jain and colleagues [5] and the principal results of their work are shown in Fig. 3. In Part (a) a weak probe beam is detuned by twenty wavenumbers from resonance. The beam passes undistorted through an aperture. In Part (b) the beam intensity is increased by about five-thousand times. The incident beam becomes filamented and distorted. In Part (c) a coupling laser is applied and a high-quality beam is again obtained.

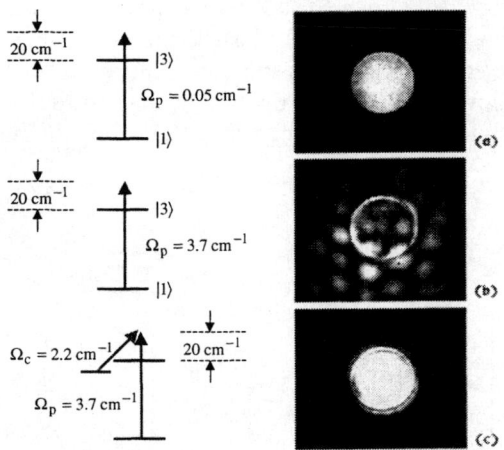

FIGURE 3. Elimination of optical self-focusing. (a) Shows the image of a weak probe beam after passing through an aperture. (b) The beam intensity is increased and filamentation is observed. (c) A coupling laser is applied and the self-focusing is eliminated. [Reprinted from Jain *et al.* [5].]

The commonality of the near-resonance experiment described in the previous paragraph with on-resonance EIT is the existence of an antiphased atomic coherence. This coherence acts in opposition to the primary dipole moment and, instead of eliminating absorption, produces a refractive index of unity.

One is therefore motivated to ask the question: Can this antiphased coherence be extended to molecular systems, such as H_2 or D_2, with detunings which approach one-hundred-thousand wavenumbers? To make this extension we require (in principle) an infinite set of Raman frequencies rather than the two frequencies of EIT. Even then, there are similarities and differences. It is not possible to set the refractive index at all frequencies equal to unity. Instead, there is a prescribed set of Raman frequencies which propagate without change in amplitude or relative phase. We also find that collinear Raman generation is significantly improved by operating off of resonance and using either a phased or antiphased molecular state.

The next section of this paper will discuss both the theory and experiment of this collinear Raman process. We then describe a new type of slow light pondermotive light scattering process. The final section of the paper will make connections to other work on EIT.

RAMAN GENERATION BY PHASED AND ANTIPHASED MOLECULAR STATES

We consider the experimental situation denoted in Fig. 4. We apply two frequencies, a pump frequency E_0 and a Stokes frequency E_{-1}. The molecular medium is deuterium which has a fundamental vibrational frequency of 2994 cm^1. We use deuterium due to the coincidence with the Ti:sapphire and Nd:YAG driving lasers. We will see in the following that these driving lasers produce a comb of collinearly-generated Raman sidebands which range from 2.94 μm in the infrared to 195 nm in the ultraviolet.

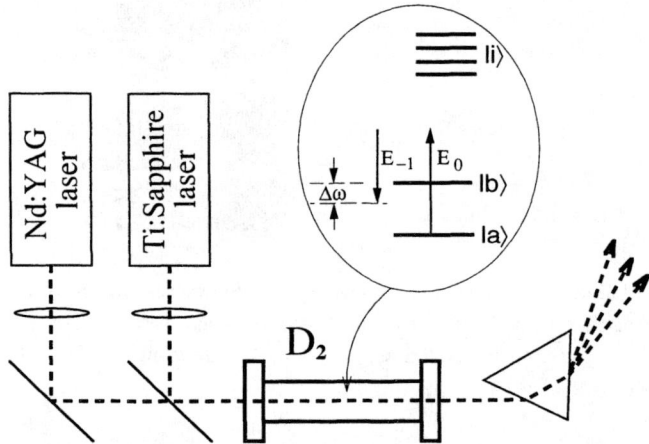

FIGURE 4. Experimental setup and energy level diagram for coherent molecular excitation and collinear Raman generation.

There is a detuning $\Delta\omega$ from the molecular resonance (Fig. 4). This detuning is important because it establishes the sign of the molecular coherence. When on-resonance and even when off-resonance and within the rotating wave approximation, the coherence of the non-allowed transition is established either by optical pumping or by first turning on the coupling laser to thus establish the quantum interference before the probe laser is applied. Here, the Stokes field E_{-1} acts as the coupling laser, but turning it on first will not select an eigenstate. The reason for this is that states $|a\rangle$ and $|b\rangle$ of Fig. 4, because they are far from the electronic states, have nearly identical Stark shifts. Instead, we choose an eigenstate by tuning

$\Delta\omega$ so that it is either above or below resonance. When below resonance, as shown, the molecular motion is in phase with the two-photon driving force. When above resonance, the molecular motion is in antiphase with the drive. The antiphased state which is established by this method is the same as the population-trapped state of on-resonance EIT.

Before proceeding further, we note pertinent prior work in this area. Several groups have studied the enhancement of stimulated Raman scattering by application of the first Stokes component collinearly with the pump laser beam [6–8]. Hakuta et al. have demonstrated collinear Raman sideband generation in solid molecular hydrogen (H_2) [9]. Nazarkin et al. have shown efficient broadband generation by impulsively exciting a coherent vibration of SF_6 [10]. Ruhman et al. have used impulsively excited Raman scattering to observe coherent molecular vibrations in the time domain [11]. Some of the work of our group is cited in Ref. [12].

Theory

We begin by following Yavuz et al. [13]. Although we apply only two frequencies, we must include all of the generated sidebands as well as the applied sidebands in the theory. The angular frequencies of the Raman sidebands are $\omega_q = \omega_0 + q(\omega_b - \omega_a - \Delta\omega) = \omega_0 + q\omega_m$. We will assume the ideal case of zero linewidth for the Raman transition. We allow for an arbitrary number of electronic states $|i\rangle$ with energies $\hbar\omega_i$ and allow for all matrix elements from states $|a\rangle$ and $|b\rangle$ (Fig. 4) to these states. By assuming that the derivatives of the probability amplitudes of the upper states $|i\rangle$ are small as compared to the detunings from these states, the problem may be written in terms of an effective Hamiltonian

$$H_{\text{eff}} = -\frac{\hbar}{2}\begin{bmatrix} A & B \\ B^* & D - 2\Delta\omega \end{bmatrix}. \tag{1}$$

The quantities $A/2$ and $D/2$ are the Stark shifts of states $|a\rangle$ and $|b\rangle$, respectively, and B is the effective Rabi frequency. With the dispersion and coupling constants defined as

$$a_q = \frac{1}{2\hbar^2}\sum_i\left[\frac{|\mu_{ai}|^2}{(\omega_i - \omega_a) - \omega_q} + \frac{|\mu_{ai}|^2}{(\omega_i - \omega_a) + \omega_q}\right],$$

$$b_q = \frac{1}{2\hbar^2}\sum_i\left[\frac{\mu_{ai}\mu_{bi}^*}{(\omega_i - \omega_a) - \omega_q} + \frac{\mu_{ai}\mu_{bi}^*}{(\omega_i - \omega_b) + \omega_q}\right],$$

$$d_q = \frac{1}{2\hbar^2}\sum_i\left[\frac{|\mu_{bi}|^2}{(\omega_i - \omega_b) - \omega_q} + \frac{|\mu_{bi}|^2}{(\omega_i - \omega_b) + \omega_q}\right], \tag{2}$$

the matrix elements of the Hamiltonian are $A = \sum_q a_q|E_q|^2$, $B = \sum_q b_q E_q E_{q-1}^*$, and $D = \sum_q d_q|E_q|^2$.

Working in local time, $\tau = t - \frac{z}{c}$, the one-dimensional propagation equation for the qth sideband is

$$\frac{\partial E_q}{\partial z} = -j\eta\hbar\omega_q N \left(a_q \rho_{aa} E_q + d_q \rho_{bb} E_q + b_q^* \rho_{ab} E_{q-1} + b_{q+1} \rho_{ab}^* E_{q+1} \right) \quad , \tag{3}$$

where N is the number of molecules per volume and $\eta = (\mu/\epsilon_0)^{1/2}$.

Using the effective Hamiltonian of Eq. (1), the equations for the density matrix elements, also in local time, are

$$\frac{\partial \rho_{ab}}{\partial \tau} - j\left(\frac{A}{2} - \frac{D}{2} + \delta\omega\right)\rho_{ab} = j\frac{B}{2}(\rho_{bb} - \rho_{aa}) \quad ,$$

$$\frac{\partial \rho_{bb}}{\partial \tau} = -\mathrm{Im}(\mathrm{B}^*\rho_{ab}) \quad . \tag{4}$$

We assume that at $\tau = 0$ all the molecules are in a single, nondegenerate ground state; therefore, $\rho_{aa}(0, z) = 1$ and $\rho_{ab}(0, z) = \rho_{bb}(0, z) = 0$.

If the elements of the effective Hamiltonian vary slowly as compared to the separation of the eigenvalues and with $B = |B| \exp j\varphi$, the solution of the density matrix which evolves smoothly from the ground state is

$$\rho_{ab} = \left(\frac{1}{2}\sin\theta\right)\exp j\varphi \quad ; \quad \rho_{bb} = \sin^2\left(\frac{\theta}{2}\right) \quad ,$$

$$\tan\theta = \frac{2|B|}{(2\Delta\omega - D + A)} \quad . \tag{5}$$

In molecular systems with large one-photon detunings $A \cong D$ and the sign of the detuning $\Delta\omega$ from Raman resonance determines the sign of the coherence ρ_{ab}.

From the previous equations we may obtain the conservation conditions for photons and power. These are

$$\frac{\partial}{\partial z}\left(\sum_q \frac{1}{\hbar\omega_q}\frac{|E_q|^2}{2\eta}\right) = 0 \quad ,$$

$$\frac{\partial}{\partial z}\left(\sum_q \frac{|E_q|^2}{2\eta}\right) = -\frac{N\hbar}{2}(\omega_b - \omega_a)\left[\frac{\partial}{\partial t}(\rho_{bb} - \rho_{aa})\right] \quad . \tag{6}$$

Even when the atomic population changes dynamically, the total number of photons remains constant with distance. Because optical power is also flowing in and out of the atomic population, it is not independent of distance.

In practice, the molecular linewidth and also the driving laser linewidths are nonzero. When this is the case, we select $\Delta\omega$ to be large as compared to these linewidths and then increase the intensity of the driving lasers so as to create as large a coherence as is possible. When the quantity θ approaches $90°$, the coherence approaches 0.5. We term this condition as maximum coherence. Instead, at fixed field, to select the antiphased state one might allow $\Delta\omega$ to chirp from an initial negative value toward zero.

In general, the previous equations must be solved numerically, but, in one limit, there is an exact solution which gives considerable insight [12]. If we make the approximation that we are tuned sufficiently far from the electronic molecular resonances that the dispersive constants a_q, b_q, c_q, and d_q are the same and we also assume that the modulation bandwidth is sufficiently small that all of the ω_q may be taken as equal, then the exact solution of both Schrödinger's and Maxwell's equations is a sum of two frequency-modulated signals. These signals center at each of the two input frequencies ω_0 and ω_{-1} and have a peak frequency deviation which is proportional to the drive strength. These eigenvectors do not change as the depth of the modulation is changed. The reason for this is that the Raman transition is driven by the envelope and a perfect FM signal has a time-invariant envelope. Either FM signal alone will not drive the transition, but the beat between these two signals does, and this beat is invariant with distance.

Experiment

In the following we show experimental results obtained by exciting either the phased or antiphased eigenstates, respectively. Since this excitation is done by selecting the detuning, we describe how we determine line center of the molecular transition. We do this three ways: First, we measure the total energy which is transmitted through the D_2 cell. Figure 5(a) shows the results of such an experiment. A second way in which we do this is to put an acoustic detector in the cell and measure heat-produced sound as energy is put into the molecular system [Fig. 5(b)]. The third way (not shown) is to measure the small signal gain as a function of a tunable probe.

Figure 6 shows recent results [14]. When we tune the driving lasers to within 1 GHz from the Raman resonance we see a bright beam of white light at the output of the D_2 cell. We disperse the spectrum with a prism and project onto a white scintilating screen about 3 m from the cell. We cool the 50-cm length of the cell to 77°K and work at pressures between about 50 and 100 torr. Up to thirteen anti-Stokes sidebands and two Stokes sidebands in addition to the driving frequencies are observed. These sidebands are spaced by 2994 cm^{-1} and range from 2.94 μm to 195 nm in wavelength. Figure 6 shows pictures of the spectrum taken with a digital color camera with a single shot exposure at a fixed aperture size. Starting from the left, the first two sidebands are the driving frequencies and the next four are anti-Stokes sidebands; beginning at the fifth anti-Stokes, the sidebands are in the ultraviolet and only fluorescence is visible. Parts (a), (b), and (c) of this figure are taken at 71 torr with the detunings noted in the figure. Part (d) is taken at a pressure of 350 torr. Here, the generation is no longer collinear and the anti-Stokes sidebands emerge in circles of increasing diameter.

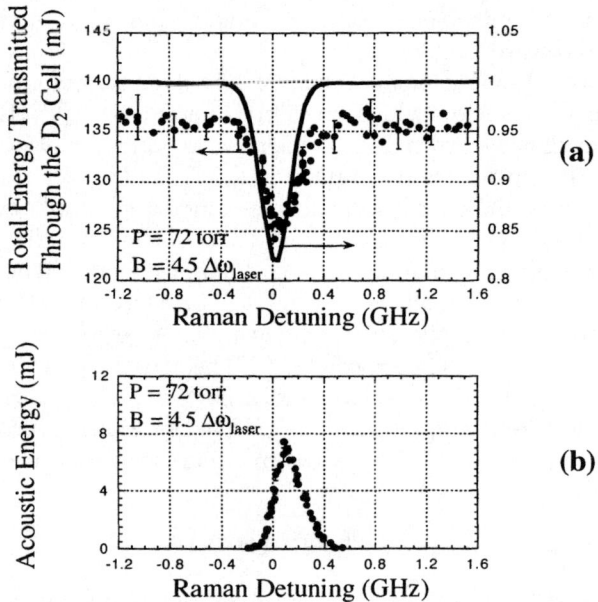

FIGURE 5. Determination of the transition line center. (a) We measure the total energy transmitted through the deuterium cell as a function of the Raman detuning. We choose zero on the x-axis as the extrapolated value of line center at zero pressure. (b) Acoustic energy as a function of the Raman detuning.

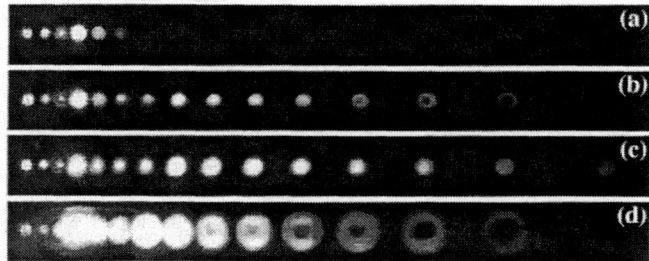

FIGURE 6. Generated spectrum. (a) $P = 71$ torr and $\Delta\omega = -400$ MHz, (b) $P = 71$ torr and $\Delta\omega = 100$ MHz, (c) $P = 71$ torr and $\Delta\omega = 700$ MHz, and (d) $P = 350$ torr and $\Delta\omega = 700$ MHz. We observe the two driving infrared fields (on the left) and multiple-generated visible and ultraviolet anti-Stokes sidebands. To reduce camera saturation for (a), (b), and (c), the first four anti-Stokes beams are attenuated by a factor of 100; the driving field at 807 nm and also the fifth anti-Stokes beam are attenuated by 10. In (d) the 807 nm beam and the first four anti-Stokes beams are attenuated by 10. [Reprinted from Sokolov *et al.* [14].]

Figure 7 shows the pulse energy as a function of order number for three different values of the Raman detuning. We observe that on-resonance generation is less efficient than off-resonance generation for all of the anti-Stokes sidebands and that generation below resonance is more efficient than above resonance. Note that we generate hundreds of μJ per pulse (at a 10-Hz repetition rate) at sidebands far into the ultraviolet.

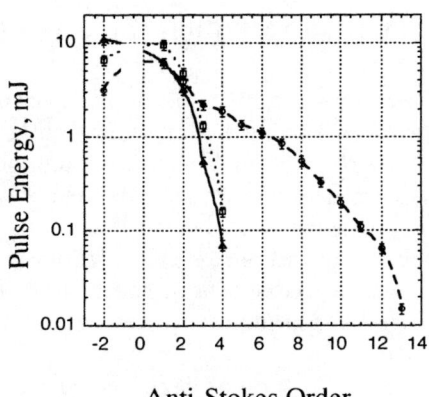

Anti-Stokes Order

FIGURE 7. Pulse energies generated at $P = 72$ torr. The triangles show on-resonance ($\Delta\omega = 0$), the circles show generation by phased ($\Delta\omega = 500$ MHz), and the squares by antiphased ($\Delta\omega = -200$ MHz) states of D_2. [Reprinted from Sokolov *et al.* [14].]

We mention two possible applications for the broad collinear spectrum which is produced by driving a single molecular eigenstate. The first of these applications is a generalization of EIT which has been shown by Yavuz *et al.* [13]. By using spectral pulse modification techniques it is possible to independently adjust the relative amplitude and phase of each spectral component of a Raman-generated spectral comb. One is therefore motivated to ask the question: Do there exist sets of Raman sidebands which will self-consistently establish the molecular coherence and propagate without change in amplitude or relative phase? Equivalently, do there exist periodic trains of femtosecond-time-scale optical pulses which establish the coherence and propagate without changing shape? Succinctly, such a set of sidebands do exist but only for the antiphased molecular state and, in effect, form an EIT-like soliton [14]. A different type soliton has been suggested earlier by Kaplan [15]. The Kaplan soliton may be thought of as an SIT-like soliton. Here, the Raman components are AM mode locked and the pulse length is chosen so that the pulse area of the two-photon Rabi frequency is 2π.

Another application of a broad Raman comb is for studies of multiphoton ionization. The spectral bandwidth of our source is about four octaves wide. Therefore, if eight photons of a long-wavelength component are required for ionization, then only two photons are required for ionization by a short-wavelength component. The relative phases of the sidebands will determine how the ionization proceeds. A first step in this direction has been made by Sokolov *et al.* [16], who have shown the generation of amplitude and frequency-modulated light with a modulation frequency equal to the fundamental vibrational frequency of molecular deuterium.

PONDERMOTIVE SCATTERING WITH SLOW LIGHT

By combining electromagnetically induced transparency (EIT) with cold-atom technology one may produce a transmission window in an otherwise optically-thick medium. Because this window is much narrower than the natural linewidth of an isolated atom, the dispersion of the medium is very steep and results in a group velocity V_g [17–19] that can be less than 10^{-7} c. As a result of this slow group velocity, an optical pulse that enters the EIT medium is spatially compressed [20], as compared to its length in free space, in the ratio of c/V_g. A test atom in this medium will experience a longitudinally-directed gradient force that is enhanced in this same ratio.

In this section we describe a type of light scattering which is based on this enhanced gradient force [21]. We are interested in conditions which are characteristic of non-condensed, trapped, cold atoms where the atom density is too low and the mean free path for collision is too long to support a propagating acoustic wave. Although this type of light scattering is in the spirit of stimulated Brillouin or Raman-Nath scattering, it differs in that it is based on local particle motion. The power efficiency of this type of scattering varies as the inverse fourth power of the group velocity. This effect therefore becomes increasingly important at slow group velocities.

Before proceeding we note that Matsko *et al.* have recently suggested that, by setting the group velocity of light equal to the phase velocity of sound, one may enhance light-sound interaction in an optical fiber [22]. Because the interaction described here is local, this synchronism condition is not required.

In the experiment which is considered, two monochromatic laser beams separated by an acoustic frequency ω_a are applied to a sample of cold atoms. The pondermotive force which results from the propagating envelope causes each atom to oscillate in the z direction. In turn, by the equation of continuity, this oscillation causes the atom density to vary sinusoidally with distance and time. As the beams propagate, the time-varying density will impose a comb of sidebands with Bessel function amplitudes onto each of the incident laser beams. For small amplitude scattering, one could then detect the beat frequency at $2\omega_a$ at the cell output.

We assume dilute, two-state test atoms with a transition energy $\omega_b - \omega_a$ immersed in an EIT medium with a group velocity of V_g. By using test atoms (for example, a different isotope) we insure that the motion of these atoms does not interact with the light-slowing EIT process. We write the propagating electromagnetic field as

$$\Omega(t, z) = \text{Re} \left[\Omega \left(t - \frac{z}{V_g} \right) \exp j(\omega t - kz) \right] . \tag{7}$$

The carrier frequency ω is tuned to the center of the EIT transparency (Fig. 8), and is detuned by $\Delta\omega = \omega - (\omega_b - \omega_a)$ from the transition of the test atom. Because the test atoms are dilute, k is the k vector in vacuum. The Fourier components of the envelope $\Omega(t - z/V_g)$ are taken to lie sufficiently within the transparency window that, to within a good approximation, the pulse propagates without change of shape. We also assume that the spectral width of the envelope is small as compared to the detuning from the test atom transition $\Delta\omega$.

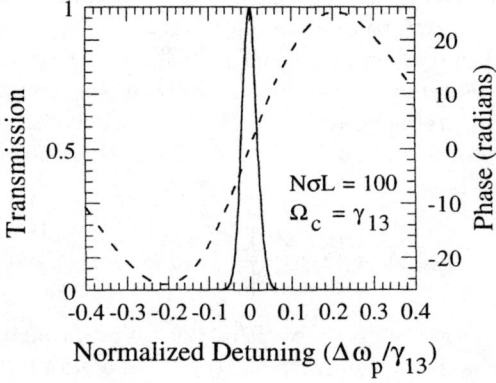

FIGURE 8. Transmission and phase vs. frequency for sub-natural linewidth EIT. The quantity $\Delta\omega_p$ is the detuning from line center of an otherwise opaque optical transition with an Einstein A coefficient of $2\gamma_{13}$. The figure assumes an optical depth of 100 and a coupling laser with a Rabi frequency equal to γ_{13}. [Reprinted from Harris and Hau [27].]

With the envelope of Eq. (7), the longitudinal gradient force is

$$F_z(t, z) = -\frac{\hbar}{4\Delta\omega} \frac{\partial}{\partial z} \left| \Omega \left(t - \frac{z}{V_g} \right) \right|^2$$

$$= \hbar k \left(\frac{c}{V_g} \right) \left(\frac{1}{4\omega\Delta\omega} \right) \frac{\partial}{\partial t} \left| \Omega \left(t - \frac{z}{V_g} \right) \right|^2 . \tag{8}$$

As the ratio (c/V_g) becomes large, many quanta may be transferred during a single pulse.

Cumulative collinear generation of a spectrum of sidebands requires that each of the generated sidebands propagate with a k vector that is equal to the k vector of its driving polarization. Because, at any z, the atom density depends only on the *electromagnetic fields at that z,* the electromagnetic nonlinearity is local in space and *k*-vector matching is insured by the linear dispersive profile of EIT. Because the second derivative of k with ω (group velocity dispersion) is zero at line center, to the extent that higher derivatives may be neglected [20], the frequencies ω_q and k vectors k_q of the generated sidebands are

$$\omega_q = \omega_0 + q\omega_a \,,$$

$$k_q = k_0 + \left(\frac{\partial k}{\partial \omega} \bigg|_{\omega_0} \right) q\omega_a \,,$$

$$= k_0 + q\frac{\omega_a}{V_g} \,, \tag{9}$$

and, irrespective of the acoustic frequency, each sideband is phasematched.

We take the driving beams to have frequencies ω_0 and $\omega_0 - \omega_a$, and term these as the pump Ω_p and Stokes Ω_s frequencies. To obtain a first-order solution we assume that the generated frequency components are weak as compared to the driving frequencies and that all frequencies lie within the transmission window and satisfy the linear dispersion condition of Eq. (9). When this is the case, the envelope $\Omega(t - z/V_g)$ of Eq. (7) is

$$\Omega(t, z) = \Omega_p + \Omega_s \exp\left[-j\omega_a \left(t - \frac{a}{V_g} \right) \right] \,. \tag{10}$$

The force on each atom is given by Eq. (8). The atom velocity follows from Newton's law. With N_0 as the number of atoms per volume, the normalized atom density $\rho(t, z)$ is defined as $N = N_0[1 + \rho(t, z)]$ and is obtained from the one-dimensional equation of continuity. These quantities are:

$$F_z(t, z) = j\frac{\hbar\omega_a}{4V_g\Delta\omega}\Omega_p\Omega_s^* \exp\left[j\omega_a \left(t - \frac{z}{V_g} \right) \right] + \text{c.c.} \,,$$

$$V_z(t, z) = \frac{\hbar}{4\Delta\omega V_g m}\Omega_p\Omega_s^* \exp\left[j\omega_a \left(t - \frac{z}{V_g} \right) \right] + \text{c.c.} \,,$$

$$\rho(t, z) = V_z(t, z)/V_g \,. \tag{11}$$

With the envelope quantity ρ defined as $\rho(t, z) = (\rho/2) \exp\left[j\omega_a \left(t - \frac{z}{V_g} \right) \right] + \text{c.c.}$, the slowly-varying envelope equation for the propagating sidebands is

$$\frac{\partial E_q}{\partial z} + \frac{1}{V_g}\frac{\partial E_q}{\partial t} = -j\frac{\omega}{c}(n - 1)\left[E_{q-1}\rho + E_{q+1}\rho^* \right] \,. \tag{12}$$

The refractive index n of the test atoms is the same for all sidebands and, to within the rotating wave approximation, is $n = 1 - (|\mu|^2 N)/(2\epsilon_0\hbar\Delta\omega)$. If ρ is sufficiently small that only the first anti-Stokes sideband has significant amplitude, the ratio of the generated anti-Stokes electric field to the pump field is

$$\left| \frac{E_{as}}{E_p} \right| = \frac{\omega_0}{c}(n-1)|\rho|L \ . \tag{13}$$

This ratio varies as $(1/V_g)^2$ and, at slow group velocities, the generated sideband should be readily observable. For example, at a detuning of 1 GHz from a test atom transition with a matrix element of 1 a.u., $V_g = 1$ m/s, an atomic mass number of 85, an atom density of 10^{12} atoms/cm^3, and a sample length of 1 cm, then at an incident power density of 1 mW/cm^2, the ratio of anti-Stokes to pump fields is $1.3 \ 10^{-5}$.

It is of interest to compare the scattering amplitude of Eq. (13) to that which is obtained with an acoustic wave which has a phase velocity equal to the optical group velocity. The atom density now satisfies the driven acoustic wave equation and the pondermotive nonlinearity is no longer local. Following Matsko and colleagues [22], with the acoustic wavelength denoted by λ_a, the generated electric field [Eq. (13)] is increased by a factor of $2\pi L/\lambda_a$. An example is provided by Bose condensation. As condensation occurs and the previously independent particles become coupled, the particle motion changes from local to wave-like. By varying the acoustic frequency within the transparency profile and thereby varying the acoustic wavelength from long to short as compared to the healing length of the condensate, it may be possible to move continuously between the local and wave-like regimes of the nonlinearity. Such experiments would be in the character of the recent work of Stamper-Kurn and colleagues [23].

As the scattering amplitude becomes large, pondermotive scattering differs in an important way from the collinear Raman scattering problem of the previous section. With two frequencies applied spaced by ω_a, the acoustic density will vary not only as ω_a, but also at harmonics of ω_a. Each spectral sideband is then coupled to all other sidebands and not only to its nearest neighbor. In the special case when there is no dispersion and all sidebands satisfy Eq. (9), we obtain a solution consisting of Bessel functions imposed on each of the two input frequencies, i.e., the same solution as Sokolov and Harris [12].

CONNECTIONS

The work of this paper should be considered in the context of a considerable on-going effort to use the unique absorptive and dispersive properties of EIT to enhance nonlinear optical processes of many types. For example, Hemmer and colleagues [24] have shown how population trapping can be used to make very efficient, low-intensity, phase conjugators. Jain and co-workers [25] introduced the idea of nonlinear optics at maximum coherence. This is the same idea which plays a large role in the collinear Raman generation. Schmidt and Imamoglu [26] have described a giant Kerr nonlinearity which is so large that it may make nonlinear optical effects possible at near the single-photon level. Harris and Hau [27] have shown how slow group velocity and the temporal slip of one pulse over another limits the maximum nonlinearity that can be obtained. Lukin and Imamoglu [28] have shown how two isotopes might be used to overcome this limitation and produced entangled states of the radiation field. An extension to dark-state polaritons has recently been discussed [29] and the combination of low-light-level nonlinear optics and nonlinear optics with entangled states opens up many possibilities [30].

The authors thank Phil Bucksbaum, Lene Hau, and Danielle Manuszak for helpful discussions. This work was supported by the U.S. Air Force Office of Scientific Research, the U.S. Army Research Office, the U.S. Office of Naval Research, and the U.S. Army Research Office OSD Multidisciplinary University Research Initiative Program (MURI).

REFERENCES

1. Harris, S. E., Field, J. E., and Imamoglu, A., *Phys. Rev. Lett.* **64**, 1107-1110 (1990).
2. Hakuta, K., Marmet, L., and Stoicheff, B., *Phys. Rev. A* **45**, 5152-5159 (1992).
3. Arimondo, E., "Coherent Population Trapping in Laser Spectroscopy," in *Progress in Optics*, edited by E. Wolf, Elsevier Science, Amsterdam, 1996, p. 257-354; Harris, S. E., *Phys. Today* **50**, 36-41 (1997); Scully, M. O., and Zubairy, M. S., *Quantum Optics*, Cambridge University Press, Cambridge, England, 1997; Harris, S. E., Yin, G. Y., Jain, M., Xia, H., and Merriam, A. J., *Philos. Trans. R. Soc. London A* **355**, 2291-2304 (1997).
4. Harris, S. E., *Opt. Lett.* **19**, 2018-2020 (1994).
5. Jain, M., Merriam, A. J., Kasapi, A., Yin, G. Y., and Harris, S. E., *Phys. Rev. Lett.* **75**, 4385-4388 (1995).
6. Schulz-von der Gathen, V., Bornemann, T., Kornas, V., and Dobele, H. F., *IEEE J. Quant. Electr.* **26**, 739-743 (1990).
7. Losev, L. L., and Lutsenko, A. P., *Quantum Electron.* **23**, 919-926 (1993); McDonald, G. S., New, G. H. C., Losev, L. L., Lutsenko, A. P., and Shaw, M., *Opt. Lett.* **19**, 1400-1402 (1994).
8. Kawano, H., Hirakawa, Y., and Imasaka, T., *IEEE J. Quant. Electr.* **34**, 260-268 (1998).
9. Hakuta, K., Suzuki, M., Katsuragawa, M., and Li, J. Z., *Phys. Rev. Lett.* **79**, 209-212 (1997).
10. Nazarkin, A., Korn, G., Wittman, M., and Elsaesser, T., *Phys. Rev. Lett.* **83**, 2560-2563 (1999).
11. Ruhman, S., Joly, A. G., and Nelson, K. A., *IEEE J. Quant. Electr.* **24**, 460-469 (1988).
12. Harris, S. E., and Sokolov, A. V., *Phys. Rev. A* **55**, R4019-R4022 (1997); Harris, S. E., and Sokolov, A. V., *Phys. Rev. Lett.* **81**, 2894-2897 (1998); Sokolov, A. V., Yavuz, D. D., and Harris, S. E., *Opt. Lett.* **24**, 557-559 (1999); Sokolov, A. V., *Opt. Lett.* **24**, 1248-1250 (1999).
13. Yavuz, D. D., Sokolov, A. V., and Harris, S. E., *Phys. Rev. Lett.* **84**, 75-78 (2000).
14. Sokolov, A. V., Walker, D. R., Yavuz, D. D., Yin, G. Y., and Harris, S. E., "Raman

Generation by Phased and Antiphased Molecular States," *Phys. Rev. Lett* (to be published).

15. Kaplan, A. E., *Phys. Rev. Lett.* **73**, 1243-1246 (1994); Kaplan, A. E., and Shkolnikov, P. L., *J. Opt. Soc. Am. B* **13**, 347-354 (1996).

16. Sokolov, A.V., Yavuz, D. D., Walker, D. R., Yin, G. Y., and Harris, S. E., "Light Modulation at Molecular Frequencies" (submitted for publication).

17. Hau, L. V., Harris, S. E., Dutton, Z., and Behroozi, C. H., *Nature* **397**, 594-598 (1999).

18. Kash, M. M., Sautenkov, V. A., Zibrov, A. S., Hollberg, L., Welch, G. R., Lukin, M. D., Rostovtsev, Y., Fry, E. S., and Scully, M. O., *Phys. Rev. Lett.* **82**, 5229-5236 (1999).

19. Budker, D., Kimball, D. F., Rochester, S. M., and Yashchuk, V. V., *Phys. Rev. Lett.* **83**, 1767-1770 (1999).

20. Harris, S. E., Field, J. E., and Kasapi, A., *Phys. Rev. A* **46**, R29-R32 (1992).

21. Harris, S. E., "Pondermotive Forces With Slow Light" (submitted for publication).

22. Matsko, A. B., Rostovtsev, Y. V., Cummins, H. Z., and Scully, M. O., *Phys. Rev. Lett.* **84**, 5752-5755 (2000).

23. Stamper-Kurn, D. M., Chikkatur, A. P., Gorlitz, A., Inouye, S., Gupta, S., Pritchard, D. E., and Ketterle, W., *Phys. Rev. Lett.* **83**, 2876-2879 (1999).

24. Hemmer, P. R., Katz, D. P., Donoghue, J., Cronin-Golomb, M., Shahriar, M. S., and Kumar, P., *Opt. Lett.* **20**, 982-984 (1995).

25. Jain, M., Xia, H., Yin, G. Y., Merriam, A. J., and Harris, S. E., *Phys. Rev. Lett.* **77**, 4326-4329 (1996).

26. Schmidt, H., and Imamoglu, A., *Opt. Lett.* **21**, 1936-1938 (1996).

27. Harris, S. E., and Hau, L. V., *Phys. Rev. Lett.* **82**, 4611-4614 (1999).

28. Lukin, M. D. and Imamoglu, A., *Phys. Rev. Lett.* **84**, 1419-1422 (2000).

29. Fleischhauer, M., and Lukin, M. D., *Phys. Rev. Lett.* **84**, 5094-5097 (2000).

30. Lukin, M. D., Yelin, S. F., and Fleischhauer, M., *Phys. Rev. Lett.* **84**, 4232-4235 (2000).

Destruction of Darkness: Optical Coherence Effects and Multi-Wave Mixing in Rubidium Vapor

A.S. Zibrov[1,2,3], L. Hollberg[1], V.L. Velichansky[2,3], M.O. Scully[2,4],
M.D. Lukin[5], H.G. Robinson[1], A.B. Matsko[2],
A.V. Taichenachev[6], and V.I. Yudin[6]

[1] NIST, 325 Broadway, Boulder, CO, USA, 80303,
[2] Texas A&M University, College Station TX,77843and
[3] Lebedev Institute of Physics, Moscow, Russia
[4] Max-Planck-Institut für Quantenoptik, 85748 Garching, Germany
5Harvard Smithsonian, Cambridge, MA 02138,
[6] Novosibirsk State University,Novosibirsk, Russia
hollberg@boulder.nist.gov, http//www.bldrdoc.gov/timefreq/ofm

Abstract. We describe several novel experimental effects resulting from optical coherences in multilevel atoms driven with coherent laser fields. Four configurations were explored using Rb vapor cells: lambda dark-line resonances, single beam double-lambda oscillations, and cascade two-photon transitions in co- and counter-propagating geometries. Experiments were performed with high spectral-resolution using low-power diode lasers sources and optically thick cells. Two of the most striking effects we observed are: (1) a self-oscillation that occurs at 3.0 GHz on the ^{85}Rb hyperfine frequency when a single laser field pumps the atoms, and (2) a coherent blue-beam emission that we observe in the co-propagating cascade two-photon case. These phenomena lead to complex lineshapes, and the effects sometimes dominate the well-known 3-level coherence effects. Additional higher-order mixing or cascaded interferences are apparent in all four experimental configurations. Many of these can be viewed as multi-wave mixing enhanced by optical coherences around closed-loop paths of atomic energy levels. The effects are particularly strong when four-photon closed-loop paths with resonant energy levels are possible, as in, the double-lambda system or as a four-wave mixing "box."

INTRODUCTION

Motivated by the desire to understand atomic coherence effects in multilevel atoms, we have been studying characteristic lineshapes in lambda and cascade 2-photon transitions with variations in experimental parameters. We continue to be surprised by the richness of phenomena that occur when even low-power laser beams interact with reasonably dense atomic vapors. Better knowledge and control of these systems should allow us to take advantage of the novel coherence effects (coherent population trapping (CPT), electromagnetic induced transparency (EIT),

CP551, *Atomic Physics 17*, edited by E. Arimondo, P. DeNatale, and M. Inguscio
2001 American Institute of Physics 1-56396-982-3

and gain without inversion, etc.) to develop advanced sensors and sources of radiation. Examples are: microwave frequency references, short-wavelength lasers, and media with high indices of refraction or tailored group velocities. Significant research efforts have been recently directed toward the study of these effects in similar systems, with interesting experimental results reported from the groups of Arimondo, Dunn, Ducloy, Hänsch, Harris, Hemmer, Lange, Wellegehausen, Wynands, and many others. Distinguishing aspects of the present work include the use of low-power cw laser sources, very high spectral resolution, and atomic vapors that are relatively dense. Optically thick atomic vapors offer potential advantages including long, effective interaction lengths, bigger signals and, sometimes, narrower linewidths. If we can achieve good transparency at high densities, then more atoms can contribute to the signal, and the spectral features could be narrowed due to exponential absorption through the cell. In addition, with long interaction lengths, dispersive and other nonlinear processes become important. These processes can significantly alter the character of the signal, sometimes interfering with, sometimes enhancing and often creating new structure within the lineshape.

This paper focuses on how these nonlinear interactions affect four different experimental configurations: the "Dark-Line" lambda resonance of Gozzini's group [1], a single-beam double-lambda oscillation, the cascade two-photon system with counter-propagating beams, and the cascade two-photon system with co-propagating beams. The underlying 3-level systems have been carefully studied over the past 20 years [2,3]. Our emphasis here is on laboratory results viewed from the experimentalist's perspective as opposed to detailed analysis of the systems. Obviously both theory and experiment are required to properly understand the results.

Perhaps the most interesting regime was when the laser field was strong (Rabi frequency large compared to the spontaneous decay rate of the atom) and the cells optically thick. This regime can give large signals with narrow linewidths, the result being improved signal-to-background and a high quality-factor, Q. All four experiments were performed with Rb vapor cells that could be heated, and with either one or two input fields from tunable diode lasers. The lasers were extended-cavity-diode-lasers (ECDL) with output powers from 10 to 35 mW, except in one case a semiconductor tapered-amplifier was used with >100 mW at 776 nm. The optical intensity applied to the atoms varied greatly, with Rabi frequencies, Ω_R, ranging from 0.01 to 10 times the spontaneous decay rate, Γ. In most cases $\Gamma < \Omega_R < \omega_D$ where ω_D is the Doppler width of the transitions involved. Cells with natural isotopic abundance were used at temperatures from room temperature up to 200 C (473 K), and lengths ranging from 3 to 10 cm, incorporating a variety of different kinds of cell windows.

DARK-LINE LAMBDA RESONANCES

The first case we consider is the "dark-line" resonance in the Λ-configuration with an optically thick cell and co-linearly propagating input beams. In some conditions, by increasing the optical thickness it is possible to retain high contrast while simultaneously narrowing the spectral linewidth. Narrower linewidths, paired with high contrast, result in larger signals and reduced background. This combination might have advantages in some applications of dark-line resonances such as compact, low-power, microwave frequency references [4,5] or for magnetometers [6,7]. Early proponents of using optically thick cells to enhance nonlinear transparencies include: work by Svanberg and Schawlow [8], using the population effects of saturated absorption, and Harris et al. who has shown spectacular results in changing transparency over orders of magnitude using the coherence effects of EIT [9,10]. In our experiments we utilize optical coherences to obtain spectrally narrow linewidths with high-contrast, and study the details of the spectral lineshapes to gain a better understanding of the interactions.

Our first experimental system is quite simple: two ECDL lasers provide the input fields to the Rb cell. The lasers were tuned near the 795 nm D1 line, and one laser was phase-locked to the other with a variable frequency offset near the ground-state hyperfine splitting of 6.8 GHz. The signal we detected was the power in the beatnote (at 6.8 GHz) between the two lasers after they passed through the Rb cell. This method allowed us to probe the structure of the dark-line resonance with a high signal-to-noise ratio, and with synthesizer precision in the offset frequency. Figure 1 shows the structure of the "dark-line" that we observed as a function of frequency offset for different cell temperatures.

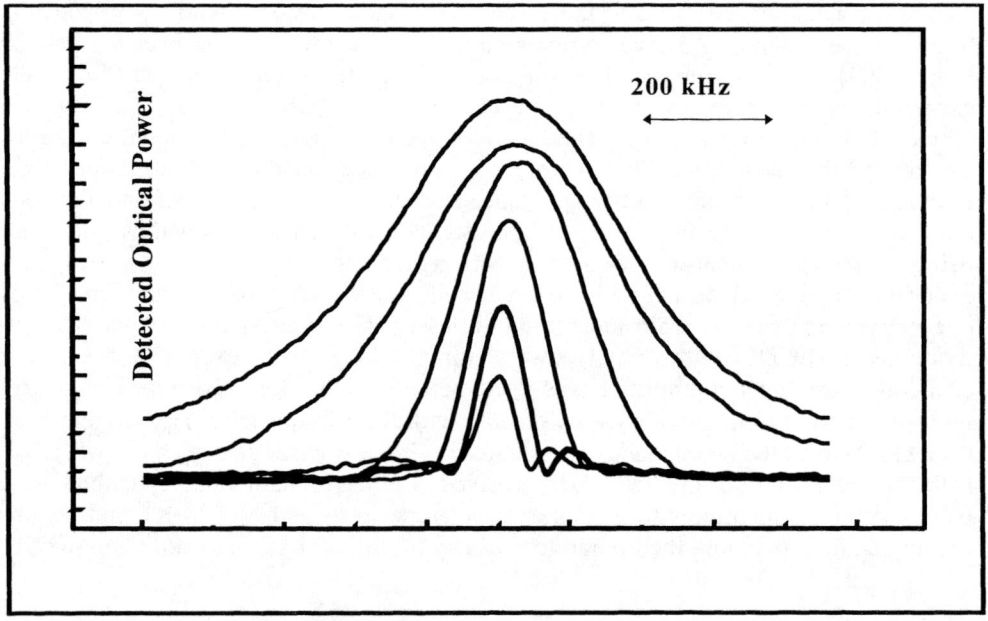

FIGURE 1. Power detected in the beatnote between the two phase-locked lasers transmitted through the Rb cell as a function of the microwave frequency offset between them. The six curves from largest to smallest correspond to the cell temperatures, 52, 57, 72, 80, 85 and 89 C. The lasers were tuned to the D1 transition at 795 nm, F=2 to F'=2, with their difference frequency centered on the ^{87}Rb hyperfine splitting of 6.8 GHz. As the cell is heated, the dark-line narrows due to exponential absorption in the wings of the resonance. The transparency (and detected signal shown here) decrease and additional interfering structure appears in the lineshape at the higher temperatures (lower curves). Laser powers were a 5.6 mW/ 3 mm dia. and 0.1 mW / 3 mm diameter, and the cell's length was 5 cm.

As the cell temperature increased the optical thickness increased and the linewidth of the "dark-line" resonance became narrower as desired and expected. With higher temperatures the transparency began to decrease and additional structure appeared in the lineshape, as seen in fig.1. As the optical depth reached the range of the two higher temperature curves, we saw the clear signature of interference effects on the lineshape. The transparency of the "dark-line" resonance seems to be destroyed by spectrally narrow dips that appear in the resonance. These features result from an additional field generated by the atoms via a double-lambda type interaction with the two input fields as diagrammed in fig. 2.

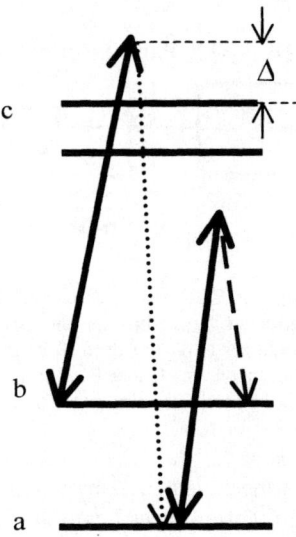

FIGURE 2. Simplified energy-level diagram of Rb showing the configuration of the input laser field(s) (indicated by the dark solid line and the dotted line) and the fields generated by the Rb atoms (dotted line plus dashed line) in the "double-lambda" interaction. The dotted line is the anti-Stokes component and the dashed line is the Stokes component.

This new field (the Stokes component in the example of fig. 2) also makes a beatnote with the strong drive-field; the beatnote is offset in frequency by the hyperfine splitting of the ground state. This additional beatnote has the same frequency offset relative to the drive (but opposite in sign) as the original beatnote generated by the dark-line resonance. Thus, the photo-currents from the two beatnotes interfered with each other in the detector, and appeared as an enhancement or a zero in the detected signal.

The effects of optical pumping and optical coherences around closed-loop double-lambda paths are critically important in understanding the details of the lineshape [11, 12]. Theoretical models including coherent-coupling between all the fields and atomic levels provide excellent agreement with the experimental lineshapes results [13].

Dark-Line Λ-Resonance With Optical Feedback

More dramatic effects become apparent in this lambda-CPT system if even a small amount of the transmitted beam travels backward through the atomic vapor. A counter-propagating beam (for example, reflection from the cell window) causes the system to "self-oscillate" at the ground state hyperfine splitting of 3.036 GHz. In this case, we observed that with a single laser-field incident on the Rb cell (tuned near the [85]Rb D2 line) the transmitted beam developed strong AM modulation at the frequency of the hyperfine splitting. The experimental apparatus is shown schematically in fig. 3.

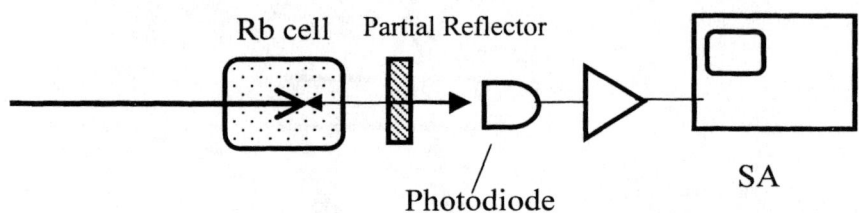

FIGURE 3. A single laser field from an ECDL is sent through a heated Rb cell, and part of the transmitted light (4 to 80 %) is reflected back through the cell by the partial reflector. A fast photodiode, amplifier and spectrum analyzer (SA) are used to display the microwave spectrum of the power fluctuations on the transmitted laser beam. Under conditions outlined in the text we observed a strong (60 dB signal to noise in 30 kHz bandwidth) and narrow (< 300 Hz) beatnote at the ground-state hyperfine splitting of [85]Rb, 3.036 GHz.

Thus, a 10 mW laser beam tuned near resonance shows large AM modulation sidebands that appear on the transmitted laser beam at the 3.036 GHz, [85]Rb hyperfine splitting (hfs). The AM sidebands have the same spectral characteristics as the input laser field, and produce a narrow linewidth beatnote with the carrier (<300 Hz FWHM at 3.036 GHz). We have observed as much as 4 % of the transmitted power contained in each of the frequency-shifted sidebands. A higher-order anti-Stokes sideband (up-shifted from the carrier by twice the ground state hyperfine splitting) was also observed under some conditions.

In contrast to our usual experience with a single laser field tuned near the D2 resonance in Rb, the atoms do not get pumped into a "dark state". In fact, the system refuses to go dark and becomes strongly interacting. The net effect is to scatter photons from the input laser field to new frequencies, shifted both up and down from the input laser frequency by the ground state hfs. With the reflected-light traveling backward through the cell the system remains in a bright-state even though there is complex hyperfine and Zeeman structure that allows many different possible dark-states. The transmitted light thus contains the input carrier-field, plus an upper and lower sideband. We studied the spectral properties of the transmitted light by making a beatnote with an independent laser, and also by analyzing the spectra with a Fabry-Perot cavity. A Fabry-Perot spectrum of the beam transmitted through the cell is shown in fig. 4.

At least for the range of experimental parameters accessible in our apparatus, the backward-going reflection of the transmitted light was critical to observe the strong and narrow self-oscillation. For example, the oscillation did not appear if the input laser was first divided into two beams and sent through the cell in opposite directions. By using an independent, tunable, diode laser we determined that the oscillation could be initiated most easily by sending a laser backward through the cell that was tuned to the anti-Stokes frequency. In the single-laser geometry the beam transmitted through the cell will contain some anti-Stokes component, which, when reflected back into the cell can initiate the self-oscillation.

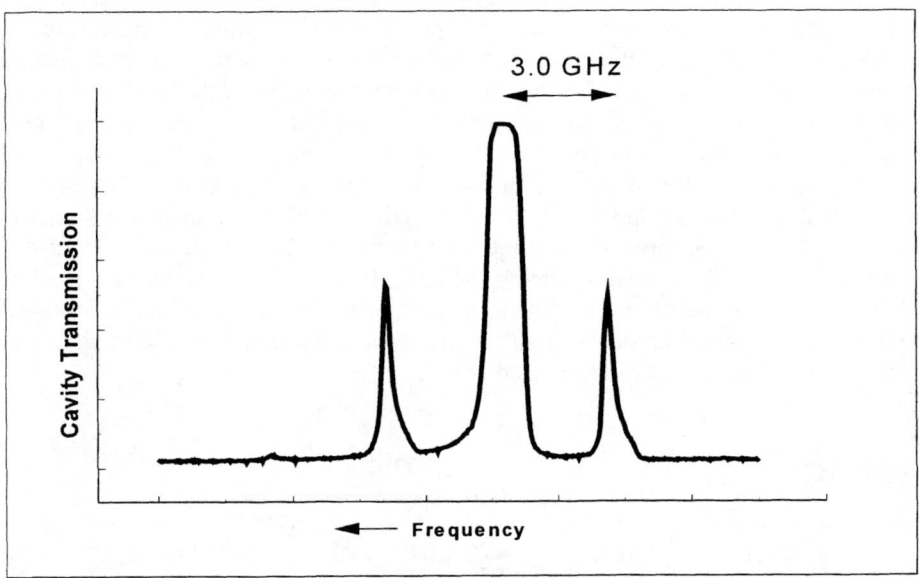

FIGURE 4. This trace shows a Fabry-Perot cavity spectrum of the light transmitted through the Rb cell in the configuration of fig. 3. and when the system is in self-oscillation at the hyperfine frequency. The oscillation is caused by the backward-going beam from the partial reflector. The smaller peaks on either side of the incident carrier were generated in the Rb vapor and correspond to the fields represented in fig. 2. by the dotted and dashed lines. The sidebands are separated from the input carrier by 3.036 GHz and each contains approximately 4 % of the power in the strong carrier, which is clipped at the top of the figure.

In the case of ^{85}Rb the oscillation only requires a single input laser field, while for ^{87}Rb some additional hyperfine re-pumping was required to see the corresponding oscillation at 6.8 GHz. Although these nonlinear oscillations and interferences are important in understanding the fundamental interaction of laser fields with alkali atoms, we have not yet found a way to use them to improve the performance of our compact microwave frequency references [4]. In fact, these interactions seem to be mostly detrimental; but this, in itself, means that it is critically important to understand the interactions and how they can affect the low intensity, low vapor-density CPT clocks that we are also studying [5]. Nonetheless, the effects may be quite useful in enhancing dispersion and altering the group velocity and index of refraction for other applications.

CASCADE 2-PHOTON COUNTER-PROPAGATING

The third experimental configuration that we studied was that of counter-propagating beams near the cascade 2-photon resonance (5S-5P-5D) in a Rb cell. The idea was to explore the potential for observing a high index-of-refraction, and/or the large changes in group-velocity associated with the coherences in near-resonance two-photon transitions when the atomic vapor is optically thick. Here we used even higher cell temperatures, a strong field on the upper part of the two-photon transition propagating in the forward direction, and a weak probe-field on the resonance line propagating in the backward direction. The probe penetrates only a short distance into the cell, but the coherent effects are detected on this beam in reflection from the cell window. Near resonance we see the well known "selective reflection" (SR) signals [14, 15]. These signals are also complicated by additional multi-wave mixing signals that appear under many conditions. Related effects have also been reported and studied in detail by Ducloy and collaborators. They have observed surprising non-phase-matched four-wave-mixing (4WM) signals in reflection from Cs vapor cells [16]. We detect additional unexplained structure in Rb as well, but we also find phase-matched 4WM in reflection from the cells. In the usual reflection geometry it is difficult to separate the effects of SR from those of multi-wave mixing. However, the SR signal can be unambiguously separated from the phase-matched 4WM signal using the experimental geometry outlined in fig. 5.

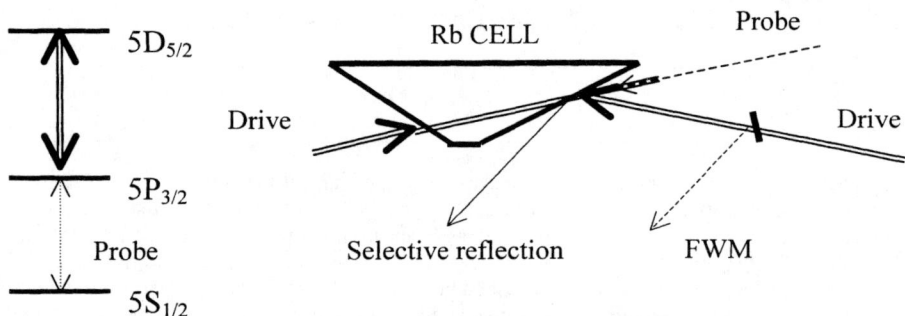

FIGURE 5. Experimental configuration used for separating the SR and 4WM signals. The two strong drive-fields (from a single amplified ECDL) are tuned to the 776 nm upper half of the two-photon cascade, and the weak probe laser is tuned to the 780 nm D2 line. The angle between the probe and the backward drive is ~ 3^0, which has been exaggerated in the figure for clarity.

We find regimes where the multi-wave mixing is much larger and much narrower than the SR signals that are induced by index-of-refraction effects. On resonance, with a heated cell, the probe-field penetrates only a very short distance into the atomic vapor; yet it was possible to have gain (as much as 130% in reflection) in the 4WM reflection signal. Examples of the signals that we observed using the reflection geometry above (fig. 5.) are shown in fig. 6. The selective-reflection signals, SR, are Doppler-free and have widths of about 60 MHz, while the 4WM resonances are larger and narrower (FWHM ~2 MHz) and are sensitive to the relative phase of the two drives-fields and the probe-laser field.

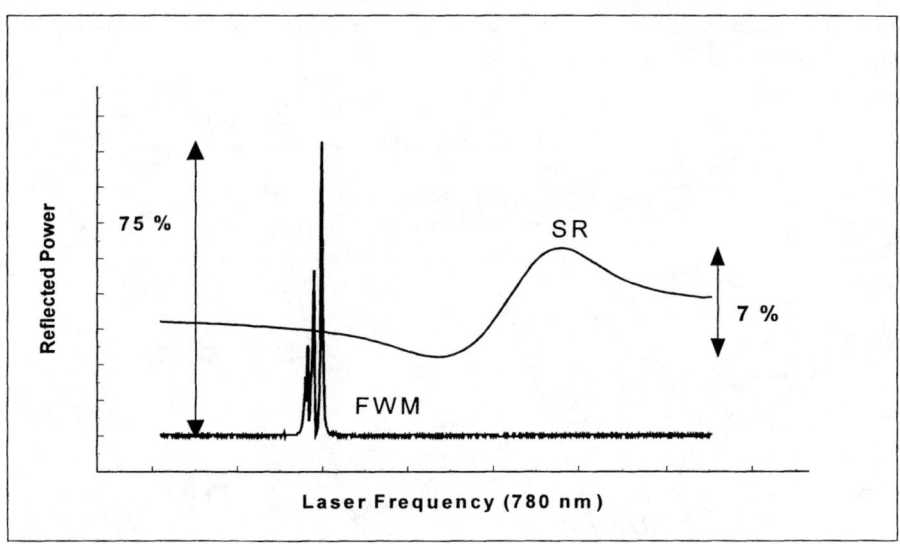

FIGURE 6. Power of the reflected signals detected simultaneously but with two separate photodetectors using the experimental arrangement of fig. 5. The SR signal is the usual Doppler-free selective-reflection signal, and 4WM is the phase-sensitive four-wave-mixing signal. The vertical axis is different for the two curves, but the inset labels show the fraction of power reflected for the individual curves. The 4WM reflection signal even shows gain under some conditions. The horizontal axis is the tuning of the 780 nm probe-laser frequency over a total range of~ 280 MHz. Other experimental parameters are: Power of the 776 nm drive laser = two x 35 mW/ 0.25 nm dia., power of 780 nm probe laser = 0.1 mW/ 0.1 mm dia., with orthogonal polarizations between drive and probe, and T_{cell} = 150 C (423 K).

CASCADE 2-PHOTON DRIVEN FORWARD-4WM "BOX"

The fourth experimental configuration consisted of two co-propagating diode laser beams near resonance with a cascade 2-photon transition in Rb (5S-5P-5D). The goal was to study experimentally the feasibility of making short-wavelength sources of coherent radiation using atomic coherences and diode laser sources. Encouraged by demonstrations of lasing-without-inversion (LWI) in systems that were experimentally convenient, and that had produced lasing at wavelengths near the wavelength of the input coherent-drive, we wanted to push toward shorter wavelengths [17,18].

Of particular interest is the V-configuration, where a strong drive-field can couple the atomic ground state to a low-lying excited state, creating gain on a shorter-wavelength transition between the ground state and a higher-lying energy level. A potential case for demonstrating this concept might be to incoherently populate the Rb 6P state by spontaneous decay from 5D, which can be pumped by a cascade two-photon transition from the ground state (fig. 7.). It should then be possible to apply a strong coherent-drive between the ground S-state and another level, for example using the D1 transition to the $5P_{1/2}$ (not shown in fig. 7.) to create the desired **V**-configuration. However, when we started to explore this in the laboratory, even without the coherent-drive on D1, we found that other multi-wave mixing effects became important in the dense Rb vapor.

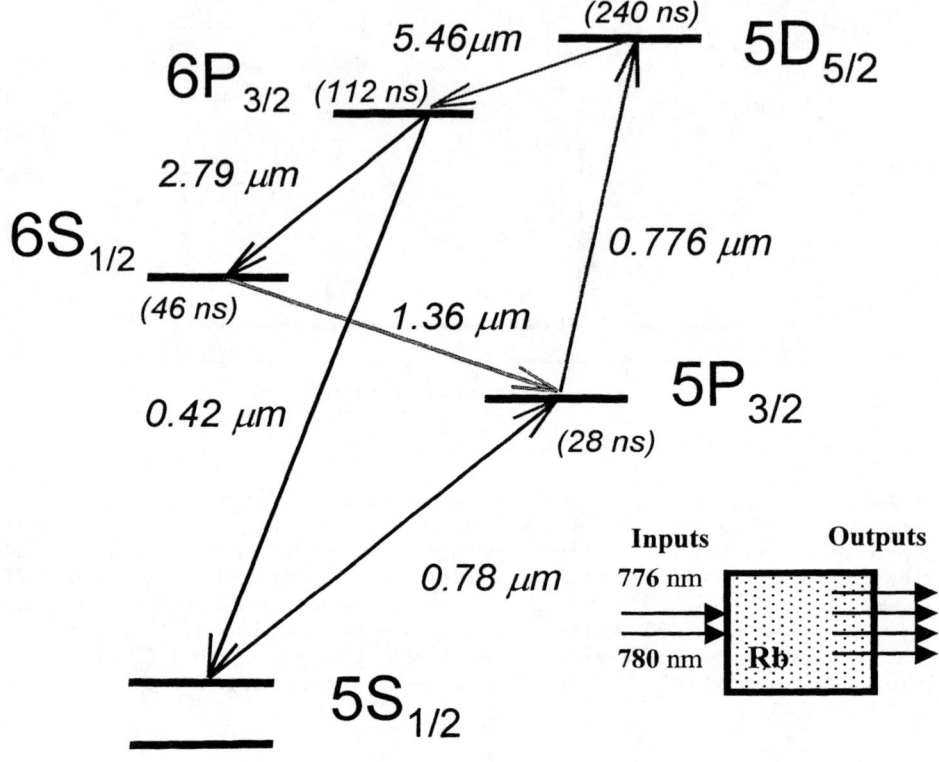

FIGURE 7. Simplified energy-level diagram of Rb showing the two input laser fields at 780 nm and 776 nm, and the coherent output beams at 420 nm, 5.4 μm, 1.36 μm, and ~2.8 μm, that we detected in the forward direction after the Rb cell.

Blue Beam Generation

With the two low-power (780 nm and 776 nm, ≈10 mW each) input laser fields we found that a coherent blue-beam was emitted in the forward direction from the Rb cell, (where "forward" means co-propagating with the two co-propagating input beams). The blue-beam is spectrally narrow, and does, in fact, come from the 420 nm 5S-6P transition. This was achieved even though the vapor cell was optically thick for the 420 nm radiation.

The configuration of two input fields in two-photon cascade (as in fig. 7) may not be an arrangement in which we expect LWI. However, the type of signals that we observe have been detected previously using high-power, short-pulse laser sources, and they have been studied in some detail with those sources [19-22]. This kind of coherent generation is attributed to forward-four-wave-mixing (F-4WM), a nonlinear parametric process. There have also been a number of experimental demonstrations of cw short-wavelength generation using F-4WM by injecting three input laser fields into an atomic vapor to produce a fourth shorter wavelength [23-25]. In a recent impressive experiment, Eikema et al. [26] have used F-4WM to generate ~1 nW of cw, 121 nm light by driving Hg vapor with three input fields. A related 4WM system using 3 input-fields but in a double-Λ configuration, has also been used to produce cw radiation using molecules [27]. Since the F-4WM interaction is quite strong, and apparently easy to initiate, we would not be surprised to learn that these signals had been observed before using just two cw input-fields; however, a preliminary search through the literature did not turn up any such examples.

With our experimental setup we detected as much as 12 microwatts of uni-directional, spectrally narrow, blue output power using two ($\approx$10 mW) laser beams in a single pass through a Rb cell (5 cm long) that was heated to about 95 C (368 K). The 420 nm beam coming out of the cell had a good, round spatial-mode that could be coupled into the TEM_{00} mode of a Fabry-Perot cavity fairly efficiently. We used the Fabry-Perot cavity to estimate the linewidth along with the tuning-rate and -range of the blue-beam. The spectral characteristics of the blue-beam corresponded directly to the spectral characteristics of the sum of the input ECDL lasers: fast linewidth about 200 kHz and low-frequency jitter of a few MHz. The frequency of the blue-light tuned at a rate equal to the tuning of either the 780 nm or 776 nm beam, as would be expected for F-4WM. The continuous tuning-range of the blue light was approximately 200 MHz, limited by our input power levels and the Doppler width of the 5D state. As long as we maintained the 776 nm laser tuning to be near resonance with the cascade two-photon transition (to the 5D state) it was possible to generate the blue-beam for a fairly wide tuning of the 780 nm laser, over a significant fraction of the ground-state hyperfine width (6.8 GHz).

Initially, it seemed surprising that the 420 nm beam could propagate through the Rb vapor cell that was optically thick for the 420 nm radiation. A better understanding of the gain on the blue transition comes from using a separate blue laser to probe the absorbance on the 420 nm, 5S – 6P transition. Figure 8 displays the probe absorption spectrum obtained when the 780 nm and 776 nm lasers are coupled to the two-photon cascade 5S – 5P – 5D. The probe shows a strong, spectrally narrow, gain peak that occurs within the Doppler-broadened absorption line.

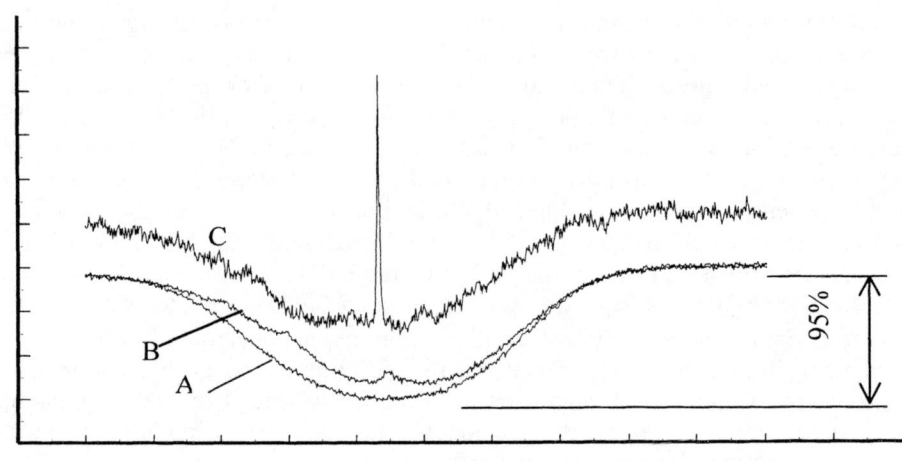

Probe Laser Frequency

FIGURE 8. An independent, tunable, frequency-doubled diode laser was used to probe the transmission spectrum of the 420 nm (5S − 5D) transition. The vertical axis shows the transmitted power of the probe beam as a function of its tuning over a range of ~500 MHz. Curve A shows the probe transmission with no other laser field present. In curve B the 780 nm laser is also present, and curve C is the probe transmission when both the 780 nm and 776 nm lasers were applied. The sharp peak near the center of the absorption line is the F-4WM gain induced by driving the 780 + 776 nm cascade transition. The increase in the background level and noise that is apparent in curve C results from scattering of the coherent blue light generated by the atoms that also finds its way to the detector. The experimental conditions were: T_{cell} = 95 C, power of 780 nm = 11 mW / 0.5 mm dia., power of 776 nm = 1.8 mW / 0.5 mm dia., and power of 420 nm probe ≈ 30 nW.

The 420 nm probe-beam (power ≈30 nW) was generated by frequency-doubling an ECDL laser in a single-pass through a LiIO₃ crystal. This beam was comparatively weak, and did not disturb the character of the blue beam emitted from the Rb cell. It is interesting to note that the Rb atoms are relatively efficient at converting near infrared diode lasers to the shorter blue wavelength. Two 10 mW lasers produced ≈12 microwatts of blue-beam in Rb, whereas similar powers in the nonlinear crystal produced tens of nanowatts. Obviously, nonlinear crystals have numerous advantages (broadly tunable, requires only one input beam, etc.) but the Rb system was simple enough to use, and was more efficient at generating a blue light.

Other Coherent Output Beams

Since the 420 nm blue-beam was strong, and if the interpretation is correct that the source is F-4WM, we expected from the energy-level diagram (fig. 7.) that a 5 μm beam would also be present. Using a new vapor cell constructed from a stainless steel vacuum tube with commercial sapphire windows we were able to detect a 5 μm beam in the forward direction (fig. 9). There was no evidence of a 5 μm beam in the backward direction. However, it was more difficult to put a stringent upper limit on the lack of a backward beam because, in both cases, the IR detector had a sizable background signal due to the heated Rb cell.

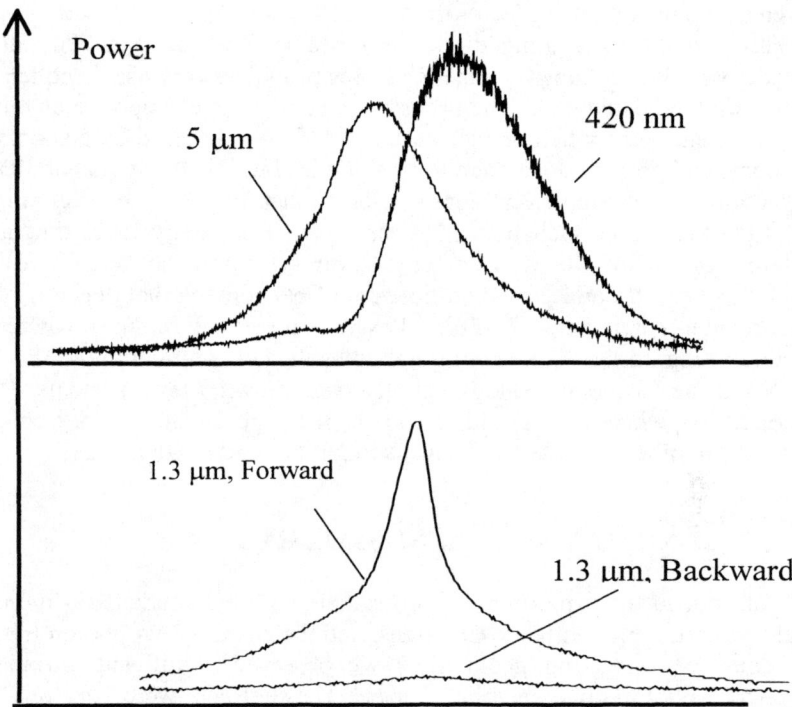

FIGURE 9. The top half of the figure displays the forward powers for the blue beam at 420 nm and the IR beam at 5 μm. These curves were taken simultaneously and are displayed as a function of tuning of the 780 nm laser frequency that was scanned over a total range of ~ 600 MHz. Peak blue power was about 12 microwatts, and the power of the 5 μm was not calibrated. For different tunings of the 776 and 780 nm beams the lower half shows the power in the forward 1.3 μm beam and the lack of a coherent 1.3 μm beam in the backward direction. For this data the experimental parameters were: 780 nm power = 12 mW/0.5 mm dia.; 776 nm = 11 mW/0.5 mm dia.; beams co-propagating with polarizations perpendicular; and T_{cell} = 85 C (358 K).

Additional coherent output beams were generated in the forward direction by adjusting the frequencies of the 780 nm and 776 nm lasers away from the tunings that produced the maximum blue-beam power. These new beams occurred at wavelengths corresponding to other Rb transitions, including: beams at 1.3 μm and ~2.8 μm. Again, we have identified these signals as resulting primarily from F-4WM mixing around closed-loop paths of four energy levels, as indicated in fig.7. The forward 1.3 μm beam was easy to detect, and we used an optical spectrum analyzer to measure its wavelength and thus identify the source as the 6S – 5P transition. The lower half of fig. 9. shows the unidirectional character of the 1.3 μm signal. A coherent beam was observed in the forward direction, while only spontaneous emission was observed in the backward direction. When the 1.3 μm beam was present, we also detected a forward beam in the wavelength range from 2 to 3.4 μm, which was probably due to the 2.79 μm transition shown in fig. 7. The wavelength discrimination in this case was limited by the responsivity of the InSb detector, combined with a quartz window that served as a long-wavelength cut-off.

With two co-propagating diode laser beams (780 nm and 776 nm) driving the cascade two-photon transition, the Rb atoms provide a very usable coherent blue-beam from a simple heated cell. The processes can be quite efficient, even with low-power cw lasers, and even when one leg of the 4WM box is closed by a spontaneous decay. The dominant process does seem to be F-4WM, but it is likely that additional coherent interactions are playing an important role. In fact, the "box" configuration of 4-levels and 4-photons (two such boxes appear in the Rb energy level diagram in fig. 7.) contains aspects of the well-known 3-level systems. The levels and fields are all coupled coherently and provide a richness of phenomena that depends on the detailed experimental parameters [28-30]. Possibly there are regimes where it would be sensible to regard the "box" configuration as the coherent-coupling of a V and a Λ, or as a V plus a cascade, in analogy with the well known 3-level systems. There may also be conditions where LWI could be found. If the goal is to generate short-wavelength sources, a number of these mechanisms might be used to advantage.

SUMMARY

In all four of the experiments discussed above (the lambda dark-line resonance, the single-beam double-lambda oscillations, and the cascade two-photon transitions in co- and counter-propagating geometries) we observed significant variations from the expected 2-photon coherence lineshapes. These effects were very obvious when the Rb atoms were strongly driven ($\Gamma < \Omega_R < \omega_D$), and when the cells were optically thick. The effects resulted mainly from multilevel, multi-photon coherent interactions that cycle around closed loops in the atomic energy levels. The additional nonlinear terms often dominated the usual 2-photon coherences even at relatively low levels of input power. Though different in outcome and manifestation, these effects can, in some sense, be viewed as multi-wave mixing (predominantly multiple 4-wave mixings) enhanced by the underlying 2-photon coherence.

ACKNOWLEDGMENTS

We thank J. Kitching, N. Vukicevic, S.Harris, T. Mossberg, M. Alegrini, E. Arimondo, R. Wynands, R. Kaiser, P. Hemmer, and M. Fleischhauer for numerous stimulating discussion and comments.

REFERENCES

1. Alzetta, G., Gozzini, A., Moi, L., and Orriols, G., *Nuovo Cimento* **36B**, 5-20 (1976).
2. Arimondo, E., *Progress In Optics* **35**, 257-354 (1996).
3. Fulton, D.J., Shepherd, S., Moseley, R.R., Sinclair, B.D., and Dunn, M.H., *Physical Review A* **52**, 2302-2311 (1995).
4. Vukicevic, N., Zibrov, A.S., Hollberg, L., Walls, F.L., Kitching, J. and Robinson, H.G., IEEE J. FCC (2000).
5. Kitching, J., Knappe, S., Vukicevic, N., Hollberg, L., Wynands, R., and Weidemann, W., proceed. Int. Freq. Cont. Symp. (2000), submitted, IEEE *Trans on Instr. and Meas* (2000).
6. Fleischhauer, M. and Scully, M.O., *Phys. Rev. A,* **49**, 1973 (1994).
7. Wynands, R. and Nagel, A., Appl. Phys. B, **68**, 1-25 (1999).
8. Svanberg, S., Yan, G.-Y., Duffey, T.P., and Schawlow, A.L., *Optics Letters* **11**, 138 (1986).
9. Field, J.E., Hahn, K.H.and Harris, S.E., *Phys. Rev. Lett.*, **67**, 3062-3035 (1999).
10. Harris, S.E., *Physics Today*, 36-42 (July, 1997) and references therein.
11. Grove, T.T., Shahriar, M.S., Hemmer, P.R., Kumar, P., Sudarshanam, V.S., and Cronin-Golomb, M., *Opt. Lett.* **22**, 769-771 (1997). Lü, B., Burkett, W.H., and Xiao, M. *Opt. Letts.* **23**, 804-806, (1998). Köster, E., Kolbe, J., Mitschke, F., Mlynek, J., and Lange, W., *Appl. Phys.* B, 201-207 (1984).
12. Fleischhauer, M., Lukin, M.D., Matsko, A.B., and Scully, M.O., *Phys. Rev. Lett.*, **84**, 3558-3561 (2000).
13. Lukin, M.D., Fleischhauer, M., Zibrov, A.S., Robinson, H.G., Velichansky, V.L., Hollberg, L., and Scully, M.O., *Phys. Rev. Lett.* **79**, 2959-2962 (1997). And Zibrov, A.S., Lukin, M..D., and Scully, M.O., *Phys. Rev. Lett.*, **83**, 4049-4052 (1999).
14. Woerdman, J.P., and Schuurmans, *Opt. Commun.*, **14**, 248 (1975).
15. Akulshin, A.M., Velichanskii, V.L., Zibrov, A.S., Nikitin, V.V., Sautenkov, V.A., Yurkin, E.K., and Senkov, N.V., Pis'ma Zh. Eksp. Teor. Fiz. **36**, 247 (1982), [JETP Lett. **36**, 303 (1982)].
16. Amy-Klein, A., Saltiel, S., Rabi, O.A., and Ducloy, M., *Physical Review A* **52**, 3101-3109 (1995). Gorris-Neveux, M., Monnot, P., Saltiel, S., Barbe, R., Keller, J.-C., and Ducloy, M., *Physical Review A* **54**, 3386-3393 (1996).
17. Zibrov, A.S., Lukin, M.D., Nikonov, D.E., Hollberg, L., Scully, M.O., Velichansky, V.L., and Robinson, H.G., *Phys. Rev. Lett.*, 75, 1499-1502 (1995).
18. Scully, M.O., Quantum Opt. 6, 203-215 1994), Mompart, J. and Corbalan, R., *J. Opt. B: Quantum Semiclass* **2**, 1-18 (2000) and references therein.
19. Zhang, P.-L., Wang, Y.-C., and Schawlow, A.L., *J. Opt. Soc. Am. B* **1**, 9-14 (1984).
20. Clark, B.K., Masters, M., and Huennekens, J., *Appl. Phys. B* **47**, 159-167 (1988).
21. Domiaty, U., Gruber, D., Windholz, L., Dinev, S.G., Allegrini, M., De Filippo, G., Fuso, F., Rinkleff, R.-H., *Appl. Phys. B* **59**, 525-531 (1994).
22. Efthimiopoulos, T., Movsessian, M.E., Katharakis, M., and Merlemis, N., *J. Appl. Phys.* **80**, 639-643 (1996).
23. Freeman, R.R., Bjorklund, G.C., Economou, N.P., Liao, P.F., and Bjorkholm, J.E., *Appl. Phys. Lett.* **33**, 739-742 (1978).
24. Wynne, J.J., Sorokin, P.P., Optical mixing in atomic vapors, in Topics in Applied Phys., edt. Shen, Y.R., Springer, Berlin (1987).
25. Nolting, J., Wallenstein, R., *Optics Communications* **79**, 437-442 (1990) and references therein.
26. Eikema, K.S.E, Walz, J., and Hansch, T.W., *Phys. Rev. Lett.* **83** 3828-3831 (1999).
27. Apolonskii, A., Baluschev, S., Hinze, U., Tiemann, E., Wellegehausen, B., *Appl. Phys. B* **64**, 435-442 (1997).
28. Shepherd, S., Fulton, D.J., and Dunn, M.H., *Phys. Rev. A* **54**, 5394-5399 (1996), and references therein.
29. Lvovsky, A.I., and Hartmann, S.R., *Phys. Rev. Lett.* **82**, 4420-4423 (1999).
30. Kosachiov, D.V., Matisov, B.G., and Rozhdestvensky, Y. V., *J. Phys. B: At Mol. Opt. Phys.* **25**, 2473-2488 (1992).

QED WITH IONS, ATOMS, AND
MUONIC SYSTEMS

The Muon Anomalous Magnetic Moment

Vernon W. Hughes

On behalf of the muon g-2 Collaboration

at Brookhaven National Laboratory

H.N. Brown[2], G. Bunce[2], R.M. Carey[1], P. Cushman[10], G.T. Danby[2], P.T. Debevec[7], H. Deng[12], W. Deninger[7], S.K. Dhawan[12], V.P. Druzhinin[3], L. Duong[10], W. Earle[1], E. Efstathiadis[1], G.V. Fedotovich[3], F.J.M. Farley[12], S. Giron[10], F. Gray[7], M. Grosse-Perdekamp[12], A. Grossmann[6], U. Haeberlen[8], M.F. Hare[1], E.S. Hazen[1], D.W. Hertzog[7], V.W. Hughes[12], M. Iwasake[11], K. Jungmann[6], D. Kawall[12], M. Kawamura[11], B.I. Khazin[3], J. Kindem[10], F. Krienen[1], I. Kronkvist[10], R. Larsen[2], Y.Y. Lee[2], I. Logashenko[1,3], R. McNabb[10], W. Meng[2], J. Mi[2], J.P. Miller[1], W.M. Morse[2], C.J.G. Onderwater[7], Y. Orlov[4], C. Özben[2], C. Polly[7], C. Pai[2], J.M. Paley[1], J. Pretz[12], R. Prigl[2], G. zu Putlitz[6], S.I. Redin[12], O. Rind[1], B.L. Roberts[1], N. Ryskulov[3], S. Sedykh[7], Y.K. Semertzidis[2], Yu.M. Shatunov[3], E. Solodov[3], M. Sossong[7], A. Steinmetz[12], L.R. Sulak[1], C. Timmermans[10], A. Trofimov[1], D. Urner[7], P. von Walter[6], D. Warburton[2], D. Winn[5], A. Yamamoto[9], D. Zimmerman[10]

[1] Department of Physics, Boston University, Boston, MA 02215, USA
[2] Brookhaven National Laboratory, Upton, NY 11973, USA
[3] Budker Institute of Nuclear Physics, Novosibirsk, Russia
[4] Newman Laboratory, Cornell University, Ithaca, NY 14853, USA
[5] Fairfield University, Fairfield, CT 06430, USA
[6] Physikalisches Institut der Universität Heidelberg, 69120 Heidelberg, Germany
[7] Department of Physics, University of Illinois at Urbana-Champaign, IL 61801, USA
[8] MPI für Med. Forschung, 69120 Heidelberg, Germany
[9] KEK, High Energy Accelerator Research Organization, Tsukuba, Ibaraki 305-0801, Japan
[10] Department of Physics, University of Minnesota, Minneapolis, MN 55455, USA
[11] Tokyo Institute of Technology, Tokyo, Japan
[12] Department of Physics, Yale University, New Haven, CT 06511, USA

CP551, *Atomic Physics 17*, edited by E. Arimondo, P. DeNatale, and M. Inguscio
© 2001 American Institute of Physics 1-56396-982-3/01/$18.00

Abstract: The muon g-2 experiment at the Brookhaven National Laboratory is described, including its motivation, goal and present status. The latest result based on 1998 data is $a_{\mu^+} = \frac{g-2}{2} = 11\ 659\ 191(59) \times 10^{-10}$ (5 *ppm*), where the error is primarily statistical . This value agrees with the present theoretical value. Data obtained thus far and now being analyzed should have a statistical error of about 0.5 ppm.

1   Introduction

The gyromagnetic ratio or spin g-value of a particle is defined as

$$g = \frac{\frac{magnetic\ moment}{e\hbar/2mc}}{\frac{angular\ momentum}{\hbar}} = 2(1+a) \tag{1}$$

in which m and e are the mass and magnitude of charge of the particle. The value $g = 2$ is that for a Dirac particle and the quantity $a = (g-2)/2$ is called the anomalous g-value of the particle.

The values of g and a for the electron and muon have played a central role in the development of modern quantum electrodynamics and particle physics. They continue to serve as fundamental quantities for testing the validity of currently established theory or that of speculative new theories.

For the electron the experimental value of a_e has been determined to the high accuracy of 3.7 ppb[1, 2] and its theoretical value has been calculated by perturbation theory as an expansion in the fine structure constant α through the α^4 term, which has an accuracy of about 1 ppb if α were known exactly. The best current value of α[2] is obtained by equating $a_e(expt)$ to $a_e(theor)$:

$$\alpha^{-1} = 137.035\ 999\ 58(52)(3.8ppb) \tag{2}$$

For the muon the experimental value for a_μ is determined with a precision of 4ppm by combining the CERN measurement[3, 4] with the recent measurements at BNL[5, 6]:

$$a_\mu(expt) = 11\ 659\ 205(46) \times 10^{-10}(4ppm) \tag{3}$$

The theoretical value is due principally to the electromagnetic interaction(QED) and as for the electron is calculated by perturbation theory as a power series in the fine structure constant α. For the muon the electron-positron field as well as the photon and muon fields contribute importantly, and the tauon field also is not negligible. Vacuum polarization contributions of hadron loops are large (60ppm) and constitute the dominant uncertainty (0.7ppm) in a_μ(theor). The contribution of the electroweak effect is significant (1.3ppm). The present theoretical value for a_μ[2] is :

$$a_\mu(theor) = 11\ 659\ 163(8)^{-10}(0.7ppm) \tag{4}$$

The theoretical and experimental values are in reasonable agreement.

$$a_\mu(expt) - a_\mu(theor) = (42 \pm 47) \times 10^{-10} \tag{5}$$

or equivalently 3.6 ± 4.0 *ppm*.

Because the muon mass m_μ is larger than th electron mass m_e, the contribution to a_μ of a heavier particle such as hadron or weak vector boson through virtual processes is greater than to a_e, indeed by the enormous factor $(m_\mu/m_e)^2 = 4 \times 10^4$. This increased sensitivity of a_μ applies generally also for postulated new particles such as additional weak vector bosons or supersymmetric particles. Hence the physics associated with a_μ is much richer than that associated with a_e even though a_μ is less well known than a_e.

The new experiment in progress at BNL with the AGS using a muon storage ring has the goal of measuring a_μ to 0.35 ppm which would represent a factor of 20 improvement over the current value. The specific scientific motivations are the following:

1. To measure the effect of the weak interactions on the predominantly electro-magnetic property a_μ of the muon. Such a measurement will provide a clean test of the renormalization prescription of the unified electroweak theory.

2. To search for physics beyond the standard model.

   In order to achieve a deeper and less arbitrary theory for particle physics than the present Standard Model, many speculative theories have been proposed, including substructure for leptons, gauge bosons, or quarks and also new particles such as gauge bosons, supersymmetric particles or excited leptons. All of these modifications of Standard Model would change $a_\mu(theor)$, and often by an amount which would be detected by a measurement of a_μ to 0.35 ppm. The muon anomalous g-value a_μ can be regarded as a standard against which any speculative new theory can be tested; such precision tests are complementary to direct searches for new particles with the highest energy accelerators.

2 The BNL Experiment

2.1 Introduction

The general method of the BNL experiment is the same as that of the CERN experiment. The new experiment, however, incorporates several major new features and advances in technology, as well as many lesser ones. These include : (1) The use of the current AGS proton beam which is a factor of about 200 greater than the CERN proton beam; (2) A superferric storage ring of high stability and homogeneity, and an NMR system capable of 0.1 ppm absolute accuracy; (3) The

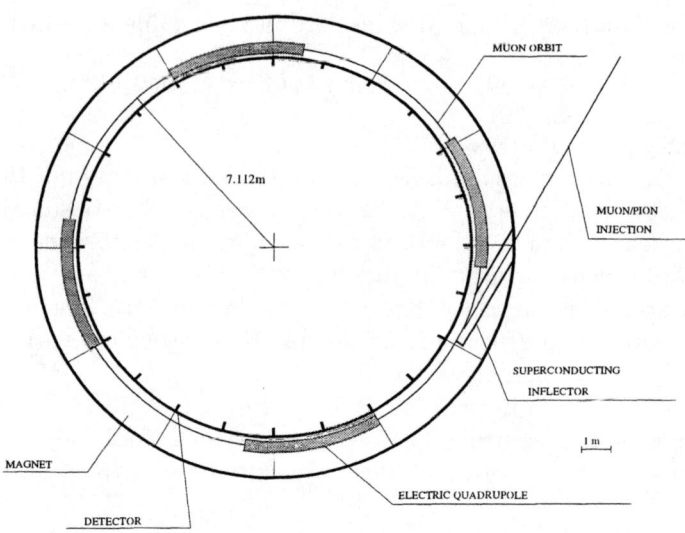

Figure 1: General schematic of the Brookhaven muon g-2 experiment, showing the muon storage ring and other elements of the experiment.

use of μ injection as well as π injection into the storage ring; (4) An improved electron spectrometer system with higher energy resolution; (5) The use of fast modern electronics; (6) The use of the dramatically increased power of modern computers for data acquisition and data analysis.

For polarized muons moving in a uniform dipole magnetic field B perpendicular to the muon spin direction and to the plane of the orbit, and with an electric quadrupole field E for vertical focusing, the difference angular frequency ω_a between the spin precession frequency ω_s and the cyclotron frequency ω_c is given by

$$\vec{\omega_a} = -\frac{e}{m_\mu}[a_\mu \vec{B} - (a_\mu - \frac{1}{\gamma_\mu^2 - 1})\vec{\beta} \times \vec{E}] \tag{6}$$

The dependence of ω_a on the electric field is eliminated by storing muons with the "magic" $\gamma_\mu = 29.3$, which corresponds to a muon momentum $p_\mu = 3.094 \, GeV/c$. Hence measurement of ω_a and of B determine a_μ from Eq. 5. The muon lifetime in the storage ring is 64.4 μs, the cyclotron period is 149 ns and the g-2 period is 4.4 μs.

2.2 Experimental arrangement

The general arrangement of the experiment is indicated in Figure 1 and a photograph of the muon storage ring is shown in Figure 2.

The AGS operated at 24 GeV and provided, per 2.5 s cycle, 8 proton bunches, each with 5×10^{12} protons and pulse width σ of 27 ns. The proton bunches

Figure 2: Photograph of the muon g-2 storage ring.

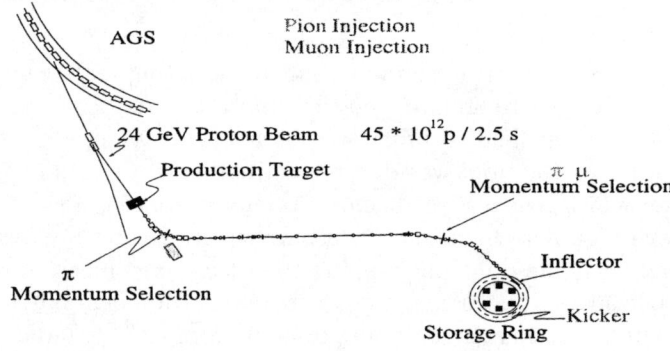

Figure 3: Secondary beam line consisting of dipole and quadrupole magnets

were individually kicked out of the AGS with the time interval between successive bunches of 33 ms and directed onto a nickel target. A positive secondary beam of 3.1 GeV/c with 10^8 particles per bunch was transported along a 116 m secondary beam line to a hole in the back of the yoke of the muon storage ring(Figure 3). The beam composition at the end of the beam line was $\pi^+/\mu^+/e^+ = 1$ with a smaller fraction of p. By momentum selection either π^+ or μ^+ could be injected into the storage ring.

A superconducting inflector magnet of 1.7 m length located in the storage ring magnet substantially cancels the 1.45 T storage ring field and delivers the beam approximately parallel to the central orbit, but 77 mm further out in radius(Figure 4)[7].

The storage ring magnet is a superferric 700 ton, 14 m diameter circular "C"-magnet, with the opening facing inward towards the circle center Figure 5[8]. The

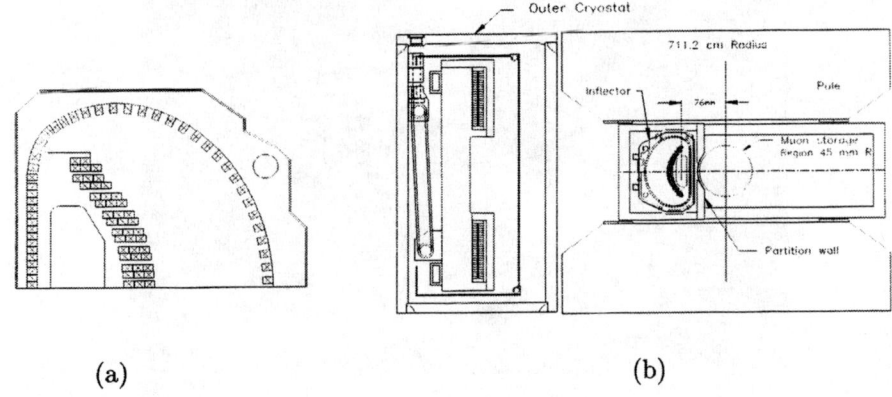

(a)   (b)

Figure 4: Superconducting inflector magnet. (a) double cosine conductor arrangement; (b) View of downstream end of the inflector in the cryostat.

field is excited by four 14 m diameter superconducting coils which carry 5200 A from a low voltage well-regulated power supply to produce the 1.45 T magnetic field. The short term field stability was better than 0.1 ppm, and the long term instability of up to 100 ppm was due to thermal expansion and contraction in the magnet yoke, which had not yet been thermally insulated.

To achieve high field homogeneity a number of features are available for shimming the magnet. These include tipping the pole pieces (each of which cover 10° in azimuth), iron wedges in the air gap between the yoke and pole pieces, edge shims, and current loops on the pole pieces running 360° around the storage ring with a radial spacing of 0.25 cm. The field was monitored by 366 fixed NMR probes placed above and below the beam vacuum chamber and was stable to a few ppm/hour[9]. Periodically the field in the storage region was mapped by an NMR trolley with 17 NMR probes, which operates in vacuum inside the beam vacuum chamber. The combined system of NMR probes permitted relative monitoring of the magnetic field to 0.1 ppm. The absolute calibration of the NMR probes in the trolley was obtained by direct comparison to a standard NMR probe which contained a spherical volume of pure water for which the ratio of the proton magnetic moment to that of a free proton is known[10, 11]. In 1998 this technique provided an absolute measurement of the storage ring field to about 0.5 ppm. For our latest data in 2000, the field averaged around the ring had a uniformity over the 9 cm diameter storage region of 3 ppm, as shown in Figure 6.

Electric quadrupoles are distributed around the ring as shown in Figure 1 and provide weak focusing with a field index n=0.135 for electrode voltages of $\pm 25 kV$.

Initially in 1997[5] we used π^+ injection into the storage ring[5]. For all subsequent data μ^+ were injected.The 1998 data are discussed in this paper[6]. The 10 mrad kick needed to put the muon beam onto a stable orbit was achieved with a

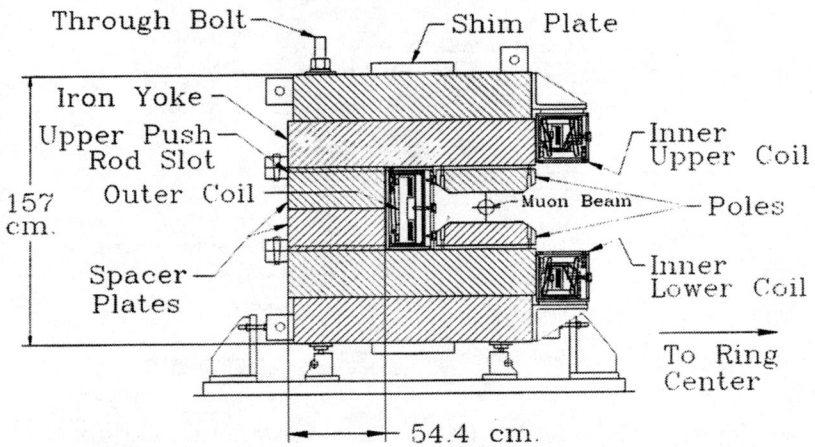

Figure 5: Cross section view of the storage ring magnet.

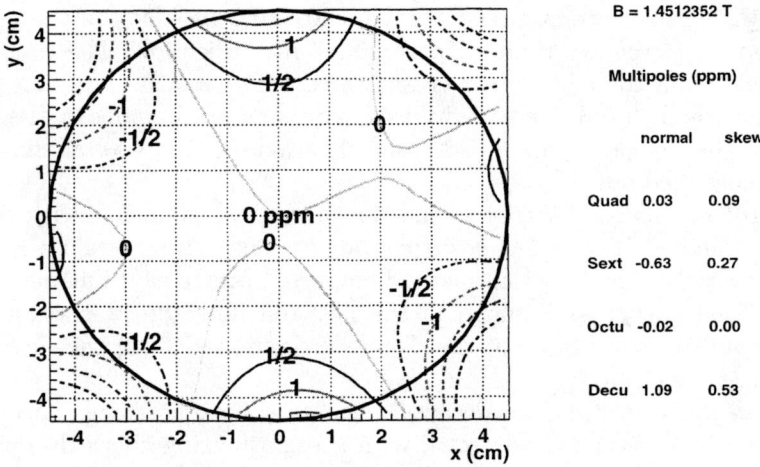

Figure 6: The magnetic field profile in the storage region averaged over azimuth. Each contour line represents a half ppm change. The multipole moments are evaluated at 4.5cm

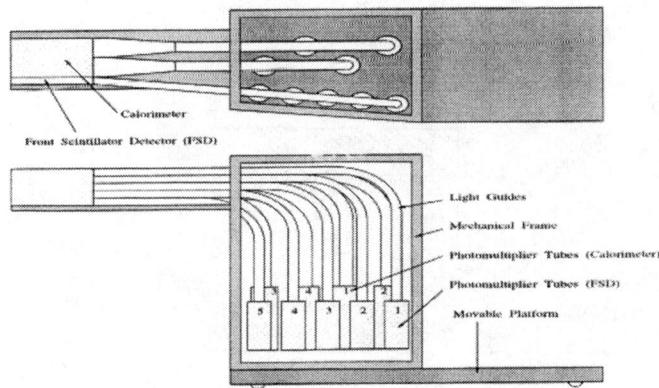

Figure 7: The scintillating fiber electromagnetic calorimeter

peak current of 4100 A and a half period of 400 ns. Three pulse-forming net works powered three identical 1.7 m long one-loop kicker sections consisting of 95 mm high parallel plates on either side of the beam. The current pulse was formed by an under-damped LCR circuit. The kicker plate geometry and composition were chosen to minimize eddy currents. The residual eddy current effect on the total field seen by the muons was less than 0.1 ppm 20 μs after injection. The time-varying magnetic field from the eddy currents was calculated with the program OPERA and was measured in a full-size straight prototype vacuum chamber by the Faraday effect. Since the muons circulate in 149 ns, they were kicked several times before the kicker pulse died out.

About 10^4 muons were stored in the ring per proton bunch. With muon injection, the number of detected positrons per hour was increased by a order of magnitude over the pion-injection method employed previously. Furthermore, the injection-related background (flash) in the positron detectors was reduced by a factor of about 50, since most of the pions were removed from the beam before entering the storage ring.

The decay positrons from the decay $\mu^+ \rightarrow e^+\nu_e\bar{\nu}_\mu$ have an energy spectrum extending up to 3.1 GeV and are detected with Pb-scintillating fiber calorimeters[12] (Figure 7) placed symmetrically at 24 positions around the inside of the storage ring. The scalloped vacuum chamber design minimized pre-showering before the electrons reach the calorimeters. Each calorimeter was read out with four photomultipliers. The observed energy spectrum is shown in Figure 8.

The calorimeter pulses were continuously sampled by custom 400 MHz waveform digitizers (WFDs), which provided both timing and energy information for the decay positrons. For each waveform digitized, the WFD provides at least four time bins before the threshold is crossed to permit pedestal monitoring and subtraction. The waveforms obtained by the WFDs were suppressed, and stored in

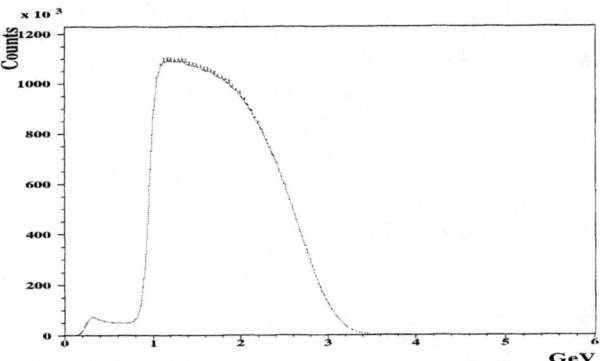

Figure 8: Observed positron energy spectrum

memory in the WFD until the end of the AGS cycle. Between AGS acceleration cycles the data were written to tape for off-line analysis, as were the spectrometer calibration data and the magnetic field data.

Data were accumulated for 8.8 muon lifetimes (567 μs) following injection. A laser and LED system were used to monitor time and gain shifts during this data-collection period. Early to late timing shifts over the first 200 μs were on-average less than 20 ps. Phototube gain and pedestal shifts were less than 1.0%. We note that a gain change over the data-taking period of a single fill will lead to an error in the determination of positron energy and also, because high energy positrons travel farther than low energy ones before striking the detectors, to a systematic timing error.

Knowledge of the muon distribution in the storage ring is important because it is the average magnetic field $\bar{B}$ weighted by the muon distribution that is needed for determining a_μ. The time spectrum of decay positrons, particularly after μ^+ injection into the storage ring, determines by Fourier analysis the distribution of rotation frequencies and hence the radial distribution of muons. This distribution was reproduced with a tracking code, and found to be 3 mm toward the outside of the central storage region. This offset was caused by the mode of operating the kicker. The calculated and measured radial distributions are shown in Figure 9. The magnetic field seen by the muon distribution was calculated by tracking a sample of muons in software through the field map measured by NMR, and by averaging the field values.

2.3   Analysis and Results

For the offline analysis the detector response (waveform shape) to positrons was determined from our data for each calorimeter. These shapes were then fit to all

229

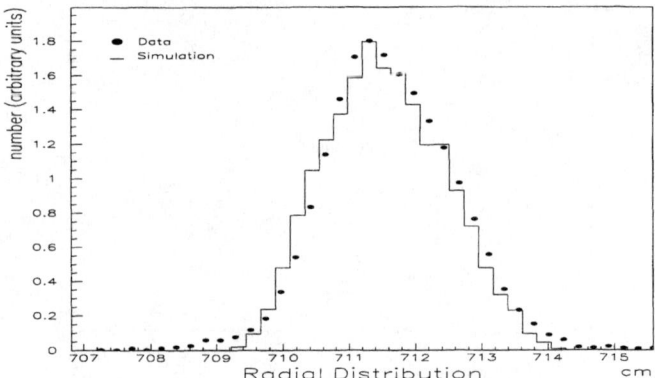

Figure 9: Muon radial distribution calculated using the tracking code (histogram) and obtained from an analysis of the beam debunching at at early times(points).

pulses in the data, and both an amplitude and a time were determined for each pulse.

The observed positron time spectrum shown in Figure 10 was adequately represented by Eq. 7.

$$N(t) = N_0(E)e^{-t/\gamma\tau_\mu}[1 + A(E)cos(\omega_a t + \phi(E))] \tag{7}$$

N is the number of decay positrons detected above some threshold energy E; $N_0(E)$ is a normalization constant; τ is the muon lifetime at rest. Because of the parity violating nature of the weak decay, the higher energy positrons are emitted preferentially along the muon spin direction and hence the muon g-2 precession is observed as a modulation with frequency ω_a and amplitude A(E). The asymmetry A(E) depends on the muon beam polarization and is 0.34 for E=1.8 GeV. The statistical error in determining ω_a is

$$\frac{\delta\omega_a}{\omega_a} = \frac{\sqrt{2}}{\omega_a\tau\gamma AN^{\frac{1}{2}}} \tag{8}$$

Values for ω_a and ω_p, the free proton NMR angular frequency in the storage ring magnetic field, were determined separately and independently. Thereafter the frequency ratio $R = \omega_a/\omega_p$ was determined. A correction of +0.9 ppm was added to R to account for the effects of the electric field and the muon vertical betatron oscillations on ω_a. From the 1998 data $R = 3.707\ 201(19) \times 10^{-3}$, where the 5 ppm error includes a 1 ppm systematic error discussed below[6].

Systematic errors associated with the 1998 data are small compared to the statistical error of 5 ppm. Measurement of the B field together with our knowledge of the muon distribution determines the average field $\bar{B}$ seen by the muons to $\pm 0.5\ ppm$.

230

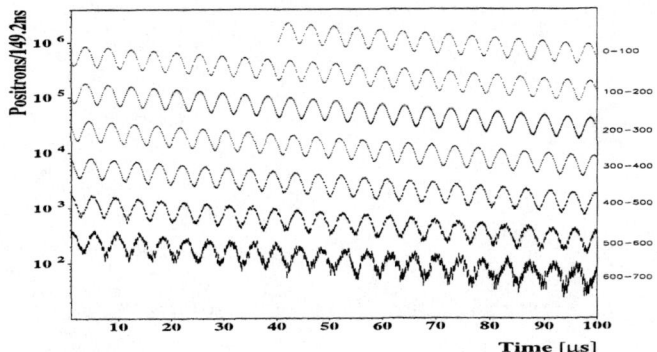

Figure 10: The positron time spectrum obtained with muon injection for the threshold energy $E = 1.8\ GeV$. These data represent 84 million positrons.

The principal systematic errors associated with the determination of ω_a from the positron data arise from pile-up and from AGS "flashlets". A pile-up error occurs when two pulses overlap within the time resolution of about 5 ns and are incorrectly identified as one, which then gives incorrect times and energies for the positrons. Pile-up is estimated to produce an effect on ω_a of less than 0.6 ppm, which we conservatively take as an error estimate. Occasionally under unstable conditions, the AGS was observed to extract beam during our data collection period of 600 μs which caused a background in our calorimeters ("flashlets"). We conservatively estimate that this effect on ω_a in the 1998 data sample is less than 0.5 ppm. Smaller errors arise from the details of the fitting procedure, rate dependent timing shifts and gain changes in the photomultipliers, uncertainties about the radial electric field, about the vertical betatron motion, and from muon losses.

Altogether, the systematic errors on ω_a and ω_p added in quadrature are less than 1 ppm.

The anomalous g-value is obtained from the frequency ratio R by

$$a_{\mu^+} = \frac{R}{\lambda - R} = 11\ 659\ 191(59) \times 10^{-10}(5\ ppm) \tag{9}$$

in which $\lambda = \mu_\mu/\mu_p = 3.183\ 345\ 39(10)[13]$. This new result[6] is in good agreement with the mean of the CERN measurements for a_{μ^+} and a_{μ^-}, and our previous measurement of $a_{\mu^+}[5]$. Assuming CPT symmetry, the weighted mean of the four measurements gives a new world average of

$$a_\mu = 11\ 659\ 205(46) \times 10^{-10}(4\ ppm) \tag{10}$$

$(\chi^2 = 2.7/3)$.

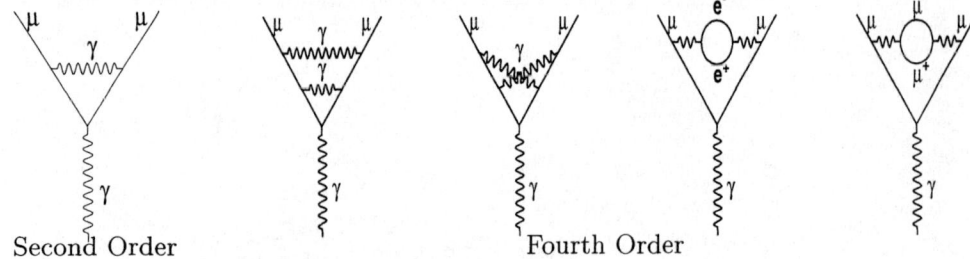

Second Order   Fourth Order

Figure 11: Some QED contributions to a_μ

Whereas for the 1998 data 84×10^6 e^+ were observed, for the 1999 data some 2×10^9 e^+ were recorded and for our 2000 data an additional 6×10^9. Hence the statistical error from all of our μ^+ data should be about 0.5 ppm. Analysis of all our data with particular attention to systematic errors is now in progress. A run with μ^- is scheduled for 2001 with the goal of obtaining a statistical accuracy of 0.5 ppm.

3   Theoretical Value for a_μ

The theoretical value for a_μ in the standard model can written

$$a_\mu(theor) = a_\mu(QED) + a_\mu(had) + a_\mu(weak) [2, 14] \tag{11}$$

3.1   QED Contribution

The contribution of the electromagnetic interaction, $a_\mu(QED)$, includes those of the photon, electron, muon and tauon fields. The result is most conveniently expressed as

$$a_\mu(QED) - a_e(QED) = 1.094\ 337\ 0(\frac{\alpha}{\pi})^2 + 22.867\ 6(33)(\frac{\alpha}{\pi})^3$$

$$+ 127.00(41)(\frac{\alpha}{\pi})^4 + 570(140)(\frac{\alpha}{\pi})^5 \tag{12}$$

Using $a_e(theor)[2]$ and the $\alpha^{-1}(a_e)$ value from Eq. 2 we obtain:

$$a_\mu(QED) = 116\ 584\ 705.7(1.8) \times 10^{-11} [16ppb] \tag{13}$$

It is interesting to note that the contribution of the tau lepton through modifying the photon propagator amounts to 40×10^{-11} or 0.36 ppm.

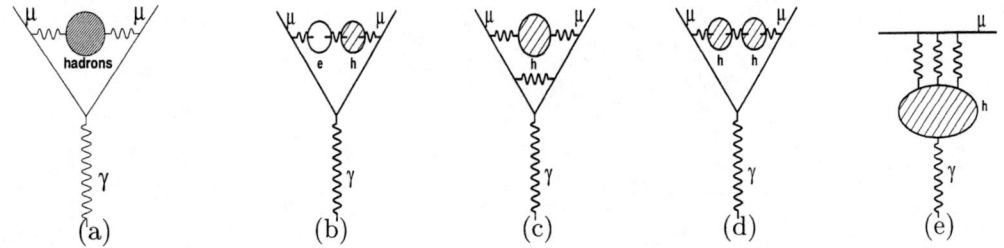

Figure 12: Hadron contribution to a_μ. (a) Lowest order vacuum polarization diagram; (b, c, d, e) Second order diagrams.

3.2   Hadronic Contribution

The hadronic contribution to a_μ is quite large, amounting to about 60 ppm, and contributes the principal uncertainty to $a_\mu(theor)$. It arises in lowest order as a vacuum polarization correction involving virtual hadrons as shown in Figure 12(a).

This diagram can not now be calculated from QCD. However, by dispersion theory this contribution can be related to $R(s) = \frac{\sigma_{total}(e^+e^- \rightarrow hadrons)}{\sigma_{total}(e^+e^- \rightarrow \mu^+\mu^-)}$, in which s is the square of the total energy in the center of mass for the colliding particles e^+, e^-. The dispersion relation of Eq. (14) involves the integral from the threshold energy for pion pair production to ∞ and contains the $1/s^2$ factor as well as the kinematic factor K(s), which increases monotonically to 1 as $s \rightarrow \infty$. The value for $a_\mu(had)$ is evaluated from the dispersion integral with measured values of R.

$$a_\mu(had\ 1) = (\frac{\alpha m_\mu}{3\pi})^2 \int_{4m_\pi^2}^\infty \frac{ds}{s^2} K(s)R(s) \tag{14}$$

Data from hadronic τ decay can also be used, assuming the validity of CVC.

The principal contribution to $a_\mu(had\ 1)$ comes from the region below $\sqrt{s} = 1\ GeV$. Extensive and accurate measurements of R have been made and are continuing at the Budker Institute of Nuclear Physics in Novosibirsk in the energy range for $\sqrt{s} = 0.3$ to 1.4 Gev, using VEPP-IIM and their CMD-II detector. New data on R are also being obtained with the e^+e^- collider in Beijing in the approximate range $\sqrt{s} = 2$ to 5 GeV and are expected also from the new Frascatti ϕ factory. Extensive data on hadronic τ decay come from Cornell and LEP.

Feynman diagrams for higher order hadronic contributions, $a_\mu(had\ 2)$, of relative order α, are shown in Figure 12. Diagrams (b),(c) and (d) can be expressed in terms of the dispersion integral, but the evaluation of diagram (e), which is designated hadronic light-by-light scattering, has not been successfully expressed in terms of experimentally accessible variables. Approximate calculations within the framework of chiral perturbation theory and the $1/N_c$ expansion give the result shown in Eq.15. The full hadron contribution is

$$a_\mu(had) = a_\mu(had\ 1) + a_\mu(had\ 2bcd) + a_\mu(had\ 2e)$$

233

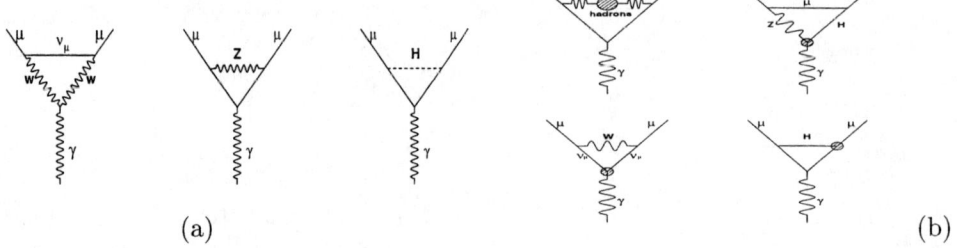

(a)   (b)

Figure 13: Weak contribution to a_μ, (a) Lowest order one loop diagrams; (b) Some higher order two loop electroweak diagrams.

$$= 6951(75) \times 10^{-11} - 101(6) \times 10^{-11} - 79(15) \times 10^{-11} \qquad (15)$$

$$= 6771(77) \times 10^{-11}$$

The ongoing experimental measurements of R mentioned above should reduce the error in a_μ(had).

3.3   Weak Contribution

The weak interaction contribution based on the standard electroweak theory is a weak radiative correction to an electromagnetic interaction and arises in lowest order from the single loop diagrams in Figure 13a which involve ν_μ, Z and H particle exchange and where two vertices are weak interaction vertices. The diagrams with the W and Z particles contribute appreciably as given in Eq.(16), but since searches establish that the mass of the Higgs particle M_H is greater than 77 GeV, the contribution from the diagram with H is relatively negligible.

$$\Delta a_\mu(W) = \frac{G_F m_\mu^2}{8\pi^2\sqrt{2}} \times \frac{10}{3} = +3.89 \times 10^{-9}$$

$$\Delta a_\mu(Z) = \frac{G_F m_\mu^2}{8\pi^2\sqrt{2}} \times \frac{1}{3}[(3 - 4cos^2\theta_W)^2 - 5] = -1.94 \times 10^{-9} \qquad (16)$$

where we take $sin^2\theta_W = 0.224$ for the weak mixing angle.

The next order electroweak contribution involving two loop diagrams (Figure 13b) has been fully calculated. The total weak contribution is

$$a_\mu(weak) = a_\mu^{EW}(1\ loop) + a_\mu^{EW}(2\ loop)$$

$$= 195 \times 10^{-11} - 44 \times 10^{-11} = 151(4) \times 10^{-11} \qquad (17)$$

Comparison of theory and experiment on a_μ(weak) will constitute a new and sensitive test of the unified electroweak theory with its prescription for renormalizability. Just as virtual electromagnetic radiative corrections were critical to the

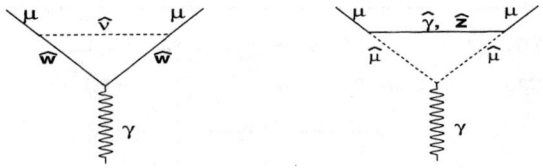

Figure 14: Supersymmetric loop comtributions to a_μ.

development of modern renormalized quantum electrodynamics so virtual radiative corrections involving both the weak and electromagnetic interactions are most important to the renormalized unified electroweak theory.

Adding the QED, hadronic and weak contributions we obtain

$$a_\mu(theor) = 116\ 591\ 628(77) \times 10^{-11}(0.66ppm) \tag{18}$$

3.4 Contributions Beyond the Standard Model

Proposed extensions of the standard model will in general contribute to a_μ and hence a comparison of a_μ(expt) with a_μ(SM) can in principle detect physics beyond the standard model[14, 15]. Two classes of extensions can be considered. One postulates compositeness or internal structure for leptons, quarks or gauge bosons and the other involves the introduction of extra groups or particles such as extra gauge bosons, leptoquarks, or supersymmetric particles.

In the standard model leptons, quarks and gauge bosons are assumed to be pointlike elementary particles with no internal structure. The magnetic moment of a particle provides a sensitive test for its compositeness, as we have learned for example in the case of the proton. If the muon is composite, the current theoretical viewpoint would imply that

$$\Delta a_\mu \sim \frac{m_\mu^2}{\Lambda^2} \tag{19}$$

in which Λ is the composite mass scale. From the present accuracy in a_μ and the agreement of a_μ(expt) and a_μ(theor) we obtain $\Lambda > 1$ TeV. A determination of a_μ to 0.35 ppm, which is the goal of the BNL experiment, would be sensitive to $\Lambda > 4$ to 5 GeV. If the muon were composite, excited muon states would be expected, and from an experimental accuracy for a_μ at 0.35 ppm a sensitivity to m_μ^* up to 400 GeV would be obtained, which is comparable to that from LEP II with $E_{cm}=200$ GeV. Compositeness of the W gauge boson would lead to an anomalous g_W value $\Delta\kappa$. Determination of a_μ to 0.35 ppm would provide a sensitivity to $\Delta\kappa=0.04$, which corresponds to $\Lambda_W \sim 2$ TeV, and exceeds considerably the sensitivity possible with LEP II or LHC.

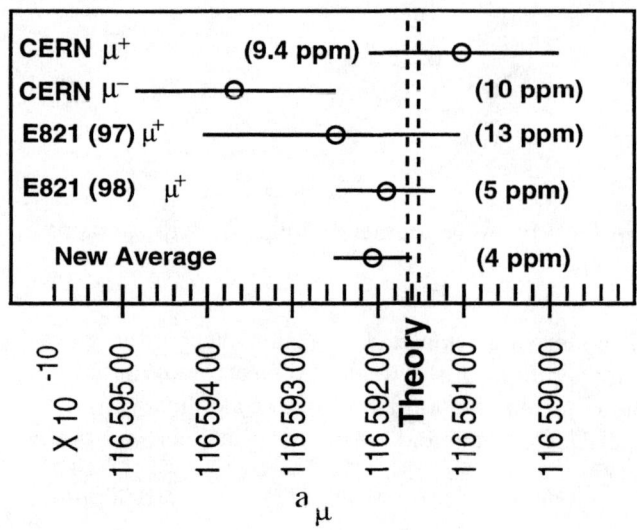

Figure 15: Experimental and theoretical values for a_μ

Supersymmetry connects fermions and bosons and introduces supersymmetric partners of known particles(sparticles). These contribute to a_μ through loop diagrams(Figure 14). A determination of a_μ to 0.35 ppm would provide a sensitivity to m_{susy} for a large value of the ratio of Higgs masses, which is comparable to that from the highest energy collider.

4   Present Comparison of Theory and Experiment

The four precise measurements of a_μ and their average are shown in Figure 15 along with the standard model prediction. The weighted mean of the experimental results agrees with the standard model, with

$$a_\mu(expt) - a_\mu(theor) = (42 \pm 47) \times 10^{-10} \tag{20}$$

or equivalently 3.6 ± 4.0 ppm. This agreement of theory and experiment further constrains new physics beyond the standard model.

5   Acknowledgements:

Preparation of this paper has been supported in part by the US Department of Energy. The author also thanks Huaizhang Deng for important assistance with this manuscript.

References

[1] Van Dyck, R.S. Jr., Schwinberg, P.B. and Dehmelt, H.G., Phys. Rev. Lett., **59**, 26 (1987); Van Dyck, R.S. Jr., in *Quantum Electrodynamics*, ed. Kinoshita, T. (*World Scientific, Singapore*, 1990), p. 322.

[2] Hughes, V.W. and Kinoshita, T., Rev. Mod. Phys. **71**, No. 2, S133(1999).

[3] Bailey, J., *et. al.*, Nucl. Phys. **B150**, 1(1979).

[4] Farley, F,J.M. and Picasso, E. in *Quantum Electrodynamics*, ed. Kinoshita, T., (*World Scientific, Singapore*, 1990), p. 479.

[5] Carey, R.M. *et al.*, Muon g-2 Collaboration, Phys. Rev. Lett. **82**, 1632(1999).

[6] Brown, H.N. *et. al.*, Muon g-2 Collaboration, "Improved Measurement of the Positive Muon Anomalous Magnetic Moment", Submitted to Phys. Rev. D.

[7] Krienen, F., Loomba, D. and Meng, W., Nucl. Instrum. Methods, **A283**, 5(1989).

[8] Danby, G. *et. al*, Muon g-2 Collaboration, "The Brookhaven Muon Storage Ring Magnet", Submitted to Nucl. Instrum. Methods.

[9] Prigl, R. *et. al.*, Nucl. Inst. Methods **A374**, 118(1996).

[10] Fei, X., Hughes, V.W. and Prigl, R., Nucl. Instrum. Methods **A394**, 349(1997).

[11] Phillips, W.D. *et. al.*, Metrologia **13**, 179(1977).

[12] Sedykh, S.A. *et. al.*, "Electromagnetic Calorimeters for the BNL Muon g-2 Experiment", Submitted to Nucl. Instrum. Methods.

[13] Liu, W. *et. al.*, Phys. Rev. Lett. **82**, 711(1999).

[14] Kinoshita, T. and Marciano, W.J. in *Quantum Electrodynamics* ed. Kinoshita T. (World Scientific, Singapore, 1990), p 419.

[15] Czarnecki, A. and Marciano, W.J., Nucl. Phys. B (Proc. Suppl.) **76**, 245 1999.

Precise Physics of Simple Atoms

Savely G. Karshenboim

D.I. Mendeleev Institute for Metrology, 198005 St. Petersburg, Russia
Max–Planck–Institut für Quantenoptik, 85748 Garching, Germany[1]

Abstract. We give a review of experimental and theoretical results on the precision study of hydrogen–like atoms with low value of the nuclear charge Z.

The simplicity of "simple" atoms has been for a while a challenge to precision theory and experiment. Are the hydrogen–like atomic systems simple enough to be calculated with an accuracy, appropriate to compete with the best experimental results? That is a question, that theorists have tried to answer. The simplest atoms are different two–body bound systems with a low value of the nuclear charge: $Z = 1$ (hydrogen, deuterium, muonium and positronium) and $Z = 2$ (ions of helium–3 and helium–4) etc. We do not try to review theoretical calculations (if necessary details can be found in Ref. [1]), but present state of art in physics of simple atoms and discuss in detail the theoretical and experimental status of studying such atoms.

LOW-ENERGY TESTS OF QED

The precise physics of simple atom is the most interesting part of the so-called low-energy tests of Quantum Electrodynamics (*QED*). Low energy tests of QED offer a number of different options:

- A study with free particles provides the possibility of testing the QED Lagrangian for free particles. The most accurate data arise from anomalous magnetic moments of the electron (*Kinoshita*[†2]) and the muon [2].

- However, one knows that the bound problem makes all calculations more complicated. Bound state QED is not a well–established theory. It involves different effective approaches to solve a two–body problem. These approaches can be essentially checked with low Z atomic systems like for e. g. hydrogen

---

[1] Summer address

[2] References marked with † correspond to presentations at a satellite meeting to ICAP named *Hydrogen atom, 2: Precision physics of simple atomic system* and its Proceedings will be published by Springer in 2001.

CP551, *Atomic Physics 17*, edited by E. Arimondo, P. DeNatale, and M. Inguscio
© 2001 American Institute of Physics 1-56396-982-3/01/$18.00

and deuterium (*Hänsch*[†]), neutral helium (*Drake*[†]) and helium ions, muonium (*Jungmann*[†]), positronium (*Conti*[†]), muonic hydrogen and helium etc. We consider here most of the low–*Z* atoms.

- Study of high–*Z* ions (*Myers*[†], *Stölker*[†]) cannot further the test of the bound state QED because of large contributions due to the nuclear structure. Rather such an investigation is useful for trying different nuclear models. However, in some particular cases, atomic systems with a not too high *Z* can give some important information on higher order terms of the QED *Zα* expansion.

- There are some other two–body atoms under investigation. They contain a hadron as an orbiting particle. Different antiprotonic (*Yamazaki*[†]) and pionic (*Nemenov*[†]) atoms provide a unique opportunity to study particle property with spectroscopic means with a high precision. In some sense it is not possible to have low precision: if a signal is detected the accuracy is granted.

The precision study of the simple atoms is not only limited by experiments with simple atoms. The theory is not able to predict anything to be comparable to the experimental data. What theory can do is to express a measurable quantity in terms of fundamental constants and particle (or nuclear) properties.

First of all we need to determine somehow the Rydberg constant (R_∞), the fine structure constant (α) and the electron mass in some appropriate units (e. g. in atomic units or in terms of the proton mass (m_e/m_p)). Uncertainties arising from these constants are sometimes compatible with other items of the uncertainty budget or they are even sometimes the most important source of inaccuracy. One should remember that the electron is the most fundamental particle for physics, chemistry, and metrology and the constants associated with its properties go through any atomic spectroscopic effects and any quantum electromagnetic effects. Due to that a number of different studies, which are very far from the spectroscopy of simple atoms (like e. g. Watt balance experiment (see *Mohr*[†] for detail)), are really strongly connected with the precision physics of simple atoms.

However, a knowledge of the universal fundamental constants is not enough for precision theoretical predictions and we need to learn also some more specific constants like for e. g. the muon mass or the proton electric charge radius. The former is important for the muonium hyperfine structure, while the latter is for calculating the hydrogen Lamb shift.

SPECTRA OF SIMPLE ATOMS

Let us discuss the spectrum of simple atoms in more detail. The gross structure of atomic levels in a hydrogen–like atom comes from the Schödinger equation with the Coulomb potential and the result is well–known[3] $E_n = -(Z\alpha)^2 m_e$. There are a number of different corrections: the relativistic ones (one can find them from

---

[3]) We use the relativistic units in which $\hbar = c = 1$.

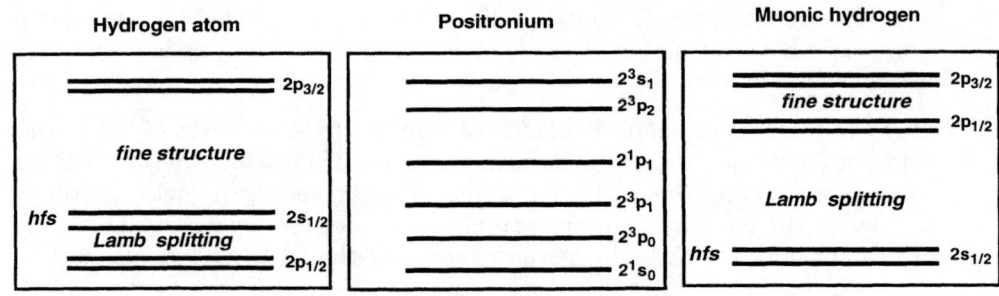

FIGURE 1. Scheme of the lowest excited levels ($n = 2$) in different simple atoms

the Dirac equation), the hyperfine structure (due to the nuclear magnetic moment) and the QED ones. A structure of levels with the same value of the principal quantum number n is a signature of any atomic system. In Fig. 1 we present three different spectra of the structure at $n = 2$. The first one is realized in "normal" (electronic) hydrogen–like atoms (hydrogen, deuterium, helium ions etc). The muonium spectrum is the same. The largest splitting, of order $(Z\alpha)^4 m_e$, is the fine structure (i. e. the splitting between levels with a different value of the electron angular momentum j), the Lamb shift arising from the electron self–energy effects is of order $\alpha(Z\alpha)^4 m_e \ln(1/(Z\alpha))$ and it splits the levels with the same j and different values of the electron orbital momentum l. Some nuclei are spinless (like e. g. in ^{4}He), while others have a non–zero spin (in hydrogen, deuterium, muonium, helium–3). In the latter case, the nuclear spin splits levels with the same electronic quantum number. The splitting are of order $(Z\alpha)^4 m_e^2/M$ or $\alpha(Z\alpha)^3 m_e^2/m_p$, where M is the nuclear mass, and the structure depends on the value of the nuclear spin. The scheme in Fig. 1 is for nuclear spin $1/2$ (hydrogen).

The structure of levels in positronium and muonic atoms is different because some other effects enter into consideration. For positronium, an important feature is a real (into two and three photons) and virtual (into one photon) annihilation. The former is responsible for the decay of the s-states, while the latter shifts triplet levels (and $2^3 s_1$ in particular). The shift is of the order of $\alpha^4 m_e$. Contributions of the same order arise from relativistic effects and hyperfine interactions. As a result the positronium level structure at $n = 2$ has no hierarchy (Fig. 1).

Another situation is that for the muonic atoms. A difference comes from a contribution due to the vacuum polarization effect (the Uehling potential). Effects of electronic vacuum polarization shift all levels to the order of $\alpha(Z\alpha)^2 m_\mu$. This shift is a nonrelativistic one and it splits 2s and 2p levels. The fine and hyperfine structures are of the same form as for the normal atoms (i. e. $(Z\alpha)^4 m_\mu$ and $(Z\alpha)^4 m_\mu^2/M$ respectively) and at low Z the Lamb shift is a dominant correction to the energy levels.

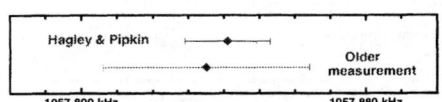

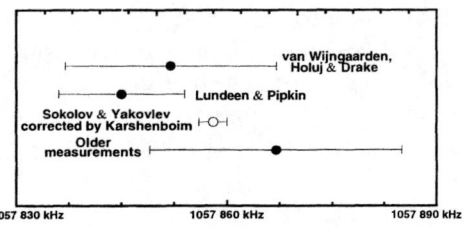

FIGURE 2. Indirect determination of the Lamb shift $(2s_{1/2}-2p_{1/2})$ in atomic hydrogen via a study of the fine structure $2p_{3/2}-2s_{1/2}$. See Ref. [3] for references.

FIGURE 3. Direct measurements of the Lamb shift in the hydrogen atom. The references can be found in Ref. [3].

HYDROGEN LAMB SHIFT

A number of different splittings have been precisely studied for about a century. Bound state QED and maybe even QED itself was essentially established after a study of the Lamb shift and the fine and hyperfine structures in hydrogen, deuterium and helium ions. In the last decades, progress with such measurements was quite slow. The results of the last twenty years are presented in Fig 2 (Lamb shift) and 3 (fine structure recalculated in terms of the Lamb shift), while the older experiments are averaged (see Ref. [3] for references). To reach the Lamb shift from the fine structure $(2p_{3/2}-2s_{1/2})$ measurement we need to use a value of the $2p_{3/2}-2p_{1/2}$ splitting which was found theoretically. The most direct results of the Lamb shift need no QED theory. A result claimed to be the most accurate one (*Sokolov[†]*) has an uncertainty of about 2 ppm. It is corrected because of a recalculation of the lifetime of the $2p_{1/2}$ state [4]:

$$\tau^{-1}(2p_{1/2}) = \frac{2^{10}\pi}{3^8}\,\alpha^3\,R_\infty\,\frac{m_R}{m}\left\{1 + \ln\left(\frac{9}{8}\right)(Z\alpha)^2 + \frac{\alpha(Z\alpha)^2}{\pi}\,(8.045...)\,\ln\frac{1}{(Z\alpha)^2}\right\}.$$

There is some criticism by E. Hinds [5] and it is not clear if this result is as accurate as claimed. We wish to note, however, that common opinion on the direct Lamb shift measurement contains two contradicting statements. Firstly, it is generally believed that a Lamb splitting of $2s_{1/2}$ and $2p_{1/2}$ (about 1 GHz), with a decay width of 2p being 0.1 GHz, cannot be measured better than 10 ppm. This means that the statistic error should be larger than 10 ppm. Secondly, it is believed that Sokolov's experiment is incorrect only because of a possible systematic error claimed by Hinds. However, nobody insists that the statistical treatment of Sokolov's data was incorrect and we can hope that traditional methods can go far beyond 10 ppm level. Measurement of the deuterium Lamb shift within the Sokolov scheme will provide a chance to test some systematics of his experiment.

Essential progress in study of the hydrogen Lamb shift comes recently from the optical two–photon Doppler–free experiments (see *Hänsch*[†] and *Schwob et al.*[†] fort detail). The Doppler–free measurement offers a determination of some transition frequency in the gross structure with a accuracy high enough to use the results to find the Lamb shift. However, two problems arise due to these experiments.

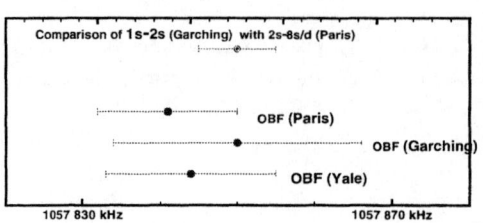

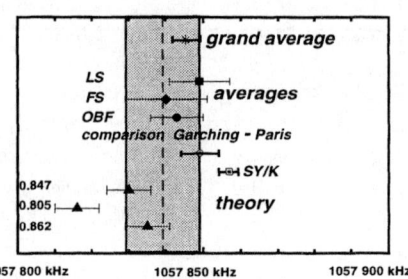

FIGURE 4. Optical determination of the lamb shift in the hydrogen atom. The references to the *optical beat frequency experiments* can be found in Ref. [3].

FIGURE 5. Comparison of experiment and theory for the hydrogen Lamb shift. The references can be found in Ref. [3].

The transition energy between different levels of the gross structure is mainly determined by the Rydberg energy: $-R_\infty/n^2$. To extract the Lamb shift we first need to find a value of the Rydberg constant. There are two ways to manage this. Following the first of them one has to measure two different frequencies within one experiment with the ratio of the frequencies being an integer number. Obtaining a beat frequency one can avoid the problem of determining the Rydberg constant. Three experiments have been performed in this way: the Garching experiment dealt with the 1s–2s transition and the 2s–4s (and 4d), at Yale 1s–2s frequency was compared with the one–photon 2s-4p transition and that was the only precision optical experiment with a one–photon transition. The recent Paris experiment worked with 1s–3s and 2s–6s (and 6d). The values derived from these experiments are collected in Fig. 4.

Another way to manage the problem with the Rydberg constant is to do two independent absolute frequency measurements (i. e. measurements in respect to the primary cesium standard) and to compare them afterwards, hence determining both the Rydberg constant and the Lamb shift. Such an approach, combining results from Garching (1s–2s) and from Paris (on 2s–8s, –8d, –12d), gave another optical value (Fig. 4). Some of the optical experiments were also performed for deuterium and that may improve the accuracy in the determination of the Rydberg constant and, thence, of the hydrogen Lamb shift.

However, the values in Fig. 4 derived from the optical measurements need further theoretical treatment. The experiments involved a number of levels (1s, 2s, 3s etc) and with optical experimental data there was also a problem of an increasing number of levels with an unknown Lamb shifts. The problem was solved with the help of a specific difference [6]:

$$\Delta(n) = E_L(1s) - n^3 E_L(ns)$$
$$= \frac{\alpha(Z\alpha)^4}{\pi} \frac{m_R^3}{m^2} \times \left\{ -\frac{4}{3} \ln \frac{k_0(1s)}{k_0(ns)} \left(1 + Z\frac{m}{M}\right)^2 + C_{Rec}\frac{Zm}{M} \right.$$
$$\left. + (Z\alpha)^2 \left[A_{61} \ln \frac{1}{(Z\alpha)^2} + A_{60}^{VP}(n) + G_n^{SE}(Z\alpha) + \frac{\alpha}{\pi} \ln^2 \frac{1}{(Z\alpha)^2} B_{62}(n) \right] \right\}, \quad (1)$$

where the coefficients A_{61}, $A_{60}^{VP}(n)$, C_{Rec} and $B_{62}(n)$ and a table for $G_n^{SE}(Z\alpha)$ can be found in Refs. [7,3]. The uncertainty was also discussed there. The difference has a better established status than that for 1s (or 2s) Lamb shifts (see Table 1). The uncertainty budget was improved recently after calculations of one–loop corrections, exactly at $Z = 1$, (*Jentschura et al.*[†]) and leading three–loop contributions (*Melnikov and van Ritbergen*[†]).

The theory of $2p_{3/2}$–$2p_{1/2}$ splitting is also well established. Perhaps, we have to clarify here the word "theoretical". A value is a *theoretical* one if it is sensitive to *theoretical* problems (like the problem of the proton radius and of higher–order QED corrections for the Lamb shift). An insensitive, sterile value is not theoretical, it is rather a *mathematical* one, and that is the case for the difference $\Delta(2)$ and the $2p_j$ energy. Details of theoretical calculations can be found in the review [1].

TABLE 1. Theoretical unceratinty of the different corrections for the Lamb shift in hydrogen. * In case of recoil term we present a value of contradiction between different calculations. * We give an estimated unceratinty of proper reevaluation of the most accurate data.

Contribution	$\delta E(2s)$	$\delta \Delta(2)$	$\delta E(2p)$
Two–loop	2 kHz	0.6 kHz	0.1 kHz
Recoil	0.9 kHz*	-	-.
Radiative-recoil	0.05 kHz	0.05 kHz	0.05 kHz
Nuclear structure	~ 10 kHz*	-	-

NUCLEAR STRUCTURE EFFECTS

Now we can compare theory and experiment for the 2s Lamb shift. We summarize them in Fig. 5, where we present average values for the Lamb shift, fine structure and optical beat frequency and comparison experiments. What is important is the influence of the nuclear charge distribution on the energy levels

$$\Delta E(nl) = \frac{2}{3} \frac{(Z\alpha)^4}{n^3} m^3 R_N^2 \delta_{l0} , \qquad (2)$$

where R_N is a mean–squared nuclear charge radius. The position of theoretical values depends on the accepted value for the proton charge radius. We label three theoretical values with the proton radius (0.847 fm – Mainz dispersion analysis paper, 0.805 fm – Stanford scattering experiment, 0.862 fm – Mainz scattering). More values for the proton radius are collected in Fig. 7 (see [3] for references). To discuss the discrepancy let us look at the most important data on electron–proton elastic scattering presented in Fig. 8. One can see that the Mainz experiment is more appropriate to precisely determine the proton radius containing more points at lower momentum transfer and with a higher precision. Due to this any compilation containing the Mainz data has to lead to a result close to the Mainz result, because the Mainz scattering points must be statistically responsible for the final result and, in particular, the dispersion analysis performed by Mainz theorists led to such a result. However it (R_p = 0.847(9) fm) differs from the empirical value (R_p = 0.862(12) fm). One problem in evaluating the data is their normalization. One can write a low momentum expansion of the form factor

$$G(q^2) = a_0 + a_1 q^2 + a_2 q^4 + \ldots \qquad (3)$$

From a theoretical point of view $G(0) = 1$ indeed. However, the normalization measurement was accurate not enough (an in particular in the Mainz case it is about 0.5%) and that means that a value tabulated from the data, as being the form factor, differed from it with some normalization. Three different fitting were performed by Wong [8] (see Fig. 8). The free fittings of a_0 led to a larger uncertainty (Wong–Mainz value in Fig. 7). Even this result must be treated as a preliminary value. It is necessary to take into account some higher–order corrections and that is not possible because of the absence of any complete description of the experiment. The reasonable estimate of the theory is presented in Fig. 5 as a filled area. All experimental values are consistent with the theory exept the corrected value from the Sokolov and Yakovlev experiment. The present status is that the computation uncertainty is about 2 ppm, the measurement inaccuracy of the grand average value is 3 ppm, while the uncertainty due to the proton size is about 10 ppm.

The problem of the nuclear size is not only a problem of the hydrogen Lamb shift: a similar situation arises with the helium–4 ion Lamb shift, where uncertainties resulting from the QED computation and the nuclear size are about the same. The comparison of theory to experiment is presented in Fig. 9. The evolution of the measured value has been due to a study of possible systematic sources (*Drake and van Wijngaarden*[†]).

PROTON–FREE HYDROGEN PHYSICS

A more difficult problem is that of the hyperfine structure, which is more sensitive to the nuclear structure. While the experimental uncertainty is below 10^{-12}, the

form factor $G(q^2)$

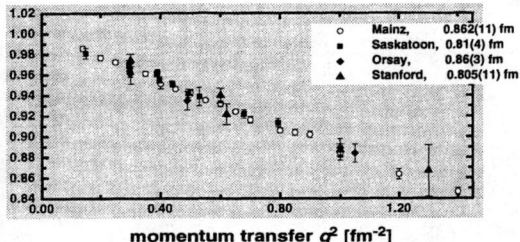

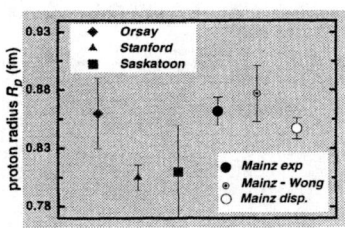

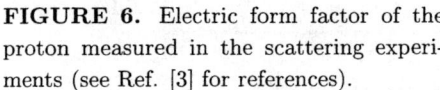

momentum transfer q^2 [fm^{-2}]

FIGURE 6. Electric form factor of the proton measured in the scattering experiments (see Ref. [3] for references).

FIGURE 7. Proton charge radius determined from the scattering experiments. The references can be found in Ref. [3].

theoretical inaccuracy is about 10 ppm. The main problem is a distribution of the magnetic moment inside the proton. It seems the scattering cannot provide accurate enough data and we need to discuss how to manage the problem of nuclear structure by means of the atomic physics. We consider here three ways to do that:

- one is based on study of muonic atoms (the muon is a lepton with a lifetime of about 2 μs and a mass of about $207\, m_e$);

- another deals with the special difference $\Delta_{hfs} = E_{hfs}(1s) - E_{hfs}(2s)$ (cf. in Eq. (1)), which can be precisily measured;

- the third is for atoms without nuclear structure. In such an atom one must substitute the proton by some more appropriate positive particle (muon or positron).

A promising way is to determine the nuclear structure with muonic atoms and in particular with muonic hydrogen. The muon orbit lies lower than the electron one. Since $m_\mu \simeq 207\, m_e$ the muon hydrogen Bohr radius is about 200 times smaller than that in hydrogen and, hence, the former is more sensitive to nuclear effects. A scheme of an experiment running now at PSI (*Pohl et al*[†]) is presented in Fig. 10. The experiment consists of the following steps: creating a metastable 2s state, exiting it to the 2p state by a laser, measuring the intensity of the X–ray decay 2p–1s. A similar scheme was used for muonic helium [9], however a recent experiment by PSI [10] showed no appropriate signal (Fig. 11). A study of the helium experiment revealed a crucial point: the creation of a metastable state, which can be destroyed by collisions. The collision rate is proportional to the target gas density as well as the rate of creating the muonic atom, and so the density cannot be varied arbitrarily. The slow muon beam at PSI allow one to use a low density gas target and creation of the 2s state has been detected. In case of success, the PSI experiment will give us the charge radius of the proton and the so–called Zemach correction

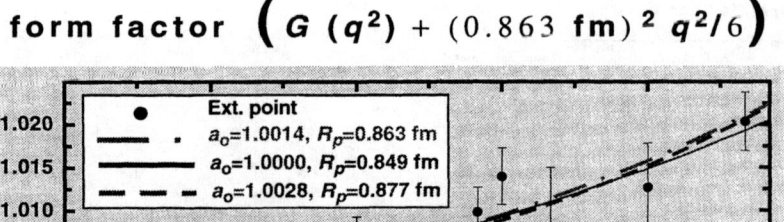

FIGURE 8. Fitting the electric form factor of the proton from the Mainz experimental data. The references can be found in Ref. [3].

to the 2s state of muonic hydrogen. Comparison of the muonic hydrogen hfs and hydrogen hfs will allow us to go farther with the study of the proton structure.

Another way to manage the problem of nuclear structure is to compare the 1s and 2s hfs. The experiments were performed for hydrogen (recently by *Rothery and Hessels*[†]), deuterium and helium ion. The recent hydrogen experiment has attracted our attention to the problem of Δ_{hfs} and it was discovered (*Karshenboim and Sapirstein*[†]) that the results (and primarily those for the helium ion) are quite sensitive to higher order corrections. All value for the hfs used to be presented in units of Fermi energy (ν_F), which is the result of a nonrelativistic interaction of the magnetic moment of an electron in the 1s state and the nucleus. The accuracy of the difference allows one to detect the fourth order corrections, namely, $\alpha(Z\alpha)^3$, $\alpha^2(Z\alpha)^2$, $\alpha(Z\alpha)^2 m/M$, and $(Z\alpha)^3 m/M$.

The same fourth order corrections are now a subject of study in the muonium ground state hfs (see Table 2). A muonium atom is a kind of hydrogen without the proton: instead the proton the nucleus is a positive muon. The present status of the muonium hyperfine splitting is as follows: the experiment at LAMPF gave 4 463 302 765(53) Hz , while the theoretical prediction is consistent with the experiment but less accurate. A computational part of the uncertainty is about 200 Hz, while a hidden experimental uncertainty in the theory is about 500 Hz. It is due to a calculation of the Fermi energy, which is proportional to the muon magnetic moment, determined from the same LAMPF experiment. Possible progress is considered by *Jungmann*[†].

Another proton–free simple atom is positronium. Its lifetime is much shorter than that of muonium, but it can be more easily produced. Different measurements in positronium are summarized in Fig. 12–16. Energy levels in positronium can be

Lamb shift 2p$_{1/2}$ - 2s$_{1/2}$ in helium ion (MHz)

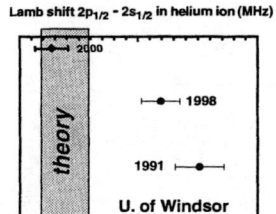

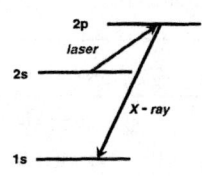

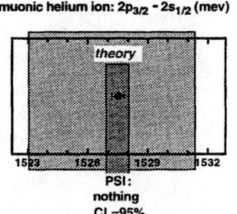

muonic helium ion: 2p$_{3/2}$ - 2s$_{1/2}$ (mev)

FIGURE 9. Experiment on the Lamb shift in ions of helium–4 at the University of Windsor (*Drake and Wijngaarden*[t]).

FIGURE 10. Scheme of the experiment on the Lamb shift in a light muonic atom

FIGURE 11. Study of the Lamb shift in muonic helium.

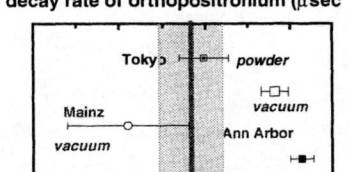

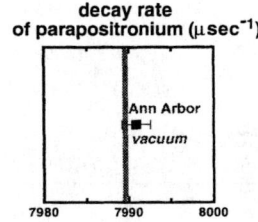

decay rate of orthopositronium (μsec^{-1})

decay rate of parapositronium (μsec^{-1})

FIGURE 12. Decay of orthopositronium. The references can be found in *Conti*[t].

FIGURE 13. Decay of parapositronium (see *Comti*[t] for reference).

presented in the form

$$R_\infty \times \left\{ C_{20} + C_{40}\alpha^2 + C_{50}\alpha^2 + (C_{61}\ln(1/\alpha) + C_{60})\,\alpha^2 + \left(C_{72}\ln^2(1/\alpha) + \ldots \right)\alpha^3 \right\}\,.$$

After two decades of intensive theoretical study we know the coefficients in the above expression up to C_{60} ([12], *Czarnecki et al.*[t]) and C_{72} ([13,14]). The decay width in positronium is known up to fractinal order α^2 (*Czarnecki et al.*[t] for parapositronium) (and for orthopositronium *Adkins*[t]) and $\alpha^3 \ln^2 \alpha$ [13]. Since the Adkins's result is a preliminary one we do not include it in Fig. 12. The Adkins's result led to a fractional correction about $4\alpha^2$ and so the theory is in contradiction to the Ann Arbor experiment. Some progress in the study of positronium is expected in the near future (*Conti*[t]).

TABLE 2. Fourth order corrections to the muonium hyperfine structure. Most of references can be found in Refs. [11] and [1].

Contribution	Numerical result	Reference
$(Z\alpha)^4$	0.03 kHz	Breit
$(Z\alpha)^2(m/M)^2\ln(1/(Z\alpha))$	-0.11 kHz	Lepage
		Bodwin et al.
$\alpha^2(Z\alpha)(m/M)\ln^3(M/m)$	-0.05 kHz	Eides and Shelyuto
$\alpha^2(Z\alpha)(m/M)\ln^2(M/m)$	0.01 kHz	Eides et al.
$\alpha(Z\alpha)^2(m/M)\ln^2(1/(Z\alpha))$	0.34 kHz	Karshenboim
$(Z\alpha)^3(m/M)\ln^2(1/(Z\alpha))$	-0.04 kHz	Karshenboim
$(Z\alpha)^3(m/M)\ln(M/m)\ln(1/(Z\alpha))$	-0.21 kHz	Karshenboim
		Kinoshita and Nio
$\alpha^2(Z\alpha)^2\ln^2(1/(Z\alpha))$	-0.04 kHz	Karshenboim
$\alpha(Z\alpha)^3\ln(1/(Z\alpha))$	-0.47 kHz	Karshenboim
$\alpha(Z\alpha)^{3+}$ (SE)	-0.04 kHz	Blundell et al.
$\alpha(Z\alpha)^3$ (VP)	0.02 kHz	Karshenboim et al.
$\alpha(Z\alpha)(m/M)^2$	-0.04	Eides et al.
$(Z\alpha)^2(m/M)^2$	0.01	Pachucki

THE STATUS OF BOUND STATE QED

After briefly reviewing the studies for hydrogen, muonium and positronium, let us discuss a problem and current trend of bound state QED. First of all we need to mention that, in our mind, the QED theory as a theory is well established. As a pure theory it is absolutely correct and absolutely useless:

- The QED theory is a theory of interaction between leptons (electrons and muons) and photons only. We need to include hadrons (such as proton) into consideration. Even for the case of pure leptonic values (like for muonium) we need to calculate a hadronic vacuum polarization contribution. So the QED theory is incomplete.

- The QED theory cannot predict anything exactly, but only in terms of expansion and the uncertainty can be presented in terms like $O(\alpha^7 m)$. It is necessary to develop an effective approach to estimate uncalculated corrections quantitively in Hz and eV.

- The QED deals rather with free particles and it is necessary to develop an effective approach to solve a bound state problem for two bodies.

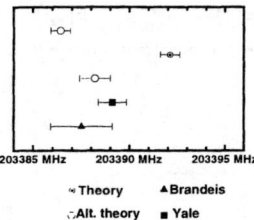

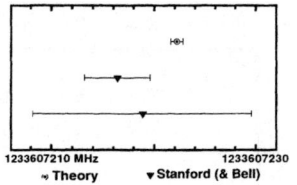

FIGURE 14. Hyperfine structure interval in the ground state of the positronium atom. The references can be found in *Conti*[†].

FIGURE 15. 1s–2s transition in positronium: comparison of experiment with theory. The references can be found in *Conti*[†].

These three problems lie beyond QED as a mathematical theory, but are an essential part of any real QED calculations. A test of different effective approaches is a real problem, as is evaluating of hadronic contributions needed for the precision theory of simple atoms.

The bound state problem has mainly three small parameters: α (associated with QED effects), $Z\alpha$ (due to binding effects) and m/M (the recoil parameter). Now, for the first time, it is necessary to try to really study effects which involve essential QED, two–body and binding effects simultaneously (the $\alpha(Z\alpha)^2 m/M$ corrections). That is a problem for the hyperfine structure of muonium, the 1s/2s hyperfine structure in hydrogen, deuterium and He$^+$ and for the positronium spectrum. Another crucial problem is that of the Lamb shift in hydrogen and light ions: this is a higher order two–loop corrections ($\alpha^2(Z\alpha)^6 m$) already known in part. We summarize all crucial terms in Table 3.

The three parameters we mention generate different expansions and it is found that all three kind of expansions are not free from problems.

- The QED expansion over α is an asymptotic one and the value of terms will decrease to some n_c and increase after it. Fortunately, that is not important for $n = 1 - 3$, which are only actual for the bound state QED.

- The $Z\alpha$ expansion involves another problem. One knows that high Z is a bad limit (strong coupling) and it is believed that low Z is a good limit. The latter is wrong. It is clear that at $Z = 0$ there is no bound system at all and the behavior of any expansion in the limit of low Z is not an analytical one. This eventually leads to logarithmic contributions. Even a cube of logarithm $(\ln^3(1/\alpha) \sim 120$ at $Z = 1)$ appears. The imaginary part of the logarithm is π and the non–leading terms have often large coefficients because of this. There is another mechanism for large coefficients. As it is well–known the Bethe logarithm $(\ln(k_0(ns)) \sim 3)$ is a logarithm of an effective energy (in atomic units) of an intermediate state in a calculation of the electron self–energy. A logarithm equal to 3 corresponds to a quite relativistic intermediate p–state $(v/c \sim 4.5(Z\alpha))$ and that also leads to large coefficients because of relativistic

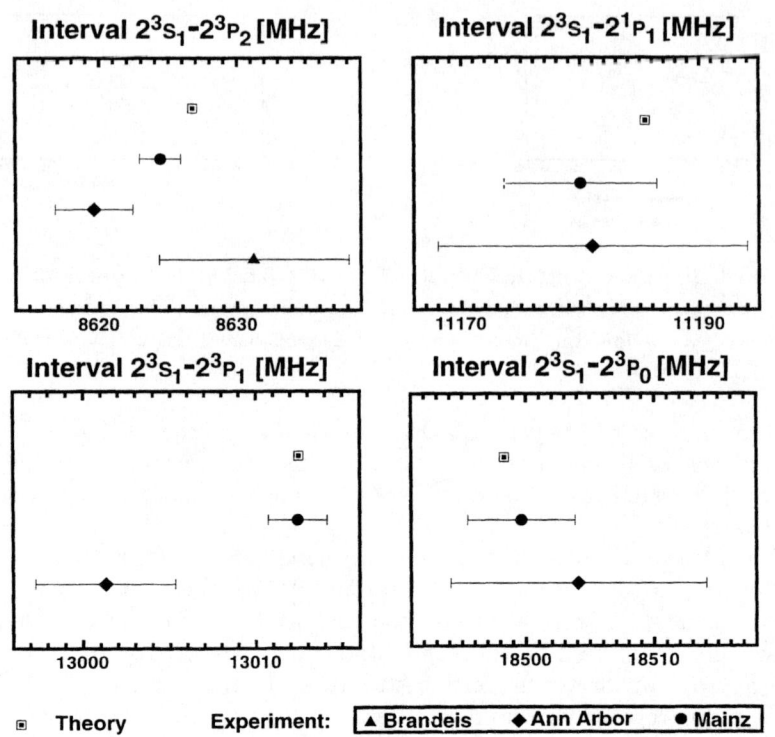

FIGURE 16. Positronium fine structure: theory and experiment. See *Conti*[†] for references.

corrections for intermediate states.

- The recoil effect with the m/M expansion also involves a non–analytical behavior. It is correct that the limit $m_1 = m_2$ (positronium) is rather complicated for a calculation, however at the opposite limit $(m/M \to 0)$ there is no bound state. Hopefully, often the logarithmic recoil corrections are not quite important numerically. Since most of the recoil effects are relativistic ones, the exchange loop generates the effective parameter $(Z\alpha)/\pi$ rather than $Z\alpha$.

Due to the increasing number of logarithmic contributions we end up with a problem of large higher–order corrections. Some higher–order logarithmic terms are compatible in comparison to a constant part of some lower–order terms.

- For the hydrogen 2s Lamb shift, non–logarithmic parts of the fifth order corrections (in unit of the Rydberg contributions) lie from 164 kHz $(\alpha(Z\alpha)^6 m)$ and 37 kHz $(\alpha^2(Z\alpha)^6 m)$ to a few kHz for recoil terms. The leading logarithmic term in the next order is $\alpha^2(Z\alpha)^6 \ln^3(Z\alpha)$, which contributs 3.6 kHz.

- The non–logarithmic parts of the third order correction (in unit of ν_F) for the ground state muonium hyperfine splitting varies from 8.8 kHz $(\alpha(Z\alpha)^2)$ to 2

TABLE 3. Crucial higher–order corrections in current studies of the simple atoms. For the g-factor of an electron we recalculated the corrections in terms of corrections to the energy.

Value	Order
hydrogen gross structure	$\alpha(Z\alpha)^7 m,\ \alpha^2(Z\alpha)^6 m$
hydrogen fine structure	$\alpha(Z\alpha)^7 m,\ \alpha^2(Z\alpha)^6 m$
hydrogen Lamb shift	$\alpha(Z\alpha)^7 m,\ \alpha^2(Z\alpha)^6 m$
He^+ Lamb shift	$\alpha(Z\alpha)^7 m,\ \alpha^2(Z\alpha)^6 m$
nitrogen fine structure	$\alpha(Z\alpha)^7 m,\ \alpha^2(Z\alpha)^6 m$
$^3He^+$ hyperfine structure	$\alpha(Z\alpha)^7 m^2/M,\ \alpha^2(Z\alpha)^6 m^2/M,\ \alpha(Z\alpha)^6 m^3/M^2,\ (Z\alpha)^7 m^3/M^2$
muonium hyperfine structure	$\alpha(Z\alpha)^7 m^2/M,\ \alpha(Z\alpha)^6 m^3/M^2,\ (Z\alpha)^7 m^3/M^2$
positronium hyperfine structure	$\alpha^7 m$
positronium gross structure	$\alpha^7 m$
positronium fine structure	$\alpha^7 m$
parapositronium decay rate	$\alpha^7 m$
orthopositronium decay rate	$\alpha^8 m$
parapositronium 4γ decay	$\alpha^8 m$
orthopositronium 5γ decay	$\alpha^8 m$
g–factor of electron in $^{40}Ca^{19+}$	$\alpha(Z\alpha)^7 m,\ \alpha^2(Z\alpha)^6 m$
g–factor of free electron	$\alpha^8 m$

kHz for recoil and radiative recoil terms and to 0.4 kHz for $\alpha^2(Z\alpha)$. Three leading logarithmic corrections are slightly below 1 kHz (see Table 2): $\alpha(Z\alpha)^3$, $\alpha(Z\alpha)^2(m/M)$ and $(Z\alpha)^3(m/M)$.

- A number of positronium levels are under study. The non-logarithmic $\alpha^6 m$ term (7.2 MHz) for the hyperfine structure is bigger than the logarithmic part of the $\alpha^7 m$ contribution (0.9 MHz), while for the 1s-2s interval the situation is different: the non-logarithmic $\alpha^6 m$ term is only 0.5 MHz and that is less than 1.2 MHZ of the $\alpha^7 m$.

These example show that an estimation of the higher order terms is extremely important and we hope that a calculation of leading logarithmic contributions provides a reasonable way to estimate uncalculated terms. We estimate the non–leading term within a half–value of the leading logarithmic contributions.

Estimation of uncalculated terms is a crucial problem in any QED calculations. Let us now mention the case of moderate Z. Study of these ions provides a unique possibility of measuring higher order corrections. In particular an experimental study of helium (*Drake*[†], *Burrows et al.*[†]) and nitrogen (*Myers*[†]) hydrogen–like ions will allow us to extract information on higher–order two–loop contributions with the help of a theoretical study of all other terms (*Ivanov and Karshenboim*[†]). Moderate–Z few–electron atoms allow us to test our understanding of higher–order electron–electron interactions which is important for high Z spectroscopy.

Large values of the higher–order terms imply a calculation without expansion. This is only possible for one parameter, either $Z\alpha$ or m/M. For the simplest corrections (like e. g. the vacuum polarization) it is possible to calculate analytically to any order of $Z\alpha$, otherwise only numerical results are possible. It is unlikely that a complete exact calculation of the two-loop self-energy can be performed soon and that means that expansion techniques is still the main approach in calculating the higher-order terms, perhaps in combination with experiments.

A new opprortunity appears due to recent measurements of a bound electron g– factor in a hydrogen–like atom (*Häffner et al.*[†], [15]). A recent result on the carbon ion is useful for indirect determination of the electron mass (*Karshenboim*[†]). Our theoretical prediction

$$g_b(e) = 2 \cdot (1 + 520795(1) \cdot 10^{-9}) \tag{4}$$

mainly based on [16] calculation has a smaller uncertainty in part because of taking into account a known $\alpha^2(Z\alpha)^2$ term. Studies of the g–factor will be very different from the Lamb shift and the hyperfine structure. In contrast to spectroscopic studies it is possible to go through all Z and to determine some unknown coefficients of the theoretical expansion if we can fix its shape (we call this *weak theory* in contrast to a real theory which can give direct numerical predictions).

SUMMARY

Concluding the paper we wish to mention briefly different applications for the study of simple atoms. These studies are important for different field of physics:

- Determination of fundamental constants (R_∞, α, m_e/m_p): some of these are important for other application, like e. g. the fine structure constant is necessary to reproduce the value of the Ohm from the quantum Hall effect.

- Development of new optical standard and tool for same: like e. g. the new frequency chain designed recently (see *Diddam et al.*[†], *Udem et al.*[†] and [17]).

- *New* physics: a study of muonium-to-antimuonium conversion (*Jungmann*[†]) and the exotic decay of positronium (*Conti*[†]) provide a possibility of looking for new particles, while the antihydrogen project and some others are expected to test some symmerties or possible variations of fundamental constants.

- Particle physics: a recent study of hydrogen is rather important to learn the proton structure, than to test the bound state QED. The theoretical methods developped recently are of use both for atoms and two-quark particles (mesons) and QED is a good opprotunity to test the methods. Exotic atoms give us accurate information on hadron-hadron interactions.

- Nuclear physics: the situation is similar to that for particles physics: a study of light atoms offers information on structure of their nuclei. One the other hand, two– and three–body atoms are an appropriate problem to testing different effective methods before to applying them to light nuclei.

Most of these questions and a more broad range of problems in physics of simple atoms were considered at a Satellite meeting to the ICAP (Hydrogen Atom, 2: *Precise Physics of Simple Atomic System*). The Proceeding will be published by Springer in 2001.

ACKNOWLEDGEMENT

I am grateful to T. Hänsch, K. Jungmann, G. Werth, J. Sapirstein and E. Myers for useful and stimulating discussions. I would like to thank S. Nic Chormaic and F. Cataliotti for useful remarks on manuscript. The work was supported in part by RFBR (grant 00-02-16718), NATO (CRG 960003) and Russian State Program *Fundamental Metrology*.

REFERENCES

1. M. I. Eides, H. Grotch and V. A. Shelyuto, Phys. Rep. *to be published.*
2. W. V. Hughes, *this volume.*
3. S. G. Karshenboim, Can. J. Phys. **76** (1998) 168.
4. S. G. Karshenboim, Phys. Sc. **57** (1998) 213.
5. E. A. Hinds, *Radiofrequency Spectroscopy* in *The Spectrum of Atomic Hydrogen: Advances.* Ed. by. G. W. Series. World Sci., Singapore, 1988.
6. S. G. Karshenboim, JETP **79**, 230 (1994).
7. S. G. Karshenboim, Z. Phys. D **39**, 109 (1997).
8. Ch. W. Wong, Int. J. Mod. Phys. **3**, 821 (1994).
9. A. Bertin *et al.*, Phys. Lett. **55B**, 411 (1975); G. Carboni *et al.*, Nuovo Cim. **34A**, 493 (1976); Nucl. Phys. **278A**, 381 (1977); Phys. Lett. **73B**, 229 (1978).
10. P. Hauser *et al.*, Phys. Rev. A**46**, 2363 (1992).
11. S. G. Karshenboim, Z. Phys. D **36**, 11 (1996).
12. G. Adkins *et al.*, Phys. Rev. Lett. **79**, 3387 (1997); A. H. Hoang *et al.*, Phys. Rev. Lett. **79**, 3383 (1997); K. Pachucki and S. G. Karshenboim, Phys. Rev. Lett. **70** 2101, (1998).
13. S. G. Karshenboim, JETP **76**, 541 (1993).
14. K. Pachucki and S. G. Karshenboim, Phys. Rev.A **60** 2792 (1999); K. Melnikov and A. Yelkhovsky, Phys. Lett. B **458**, 143 (1999).
15. H. Häffner *et al.*, *this volume.*
16. T. Beier *et al.*, Hyp. Inter. *to be published.*
17. R. Holzwarth *et al.*, *this volume.*

The impact of atomic precision measurements in high energy physics

Roberto Casalbuoni

Dept. of Physics, University of Florence and I.N.F.N., Florence
e-mail: casalbuoni@fi.infn.it

Abstract. In this talk I discuss the relevance of atomic physics in understanding some important questions about elementary particle physics. A particular attention is devoted to atomic parity violation measurements which seem to suggest new physics beyond the Standard Model. Atomic physics might also be relevant in discovering possible violations of the CPT symmetry.

INTRODUCTION

The aim of this talk is to review some of the atomic precision measurements in atomic physics leading to precious informations in the realm of high-energy physics. The idea of atomic physics bringing light on the high-energy physics world requires some qualification due to the very different scales of energy involved in the two cases. In fact, typically one has a separation of about six or seven order of magnitude between the two scales and one expects the two physics being almost decoupled. In fact, if we look at some observable, A, at a scale $\Lambda_1 \ll \Lambda_2$, we expect that the observable can be represented in the form

$$A(\Lambda_1, \Lambda_2) = A(\Lambda_1) + \mathcal{O}\left(\left(\frac{\Lambda_1}{\Lambda_2}\right)^n\right) \qquad (1)$$

In order to be able to derive informations about the physics at the scale Λ_2, being at the scale Λ_1, one starts considering a combination of observables corresponding to the corrections coming from the higher scale

$$B = c\left(\frac{\Lambda_1}{\Lambda_2}\right)^n \qquad (2)$$

In order to measure B one needs either the coefficient c being very large in such a way to partially compensate the scale factor, or having an extremely good experimental sensitivity. In this talk I will consider two particular examples of situations where atomic physics can be relevant to high-energy physics, namely Atomic Violation of Parity (APV) and possible violations of the discrete symmetry CPT,

CP551, *Atomic Physics 17*, edited by E. Arimondo, P. DeNatale, and M. Inguscio
© 2001 American Institute of Physics 1-56396-982-3/01/$18.00

that is the product of charge-conjugation, parity and time-reversal. In fact, using heavy atoms like cesium in APV measurements one can get a good enhancement factor. On the other hand, CPT symmetry can be tested by using the extraordinary opportunities offered by the atomic traps in order to obtain a very accurate determination of frequencies.

ATOMIC PARITY VIOLATION IN ATOMS

In this Section I will discuss mainly the latest determination of the weak charge in atomic cesium and some of its implications in models of physics beyond the Standard Model (SM). The SM has been tested very precisely at machines such as LEP and SLC, where, working at an energy around the Z mass, one is mainly testing the property of the Z itself. Therefore, the physics beyond the SM that can be looked for at these machines is the one giving corrections to the Z-propagator and/or to the couplings of the Z with fermion-antifermion pairs. Namely, new massive vector bosons, Z', which mix to the Z, or new particles running in loops and contributing to the Z self-energy or to vertex corrections. But consider, for instance, the case of a massive vector boson which does not mix to the Z, and therefore invisible at LEP (except for tiny radiative corrections). If the Z' is coupled to fermions, in the low-energy limit it gives rise to an effective four-fermi interaction. Therefore, low-energy experiments are complementary to the high-energy ones, and furthermore they are able to measure directly the couplings of the Z to light quarks; something that at LEP and SLC can be done only in an indirect way. Among the low-energy experiments a particular role is played by the APV experiments, due to the precision almost at the level of the one reached at LEP/SLC.

Let us now recall some feature of APV in atoms. First of all, within the SM the four-fermi parity violating hamiltonian density for nucleons is given by

$$\mathcal{H}^{PV} = \frac{G_F}{\sqrt{2}} \left[(\bar{e}\gamma_\mu\gamma_5 e) \sum_{N=p,n} c_{1N} \bar{N}\gamma^\mu N + (\bar{e}\gamma_\mu e) \sum_{N=p,n} c_{2N} \bar{N}\gamma^\mu\gamma_5 N \right] \quad (3)$$

where

$$c_{ip} = -2c_{iu} - c_{id}, \quad c_{in} = -c_{iu} - 2c_{id}, \quad i = 1,2 \quad (4)$$

and

$$c_{1q} = -8a_e v_q = -(T_3^q - 2s_\theta^2 Q^q), \quad c_{2q} = -8v_e a_q = -T_3^q(1 - 4s_\theta^2), \quad q = u, d \quad (5)$$

Here v_e, v_q, a_e and a_q are the vector and vector-axial couplings of the Z to the electrons and quarks. For a point-like nucleus with Z protons and N neutrons, the hamiltonian density, in the non-relativistic limit, is given by

$$\mathcal{H}_{PV} = \frac{G_F}{4\sqrt{2}m_e} \Big[Q_W(Z,N)\vec{\sigma}_\ell \cdot [\vec{p}, \delta^3(\vec{r})]_+ + 2(c_{2p}\vec{S}_p + c_{2n}\vec{S}_n) \cdot [\vec{p}, \delta^3(\vec{r})]_+ \quad (6)$$

$$- 2i\vec{\sigma}_\ell \wedge (c_{2p}\vec{S}_p + c_{2n}\vec{S}_n) \cdot [\vec{p}, \delta^3(\vec{r})]_+ \Big] \quad (7)$$

255

where $\vec{p}$ is the momentum of the electron, $\vec{S}_{p(n)}$ the total spin of the protons (neutrons) and m_e the electron mass. I have also defined the *weak charge* of the atom as

$$Q_W(Z,N) = 2\left[c_{1p}Z + c_{1n}N\right] \tag{8}$$

Notice that for a heavy atom (large values of Z) the matrix element of the first term in $\mathcal{H}_{PV}$ is roughly proportional to Z^3, one factor coming from Q_W, one from the momentum of the electron and the third one from the wave function evaluated at the origin. This coherence effect was noticed by Bouchiat and Bouchiat [1] and it provides, in the case of cesium ($Z = 55$) an enhancement factor of about 10^5, more or less what is necessary in order to compensate for the decoupling factor from the scales mentioned in the Introduction.

In order to get a rough idea of the bounds on new physics that can be obtained by a measurement of Q_W with a given sensitivity, we parametrize the new physics contribution to Q_W by a four-fermi effective interaction [2]

$$\mathcal{L}_{NP}^{PV} = \frac{g_{NP}^2}{\Lambda^2}\,\bar{e}\gamma_\mu\gamma_5\,e \sum_{q=u,d} h_{1q}\,\bar{q}\,\gamma^\mu\,q \tag{9}$$

If we assume $h_{1q} \approx c_{1q}$, for a sensitivity $\Delta Q_W/Q_W \approx 1\%$ one gets a bound

$$\Lambda \approx (5\,g_{NP})\,TeV \tag{10}$$

If new physics is strongly interacting ($g_{NP}^2 \approx 4\pi$), then $\Lambda \approx 17\,TeV$, whereas in the weakly interacting case ($g_{NP}^2 \approx 4\pi\alpha$) we get $\Lambda \approx 1.5\,TeV$. In any case we see that at 1% level of sensitivity, Q_W is able to test new physics for scales greater than 1 TeV.

In APV measurements one looks at optical transitions between a pair of states $|\psi_\pm\rangle$ mixed by $\mathcal{H}_{PV}$ and a state $|\psi_0\rangle$ of the same nominal parity as $|\psi_+\rangle$. The mixing of the two eigenstates of parity is given by

$$\eta = \frac{\langle\psi_-|H_{PV}|\psi_+\rangle}{\Delta E} \tag{11}$$

where ΔE is the splitting between the two levels. If I denote by M_1 and E_1^{PV} the amplitudes for the two unperturbed transitions $|\psi_+\rangle \to |\psi_0\rangle$ and $|\psi_-\rangle \to |\psi_0\rangle$, the transition probability, after the mixing, is given by

$$W = M_1^2 + |E_1^{PV}|^2 \pm 2\,Im\,(E_1^{PV})M_1 \tag{12}$$

The choice of the sign depends on the helicity of the photon which is emitted or absorbed in the transition. In the actual experiment on cesium one measures the *circular dichroism*, that is the asymmetry for the absorption cross-section

$$\delta = \frac{\sigma_+ - \sigma_-}{\sigma_+ + \sigma_-} \approx 2\,\frac{Im\,(E_1^{PV})}{M_1} \tag{13}$$

Of course, the PV amplitude E_1^{PV} is proportional to the mixing parameter η and therefore measuring δ one can get the matrix element of the PV hamiltonian. These ideas have been applied in particular to the transition $6S \rightarrow 7S$ in atomic cesium $^{133}_{55}Cs$ [3–5], but also to other atoms as thallium [6]. The typical value of δ is $10^{-4} \div 10^{-5}$, but there is a strong background which can be overcomed by letting the PV amplitude to interfere with a large electro-induced (Stark) transition. Eventually one extracts from the experiment the matrix element of H^{PV} which is proportional to Q_W times an atomic form factor κ_{PV} which must be evaluated theoretically in order to extract the value of the weak charge. Therefore the measurement must be coupled with theoretical calculations of similar accuracy in order to get a precise determination of Q_W. In the case of atomic cesium the calculation of κ_{PV} was performed independently by two groups [7,8]. This calculation is not an easy task, as one has to use many-body perturbation theory coupled with Hartree-Fock techniques. The theoretical errors are quite difficult to estimate. The authors of Refs. [7,8] did their estimate by looking at the differences between the theoretical and the experimental values of parity conserving quantities as dipole matrix elements and hyperfine splittings for the $6S_{1/2}$, $7S_{1/2}$, $6P_{1/2}$ and $7P_{1/2}$ states. In this way the error $\Delta\kappa_{PV}/\kappa_{PV} \approx 1\%$ was obtained. After the new measurement of the weak charge of the cesium by the Boulder group [5], which improved the accuracy of the previous experiment [4] by more than a factor five, Bennett and Wieman [9] re-examined the theoretical errors on κ_{PV}. In fact, since the time of the previous estimate there have been a number of new and more precise measurements of the quantities of interest. The result is that now the agreement is much better than before, and as a consequence Bennett and Wieman got the estimate $\Delta\kappa_{PV}/\kappa_{PV} \approx 0.4\%$. It should be noticed that there is a third element which contributes to the extraction of Q_W from the data. This is the Stark mixing-induced electric dipole moment amplitude, β. The experiments in Refs. [3,4] were using a theoretical determination of β. In [5] the ratio M_{hf}/β has been measured. The off-diagonal magnetic dipole moment induced by the hyperfine interaction is well known empirically and it is possible to extract a precise value for β. However, in a contribution to this Conference [10], the matrix element M_{hf} has been accurately calculated with the result that the empirical formula for it should be corrected by a factor of 0.24% increasing the discrepancy with the SM (see later). I would like also to comment about some possible neglected contribution in the evaluation of the atomic form factor. It has been pointed out in ref. [11] that there could be a contribution arising from the difference of neutron and proton spatial distributions inside the nucleus. This contribution turns out to be very difficult to estimate, in fact it is quite model dependent. Most probably it could introduce a further error on $Q_W(Cs)$ of about 0.3. This would not change the conclusions in a very significant way. Another point has been raised recently in ref. [12]. This author argues that the contribution from the Breit interaction (exchange of a transverse photon between two electrons) could have been underestimated. The Breit interaction contribution to the atomic form factor was estimated in [7] and it was found to be very small. However in ref. [12] it is found that the total effect, taking into account also second and third order contributions, is about twice the first order effect. As a consequence, if "all" the higher order contributions could be shown to be negligible, the experimental measure would reconcile with the SM expectation for $Q_W(Cs)$. However, see also ref. [13].

To conclude this analysis I think that an evaluation of the atomic form factor by taking into account the next order in the many-body perturbative theory is highly desirable in order to settle the question. In any case I find of some interest to assume that the theoretical error is indeed at the level of 0.4% in order to see which are the possible implications of the APV in high-energy physics.

I can start now to discuss the experimental results on $Q_W(Cs)$. It is interesting to recall the value obtained in [4] combined with the theoretical determination of κ_{PV} [7,8]

$$Q_W(Cs) = -71.04 \pm (1.58)_{\text{exp}} \pm (0.88)_{\text{th}} \tag{14}$$

The total error of these measurement on $Q_W(Cs)$ is at 2.5% level of accuracy that, at that time, was comparable with the sensitivity obtained at LEP1. In fact, this determination of $Q_W(Cs)$ lead to the first indication that technicolor models, in their most simple version obtained from scaling of QCD, could not possibly fit the data. The new experimental result on $Q_W(Cs)$ [5] combined with the new determination of the theoretical error [9] gives

$$Q_W(Cs)^{\text{exp}} = -72.06 \pm (0.28)_{\text{exp}} \pm (0.34)_{\text{th}} \tag{15}$$

A result at 0.6% level of accuracy. On the theoretical side, Q_W can be expressed as [14]

$$Q_W(Cs)^{\text{th}} = -72.72 \pm 0.13 - 102\epsilon_3^{\text{rad}} + \delta_N Q_W \tag{16}$$

including hadronic-loop uncertainty. I use here the variables ϵ_i (i=1,2,3) of ref. [15], which include the radiative corrections, in place of the set of variables S, T and U originally introduced in ref. [16]. In the above definition of $Q_W^{\text{th}}(Cs)$ I have explicitly included only the Standard Model (SM) contribution to the radiative corrections. New physics (that is physics beyond the SM) contributions are represented by the term $\delta_N Q_W$. Also, I have neglected a correction proportional to ϵ_1^{rad}. In fact, as well known [14], due to the particular values of the number of neutrons ($N = 78$) and protons ($Z = 55$) in cesium, the dependence on ϵ_1 almost cancels out. For a top mass of 175 GeV and $m_H = 100(300)$ GeV the value of ϵ_3^{rad} is given by [17]

$$\epsilon_3^{\text{rad}} = 5.110(6.115) \times 10^{-3} \tag{17}$$

For $m_H = 100$ GeV, corresponding roughly to the lower experimental bound from direct search at LEP2 [18], one gets

$$Q_W(Cs)^{\text{exp}} - Q_W(Cs)^{SM} = 1.18 \pm 0.46 \tag{18}$$

giving rise to a deviation of about 2.57 SD. Furthermore, for increasing mass of the Higgs the discrepancy increases. Therefore, if we assume as being correct the experimental result, the theoretical evaluation of κ_{PV} and the evaluation of the theoretical errors, we are forced to conclude that the SM is disfavored at 99% CL.

We can draw another conclusion, that is, that in order to explain the data on $Q_W(Cs)$ we need new physics not constrained by the LEP and SLC data. In fact, as an example let me consider a type of new physics visible at LEP as, for instance, contributing to the self-energy of the Z, the so called oblique corrections. In such a case one can write $\delta_N Q_W(\text{oblique}) = -102\epsilon_{3N}$, and in order to compensate for the discrepancy on $Q_W(Cs)$ one needs

$$\epsilon_{3N} = (-11.6 \pm 4.5) \times 10^{-3} \tag{19}$$

whereas from LEP and SLC data one can determine the sum

$$\epsilon_3^{\text{exp}} = \epsilon_3^{\text{rad}} + \epsilon_{3N} = (4.19 \pm 1) \times 10^{-3} \tag{20}$$

Therefore one gets $\epsilon_{3N} \approx 10^{-3}$, one order of magnitude too small to explain the data on $Q_W(Cs)$.

I would like also recall the experimental result of APV on Thallium [6]

$$Q_W(Tl)^{\text{exp}} = -114.8 \pm (1.2)_{\text{exp}} \pm (3.4)_{\text{th}} \tag{21}$$

This result is not as precise as the one on Cs, and in fact the total error is about 3%. At this level it is perfectly compatible with the SM prediction

$$Q_W(Tl)^{\text{SM}} = -116.7 \pm 0.1 \tag{22}$$

A new experiment on cesium is being planned in Paris but the experimental sensitivity is going to be lower than the one obtained in Boulder.

In Berkeley and Seattle there are plans for isotope ratio measurements. In this case the dependence on the atomic form factor would go away eliminating the theoretical error. However these ratios depend on the variation of the neutron density along the isotope chain. This would introduce errors at least twice as big as the experimental ones [19].

We are now in the position of discussing the implications of eq. (18) on new physics. Assuming that the contribution of new physics, $\delta_N Q_W$, is such to reproduce the experimental results, we can make use of eqs. (15) and (16) to write [20]

$$Q_W(Cs)^{\text{exp}} - Q_W(Cs)^{\text{th}}(m_H) = 0.66 + 102\epsilon_3^{\text{rad}}(m_H) - \delta_N Q_W \pm 0.46 \tag{23}$$

For $m_H = 100\ GeV$ at a 95% CL we find

$$0.28 \le \delta_N Q_W \le 2.08 \tag{24}$$

Notice that the lower positive bound arises since the SM (corresponding to $\delta_N Q_W = 0$) does not fit the experimental value of $Q_W(Cs)$ at this CL value. This is quite important since it implies an upper bound on the scale of new physics. For the same reason new physics with a contribution $\delta_N Q_W < 0$ is not allowed. Also notice that lower and upper bounds both increase for increasing Higgs mass.

Contact interactions from compositeness. A typical four-fermi operator in composite models contributing to the PV lagrangian is [21,20]

$$\pm \frac{g^2}{\Lambda^2} \, \bar{e} \, \gamma_\mu \frac{1-\gamma_5}{2} \, e \bar{q} \, \gamma^\mu \frac{1-\gamma_5}{2} \, q \tag{25}$$

The effect of this interaction is to modify the coefficients $c_{1u,1d}$

$$c_{1u,1d} \to c_{1u,1d} \mp \frac{\sqrt{2}\pi}{G_F \Lambda^2} \tag{26}$$

where, since composite models correspond to strongly interacting new physics, we have assumed $g^2 = 4\pi$. From

$$Q_W = -2[(2Z+N)c_{1u} + (Z+2N)c_{1d}] \tag{27}$$

we see that the negative sign for the operator (25) is excluded. For the positive sign we get the bounds

$$12.1 \le \Lambda(TeV) \le 32.9 \tag{28}$$

The typical lower bound from high energy physics is about 3.5 TeV [22].

Extra-dimension models. In ref. [23] a minimal extension to higher dimensions of the SM, with extra dimensions compactified, was considered. In this model the fermions live in a 4-dimensional subspace, the wall, whereas the gauge bosons live in the full D-dimensional space, the bulk. In general, there might be two Higgs fields, one living in the bulk, ϕ_1, and the other living on the wall, ϕ_2. The propagation of the gauge fields in the bulk is equivalent to the exchange of an infinite tower of Kaluza-Klein (KK) excitations with increasing mass. For example, for $D = 5$, $M = n/R$, $n = 1, \cdots, \infty$, with R the compactification radius. If only the Higgs field ϕ_2 is present, the ordinary gauge bosons do not mix with the KK resonances and it is easy to see that the contribution of these modes to Q_W is negative [24]. Therefore the model does not fit the data on $Q_W(Cs)$. For the more general case of both Higgs fields present it has been shown [24] that the LEP/SLC and $Q_W(Cs)$ experimental data are not compatible among them at 95% CL.

Extra Z' models. The implications of models with an extra neutral vector boson Z' for APV have been considered in the literature for quite a long time [25,24,26]. The Z' has couplings comparable to the ones of the Z in the SM and therefore this is an example of weakly interacting new physics. There is a continuum of such models characterized by an angle $0^0 \le \theta_6 \le 90^0$. To any value of θ_6 it corresponds a different model. The 95% CL regions allowed by Q_W, in the plane $(\theta_6, M_{Z'})$, for different values of the Higgs mass, are shown in Fig. 1. In deriving these Figures the assumption of zero mixing between Z and Z' has been made. In the Figure are also shown three popular models: η ($\theta_6 \approx -52^0$), χ ($\theta_6 = 0^0$), ψ ($\theta_6 = 90^0$). We see that the η and the ψ models are not allowed by the data. The

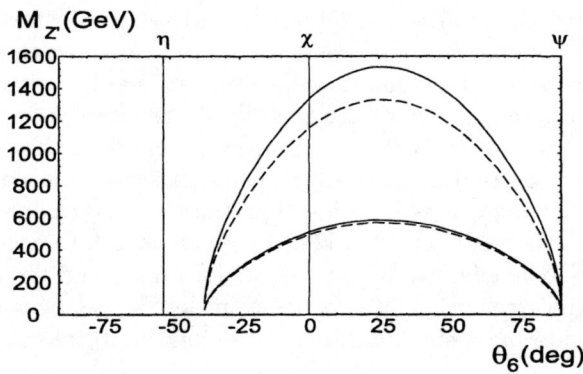

FIGURE 1. The Figure shows the 95% CL regions allowed by Q_W for the Z' models. The solid contour corresponds to $m_H = 100\,GeV$, and the dashed one to $m_H = 300\,GeV$.

direct search at the Tevatron for a Z' within the χ model gives a direct lower bound at 95% CL, $M_{Z'} \geq 590\,GeV$ (a similar bound holds for all these models) . Therefore this model is compatible with the data. A recent best fit to all the data (including APV) gives for the χ model the following results [26], $M_{Z'} = 812^{+339}_{-152}\,GeV$ and a mixing angle compatible with zero, $\theta_M = (-1.12 \pm 0.80) \times 10^{-3}$.

ATOMIC PHYSICS AND CPT VIOLATION

The CPT theorem is one of the fundamental results in local relativistic field theories. Therefore the idea of possible violations of this theorem implies that some of the axioms of these theories should be reviewed. Let me recall here the exact statement of the theorem [27]: *In a field theory satisfying*

1. Locality

2. Lorentz invariance

3. Analiticity of the Lorentz group representations in the boost parameters

the CPT transformation is a symmetry of the theory itself.

The first two conditions say that one is dealing with a local relativistic field theory, whereas the third one is satisfied in any finite-dimensional representation of the Lorentz group. It is interesting to notice that unitary representations fail to be analytic and as a consequence the CPT theorem can be violated in this case. The first example of this situation dates back to Majorana [28] when he formulated a first order wave equation without negative-energy solutions. He was able to do that by making use of a unitary infinite-dimensional representation of the Lorentz group. Since this theory does not contain antiparticles the CPT symmetry is broken. However, the quarks and leptons described by the SM belong to finite-dimensional representation of the Lorentz group and therefore this does not seem a possible way to break the theorem. It seems also very hard to give up locality, since it guarantees the microcausality of the theory. Therefore, the only sensible way to avoid the consequences of the CPT theorem in a local field theory seems to break Lorentz invariance. A situation of this type could arise at a more fundamental level as in string theory, where it is possible that Lorentz invariance is spontaneously broken around the Planck mass, M_P [29]. One can take into account these effects by writing down a local effective lagrangian with Lorentz and CPT breaking terms. These terms can be written as an expansion in derivatives over the Planck mass. For instance, considering a single fermion, the violating term can be written as

$$\mathcal{L}_v = \sum_n \frac{g_n}{M_P^n} T \bar{\psi} \Gamma (i\partial)^n \psi \tag{29}$$

I have used a somewhat symbolic notation where Γ stays for a generic combination of Dirac matrices and T is a constant tensor and I take the mass dimensions of g_n as $[g_n] = 1$. Furthermore I will assume the same internal symmetries as in the SM, that is $SU(3) \otimes SU(2) \otimes U(1)$ [30,31]. Since the breaking terms should vanish in the limit $M_P \to \infty$ also for $n = 0$, I will require

$$g_0 = c_o \frac{m^2}{M_P} \tag{30}$$

where m is some low-energy mass scale parameter. We see that the relevant terms are the ones with $n = 0$ and $n = 1$, and therefore the resulting theory preserves the renormalizability property.

Let me now consider a single fermion interacting with the electromagnetic field. One adds to the standard QED lagrangian the following two terms

$$\mathcal{L}_v^{(n=0)} = \bar{\psi}[-a_\mu \gamma^\mu - b_\mu \gamma_5 \gamma^\mu - \frac{1}{2} H_{\mu\nu} \sigma^{\mu\nu}]\psi \tag{31}$$

and

$$\mathcal{L}_v^{(n=1)} = \bar{\psi}[ic_{\mu\nu} \gamma^\mu D^\nu + id_{\mu\nu} \gamma_5 \gamma^\mu D^\nu]\psi \tag{32}$$

where $D_\mu = \partial_\mu - iqA_\mu$, with q the electric charge of the fermion. There are other possible terms with $n = 1$, but they are not compatible with the symmetries of the SM and therefore they should be suppressed. The following orders of magnitude are expected

$$a_\mu, b_\mu, H_{\mu\nu} \approx \mathcal{O}\left(m^2/M_P\right), \quad c_{\mu\nu}, d_{\mu\nu} \approx \mathcal{O}\left(m/M_P\right) \tag{33}$$

The terms in $\mathcal{L}_v^{(n=0,1)}$ violate Lorentz invariance, since all the tensors in eq. (33) are constant ones. However only the terms proportional to a_μ and b_μ violate CPT symmetry since γ_μ, $\gamma_\mu \gamma_5$ and D_μ are CPT odd, whereas the other covariant terms are CPT even. Therefore, in the following I will take into consideration only $\mathcal{L}_v^{(n=0)}$. Notice also that when dealing with a single fermion the term in a_μ does not have physical meaning since we can write $a_\mu = \partial_\mu(a \cdot x)$, showing that a_μ is a trivial gauge background field. Of course, the situation changes when dealing with different fermions having different a_μ's. From eq. (33) we expect that the order of magnitude of the CPT and Lorentz breaking terms is given by $m/M_P \approx 10^{-22} \div 10^{-17}$ for $m = m_e \div v$, where m_e is the electron mass and $v \approx 250\ GeV$ is the electroweak symmetry breaking scale. Lorentz and CPT breaking terms could appear also in the photon part of the total lagrangian. This instance is discussed thoroughly in the second paper of ref. [30], but I will not consider it in this talk.

Here I want to illustrate some atomic physics experiment about CPT violation. But before doing that let me just give a list of other existing or planned experiments about the violation of this fundamental symmetry

- $K - \bar{K}$ mass difference. This experiment gives the best high-energy result [22]

$$\frac{|m_K - m_{\bar{K}}|}{m_K} \lesssim 10^{-18} \tag{34}$$

- Experiments on neutral meson oscillations to be done at meson factories [32].

- Experiments on muons [33].

- Experiments with spin-polarized solids [34].

- Experiments from clock-comparison [35].

CPT violation may have also some relevance for baryogenesis and this subject has been discussed in ref. [36].

Let me now consider atomic physics experiments for testing CPT using atomic traps. Several of these experiments have been performed by confining single particles or antiparticles in a Penning trap for a long time. These experiments have a very high precision, of order 10^{-9} or better, whereas the precision in experiments about mesons (see eq. (34)) is much lower, of order 10^{-3}. I recall here the comparison of the electron and positron gyromagnetic ratios, $g_\mp$, obtained measuring their cyclotron and anomaly frequencies (see later), which gives the figure of merit [37]

$$\left| \frac{g_- - g_+}{g_{av}} \right| \lesssim 2 \times 10^{-12} \tag{35}$$

Measuring the proton and antiproton cyclotron frequencies, one can get their charge-to-mass ratios. $r_{p.\bar{p}}$ [38]

$$\left| \frac{r_p - r_{\bar{p}}}{r_{av}} \right| \lesssim 9 \times 10^{-11} \tag{36}$$

Analogously, from the charge-to-mass ratio for electron and positron [39]

$$\left|\frac{r_{e^-} - r_{e^+}}{r_{av}}\right| \lesssim 1.3 \times 10^{-7} \tag{37}$$

As we see the relevant figures of merit are much bigger than the one for the mass difference $K - \bar{K}$, although, as noticed, these measurements are about six order of magnitude more sensitivity than the one leading to (34). In ref. [40] it has been argued that these figures of merit could not be the relevant ones in testing CPT breaking. In fact, within the approach presented here, at the lowest order in the CPT violating parameters, one has $g_- = g_+$, and similarly the charge-to-mass ratios do not depend on these parameters [40]. To review this point, let me start by the Dirac equation for an electron or a proton including the breaking terms contained in $\mathcal{L}_v^{(n=0)}$ (of course, the breaking parameters may depend on the type of particle one is considering)

$$\left(i\gamma^\mu D_\mu - m - b_\mu \gamma_5 \gamma^\mu - \frac{1}{2} H_{\mu\nu} \sigma^{\mu\nu}\right) \psi = 0 \tag{38}$$

In a Penning trap the radial confinement is obtained through a strong axial magnetic field, whereas the axial confinement is obtained by a quadrupole electric field. The main corrections due to the CPT and Lorentz breaking parameters are obtained by taking A_μ as the four-potential for a constant magnetic field. Then, to obtain the energy shifts generated by the breaking parameters one makes use of the relativistic Landau levels wave functions and the expressions containing the full QED corrections for the unperturbed levels [40,41]. However, the underlying physics can be understood quite simply recalling the expression for the non-relativistic Landau levels

$$E_{n,\sigma} = \left(n + \frac{1}{2} + \frac{g}{2}\right)\frac{Be}{m}, \quad \sigma = \pm\frac{1}{2} \tag{39}$$

The cyclotron and anomalous frequencies are obtained comparing two Landau levels with different quantum number n and with the same and opposite spin configurations respectively

$$\omega_c = E_{1,-1/2} - E_{0,-1/2} = \frac{Be}{m}$$
$$\omega_a = E_{0,+1/2} - E_{1,-1/2} = \frac{g-2}{2}\frac{Be}{m} \tag{40}$$

The relevant CPT and Lorentz breaking corrections to the energy levels are given by [41]

$$\delta E_{n,\pm 1/2}^{e^-} = \mp b_3 \pm H_{12}, \quad \delta E_{n,\pm 1/2}^{e^+} = \mp b_3 \mp H_{12} \tag{41}$$

where we have taken the third axis along the magnetic field of the trap. The frequencies for the antiparticles that we need according to the CPT theorem are the ones with inverted spin, therefore

264

$$\omega_c^{e^-} = \omega_c^{e^+} = \omega_c, \quad \omega_a^{e^\mp} = \omega_a \mp 2b_3 + 2H_{12} \tag{42}$$

We get

$$\Delta\omega_c \equiv \omega_c^{e^-} - \omega_c^{e^+} = 0, \quad \Delta\omega_a \equiv \omega_a^{e^-} - \omega_a^{e^+} = -4b_3 \tag{43}$$

We recall that these equations hold only at the first order in the breaking parameters and also that the usual relation $(g-2)/2 = \omega_a/\omega_c$ does not hold here since, as noted before, the gyromagnetic ratios do not change at the lowest order.

Since the observables that are measured in a Penning trap are the anomalous and cyclotron frequencies, it seems natural to introduce figures of merit related to these observables. A such figure of merit for CPT violation is [40]

$$r_{\omega_a}^e = \frac{|\mathcal{E}_{n,\sigma}^{e^-} - \mathcal{E}_{n,-\sigma}^{e^+}|}{\mathcal{E}_{n,\sigma}^{e^-}} = \frac{|\delta E_{n,\sigma}^{e^-} - \delta E_{n,-\sigma}^{e^+}|}{\mathcal{E}_{n,\sigma}^{e^-}} \tag{44}$$

where $\mathcal{E} = E + \delta E$. For a weak magnetic field one gets

$$r_{\omega_a}^e = \frac{|\Delta\omega_a|}{2m} = 2\frac{|b_3|}{m} \tag{45}$$

A new analysis of the 1987 experiment by Dehmelt et al. [37] has been done recently in ref. [42] obtaining the following bound

$$r_{\omega_a}^e \lesssim 1.2 \times 10^{-21} \tag{46}$$

However, the vector b_μ is absolutely constant and as such it rotates with a diurnal period of 23 h and 56 m, when seen in the laboratory frame wich is fixed with respect to the earth. This effect might have given rise to non favorable situations during the observation, and therefore the bound has been a bit relaxed [42]

$$r_{\omega_a}^e \lesssim 3 \times 10^{-21} \div 2 \times 10^{-20} \tag{47}$$

In the case of proton and atiproton there is no experiment at the moment. Assuming an experimental sensitivity analogous to the electron positron case (meaning $\delta\omega_a \approx 2\ Hz$) one gets [41]

$$r_{\omega_a}^p = 2\frac{|b_3^p|}{m_p} \lesssim 10^{-23} \tag{48}$$

The last case I consider is the spectroscopy of free or magnetically trapped hydrogen (H) and antihydrogen ($\bar{H}$). This is interesting since the two-photon $1S-2S$ transition has been measured with a precision of 3.4×10^{-14} [43] in a cold atomic beam of H and with a precision of 10^{-12} in trapped H [44]. However for the free case the dependence of the $1S - 2S$ transition on the CPT and Lorentz breaking parameters is suppressed by a factor $\alpha^2/8\pi$, since the $1S$ and $2S$ levels shift by the same amount at the leading order in the breaking [45]. Consider now the spectroscopy of H and $\bar{H}$ in a magnetic field B. In the basis $|m_J, m_I\rangle$ the four $1S$ and $2S$ hyperfine Zeeman levels are, for $n = 1, 2$

$$
\begin{aligned}
|b_n\rangle &= |-1/2, -1/2\rangle, \quad |d_n\rangle = |1/2, 1/2\rangle \\
|a_n\rangle &= \cos\theta_n |-1/2, 1/2\rangle - \sin\theta_n |1/2, -1/2\rangle \\
|c_n\rangle &= \sin\theta_n |-1/2, 1/2\rangle + \cos\theta_n |1/2, -1/2\rangle
\end{aligned}
\tag{49}
$$

with $\tan 2\theta_n = (51\ \mathrm{mT})/n^3 B$. Transitions of the type $|c_1\rangle \rightarrow |c_2\rangle$ have leading-order sensitivity to Lorentz and CPT violation, but they are field-dependent. As a consequence there is a problem connected with the broadening of the lines due to trapping field inhomogeneities.

Consider now hyperfine transitions in the ground state. Again there is the problem of the Zeeman broadening. However one can try to eliminate the frequency dependence on B (at lowest order) by choosing a field independent transition point [45]. For $B \approx 0.65\ T$ the state $|c_1\rangle$ is highly polarized ($|1/2, -1/2\rangle$). Then the effect on the transition $|c_1\rangle \rightarrow |d_1\rangle$ of the CPT and Lorentz violating parameters is $\delta\omega_{c\rightarrow d}^{H,\bar{H}} = 2(\mp b_3^p + H_{12}^p)$. Therefore by putting $\Delta\omega_{c\rightarrow d} = \omega_{c\rightarrow d}^{H} - \omega_{c\rightarrow d}^{\bar{H}}$ the corresponding figure of merit can be defined as

$$
r_{c\rightarrow d}^{H} = \frac{|\Delta\omega_{c\rightarrow d}|}{m_H} = 4\frac{|b_3^p|}{m_H}
\tag{50}
$$

Attaining a resolution of $1\ mHz$, one would get [45]

$$
r_{c\rightarrow d}^{H} \lesssim 5 \times 10^{-27}
\tag{51}
$$

CONCLUSIONS

In this talk I have reviewed some important consequences of atomic physics measurements in the domain of high-energy physics. In particular APV in cesium could be the first real indication of new physics beyond the SM. The atomic physics tests of the CPT symmetry are already at a spectacular level of sensitivity, and the future experiments on H and $\bar{H}$ could give bounds well below the one expected from string theory.

REFERENCES

1. Bouchiat M.A. and Bouchiat C.C., *Phys. Lett.* **B48**, 111 (1974); *J. Phys.* **35**, 899 (1974).
2. Ramsey-Musolf M.J., physics/0001250, (2000).
3. Bouchiat M.A., Guena J., Hunter L. and Pottier L., *Phys. Lett.* **B117**, 358 (1982).
4. Noecker M.C., Masterson B.P. and Wieman C.E., *Phys. Rev. Lett.* **61**, 310 (1988).
5. Wood C.S., Bennett S.C., Cho D., Masterson B.P., Roberts J.L., Tanner C.E. and Wieman C.E., *Science* **275**, 1759 (1999).
6. Vetter P.A., Meekhof D.M., Majumder P.K., Lamoreaux S.K. and Fortson E.N., *Phys. Rev. Lett.* **71**, 3442 (1993); Edwards N.H., Phipp S.J., Baird E.G., Nakayama S., *Phys. Rev. Lett.* **74**, 2654 (1995).
7. Dzuba V.A., Flambaum V.V., Silvestrov P. and Sushkov O., *Phys. Lett.* **A141**, 147 (1989).
8. Blundell S.A., Johnson W.R. and Sapirstein J., *Phys. Rev. Lett.* **65**, 1411 (1990).
9. Bennett S.C. and Wieman C.E., *Phys. Rev. Lett.* **82**, 2484 (1999).
10. Dzuba V.A., Flambaum V.V. and Ginges J.S.M., contribution A11 to this meeting, *Conference Abstracts*, eds. F. Fusi and F. Cervelli. See also Dzuba V.A. and Flambaum V.V., physics/0005038, (2000).
11. Pollock S.J. and Welliver M.C., *Phys. Lett.*, **B464**, 177 (1999).
12. Derevianko A. physics/0001046, (2000); Derevianko A., hep-ph/0005274, (2000).
13. Kozlov M.G., Porsev S.G. and Tupitsyn I.I., physics/0004076, (2000) and contribution A15 to this meeting, *Conference Abstracts*, eds. F. Fusi and F. Cervelli.
14. Marciano W.J. and Rosner J.L., *Phys. Rev. Lett.* **65**, 2963 (1990); Altarelli G., Lectures given at the *Les Houches Summer School: Particles In The Nineties*, 30 Jun - 26 Jul 1991, Les Houches, France.
15. Altarelli G., Barbieri R. and Jadach S., *Nucl. Phys.* **B369**, 3 (1992); Altarelli G., Barbieri R. and Caravaglios F. *Nucl. Phys.* **B405**, 3 (1993); *ibidem Phys. Lett.* **B349**, 145 (1995).
16. Peskin M.E. and Takeuchi T., *Phys. Rev. Lett.* **65**, 964 (1990); *ibidem Phys. Rev.* **D46**, 381 (1991).
17. Altarelli G., Barbieri R. and Caravaglios F., *Int. J. Mod. Phys.* **A13**, 1031 (1998), and updating of these results as communicated to me by Dr. Caravaglios.
18. The actual experimental lower bound on the Higgs mass is 107.9 *GeV* at 95% CL, see: ALEPH, DELPHI, L3 and Opal Collaborations. The LEP Working group for Higgs boson searches, CERN-EP-2000-055, April 2000.
19. Pollock S.J., Fortson E.N. and Wilets L., *Phys. Rev.* **C46**, 2587 (1992); Chen B.Q. and Vogel P., *Phys. Rev.* **C48**, 1392 (1993).
20. Casalbuoni R., De Curtis S., Dominici D. and Gatto R., *Phys. Lett.* **B460**, 135 (1999).
21. Langacker P., *Phys. Lett.* **B256**, 277, (1991).
22. Particle Data Group, Caso C. *et al.*, *Eur. Phys. J.* **C3**, 1 (1998).
23. Pomarol A. and Quiros M., *Phys. Lett.* **B438**, 255 (1998); Delgado A., Pomarol A. and Quiros M., *Phys. Rev.* **D60**, 95008 (1999); Masip M. and Pomarol A. *Phys. Rev.* **D60**, 96005 (1999).

24. Casalbuoni R., De Curtis S., Dominici D. and Gatto R., *Phys. Lett.* **B462**, 48 (1999).
25. Amaldi U. *et al.*, *Phys. Rev.* **D36**, 1385 (1987); Marciano W.J. and Rosner J.L., *Phys. Rev. Lett.* **65**, 2963 (1990); Altarelli G. *et al. Phys. Lett.* **B261**, 146 (1991); Mahantappa K.T. and Mohapatra P.K., *Phys. Rev* **D43** 3093 (1991); Rosner J.L., *Phys. Rev.* **D61**, 016006 (2000). In these papers it is also possible to find a complete list of references to the Z' models.
26. Erler J. and Langacker P., *Phys. Rev. Lett.* **84**, 212 (2000).
27. Streater R.F. and Wightman A.S., *PCT, spin and statistics and all that*, edited by W.A. Benjamin, Inc., New York, Amsterdam, 1964.
28. Majorana, E., *Il Nuovo Cimento* **9**, 335 (1932). This paper is in italian and it was translated in english by Fradkin E.S., *AJP* **34**, 314 (1966).
29. Kostelecký V.A. and Potting R., *Nucl. Phys.* **B359**, 545, (1991); *ibidem*, *Phys. Lett.* **B381**, 389, (1996); Kostelecký V.A. and Samuel S., *Phys. Rev. Lett.* **63**, 224, (1989); *ibidem*, *Phys. Rev. Lett.* **66**, 1811, (1991); *ibidem*, *Phys. Rev.* **D39**, 683, (1989); *ibidem*, *Phys. Rev.* **D40**, 1886, (1989).
30. Colladay D. and Kostelecký V.A., *Phys. Rev.* **D55**, 6760, (1997); *ibidem*, **D58**, 116002, (1998).
31. For a recent review of the results on this subject, see Kostelecký V.A., hep-ph/0005280, (2000).
32. Kostelecký V.A., *Phys. Rev. Lett.* **80**, 1818, (1998).
33. Bluhm R., Kostelecký V.A. and Lane C.D., *Phys. Rev. Lett.* **84**, 1098, (2000).
34. Bluhm R. and Kostelecký V.A., *Phys. Rev. Lett.* **84**, 1381, (2000).
35. Kostelecký V.A. and Lane C.D., *Phys. Rev.* **D60**, 116010, (1999).
36. Bertolami O. *et al.*, *Phys. Lett.* **B395**, 178 (1997).
37. Van Dyck R.S. Jr., Schwinberg P.B. and Dehmelt H.G., *Phys. Rev. Lett.* **59**, 26 (1987); *ibidem Phys. Rev.* **D34**, 722 (1986). For a review of the principles of the Penning trap, see: Brown L.S. and Gabrielse G. *Rev. Mod. Phys.* **58**, 233 (1986).
38. Gabrielse G. *et al.*, *Phys. Rev. Lett.* **82**, 3198 (1999).
39. Schwinberg P.B., Dyck R.S. Jr. and Dehmelt H.G., *Phys. Lett.* **A81**, 119 (1981).
40. Bluhm R., Kostelecký V.A. and Russell N., *Phys. Rev. Lett.* **79**, 1432 (1997).
41. Bluhm R., Kostelecký V.A. and Russell N., *Phys. Rev.* **D57**, 3932 (1998).
42. Dehmelt H., Mittleman R., Van Dyck R.S. Jr. and Schwinberg P., *Phys. Rev. Lett.* **83**, 4694 (1999).
43. Udem T. *et al.*, *Phys. Rev. Lett.* **79**, 2646 (1997).
44. Cesar C.L. *et al.*, *Phys. Rev. Lett.* **77**, 255 (1996).
45. Bluhm R., Kostelecký V.A. and Russell N., *Phys. Rev. Lett.* **82**, 2254 (1999).

The g-Factor of the Bound Electron in Hydrogenic Ions

Wolfgang Quint[1] and the g-Factor Collaboration

Gesellschaft für Schwerionenforschung
Planckstr. 1, D-64291 Darmstadt, Germany

Abstract. We report on the measurement of the g-factor of the electron bound in an atomic ion. A single hydrogenic ion ($^{12}C^{5+}$) is stored in a Penning trap. The electronic spin state of the ion is monitored via the continuous Stern-Gerlach effect in a quantum non-demolition measurement. Quantum jumps between the two spin states (spin *up* and spin *down*) are induced by a microwave field at the spin precession frequency of the bound electron. The g-factor of the bound electron is obtained by varying the microwave frequency and counting the number of spin flips for a fixed time interval. Applications of the continuous Stern-Gerlach effect include high-accuracy tests of bound-state quantum electrodynamics (QED), the measurement of the atomic mass of the electron, the determination of the fine structure constant α, and the measurement of nuclear g-factors.

THE G-FACTOR OF THE ELECTRON

The g-factor of the electron relates its magnetic moment μ, in units of the Bohr magneton μ_B, to its spin angular momentum s, in units of Planck's constant $\hbar$.

$$\frac{|\mu|}{\mu_B} = g\frac{|s|}{\hbar} \tag{1}$$

In the Dirac theory the g-factor of the free electron is

$$g_{Dirac} = 2 \tag{2}$$

The g-factor of the free electron differs slightly from this Dirac value due to the presence of the virtual radiation field. The different contributions to the g-factor bound electron in hydrogenic ions are the relativistic correction, radiative corrections (QED), and nuclear corrections [1].

---

[1] supported by Deutsche Forschungsgemeinschaft.

CP551, *Atomic Physics 17,* edited by E. Arimondo, P. DeNatale, and M. Inguscio
© 2001 American Institute of Physics 1-56396-982-3/01/$18.00

- *Relativistic correction (Breit term):*

The g-factor of the electron in the Coulomb field of a point-like nucleus with charge Ze was first calculated by Breit in 1928 who solved the Dirac equation for hydrogenic ions in the presence of a magnetic field (Zeeman level splitting) [2]. According to Breit the g-factor of the electron in the $1s_{1/2}$ ground state of a hydrogenic ion is reduced to

$$g = \frac{2}{3}\left(1 + 2\sqrt{1 - (Z\alpha)^2}\right) = 2 - \frac{2}{3}(Z\alpha)^2 - \frac{1}{6}(Z\alpha)^4 + 0(Z\alpha)^6. \qquad (3)$$

For hydrogenic ions the Breit term is the largest bound-state correction of the g-factor of the electron. In the case of hydrogenic uranium (U^{91+}) the Breit term reduces the g-factor of the electron by as much as 15%.

- *Radiative corrections:*

The first calculation of a radiative correction to the g-factor of the free electron was done by Schwinger. The Schwinger term α/π describes the virtual emission and absorption of a photon by the electron (Fig. 1). The higher-order QED terms account for the virtual emission and absorption of several photons and for the vacuum polarization. Meanwhile, the radiative corrections were calculated up to the fourth order in α [3].

$$g_{free} = 2 + C_1\left(\frac{\alpha}{\pi}\right) + C_2\left(\frac{\alpha}{\pi}\right)^2 + C_3\left(\frac{\alpha}{\pi}\right)^3 + C_4\left(\frac{\alpha}{\pi}\right)^4 + 0\left(\frac{\alpha}{\pi}\right)^5 \qquad (4)$$

The precision of the quantum electrodynamical calculations of the g-factor of the free electron is limited by our knowledge of the fine structure constant $\alpha \approx 1/137$. The theoretical value for the g-factor of the free electron is experimentally confirmed on a level of 10^{-11} [4,5].

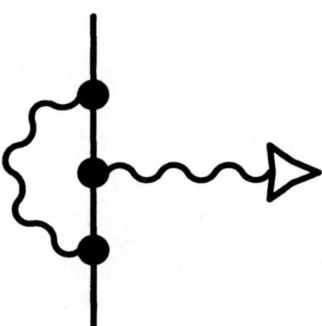

FIGURE 1. Feynman diagram describing the QED contribution to the g-factor of the free electron of order α/π (Schwinger term). The straight line is the propagator of the free electron, the wavy line is the photon propagator, and the triangle denotes the magnetic field.

- *Bound-state radiative corrections:*

  In addition to the radiative corrections of the free electron, the g-factor of the bound electron in the $1s_{1/2}$ state is modified by bound-state QED corrections. Six Feynman graphs contribute to the g-factor of the bound electron on the one-loop level (of order α). The three graphs a, c, and e in Fig. 2 describe the self-energy corrections, and the graphs b, d, and f represent the vacuum-polarization corrections to the bound-state g-factor. The Feynman graphs were evaluated for hydrogenic ions with different nuclear charge Z in non-perturbative calculations which include all orders in $Z\alpha$ [6,7]. These new theoretical calculations are an exciting advancement in the field of bound-state quantum electrodynamics. The comparison of the new theoretical results with our high-accuracy measurements of the g-factor of the bound electron in hydrogenic ions will provide one of the most stringent tests of the theory of quantum electrodynamics in very strong electromagnetic fields.

  Before these new theoretical developments the bound-state QED contributions to the g-factor of the bound electron were perturbatively evaluated in a $Z\alpha$ expansion [8]. The leading term of the $Z\alpha$ expansion is $\Delta g_{bound-QED} = (\alpha/\pi)(Z\alpha)^2/6$. Our recent measurements on hydrogenic carbon ($^{12}C^{5+}$) are sensitive to the terms of higher order in $Z\alpha$ and clearly discriminate between the recent non-perturbative evaluation of the bound-state QED contributions and the previous calculations which yielded only the leading term of the $Z\alpha$ expansion.

- *Nuclear corrections:*

  The following nuclear effects influence the g-factor of the bound electron in hydrogenic ions and have to be considered in the theoretical calculations:

  i) nuclear recoil correction due to the finite nuclear mass,

  ii) nuclear size effect, and

  iii) nuclear polarization, i.e. the virtual excitation of nuclear energy levels.

  Compared to other bound-state QED tests, for example the Lamb shift and the hyperfine splitting in hydrogenic ions, the g-factor of the bound electron in hydrogenic ions is less sensitive to the details of the nuclear structure. The nucleus influences the g-factor of the bound electron only via the dependence of the electronic wave function on the nuclear properties.

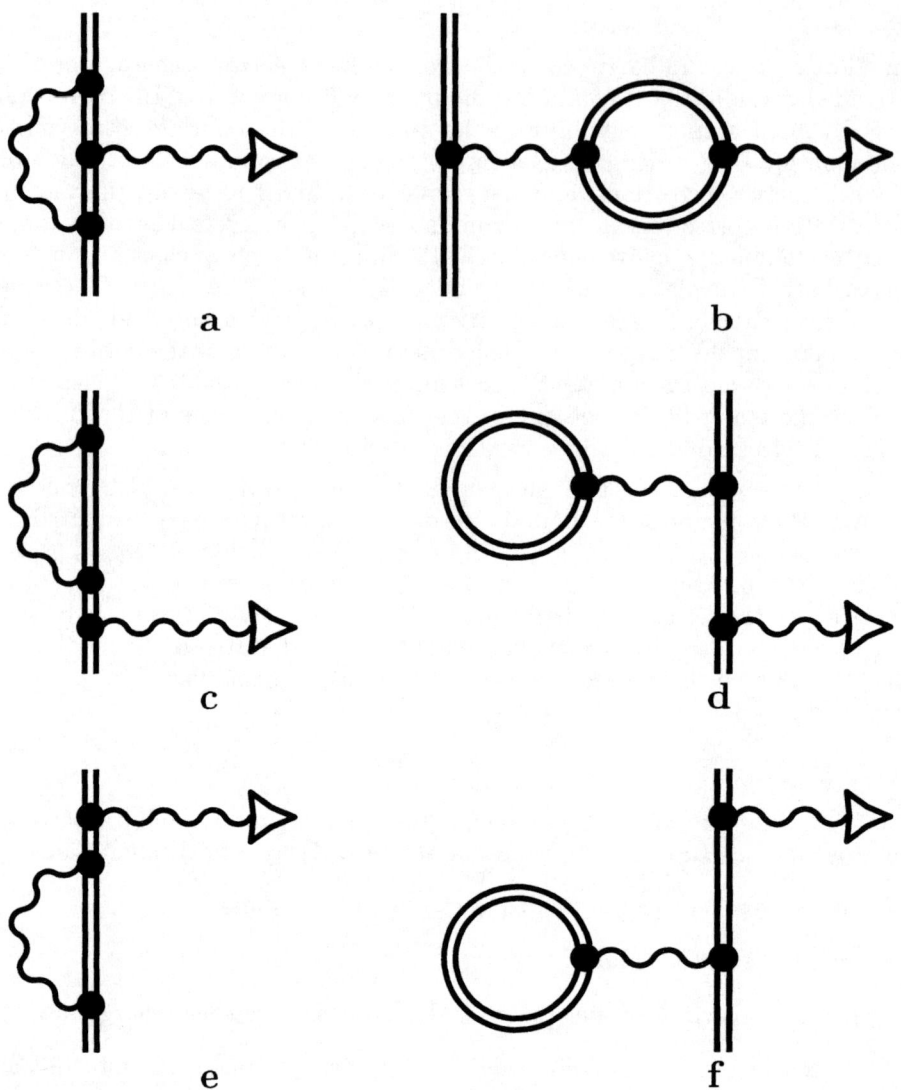

FIGURE 2. Feynman diagrams describing the bound-state QED contributions to the g-factor of the electron of order α/π. Graphs a, c, and e: self-energy corrections; graphs b, d , and f: vacuum-polarization corrections. The double line represents the propagator of a bound electron in a hydrogenic ion.

PENNING TRAP APPARATUS

In a Penning trap a charged particle is stored in a combination of a homogeneous magnetic field B_0 and an electrostatic quadrupole potential [5]. The magnetic field confines the particle in the plane perpendicular to the magnetic field lines, and the electrostatic potential in the direction parallel to the magnetic field lines. The three eigenmotions (Fig. 3) that result are the trap-modified cyclotron motion (frequency ω_+), the magnetron motion (frequency ω_-), which is a circular E × B drift motion perpendicular to the magnetic field lines, and the axial motion (parallel to the magnetic field lines, frequency ω_z). The free-space cyclotron frequency $\omega_c = (Q/M)B_0$ of an ion with charge Q and mass M in a magnetic field B_0 can be determined from a combination of the trapped ion's three eigenfrequencies ω_+, ω_z and ω_- with the formula [9]

$$\omega_c^2 = \omega_+^2 + \omega_z^2 + \omega_-^2. \tag{5}$$

The magnetron frequency ω_- can be calculated from the axial frequency ω_z and the trap-modified cyclotron frequency ω_+ through the relation $\omega_- = \omega_z^2/2\omega_+$. Thus, the free-space cyclotron frequency ω_c can be determined from a measurement of the axial frequency ω_z and the trap-modified cyclotron frequency ω_+.

In our experiment, the magnetic field ($B_0 = 3.8$ Tesla) is provided by a super-conducting solenoid. The electrostatic quadrupole potential is produced by a stack of five cylindrical electrodes with the same inner diameter (7 mm): the so-called ring electrode at the center and the compensation electrodes and endcaps placed on either side of the ring electrode [10,11]. With a positive voltage U_0 applied between the two endcaps and the ring electrode, a potential well $V_{el}(z) = QU_0z^2/d^2$ is created along the magnetic field lines for a positively charged ion, where d is a

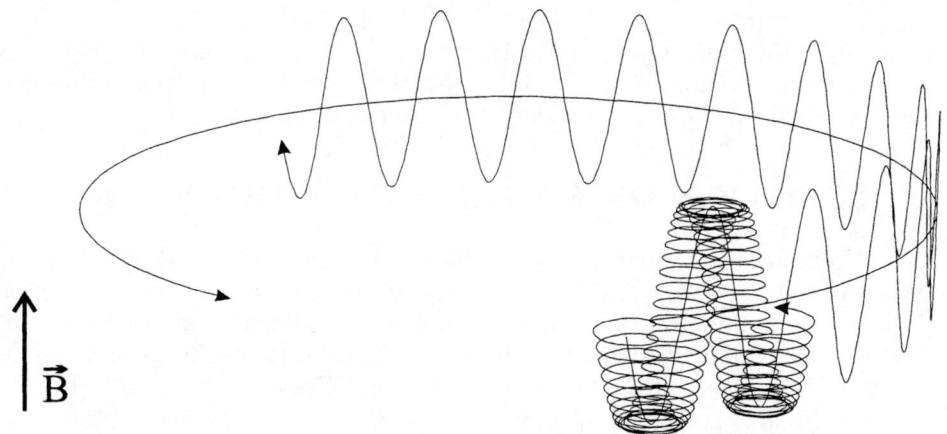

FIGURE 3. Three eigenmotions of a particle in a Penning trap: axial oscillation, cyclotron motion (small circles), and magnetron motion.

characteristic size of the trap electrodes. The voltage at the compensation electrodes is tuned to minimize anharmonic terms of the electric potential. The axial motion of the trapped ion is a harmonic oscillation with frequency

$$\omega_z = \sqrt{\frac{Q}{M}\frac{U_0}{d^2}}. \tag{6}$$

CONTINUOUS STERN-GERLACH EFFECT

The principle of the continuous Stern-Gerlach effect is based on a coupling of the magnetic moment μ of the particle to its axial oscillation frequency ω_z in the Penning trap [12,13]. This coupling is achieved by a quadratic magnetic field component ("magnetic bottle") superimposed on the homogeneous magnetic field B_0 of the Penning trap.

$$B(z) = B_0 + \beta_2 z^2 \tag{7}$$

Due to the interaction of the z-component μ_z of the magnetic moment with the "magnetic bottle" term the trapped ion possesses a position-dependent potential energy $V_m = -\mu_z(B_0 + \beta_2 z^2)$, which adds to the potential energy V_{el} of the ion in the electrostatic well. Therefore, the effective trapping force is modified by the magnetic interaction, and the axial frequency of the trapped ion is shifted upwards or downwards, depending on the sign of the z-component μ_z of the magnetic moment. This axial frequency shift is given by

$$\delta\omega_z = \frac{\beta_2 \mu_z}{M\omega_z} \tag{8}$$

where ω_z is the unshifted axial frequency. The determination of the spin orientation via the continuous Stern-Gerlach effect is a quantum non-demolition measurement because the spin state is measured non-destructively.

DETECTION OF A SINGLE HYDROGENIC ION

In our Penning trap apparatus, which has been described in [11], there are two positions in the stack of cylindrical electrodes where ions can be trapped (Fig. 4).

(i) In the precision trap ions are created in different charge states by electron impact ionization of neutral atoms. An electron beam with an energy of 2 keV and a current of 5 nA is emitted from a tungsten field emission point close to the trap electrodes. In the precision trap unwanted ion species are removed by selectively heating their axial motion; consecutively, the number of $^{12}C^{5+}$ ions is reduced to one particle by lowering the electrostatic potential well. In the precision trap the spin-flip transition of the bound electron is excited by applying a microwave field.

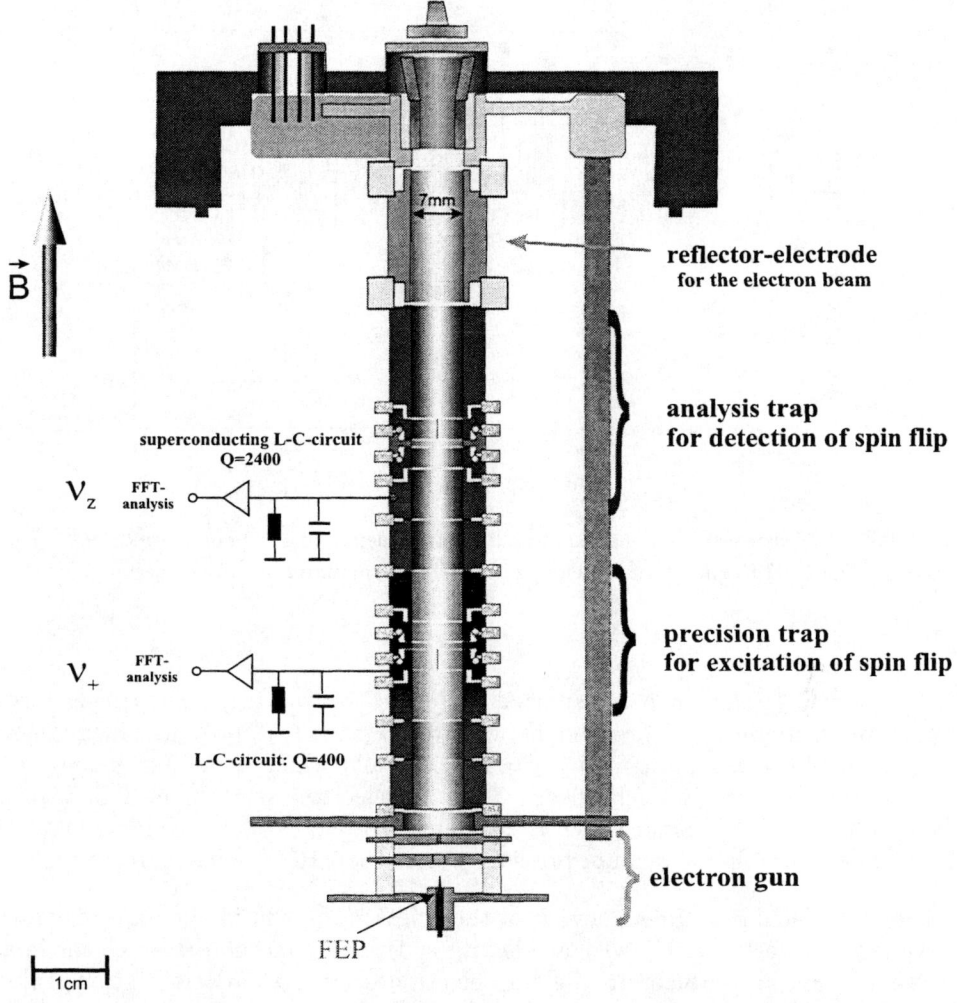

FIGURE 4. Penning trap set-up with the cylindrical electrodes for trapping, the electron gun for production of hydrogenic ions by electron ionization, and L-C-circuits for detection of trapped ions.

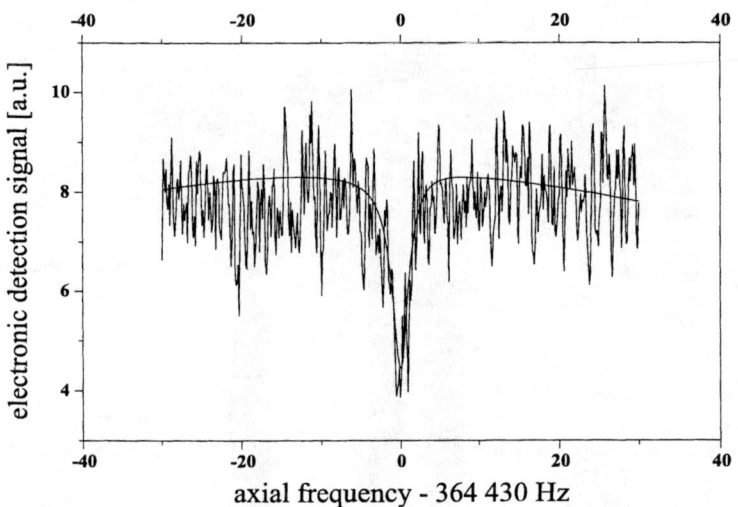

FIGURE 5. Measurement of the axial oscillation frequency of a single trapped $^{12}C^{5+}$ ion at T = 4 K. The axial frequency of the ion is observed as a minimum in this spectrum.

The single $^{12}C^{5+}$ ion is then transported to the *(ii)* analysis trap. The ring electrode of this trap is made out of ferromagnetic material (nickel) to produce the quadratic component of the magnetic field ($\beta_2 = 1$ Tesla/cm^2) which is necessary to observe the continuous Stern-Gerlach effect. The trap electrodes are housed in a sealed ultra-high vacuum chamber which is kept at a temperature of 4 Kelvin. With the effect of cryopumping a vacuum pressure better than 10^{-16} mbar is reached.

The axial oscillation frequency ω_z of the single $^{12}C^{5+}$ ion in the analysis trap is measured non-destructively with an electronic detection method through the image currents which are induced in the trap electrodes by the particle motion [14]. A *LCR* circuit resonant at $\omega_z = 2\pi \times 364$ kHz (with quality factor $Q = 2400$) is attached to one of the trap electrodes to optimise the detection sensitivity. Through the image currents dissipated in the *LCR* circuit the trapped ion is cooled to the ambient temperature of 4 K with a time constant of $\tau_z = 100$ ms. The ion's axial frequency is determined in a frequency analysis of the signal across the resonance circuit. Such a frequency spectrum is shown in Fig. 5. When the $^{12}C^{5+}$ ion is in thermal equilibrium at $T = 4$ K its axial frequency is observed as a minimum in this frequency spectrum [14]. The elegance of the described measurement technique lies in the fact that it allows for the detection of a single trapped ion while it is at a temperature of 4 K. At this low temperature its oscillation amplitudes in the trap are strongly reduced ($< 50\,\mu$m), and systematic errors due to inhomogeneities of the magnetic field or anharmonicities of the electrostatic trapping field are minimized.

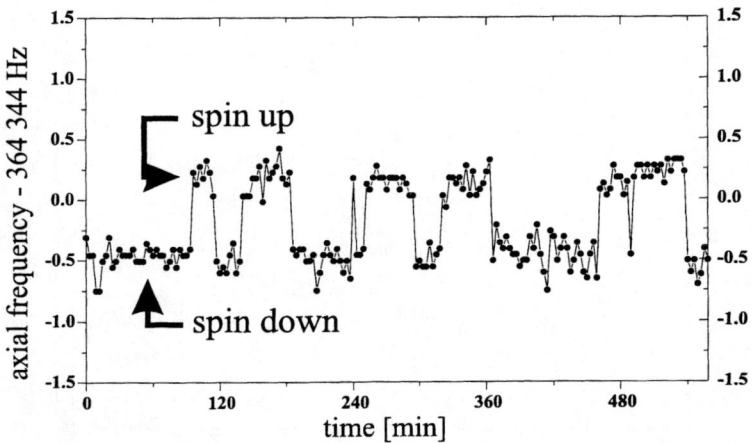

FIGURE 6. Quantum non-demolition measurement of the spin state: the spin-flip transitions are observed as small discrete changes of the axial frequency of the stored $^{12}C^{5+}$ ion.

G-FACTOR MEASUREMENT

The quantum state of the $^{12}C^{5+}$ ion, i.e. the magnetic quantum number m_s of the bound electron in the $1s_{1/2}$ ground state, can be monitored non-destructively in the analysis trap in repeated measurements of the axial frequency of the trapped ion. Transitions between the two spin states $m_s = \pm 1/2$ are induced by a microwave field (at 104 GHz) resonant with the Larmor precession frequency ω_L of the bound electron.

$$\hbar \omega_L = g \frac{e\hbar}{2m_e} B = g \mu_B B \tag{9}$$

Here, g is the g-factor of the bound electron and $\mu_B = e\hbar/2m_e$ is the Bohr magneton. The spin-flip transitions are observed as discrete changes of the ion's axial frequency. Fig. 6 shows a clear demonstration of such quantum jumps observed via the continuous Stern-Gerlach effect. The measured axial frequency shift for a transition between the two quantum levels is $\omega_z(\uparrow) - \omega_z(\downarrow) = 2\pi \times 0.7$ Hz, in excellent agreement with the expected value calculated from Equ. 8.

The observation of the continuous Stern-Gerlach effect on the hydrogen-like carbon ion $^{12}C^{5+}$ makes it possible to measure its electronic g-factor to high accuracy. Using the cyclotron frequency ω_c of the ion (Equ. 5) for the calibration of the magnetic field B, the g-factor of the bound electron can be expressed as the ratio of the Larmor precession frequency ω_L of the electron and the cyclotron frequency ω_c of the $^{12}C^{5+}$ ion.

$$g = 2 \cdot \frac{\omega_L}{\omega_c} \cdot \frac{Q/M}{e/m_e} \tag{10}$$

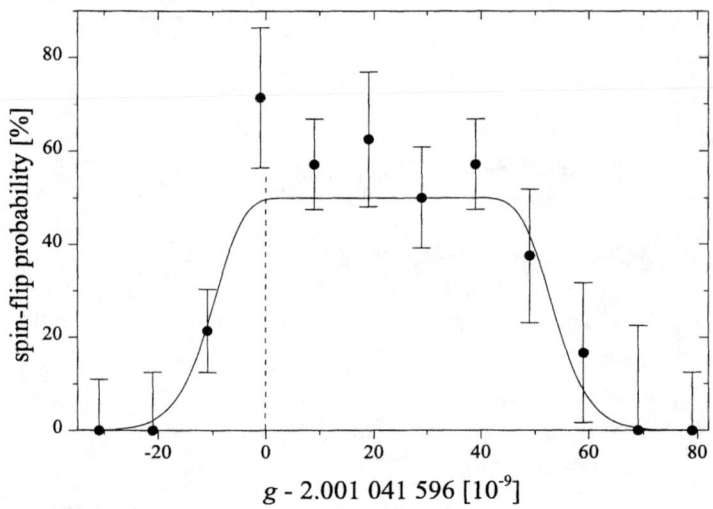

FIGURE 7. g-Factor resonance measured in the precision trap. The resonance curve broadened and shifted due to saturation effects.

The ratio of charge-to-mass ratios of the carbon ion (Q/M) and of the electron (e/m_e) was measured in a Penning trap to an accuracy of $2 \cdot 10^{-9}$ [15].

A resonance spectrum of the Larmor precession frequency ω_L of the bound electron is obtained in the following way. In the precision trap a microwave field at ω_L is applied to excite a spin-flip transition. Then the $^{12}C^{5+}$ ion is transferred to the analysis trap where the spin state is analysed via the continuous Stern-Gerlach effect. The ion is moved back to the precision trap, and the measurement cycle is starts again. The number of spin-flip transitions which occurred in the precision trap is counted for a fixed time interval. Then the microwave frequency is varied and the measurement is repeated at different excitation frequencies.

Finally, the plot of the quantum jump rate versus excitation frequency yields the resonance spectrum. A g-factor resonance is obtained by dividing the Larmor frequency by the cyclotron frequency which is simultaneously measured (Fig. 7). The experimental data points are fitted to a theoretical lineshape. In the specific example shown in Fig. 7 the resonance curve is shifted and broadened due to saturation effects. A reduction of the intensity of the driving microwave field results in a substantially smaller line width. The trap-modified cyclotron frequency ω_+ of the stored $^{12}C^{5+}$ ion is determined in a Fourier transform of the image currents induced in the trap electrodes by the ion motion (Fig. 8).

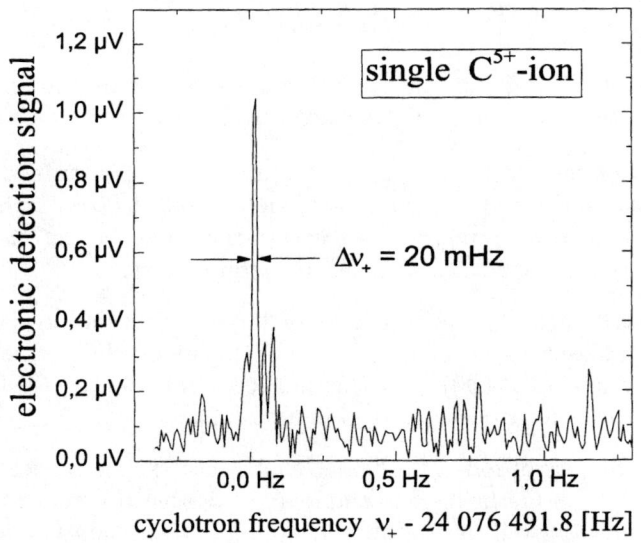

FIGURE 8. Mesurement of the trap-modified cyclotron frequency ω_+ of a $^{12}C^{5+}$ ion. The fractional line width is smaller than 10^{-9}.

EXPERIMENTAL RESULTS

In our first measurement [16] we obtained values of $\omega_L = 2\pi \times 103\,958.105\,(10)$ MHz for the Larmor precession frequency of the bound electron and $\omega_c = 2\pi \times 23.755\,285\,(20)$ MHz for the free-space cyclotron frequency of the $^{12}C^{5+}$ ion. From these frequencies we calculated using Equ. 10 the g-factor of the bound electron in $^{12}C^{5+}$ with an accuracy of 10^{-6}

$$g_e(C^{5+}) = 2.001\,042\,(2) \tag{11}$$

In our most recent measurements we improved the measurement accuracy by a factor of 1000. This was achieved by spatially separating the functions of inducing and detecting the spin-flip transitions, as described above. For a detailed discussion we refer to a forthcoming publication [17]. Our new experimental value for the g-factor of the bound electron in $^{12}C^{5+}$ is $g_{\text{exp}} = 2.001\,041\,596\,(5)$. Our measurement is in excellent agreement with the theoretical value $g_{\text{th}} = 2.001\,041\,591\,(7)$. Together experiment and theory test the bound-state QED contributions to the g-factor of the bound electron with high precision. Future measurements of the g-factor of the bound electron in heavier hydrogenic ions will provide even more stringent tets of bound-state quantum electrodynamics.

OUTLOOK

A couple of interesting experiments are made possible by the application of the continuous Stern-Gerlach effect to atomic ions. For example:

- The mass of the electron in atomic units can be determined using Equ. 10 from measurements of the g-factor of the bound electron in light hydrogenic ions. In such measurements the electron mass can be determined by nearly one order of magnitude more accurately than today.

- The fine structure constant α can be deduced with an accuracy of a few times 10^{-8} from measurements of the g-factor of the bound electron in medium-heavy hydrogenic ions ($Z \approx 30$). For this purpose, bound-state QED contributions of second order in α need to be evaluated [18].

- Nuclear g-factors of ions in different charge states can be determined via the continuous Stern-Gerlach effect employing a double-resonance technique where the electronic as well as the nuclear spin-flip transition is excited. Also the effect of diamagnetic shielding can be investigated in such measurements.

We are grateful to S. Karshenboim, I. Lindgren, A.-M. Mårtensson-Pendrill, V. Natarajan, K. Pachucki, H. Persson, S. O. Salomonson, V. M. Shabaev, and G. Soff for many stimulating discussions. This work is financially supported by the European Union under the contract number ERB FMRX CT 97-0144 within the EUROTRAPS network.

REFERENCES

1. W. Quint, Phys. Scr. **T59**, 203 (1995).
2. G. Breit, Nature **122**, 649 (1928).
3. V.W. Hughes and T. Kinoshita, Rev. Mod. Phys. **71**, 133 (1999).
4. R. S. Van Dyck, Jr., P. B. Schwinberg and H. G. Dehmelt, Phys. Rev. Lett. **59**, 26 (1987).
5. Hans Dehmelt, Rev. Mod. Phys. **62**, 525 (1990).
6. H. Persson, S. Salomonson, P. Sunnergreen, I. Lindgren, Phys. Rev. A **56**, R2499 (1997).
7. T. Beier, I. Lindgren, H. Persson, S. Salomonson, P. Sunnergren, H. Häffner, and N. Hermanspahn, Phys. Rev. A, accepted.
8. H. Grotch and R.A. Hegstrom, Phys. Rev. **A 4**, 59 (1971).
9. L.S. Brown and G. Gabrielse, Phys. Rev. A **25**, 2423 (1982).
10. G. Gabrielse, L. Haarsma and S.L. Rolston, Intl. J. of Mass Spec. and Ion Proc. **88**, 319 (1989); ibid. **93**, 121 (1989).
11. M. Diederich, H. Häffner, N. Hermanspahn, M. Immel, W. Quint, S. Stahl, H.-J. Kluge, R. Ley, R. Mann, G. Werth, Hyp. Int. **115**, 185 (1998).
12. H. Dehmelt, Proc. Natl. Acad. Sci. U.S.A. **83**, 2291 (1986).

13. H. Dehmelt, Z. Phys. D **10**, 127 (1988).
14. D.J. Wineland, H.G. Dehmelt, J. Appl. Ph. **46**, 919 (1975).
15. R.S. Van Dyck Jr., D.L. Farnham, P.B. Schwinberg, Physica Scripta **T59**, 134 (1995).
16. N. Hermanspahn, H. Häffner, H.-J. Kluge, S. Stahl, W. Quint, J. Verdú, and G. Werth, Phys. Rev. Lett. **84**, 427 (2000).
17. H. Häffner, T. Beier, N. Hermanspahn, H.-J. Kluge, S. Stahl, W. Quint, J. Verdú, and G. Werth, submitted to Phys. Rev. Lett.
18. T. Beier, private communication.

Tests of Quantum Electrodynamics in Hydrogenic Ions

Joshua Silver

Oxford University
Clarendon Laboratory
Parks Road
Oxford OX1 3PU

Abstract. This paper discusses measurements of the Lamb shift in hydrogenic ions and how they have been used to test quantum electrodynamics. There is a discussion of the validity of renormalisation and also a consideration of Lamb shift measurements for hydrogenic ions of medium Z elements, with a description of a novel experiment on slow ions and a discussion of what the results of this experiment should tell us.

The Dirac equation is believed to properly describe the relativistic quantum mechanics of the one-electron system but of course it does not completely describe such a system since it ignores certain quantum electrodynamic effects – as well as small effects due to nuclear size, shape and spin. In this paper I will restrict myself to discussing Lamb shifts in hydrogenic ions since measurement of these shifts, which are quantum electrodynamic in origin, have sometimes been compared to calculation and the comparison used as a "test of quantum electrodynamics". It is interesting to ask how good a test this really provides, and to examine the case for measurements in hydrogenic ions when compared to atomic hydrogen.

Firstly, perhaps, it is appropriate to examine the case for testing quantum electrodynamics in general, and then looking at the specific case for measuring Lamb shifts in both hydrogen and the hydrogenic ions. I hope I will be excused as an experimenter by taking a somewhat non-technical approach here. Feynman wrote an excellent though non-technical text on Quantum Electrodynamics in 1985[1] and in this text he gives a nice justification for testing quantum electrodynamics as follows:
" if we arrange in the laboratory an experiment involving just a *few* electrons in *simple* circumstances, then we can calculate what might happen very accurately, and we can measure it very accurately, too. Whenever we do such experiments, the theory of quantum electrodynamics works very well. We physicists are always checking to see if there is something the matter with the theory. That's the game, because if there *is* something the matter, it's interesting. But so far, we have found nothing wrong with the theory of quantum electrodynamics. It is, therefore, I would say, the jewel of physics – our proudest possession." Feynman goes on to explain that quantum electrodynamics is also the prototype for newer theories, implying that this adds to the importance that it be tested.

CP551, *Atomic Physics 17,* edited by E. Arimondo, P. DeNatale, and M. Inguscio
© 2001 American Institute of Physics 1-56396-982-3/01/$18.00

Later in the same reference[2], Feynman describes the technique of renormalisation which is needed to calculate experimentally observable effects such as the Lamb shift, and makes the comment " The shell game we play to find n and j is technically called "renormalisation". But no matter how clever the word, it is what I would call a dippy process! Having to resort to such hocus-pocus has prevented us from proving that the theory of quantum electrodynamics is mathematically self-consistent. It's surprising that the theory still hasn't been proved self-consistent one way or the other by now; *I suspect that renormalisation is not mathematically legitimate* (my italics). What *is* certain is that we do not have a good mathematical way to describe the theory of quantum electrodynamics: such a bunch of words to describe the connection between n and j and m and e is not good mathematics. So far as I am aware[3] the situation described by Feynman in 1985 regarding the mathematical validity of renormalisation still holds at the time of writing. This can be viewed as one underlying justification for the measurement of Lamb shifts in hydrogenic ions – these are systems which involve a few electrons – in fact only one electron – in simple circumstances. However, it has been stated with some authority that the theoretical method used to calculate the measured quantities has not been shown to be mathematically self-consistent or "legitimate". Questions about the mathematical legitimacy of infinite renormalisation might be better considered from the perspective of a philosopher of science and are for example considered by Teller [4,5] and a discussion of the problem of infinities in quantum electrodynamics is also given by Weinberg[6].

So we now have a general case for testing quantum electrodynamics, and a case for testing quantum electrodynamic quantities which are calculated using renormalisation. One obvious candidate (smallest number of electrons, simple system) is the hydrogen atom and indeed Lamb shift measurements in the hydrogen atom itself have long been regarded as a means of testing quantum electrodynamics. As is well known however [7] the charge radius of the proton is at present quite poorly known[8] and this limits the interpretation of measurements of the Lamb shift in atomic hydrogen. So what is the case for measuring Lamb shifts in hydrogenic ions (by which I mean systems which consist of a nucleus and a single electron) and, given the astonishing accuracy recently achieved for the measurement of the 1s – 2s separation in hydrogen, [9] can such measurements tell us anything which we cannot learn by a study of hydrogen itself – so are they really worth pursuing?

Perhaps the first point to consider is Z scaling. The Schrödinger energies in the hydrogenic ion scale of course as Z^2 whilst the order alpha self-energy, the largest contribution to the s state Lamb shift, scales essentially as Z^4. This means that, relatively, the Lamb shift can be much larger when compared to the s electron binding energy for a high Z hydrogenic ion that for hydrogen itself. On the face of it, therefore, more sensitive Lamb shift measurements should be possible at higher Z – and indeed, the power of this argument is very nicely demonstrated by the little-known historic fact that the ground state (1s) Lamb shift in a one-electron ion was actually observed and measured as early as 1933 [10] long before the development of modern quantum electrodynamics, though this went unrecognised until some 16 years later. The 1933 measurement was carried out in hydrogenic lithium ions, and the Z^4 scaling of the shift meant that it could be measured using relatively low resolution grating spectroscopy, rather than the radio frequency spectroscopy which was first applied in hydrogen by Lamb. The fact that the Lamb shift grows fast with Z means that, in the fullness of time, we might perhaps expect measurements on high Z hydrogenic ions to become a very sensitive way to test quantum electrodynamics, as long as appropriate experimental techniques can be found, and methods of measuring nuclear size are improved – in this context it might be helpful to bear in mind that the spectroscopic methods which have now worked so very well in atomic hydrogen have arguably been under development since the 19th century, whilst studies of systems such as U^{91+} are still, by comparison, in their infancy.

The energy levels of interest for a measurement of the 1s or 2s Lamb shift are shown in figure 1, which is for the hydrogenic silicon ion.

The most straightforward method of measuring the ground state (1s) shift is to follow the method originated by Edlen, namely to make an accurate measurement of the 1s – 2p Lyman alpha wavelengths, or equivalently, energies. The energy of the transition may be thought of as the sum of Dirac and quantum electrodynamic contributions, with small nuclear corrections, and differencing the measured values from the Dirac contribution yields the 1s shift when allowance is made for the effect of nuclear mass, size and spin. This technique is limited in accuracy essentially because the 1s shift is a small fraction of the 1s –2p energy[11]. Very accurate measurements of this energy in systems other than the neutral hydrogen atom are difficult at the present time for several reasons, including the spectral region in which the transition lies, and the fact that it is difficult to produce highly charged ions with low velocities, so that Doppler shifts must be dealt with.

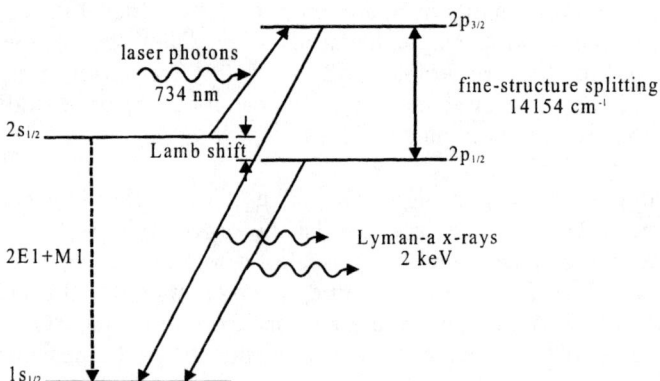

Figure 1. Term diagram showing energy levels relevant for 1s and 2s Lamb shift measurement.

One method of dealing with Doppler shifts in these measurements is to "overlap" Lyman and Balmer transitions using Bragg crystal spectroscopy[12], a method which is in some loose sense analogous to Doppler-free two photon spectroscopy of the 1s-2s interval in atomic hydrogen and simultaneous comparison with 2 – 4 transitions, as pioneered by Wieman and Hansch[13]. This method was shown by Laming et al[14] to be capable of ppm statistical accuracy for the relative measurement of Lyman alpha wavelengths in hydrogenic Ge^{31+} ions some years ago, however the statistical errors associated principally with the comparison of different orders of diffraction[15] have so far limited the actual accuracy which may be achieved, even though this approach still has promise. Figure 2 shows a comparison of uncertainties between calculated and measured values for the 1s Lamb shift for hydrogenic systems from hydrogen to hydrogenic uranium. It will be seen that so far, the accuracy of measurement is not sufficient to provide a critical test of the theory. A most interesting beautiful recent measurement to which I would like to draw attention is the 1s Lamb shift in hydrogenic uranium ions[16], in a sense a direct descendant of the very first work by Edlen. This recent work, which uses very fast ions, is essentially limited by Doppler effects, but there is hope that improved instrumentation will reduce the error noticeably[17]. There are also recent proposals due to Tarbutt[18] for differential comparison of Lamb shifts between different hydrogenic ions which offer the promise of improved experimental precision.

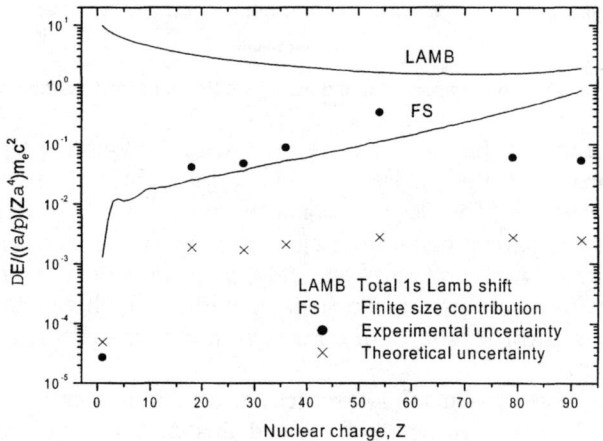

Figure 2. Experimental and theoretical uncertainties for 1s Lamb shifts, also shown is the relative contribution of the finite nuclear size.

The "classical" 2s state Lamb shift has been studied in hydrogenic ions using a number of approaches, some directly spectroscopic, and others less direct[19]. I will concentrate here on spectroscopic measurements. The publication of a paper by Frank Pipkin[20] which applied rf spectroscopy and fast beam techniques to make an accurate measurement of the Lamb shift in atomic hydrogen could be seen as suggesting a route to Lamb shift measurement in higher Z hydrogenic ions[21,22] in that the Z^4 scaling of the 2s Lamb shift and the 2p finestructure means that for hydrogenic ions in the approximate range $Z = 1 \sim Z=20$, rf or laser spectroscopy may be used to drive resonances between the 2s and 2p levels and this resonance spectroscopy may be used, in principle, to measure accurately fine-structures and Lamb shifts. An very thorough review of the work to date in this field was given by Myers[23] at the satellite meeting to this ICAP, Hydrogen II, and the results of measurements of the 2s Lamb shift for the ions of phosphorus, sulphur, chlorine and argon[24] are given in figure 3.

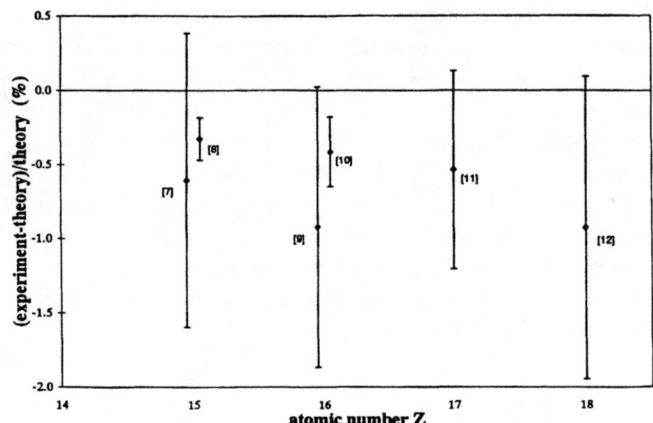

Figure 3. Comparison of experiment and theory for 2s Lamb shift in the range Z=15-18.

It will be seen from the figure that the experimental values are all consistently and intriguingly lower than the theoretical values, but that the discrepancies are not statistically significant. The laser resonance experiments – and the quenching experiment on hydrogenic argon – were all carried out with a fast atomic beam source which, at the time, was probably the only practicable source of ions for such measurements, and the uncertainties associated with the high velocity of the ions in a fast beam form a significant source of error in all these measurements.

Relatively recent developments in ion sources mean that it is now possible to generate useful numbers of *slow* hydrogenic ions and this offers the promise of improved precision for laser resonance Lamb shift measurements in these ions[25]. The source of ions chosen for the first experiment of this sort is an electron beam ion trap (ebit), a device which is principally due to Levine[26] but which derives from the earlier electron beam ion source due principally to Donets[27] which, in their turn, seem to originate from a suggestion by Bleakney[28] Given the situation shown in figure 3, it was decided initially to try to use the laser resonance method to measure the Lamb shift in hydrogenic silicon ions. The choice of ion is dictated partly by the availability of a suitable laser, and partly by the laser power requirements. Careful analysis suggests that hydrogenic silicon ions may be an appropriate choice, since a suitable solid state laser system (ti-sapphire) exists, and the high powers required to obtain good statistics for the resonance – and high statistical accuracy of the final measurement – can be achieved with the use of a very high finesse build-up cavity. The laser system needed for this experiment is presently under development, and the experiment is described by Klein et al[29] in a presentation to the Hydrogen II satellite. A layout of the experiment, is given in figure 4.

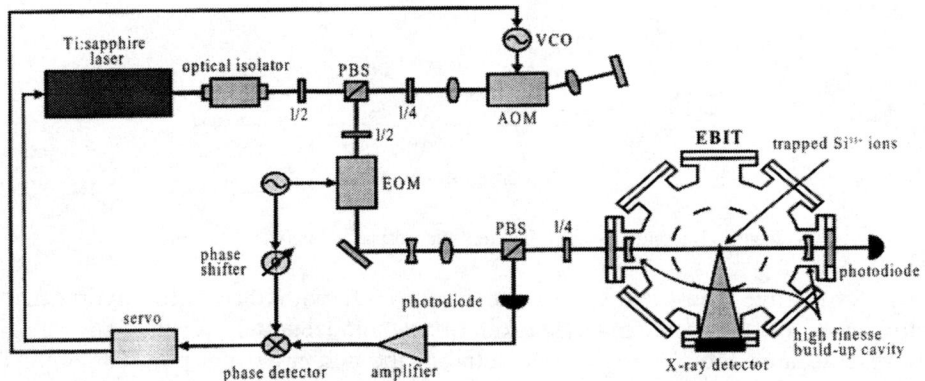

Figure 4. Layout of the Si^{13+} 2s Lamb shift experiment.

It is hoped that first results from this very difficult experiment will be available by the time of the next ICAP, and that eventually this measurement will achieve an accuracy for the 2s Lamb shift of better than +/- 1000ppm.

Let us speculate that, over the next several years, we will have reliable measurements of the 2s Lamb shift for the medium Z hydrogenic ions such as Silicon and perhaps phosphorus and sulphur with an accuracy of 1000ppm. What will a comparison of these measurements with theory actually tell us? In order to answer this question, we must look critically at the theory. Following Mallampalli and Sapirstein[30], the Lamb shift may be written as

$$\Delta E_n = m \frac{\alpha}{\pi} \frac{(Z\alpha)^4}{n^3} F_n(Z\alpha) + m \left(\frac{\alpha}{\pi}\right)^2 \frac{(Z\alpha)^4}{n^3} G_n(Z\alpha)$$

where the function F_n is associated with the one loop Lamb shift and G_n is associated with the two loop Lamb shift. The one loop Lamb shift has been calculated by Mohr for high Z hydrogenic ions and for hydrogen with high accuracy[31]. However, calculations of the two loop Lamb shift are presently incomplete, and there is also some disagreement between Mallampalli and Sapirstein[30] and Goidenko et al[32] as to the numerical value of this contribution at low and intermediate Z. The situation is shown graphically in figure 5[33].

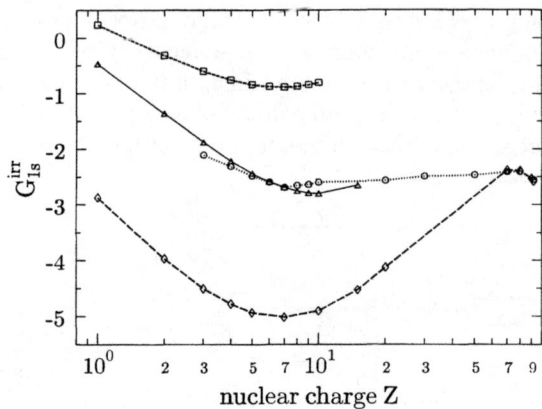

Figure 5. Comparison of different theoretical values of G_{1s}^{irr} from [32].

When comparing measurements of the Lamb shift with theory for hydrogen and hydrogenic ions, great care must be taken to account accurately for the 1s level shifts that arise because the electrostatic potential of the nucleus is not pure coulomb. The situation is complicated for hydrogen itself because the proton size is not very well known[8]. However, the Z scaling of the two loop contributions to the Lamb shift means that for intermediate Z, these two loop terms grow relative to nuclear size contributions. This means that where there is disagreement between different theoretical values, as shown in fig 5, these terms may in principle be tested experimentally by measurement - for example it has been suggested [34] that the uncertainty in the theoretical values of the Lamb shift for hydrogenic silicon arising from the two loop terms could be at the level of 1000ppm or larger. In this range of Z, then, accurate measurements of the Lamb shift could show up discrepancies with quantum electrodynamic contributions – i.e. the two loop Lamb shift - which would be masked by proton size uncertainty in atomic hydrogen.

REFERENCES

1. Feynman, R. P., *QED The Strange Theory of Light and Matter,* Princeton University Press, Princeton N.J. 1985
2. See the discussion on pages 126-9 of ref 1.
3. Mohr, P.J., Private Communication, 2000
4. Teller, P., *An Interpretive Introduction to Quantum Field Theory,* Princeton University Press, Princeton N.J. 1995
5. Cao, T.,Y., *Conceptual Foundations of Quantum Field Theory,* Cambridge University Press, Cambridge
6. Weinberg, S., *The Quantum Theory of Fields,* Cambridge University Press, Cambridge 1995 pp 31 – 38

7. Karshenboim, S. G., Precise Physics of Simple Atoms – ICAP Proceedings 2000-
8. Karshenboim, S.G., *Can. J. Phys.* **77** 241 - 266 (1999)
9. Niering, M., et al *Phys. Rev. Lett.* **84,** 5496 – 5499 (2000)
10. Edlen, B., *Nov. Acta. Regiae. Soc. Sci. Ups Ser.* **C9** 28 (1934)
11. Silver, J. D., *Physica Scripta* **37** 720 – 727 (1988)
12. McClelland, A. F. et al, *Nucl. Instrum. Methods* **B9** 706 – 709 (1985), see also Silver, J.D. et al *Phys. Rev. A* **36** 1515 – 1518 (1987)
13. Wieman, C. E., and Hansch, T.W., *Phys. Rev. A* **22** 192 (1980)
14. Laming, J.M., et al *Nucl. Instrum.Methods.* **B 31** 21 – 23 (1988)
15. C. T. Chantler, D. Phil Thesis, University of Oxford, 1990.
16. Stohlker, T., et al *Phys Rev Lett* **85** 3109 – 3112 (2000)
17. Stohlker, T., Private Communication 2000
18. Tarbutt, M. R., et al *Hyperfine Interactions* **127** 333 – 337 (2000)
19. Indirect methods include anisotropy of radiation following mixture of the 2s and 2p states, and the reduction of the 2s lifetime caused by state mixture.
20. Fabjan, C.W., and Pipkin, F.M., *Phys. Rev. Lett.* **25** 421 (1970)
21. Churassy, S., Gaillard, M.L., and Silver, J.D., *Phys. Rev. Lett.* **33** 184 (1974)
22. Kugel, H.W., and Murnick, D.E., *Rep Prog Phys* **40** 297 (1977)
23. Myers, E. G., in *The Hydrogen Atom II* (2000)
24. Margolis, H.S., et al *Hyperfine Interactions* **99** 169 – 174 (1996)
25. Silver, J.D., et al, *Rev. Sci. Instr.* **65** 1072 – 1074 (1994)
26. Levine, M.A., et al, *Physica Scripta* **T22** 157 – 163 (1988)
27. Donets, E.D. Soviet Patent 248860 1967 but see also Baker, F.A., and Hasted, J.B., *Phil. Trans. Roy. Soc.* **261** 33 – 65 (1966)
28. Bleakney, W., *Phys Rev* **34** 157 – 160 (1929)
29. Klein, H.A., et al in *The Hydrogen Atom II* (2000)
30. Mallampalli, S., and Sapirstein, J., *Phys. Rev. Lett.* **80** 5297 – 5300 (1998)
31. Mohr, P.J., *Phys. Rev. A* **46** 4421 – 4424 (1992), and Jentschura, U. D., Mohr, P.J., and Soff, G., *Phys Rev Lett* **82** 53 – 56 (1999)
32. Goidenko et al, *Phys. Rev. Lett.* **83** 2312 – 2315 (1999) – see also Yerokhin, *Phys. Rev. A* **62** 2508 – 2514 (2000)
33. Labzowsky, L., Private Communication 2000, see also ref 32
34. Karshenboim, S. Private Communication, 2000

GRAVITATIONAL WAVES
AND DETECTORS

Gravitational Wave Detectors on the Ground and in Space

Karsten Danzmann*

*Institut für Atom- und Molekülphysik, Universität Hannover,
and Max-Planck Institut für Quantenoptik, Außenstelle Hannover,
Callinstr. 38, D-30167 Hannover, Germany*

Abstract. Small prototypes of gravitational wave detectors have been under development for over 30 years. But it is only now that we have the necessary technology available to build large instruments with good sensitivity. After several years of construction, the first ground-based interferometers will go into operation in 2001 and a space-based detector is expected to be launched in 2010. These instruments will complement each other because the gravitational wave spectrum extends over many decades in frequency. Ground-based detectors can only observe the audio-frequency regime above 1 Hz, while sources in the low-frequency regime are only accessible from space because of the unshieldable background of local gravitational noise on the ground.

I INTRODUCTION

If a body changes its shape, the resulting change in the force field will make its way outward at the speed of light. As early as 1805, Laplace, in his famous *Traité de Mécanique Céleste* stated that, if Gravitation propagates with finite speed, the force in a binary star system should not point along the line connecting the stars, and the angular momentum of the system must slowly decrease with time. In modern language we would say the binary star is losing energy and angular momentum by emitting gravitational waves. In 1993, Hulse and Taylor were awarded the Nobel prize in physics for the indirect proof of the existence of Gravitational Waves using exactly this kind of observation on the binary pulsar PSR 1913+16. A direct detection of gravitational waves has not been achieved up to this day.

Einstein's paper on gravitational waves was published in 1916, but it was not before the late 1950s that it was rigorously proven that gravitational radiation was in fact a physically observable phenomenon, that gravitational waves carry energy and that, as a result, a system that emits gravitational waves should lose energy.

General Relativity replaces the Newtonian picture of Gravitation by a geometric one that is very intuitive if we are willing to accept the fact that space and time do not have an independent existence but rather are in intense interaction with the physical world. Massive bodies produce "indentations" in the fabric of spacetime,

CP551, *Atomic Physics 17,* edited by E. Arimondo, P. DeNatale, and M. Inguscio
© 2001 American Institute of Physics 1-56396-982-3/01/$18.00

and other bodies move in this curved spacetime taking the shortest path, much like a system of billard balls on a springy surface. In fact, the Einstein field equations relate mass (energy) and curvature in just the same way that Hooke's law relates force and spring deformation, or phrased somewhat poignantly: spacetime is an elastic medium.

If a mass distribution moves in an asymmetric way, then the spacetime indentations travel outwards as ripples in spacetime called gravitational waves. Gravitational waves are fundamentally different from the familiar electromagnetic waves. While electromagnetic waves, created by the acceleration of electric charges, propagate IN the framework of space and time, gravitational waves, created by the acceleration of masses, are waves of the spacetime fabric ITSELF.

Unlike charge, which exists in two polarities, mass always come with the same sign. This is why the lowest order asymmetry producing *electro-magnetic* radiation is the dipole moment of the charge distribution, whereas for *gravitational* waves it is a change in the quadrupole moment of the mass distribution. Hence those gravitational effects which are spherically symmetric will not give rise to gravitational radiation. A perfectly symmetric collapse of a supernova will produce no waves, a non-spherical one will emit gravitational radiation. A binary system will always radiate.

Gravitational waves distort spacetime, in other words they change the distances between free macroscopic bodies. A gravitational wave passing through the Solar System creates a time-varying strain in space that periodically changes the distances between all bodies in the Solar System in a direction that is perpendicular to the direction of wave propagation. The main problem is that the relative length change due to the passage of a gravitational wave is exceedingly small. This is not to mean that gravitational waves are weak in the sense that they carry little energy. On the contrary, a supernova in a not too distant galaxy will drench every square meter here on earth with kilowatts of gravitational radiation intensity. The resulting length changes, though, are very small because spacetime is an extremely stiff elastic medium so that it takes extremely large energies to produce even minute distortions.

II SOURCES OF GRAVITATIONAL WAVES

The gravitational wave sources observable from the ground and from space space are profoundly different, due to the very different frequency ranges in which they emit.

A Sources for ground-based detectors

Strong bursts from supernovae have always been a prime target for ground-based detectors. It is becoming increasingly clear that there is a large variety of supernovae, and many low-luminosity ones (like SN 1987 a) are missed in surrounding galaxies. Computer simulations are still not able to predict realistically what will happen in a gravitational collapse with high angular momentum, which is the situation likely to lead to gravitational radiation. Pulsar evidence [1] now sugggests that the mean space velocity of pulsars is three times higher than had previously been estimated. This linear velocity must come from some non-axisymmetric asymmetry in the gravitational collapse, and this would also enhance one's expectations of gravitationla radiation

Coalescing binary systems have long been among the best understood sources of gravitational radiation. Observations of pulsars like the Hulse-Taylor pulsar PSR 1913+16 are suggesting that in order to see an event rate of coalescences of such systems of more than one per year, one has to observe out to a distance of more than 100 Mpc. Recently though, theoretical studies of binary evolution [2] have suggested that there should be a large population of very tight neutron star binaries that have such short inspiral times that the chances of seeing one at any particular time in our galaxy are very small, but the coalescence rate integrated over time in our galaxy could be factors of ten or more larger than before.

Pulsars and accreting neutron stars are promising sources that would produce nearly monochromatic gravitational wave spectra. The spindown rate of observed pulsars sets an upper limit to their emission of gravitational radiation. The assumption that gravitational radiation is the dominant source of energy loss is not unreasonable at least for young pulsars with significant irregularities in their shape. In this case there might well be several in our neighborhood.

B Sources for space-based detectors

The two main categories of gravitational waves sources for space-based detectors like LISA are the galactic binaries and the massive black holes (MBHs) expected to exist in the centres of most galaxies.

Because the masses involved in typical binary star systems are small (a few solar masses), the observation of binaries is limited to our Galaxy. Galactic sources that can be detected by LISA include a wide variety of binaries, such as pairs of close white dwarfs, pairs of neutron stars, neutron star and black hole (5 − −20 $M_{\odot}$) binaries, pairs of contacting normal stars, normal star and white dwarf (cataclysmic) binaries, and possibly also pairs of black holes. It is likely that there are so many white dwarf binaries in our Galaxy that they cannot be resolved at frequencies below 10^{-3} Hz, leading to a confusion-limited background. Some galactic binaries are so well studied, especially the X-ray binary 4U1820-30, that it is one of the most reliable sources. If LISA would not detect the gravitational waves from known binaries with the intensity and polarisation predicted by General Relativity, it will shake the very foundations of gravitational physics.

The main objective of the LISA mission, however, is to learn about the formation, growth, space density and surroundings of massive black holes (MBHs). There is now compelling indirect evidence for the existence of MBHs with masses of 10^6

to $10^8 \, M_\odot$ in the centres of most galaxies, including our own. The most powerful sources are the mergers of MBHs in distant galaxies, with amplitude signal-to-noise ratios of several thousand for $10^6 \, M_\odot$ black holes. Observations of signals from these sources would test General Relativity and particularly black-hole theory to unprecedented accuracy. Not much is currently known about black holes with masses ranging from about $100 \, M_\odot$ to $10^6 \, M_\odot$. LISA can provide unique new information throughout this mass range.

III COMPLEMENTARITY OF DETECTION ON THE GROUND AND IN SPACE

Astronomical observations of electromagnetic waves cover a range of 20 orders of magnitude in frequency, from ULF radio waves to high-energy gamma-rays. Almost all of these frequencies (except for visible and radio) cannot be detected from the Earth, and therefore it is necessary to place detectors optimised for a particular frequency range (e.g. radio, infrared, ultraviolet, X-ray, gamma-ray) in space.

The situation is similar for gravitational waves. The range of frequencies spanned by ground- and space-based detectors, as shown schematically in Figure 1, is comparable to the range from high frequency radio waves up to X-rays. Ground-based detectors will never be sensitive below about 1 Hz, because of terrestrial gravity-gradient noise. A space-based detector is free from such noise and can be made very large, thereby opening the range from 10^{-4} Hz to 1 Hz, where both the most certain and the most exciting gravitational-wave sources radiate most of their power.

The importance of low frequencies is a consequence of Newton's laws. For systems involving solar-mass objects, lower frequencies imply larger orbital radii, and the range down to 10^{-4} Hz includes sources with the typical dimensions of many galactic neutron star binaries, cataclysmic binaries, and some known binaries. These are the most certain sources. For highly relativistic systems, where the orbital velocities approach the speed of light, lower frequencies imply larger masses ($M \propto 1/f$), and the range down to 10^{-4} Hz reaches masses of $10^7 \, M_\odot$, typical of the black holes that are believed to exist in the centres of many, if not most, galaxies. Their formation and coalescences could be seen anywhere in the Universe and are among the most exciting of possible sources. Detecting them would test the strong-field limit of gravitational theory and illuminate galaxy formation and quasar models.

For ground-based detectors, on the other hand, their higher frequency range implies that even stellar-mass systems can last only for short durations, so these detectors will mainly search for short-lived catastrophic events (supernovae, coalescing neutron-star binaries). Normally, several detectors are required for directional information. If such events are not detected in the expected way, this will upset the astrophysical models assumed for such systems, but not necessarily contradict gravitation theory.

By contrast, if a space-based interferometer does not detect the gravitational waves from known binaries with the intensity and polarisation predicted by Gen-

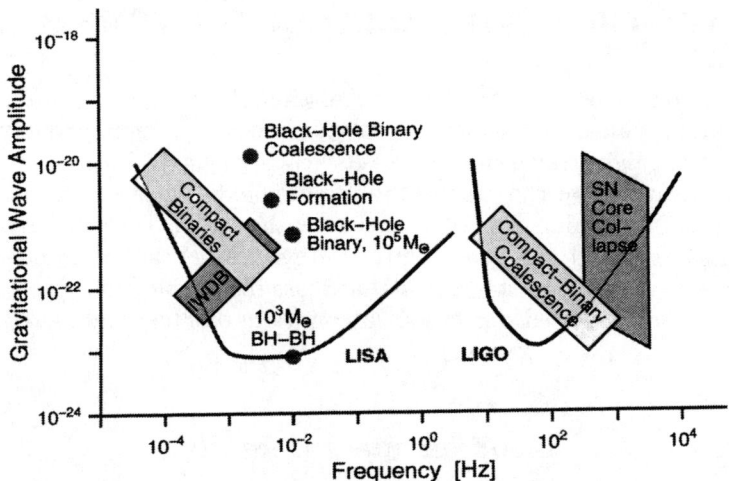

FIGURE 1. Comparison of frequency range of sources for ground-based and space-based gravitational wave detectors. Only a few typical sources are indicated, ranging in frequency from the kHz region of supernovae and final mergers of binary stars down to mHz events due to formation and coalescence of supermassive black holes, compact binaries and interacting white dwarf binaries. The sources shown are in two clearly separated regimes: events in the range from, say, 10 Hz to several kHz (and only these will be detectable with terrestrial antennas), and a low-frequency regime, 10^{-4} to 10^{-1} Hz, accessible only with a space project. Sensitivities of LISA for periodic sources, and of (the "Advanced") LIGO for burst sources, are indicated.

eral Relativity, it will undermine the very foundations of gravitational physics. Furthermore, even some highly relativistic events, such as massive black hole coalescences with masses below $10^5 M_\odot$, last roughly a year or longer. This allows a single space-based detector to provide directional information as it orbits the Sun during the observation.

Both ground- and space-based detectors will also search for a cosmological background of gravitational waves. Since both kinds of detectors have similar energy sensitivities, their different observing frequencies are ideally complementary: observations can supply crucial spectral information.

The space-based interferometer proposal has the full support of the ground-based detector community. Just as it is important to make observations at radio, optical, X-ray, and all other electromagnetic wavelengths, so too is it important to cover different gravitational-wave frequency ranges. Ground-based and space-based observations will therefore complement each other in an essential way.

IV GROUND-BASED DETECTORS

The highest frequencies expected for the emission of strong gravitational waves are around 10 kHz because a gravitational wave source cannot emit strongly at periods shorter than the light travel time across its gravitational radius. At frequencies below 1 Hz, observations on the ground are impossible because of an unshieldable background due to Newtonian gravity gradients on the earth. These two frequencies define the limits of the high-frequency band of gravitational radiation, mainly populated by signals from neutron star and stellar mass black hole binaries. This band is the domain of ground-based detectors: laser interferometers and resonant-mass detectors.

A Resonant-mass detectors

The history of attempts to detect gravitational waves began in the 1960s with the famous bar experiments of Joseph Weber [7]. A resonant-mass antenna is, in principle, a simple object. It consists of a solid body that during the passage of a gravitational wave gets excited similarly to being struck with a hammer, and then rings like a bell.

The solid body traditionally used to be a cylinder, that is why resonant-mass detectors are usually called bar detectors. But in the future we may see very promising designs in the shape of a sphere or sphere-like object like a truncated icosahedron. The resonant mass is usually made from an aluminum alloy and has a mass of several tons. Occasionally, other materials are used, e.g. silicon, sapphire or niobium.

The first bar detectors were operated at room temperature, but the present generation of bars is operating below liquid-helium temperature. The next generation, which is already under construction (NAUTILUS in Frascati, AURIGA in Legnaro), will operate at a temperature around 50 mK.

Resonant-mass detectors are equipped with transducers that monitor the complex amplitudes of one or several of the bar's vibrational modes. A passing gravitational wave changes these amplitudes due to its frequency content near the normal mode frequencies. Present-day resonant mass antennas are fairly narrow-band devices, with bandwidths of only a few Hz around centre-frequencies in the kHz range. With improved transducer designs in the future, we may see the bandwidth improve to 100 Hz or better.

The sensitivities of bar antennas have steadily improved since the first experiments of Joe Weber. Currently, we see a network of antennas at CERN, Frascati, Louisiana State, Legnaro and Perth operating at an rms noise level equivalent to $h_{\rm rms} \approx 6 \times 10^{-19}$. In the first decade of the next millennium, planned sphere-like detectors operating near the standard quantum limit may reach burst sensitivities below 10^{-21} in the kHz range [8].

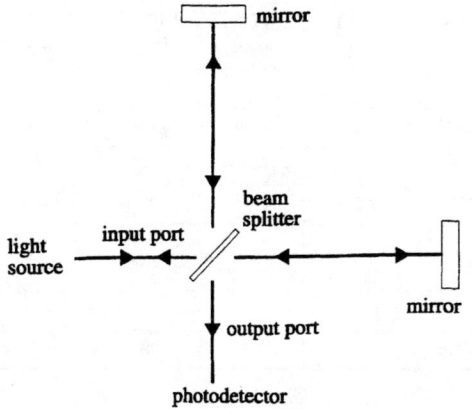

FIGURE 2. Schematic of a two-arm Michelson interferometer. Interference of the two returning beams on the photodetector gives a measure of their relative phase. Any variation in the mirror distances caused by a passing gravitational wave will modulate this phase signal. By having both arms fed from the same light source, the phase noise inherent to a non-ideal source is the same in each arm and cancels.

B Laser Interferometers

Although the seeds of the idea can be found in early papers by Pirani [9] and Gertsenshtein and Pustovoit [10], it was really in the early 1970s that the idea emerged that laser interferometers might have a better chance of detecting gravitational waves, mainly promoted by Weiss [11] and Forward [12].

A Michelson interferometer measures the phase difference between two light fields having propagated up and down two perpendicular directions, i.e. essentially the length difference between the two arms. This is the quantity that would be changed by a properly oriented gravitational wave. The phase difference measured can be increased by increasing the armlength, or, equivalently, the interaction time of the light with the gravitational wave, up to an optimum for an interaction time equal to half a gravitational wave period. For a gravitational wave frequency of 100 Hz this corresponds to five milliseconds or an armlength of 750 km. On the ground it is clearly impractical to build such large interferometers, but there are ways to increase the interaction time without increasing the physical armlength beyond reasonable limits. Several variants have been developed, all of them relying on storing and enhancing the laser light, or the gravitational-wave induced sidebands, or both. The technology and techniques for such interferometers have now been under development for nearly 30 years. Figure 3 gives an impression of the wide international scope of the interferometer efforts. After pioneering work at MIT, other groups at Munich/Garching, at Glasgow, then Caltech, Paris/Orsay, Pisa, and later in Japan, also entered the scene. Their prototypes range from a few meters up to 30, 40, and even 100 m.

Today, these prototype interferometers are routinely operating at a displacement noise level of the order $10^{-19}\,\mathrm{m}/\sqrt{\mathrm{Hz}}$ over a frequency range from 200 Hz to 1000 Hz, corresponding to an rms gravitational-wave amplitude noise level of $h_{\mathrm{rms}} \approx 10^{-19}$.

Plans for kilometer-size interferometers have been developed for the last 15 years. All of these large-scale projects will use low-noise Nd:YAG lasers (wavelength $1.064\,\mu\mathrm{m}$), pumped with laser diodes, just as is intended for the space detector

Country:	USA	USA	GER	GBR	FRA	ITA	JPN	JPN
Institute:	MIT	Caltech	MPQ	Glasgow	CNRS	INFN	ISAS	NAO
Prototypes:								
Start:	1972	1980	1975	1977	1983	1986	1986	1991
Laser:	Ar$^+$	Ar$^+$	Ar$^+$	Ar$^+$	(Ar$^+$)	(Ar$^+$)	Ar$^+$	YAG
Arm length ℓ:	40 m		30 m	10 m	0.5 m		100 m	20 m
Strain sensitivity $\tilde{h}$ [Hz$^{-\frac{1}{2}}$] :	$1 \cdot 10^{-20}$ 1995		$11 \cdot 10^{-20}$ 1986	$6 \cdot 10^{-20}$ 1992			$8 \cdot 10^{-20}$ 1996	$2 \cdot 10^{-18}$ 1996
Large Interferometric Detectors:								
Planning (start):	1982	1984	1985	1986	1986	1986	1987	1994
Arm length ℓ:	4 km 2 km	4 km	600 m		3 km		300 m	
Site (State)	Hanford (WA)	Livingston (LA)	Hannover GER		Pisa ITA		Mitaka JPN	
Cost (10^6 US$):	292		7		90		15	
Project name:	LIGO		GEO 600		VIRGO		TAMA 300	

FIGURE 3. Funded ground-based interferometric gravitational wave detectors: List of prototypes (upper part) and long-baseline projects (lower part).

LISA, which will greatly benefit from their efforts for achieving extreme stability and high overall efficiency. In contrast to LISA, all ground-based detectors will employ a technique called power recycling to enhance the laser power circulating in the interferometer over the power available out of the laser. To achieve this, the interferometer output is kept dark by a control system, so that all the laser power is going back to the input, where it is sent back into the onterferometer by a power recycling mirror, forming a resonant Fabry-Perot cavity that consists of the power recycling mirror as incoupling mirror and the whole locked interferometer as end mirror.

The US project LIGO calls for *two* facilities at two widely separated sites [13], one in Hanford, Washington, and one in Livingston, Louisiana. Both will house a 4 km interferometer, Hanford an additional 2 km interferometer. At both sites ground-work and construction have been finished, and vacuum tests (of the "world's largest vacuum chamber") were successful. The installation of optics and suspensions has been completed and laser light is on site. In Hanford, both 2 km arm cavities were locked to the laser for extended periods, and currently the power-recycled vertex interferometer without the long arms is resonating. Louisiana is scheduled to follow with a certain time delay and commissioning of both interferometers is expected to begin in 2001 with routine data taking in 2002.

In the French-Italian project VIRGO, being built near Pisa, an elaborate seismic isolation system will allow this project to measure down to a frequency of 10 Hz or even below [14]. Construction of the center building and the mode cleaner is finished and the ground-work for the long arms has begun. A laser is on site, the central vacuum system is in place, and currently the mode-cleaner cavity is going into operation. VIRGO will follow about a year after LIGO, with routine data taking planned to commence in 2003.

A British-German collaboration has de-scoped the project of a 3 km antenna to a length of only 600 m: GEO 600 [15]. It will employ advanced optical techniques to make up for the shorter arms. Ground work and construction at the site near Hannover are completed, the vacuum system tested. Most of the optics have been installed and both mode-cleaners are locking. Currently, the east-west arm of GEO 600 is going into operation. The schedule is comparable to LIGO, with routine data taking planned for 2002.

In Japan, after a merger of efforts at ISAS and other institutions, construction and vacuum verification of a common 300 m project called TAMA 300 [16] is completed. TAMA is now operating as an interferometer, although not with final sensitivity yet, because power recycling has not yet been installed.

Not included in Figure 3 is the (not yet funded) Australian project of a 500 m detector to be built near Perth. The site would allow later extension to 3 km arms.

LIGO, VIRGO, GEO 600 and TAMA 300 are scheduled to be completed by the end of this century. Observations may begin in 2001 or 2002, although the sensitivity of the first stage detectors may be only marginally sufficient to detect gravitational waves. However, step-by-step improvements will be made, until the network finally reaches the advanced detector sensitivity sometime between 2005 and 2010. At that point, one can be confident that signals will be observed from sources such as supernovae, compact binary coalescences and pulsars, unless something is fundamentally wrong with our current estimates of their strength and distribution.

V PULSAR TIMING

Man-made gravitational wave detectors operate by detecting the effect of gravitational waves on the apparatus. It is also possible to detect gravitational waves by observing their effect on electromagnetic waves as they travel to us from astronomical objects. Such methods of detection are like "one-arm interferometers" – the second arm is not needed if there is another way to provide a reference clock stable enough to sense the changes in propagation time produced by gravitational waves.

Pulsar timing makes use of the fact that the pulsar is a very steady clock. If we have a clock on the Earth that is as stable as the pulsar, then irregularities in the arrival times of pulses that are larger than expected from the two "clocks" can be attributed to external disturbances, and in particular possibly to gravitational waves. Since the physics near a pulsar is poorly known, it might be difficult to prove that observed irregularities are caused by gravitational waves. But where irregularities are absent, this provides an upper limit to the gravitational wave field. This is how such observations have been used so far.

All pulsars slow down, and a few have shown systematic changes in the slowing down rate. Therefore, it is safer to use random irregularities in the pulsar rate as the detection criterion, rather than systematic changes. Such random irregularities set limits on random gravitational waves: the stochastic background.

The arrival times of individual pulses from most pulsars can be very irregular. Pulsar periods are stable only when averaged over considerable times. The longer the averaging period, the smaller are the effects of this intrinsic irregularity. Therefore, pulsar timing is used to set limits on random gravitational waves whose period is of the same order as the total time the pulsar has been observed, from its discovery to the present epoch. Millisecond pulsars seem to be the most stable over these long periods, and a number of them are being used for these observations.

The best limits come from the first discovered millisecond pulsar, PSR 1937+21. At a frequency of approximately 1 per 10 years the pulsar sets an upper limit on the energy density of the gravitational wave background of $\Omega_{\mathrm{GW}} < 10^{-7}$ [17]. This is in an ultra-low frequency range that is 10^5 times lower than the LISA band and 10^{10} times lower than the ground-based band. If one believes a theoretical prediction of the spectrum of a cosmic gravitational wave background, then one can extrapolate this limit to the other bands. But this may be naive, and it is probably wiser to regard observations in the higher-frequency bands as independent searches for a background.

More-recently discovered millisecond pulsars are also being monitored and will soon allow these limits to be strengthened. If irregularities are seen in all of them at the same level, and if these are independent of the radio frequency used for the observations, then that will be strong evidence that gravitational waves are indeed responsible.

These observations have the potential of being extended to higher frequencies by directly cross-correlating the data of two pulsars. In this way one might detect a correlated component caused by gravitational waves passing the Earth at the moment of reception of the radio signals from the two pulsars. Higher frequencies are accessible because the higher intrinsic timing noise is reduced by the cross-correlation. Again, seeing the effect in many pairs of pulsars independently of the radio frequency would be strong evidence for gravitational waves.

VI SPACECRAFT TRACKING

Precise, multi-frequency transponding of microwave signals from interplanetary probes, such as the ULYSSES, GALILEO and CASSINI spacecraft, can set upper limits on low-frequency gravitational waves. These appear as irregularities in the time-of-communication residuals after the orbit of the spacecraft has been fitted. The irregularities have a particular signature. Searches for gravitational waves have produced only upper limits so far, but this is not surprising: their sensitivity is far short of predicted wave amplitudes. This technique is inexpensive and well worth pursuing, but will be limited for the forseeable future by some combination of measurement noise, the stability of the frequency standards, and the uncorrected parts of the fluctuations in propagation delays due to the interplanetary plasma and the Earth's atmosphere. Consequently, it is unlikely that this method will realise an *rms* strain sensitivity much better than 10^{-17}, which is six orders of magnitude worse than that of a space-based interferometer.

VII   THE LISA MISSION

The LISA mission comprises three identical spacecraft located 5×10^6 km apart forming an equilateral triangle. LISA is basically a giant Michelson interferometer placed in space, with a third arm added to give independent information on the two gravitational wave polarizations, and for redundancy. The distance between the spacecraft – the interferometer arm length – determines the frequency range in which LISA can make observations; it was carefully chosen to allow for the observation of most of the interesting sources of gravitational radiation. The centre of the triangular formation is in the ecliptic plane, 1 AU from the Sun and 20° behind the Earth. The plane of the triangle is inclined at 60° with respect to the ecliptic. These particular heliocentric orbits for the three spacecraft were chosen such that the triangular formation is maintained throughout the year with the triangle appearing to rotate about the centre of the formation once per year.

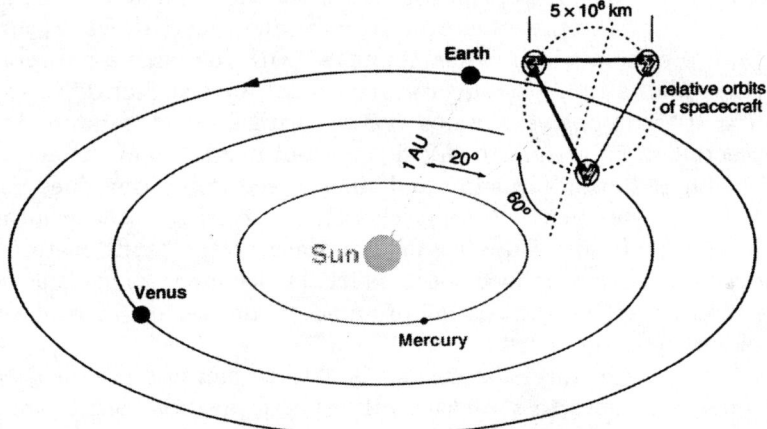

FIGURE 4. LISA configuration: three spacecraft in an equilateral triangle. Drawing not to scale: the LISA triangle is drawn one order of magnitude too large.

While LISA can be described as a big Michelson interferometer, the actual implementation in space is very different from a laser interferometer on the ground and is much more reminiscent of the technique called spacecraft tracking, but here realized with infrared laser light instead of radio waves. The laser light going out from the center spacecraft to the other corners is not directly reflected back because very little light intensity would be left over that way. Instead, in complete analogy with an RF transponder scheme, the laser on the distant spacecraft is phase-locked to the incoming light providing a return beam with full intensity again. After being transponded back from the far spacecraft to the center spacecraft, the light is superposed with the on-board laser light serving as a local oscillator in a heterodyne detection. This gives information on the length of one arm modulo the laser frequency. The other arm is treated the same way, giving information on the length of the other arm modulo the same laser frequency. The difference between these two signals will thus give the difference between the two arm lengths (i.e. the gravitational wave signal). The sum will give information on laser frequency fluctuations.

Each spacecraft contains two optical assemblies. The two assemblies on one spacecraft are each pointing towards an identical assembly on each of the other two spacecraft to form a Michelson interferometer. A 1 W infrared laser beam is transmitted to the corresponding remote spacecraft via a 30-cm aperture $f/1$ Cassegrain telescope. The same telescope is used to focus the very weak beam (a few pW) coming from the distant spacecraft and to direct the light to a sensitive photodetector where it is superimposed with a fraction of the original local light. At the heart of each assembly is a vacuum enclosure containing a free-flying polished platinum-gold cube, 4 cm in size, referred to as the proof mass, which serves as an optical reference ("mirror") for the light beams. A passing gravitational wave will change the length of the optical path between the proof masses of one arm of the interferometer relative to the other arm. The distance fluctuations are measured to sub-Ångstrom precision which, when combined with the large separation between the spacecraft, allows LISA to detect gravitational-wave strains down to a level of order $\Delta\ell/\ell = 10^{-23}$ in one year of observation, with a signal-to-noise ratio of 5.

The spacecraft mainly serve to shield the proof masses from the adverse effects due to the solar radiation pressure, and the spacecraft position does not directly enter into the measurement. It is nevertheless necessary to keep all spacecraft moderately accurately ($10^{-8}\,\text{m}/\sqrt{\text{Hz}}$ in the measurement band) centered on their respective proof masses to reduce spurious local noise forces. This is achieved by a "drag-free" control system, consisting of an accelerometer (or inertial sensor) and a system of electrical thrusters.

Capacitive sensing in three dimensions is used to measure the displacements of the proof masses relative to the spacecraft. These position signals are used in a feedback loop to command micro-Newton ion-emitting proportional thrusters to enable the spacecraft to follow its proof masses precisely. The thrusters are also used to control the attitude of the spacecraft relative to the incoming optical wavefronts, using signals derived from quadrant photodiodes. As the three-spacecraft constellation orbits the Sun in the course of one year, the observed gravitational waves are Doppler-shifted by the orbital motion. For periodic waves with sufficient signal-to-noise ratio, this allows the direction of the source to be determined to arc minute or degree precision, depending on source strength.

Each of the three LISA spacecraft has a launch mass of about 400 kg (plus mar-

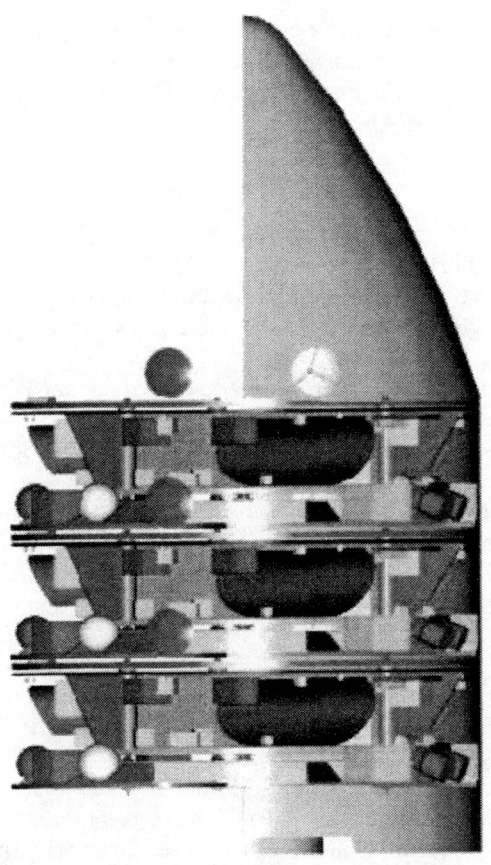

FIGURE 5. 3 LISA composite spacecraft in the Delta II 10ft fairing.

gin) including the payload, ion drive, all propellants and the spacecraft adapter. The ion drives are used for the transfer from the Earth orbit to the final position in interplanetary orbit. All three spacecraft can be launched by a single Delta II 7925H. Each spacecraft carries a 30 cm steerable antenna used for transmitting the science and engineering data, stored on board for two days, at a rate of 7 kBps in the X-band to the 34-m network of the DSN. Nominal mission lifetime is two years, but consumables are sized for an extended mission of more than 10 years.

LISA is envisaged as a NASA/ESA collaborative project, with NASA providing the launch vehicle, mission and science operations and about 50 % of the payload, ESA providing the three spacecraft including the ion drives, and European institutes, funded nationally, providing the other 50 % of the payload. The collaborative NASA/ESA LISA mission is aimed at a launch in the 2010 time frame. LISA is a Cornerstone mission in ESA's Future Science Program Horizons 2000 and has recently been included in NASA's Roadmap.

REFERENCES

1. A.G. Lyne and D.R. Lorimer, Nature **369** (1994) 127.
2. A.V. Tutukov and L.R. Yungelson, MNRAS **260** (1993) 675.
3. C.W. Misner, K.S. Thorne, and J.A. Wheeler, *Gravitation* (Freeman & Co., San Francisco, 1973).
4. P.R. Saulson, *Fundamentals of Interferometric Gravitational Wave Detectors* (World Scientific, Singapore, 1994).
5. B.F. Schutz, *A First Course in General Relativity* (Cambridge University Press, Cambridge, 1985).
6. K.S. Thorne, *Gravitational Radiation*, in: S.W. Hawking and W. Israel, eds., *300 Years of Gravitation* (Cambridge University Press, Cambridge, 1987) 330–458.
7. J. Weber, Phys. Rev. **117** (1960) 306.
8. M. Bassan, Class. Quant. Grav. Supplement A **39** (1994) 11.
9. F.A.E. Pirani, Acta Physica Polonica **15** (1956) 389.
10. M.E. Gertsenshtein and V.I. Pustovoit, JETP **16** (1963) 433.
11. R. Weiss, Quarterly Progress Report of RLE, MIT **105** (1971) 54.
12. G.E. Moss, L.R. Miller, and R.L. Forward, Appl. Opt. **10** (1971) 2495.
13. A. Abramovici *et al.*, Science **256** (1992) 325.
14. G. Bradaschia *et al.*, Nucl. Instrum. and Methods A **289** (1990) 518.
15. K. Danzmann *et al.*, *GEO 600 – A 300 m laser-interferometric gravitational wave antenna*, Proc. 1st Edoardo Amaldi Conference, Frascati, June 1994; and also: J. Hough *et al.*, Proc. MG7, Stanford, July 1994.
16. K. Tsubono and TAMA collaboration, TAMA *Project* in: K. Tsubono, M.-K. Fujimoto, K. Kuroda, (Eds.), *Gravitational Wave Detection, Proc. TAMA Intern. Workshop, Nov. 1996*, p. 183–191, Universal Academy Press (Tokyo, 1997).
17. V.M. Kaspi, J.H. Taylor, M.F. Ryba, Astrophys. J. **428** (1994) 713.

Thermal noise and radiation pressure effects in interferometric measurements

A. Heidmann, P.F. Cohadon, Y. Hadjar, T. Briant and M. Pinard
Laboratoire Kastler Brossel [1], 4 place Jussieu, F75252 Paris, France

Abstract. Sensitivity in gravitational-wave interferometers is limited by the Brownian motion of the mirrors. This thermal noise is a consequence of the energy of the mirror at thermal equilibrium and it induces small displacements and deformations of the mirror surface. This noise can be studied experimentally with a high-finesse optical cavity since the phase of the reflected field is very sensitive to changes of the cavity length. We have observed the Brownian motion of a mirror and we have cooled the mirror by the radiation pressure of light. Such an active control of the thermal noise may be useful to increase the sensitivity of gravitational-wave detectors.

INTRODUCTION

Observation of gravitational waves has become an important challenge and several projects of interferometers are now under construction [1,2]. These detectors are based on very sensitive laser interferometry to measure the relative motions of mirrors placed at the ends of two perpendicular arms. A gravitational wave would produce a differential change of the arms length, with a relative variation less than 10^{-21} for gravitational-wave frequencies between 10 Hz and 1 kHz. There are thus many noise sources which can affect the sensitivity of gravitational-wave detectors and a great effort has been done to reduce these noises. For example, effects of ground motion are filtered by suspending the mirrors to very complex super-attenuator devices. The sensitivity is now mainly limited by the thermal noise of the mirrors at low frequency, and by the photon noise of the laser beam at higher frequency [3].

The Brownian motion of the mirrors can be decomposed into suspension and internal thermal noises. The former corresponds to the motion of the center of mass of the mirror induced by the thermal excitation of the pendulum suspension. The latter is due to thermally induced deformations of the mirror surfaces. It constitutes the main limitation of interferometric detectors in the intermediate frequency domain. Experimental observation of this noise is of particular interest since it strongly depends on the spatial matching between light and internal

[1] Laboratoire de l'Université Pierre et Marie Curie et de l'Ecole Normale Supérieure associé au Centre National de la Recherche Scientifique

CP551, *Atomic Physics 17,* edited by E. Arimondo, P. DeNatale, and M. Inguscio

acoustic modes of the mirror [4–7]. It also depends on the mechanical dissipation mechanisms which are not well known in solids.

Observation and control of this Brownian motion is thus an important issue for very sensitive interferometric measurements [8]. Mirror displacements induced by thermal noise are however very small, on the order of 10^{-19} m$/\sqrt{\text{Hz}}$ at room temperature. This corresponds to displacements of a billionth of an Angstrom for an analysis frequency bandwidth of 1 Hz. Very sensitive displacement sensors have thus to be developed, and one promising technique consists in optical transducers based on a high-finesse Fabry-Perot cavity. Near an optical resonance, the phase of the beam reflected by a single-ended Fabry-Perot cavity strongly depends on the cavity length, and then on the mirror motion. By using a homodyne detection of the reflected field, one can expect to measure small mirror displacements with a sensitivity only limited by the shot-noise of the light beam.

Another interest of the measurement of small displacements is the study of quantum effects of radiation pressure. When a movable mirror is exposed to a laser beam, its motion is coupled to the laser intensity via radiation pressure. This optomechanical coupling plays an important role in interferometric measurements since it induces a Standard Quantum Limit of the sensitivity [9,10]. This quantum effect can be studied by using a single-port high-finesse cavity with a movable mirror for which the effects of the mirror motion are enhanced by the cavity finesse. Such a device has other interesting quantum properties since it can be used to generate squeezed states of light [11] or to perform a quantum nondemolition measurement of the intensity of light [12].

In this paper we report the results of our experiment based on a compact optical cavity with a mirror coated on a mechanical resonator. We first describe our experimental setup and the observation of the Brownian motion [13]. We have observed the internal thermal noise of the mirror, both near the fundamental resonance frequency of the resonator and at low frequency. At resonance, the thermal noise spectrum has a lorentzian shape, as expected from the fluctuation-dissipation theorem. The minimum observable displacement corresponds to the shot-noise limit and is large enough to observe the background thermal noise of the mirror at low frequency.

We then present the cooling of the mirror by radiation pressure [14]. A feedback mechanism controls the Brownian motion via the intensity modulation of a laser beam reflected on the mirror. This allows to reduce the effective temperature of the mirror and its thermal noise, both at the mechanical resonance frequency and at low frequency. At resonance, this cooling mechanism corresponds to a cold damping process, the radiation pressure exerted by the light corresponding to a viscous force applied to the mirror without any additional thermal noise. We have also studied the transient evolution of the thermal noise when the cold damping is applied or stopped. The results show that the relaxation time towards the cooled regime can be much shorter than the relaxation time towards the thermal equilibrium.

We finally present some approaches that are currently studied in order to reduce the thermal noise in gravitational-wave interferometers. We examine in particular the possible application of the cooling mechanism to control the thermal noise. One may envision to measure the thermal noise by a high-finesse cavity and to use this signal to cool the mirror by radiation pressure. We also propose a cycling cooling mechanism which takes advantage of the transient characteristics of the cold damping mechanism to reduce the thermal noise associated with the violin modes of the mirror suspension.

OBSERVATION OF THERMAL NOISE

In this section we present the experimental results concerning the observation of thermal noise, both for a small plano-convex mirror and for a cylindrical one. We first recall some well known results concerning the thermal noise and the fluctuation-dissipation theorem.

Thermal noise of mirrors

In a gravitational-wave interferometer, the Brownian motion of the mirrors induces changes of the arms length and leads to a noise at the output of the interferometer. The suspended mirrors have many degrees of freedom, such as the pendulum modes of the suspension, the violin modes of the suspension wires, and the internal acoustic modes of the mirrors. The thermal motion of each degree of freedom is a consequence of the energy $\frac{1}{2}k_B T$ stored at thermal equilibrium, where k_B is the Boltzmann constant and T is the temperature. It can also be seen as the result of the mechanical response to a thermal random force. If we consider only one degree of freedom, the displacement $\delta x [\Omega]$ of the mirror at frequency Ω can be written as

$$\delta x [\Omega] = \chi [\Omega] F_T [\Omega], \qquad (1)$$

where $\chi [\Omega]$ is the mechanical susceptibility of the mode and $F_T [\Omega]$ is a Langevin force describing the coupling of the mode with the thermal bath. Each mode is equivalent to a harmonic oscillator characterized by a lorentzian mechanical susceptibility

$$\chi [\Omega] = \frac{1}{M (\Omega_M^2 - \Omega^2 - i\Omega_M^2 \Phi [\Omega])}, \qquad (2)$$

with a mass M, a resonance frequency Ω_M, and a loss angle $\Phi [\Omega]$. The spectral dependence of the loss angle is related to the dissipation mechanisms. In the following, we will consider for simplicity the case of a viscous damping for which the loss angle is a linear function of the frequency Ω and is related to the mechanical damping Γ of the mode by

$$\Phi\left[\Omega\right] = \frac{\Gamma\Omega}{\Omega_M^2}. \tag{3}$$

From the fluctuation-dissipation theorem [15], the spectrum $S_T\left[\Omega\right]$ of the Langevin force is related to the imaginary part of the susceptibility by

$$S_T\left[\Omega\right] = -\frac{2k_B T}{\Omega} \mathrm{Im}\left(\frac{1}{\chi\left[\Omega\right]}\right). \tag{4}$$

According to eqs. (1) to (4), the noise spectrum $S_x\left[\Omega\right]$ of the displacement $\delta x\left[\Omega\right]$ has a lorentzian shape. The total noise spectrum, taking into account all the mechanical modes, is the sum of these lorentzian contributions. Near one of the mechanical resonances, the thermal noise is mainly ruled by the resonant component, with a background noise related to all other modes and approximately flat in frequency. At low frequency, the noise spectrum is the sum of the low-frequency tails of the noise spectra of each mode. The spectral dependence of this background noise however depends on the loss angle $\Phi\left[\Omega\right]$ which is not well known in solids. In particular, the dependence of the loss angle with frequency may not be appropriately described by a viscous damping for internal acoustic modes of the mirrors.

For the internal thermal noise, the situation is even complicated by the fact that the interferometer is not sensitive to a global motion δx of the mirror, but to the deformation of the mirror averaged over the beam profile. As a consequence, the effect of the thermal noise may depend on the spatial matching between the internal acoustic modes and the light, which can be taken into account by including a spatial overlap with the light in the effective mass M of each acoustic mode [4,6,11].

Observation of thermal noise

This thermal noise can be observed experimentally by using a high-finesse Fabry-Perot cavity. A displacement of a mirror of the cavity changes the detuning between the laser frequency and the optical resonance frequency of the cavity. Near such an optical resonance, the phase of the field reflected by a single-ended cavity strongly depends on the cavity length, and then on the displacement of the mirrors: a displacement δx induces a phase shift of the reflected field on the order of

$$\delta\varphi_x \simeq 8\mathcal{F}\frac{\delta x}{\lambda}, \tag{5}$$

where $\mathcal{F}$ is the cavity finesse and λ is the optical wavelength. If all the technical noise is suppressed, this signal has to be compared to the quantum noise $\delta\varphi_n$ of the reflected field, which is on the order of the shot noise of the incident beam,

$$\delta\varphi_n \simeq 1/\sqrt{\overline{I}_{in}}, \tag{6}$$

where $\overline{I}_{in}$ is the mean incident intensity. One then expects to be able to detect a displacement corresponding to a small fraction of the optical wavelength: for a

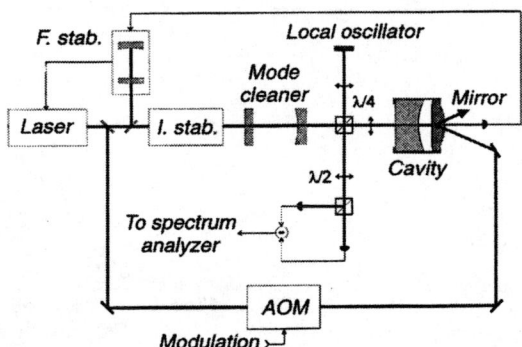

FIGURE 1. Experimental setup. The Brownian motion of the mirror is measured with a high-finesse cavity. A frequency (F. stab.) and intensity (I. stab.) stabilized laser beam is sent into the cavity and the phase of the reflected field is measured by homodyne detection. An auxiliary beam with modulated intensity (AOM) can optically excite the acoustic modes of the mirror.

cavity finesse of 10^5 and an incident power of 1 mW, the sensitivity is on the order of 10^{-20} m/$\sqrt{\text{Hz}}$.

Our experimental setup is shown in figure 1. The 1 mm long high-finesse cavity is composed of a coupling mirror with a 50 ppm transmission, and a totally reflecting back mirror coated on the plane side of a small plano-convex mechanical resonator made of fused silica. The substrate of the back mirror has a thickness of 1.5 mm and a curvature radius of the convex side of 100 mm. Internal acoustic modes of this mechanical resonator can be described as Gaussian modes confined near the central axis of the resonator [11,16]. The fundamental mode has a resonance frequency close to 2 MHz.

The light entering the cavity is provided by a titane-sapphire laser working at 810 nm and frequency locked to an optical resonance of the high-finesse cavity. The beam is also intensity-stabilized and spatially filtered by a mode cleaner. The phase of the field reflected by the cavity is measured by a homodyne detection.

Figure 2 shows the phase noise spectrum of the reflected beam for frequencies around the fundamental resonance frequency of the mirror. Curve (a) reflects the Brownian motion of the mirror which is peaked around the resonance frequency. The noise spectrum is calibrated in displacements of the mirror (scale on the left in figure 2). The Brownian motion corresponds to very small displacements of the mirror since they are only on the order of 10^{-17} m/$\sqrt{\text{Hz}}$ at the top of the resonance. We have also determined the sensitivity of our device. It is limited by the quantum noise of the light and the minimum observable displacement is equal to 2×10^{-19} m/$\sqrt{\text{Hz}}$. This sensitivity is in excellent agreement with the theoretical value which can be deduced from the experimental parameters of our cavity (cavity finesse 37000, incident power 100 μW) [13].

We can also optically excite the mechanical motion of the mirror. We use an

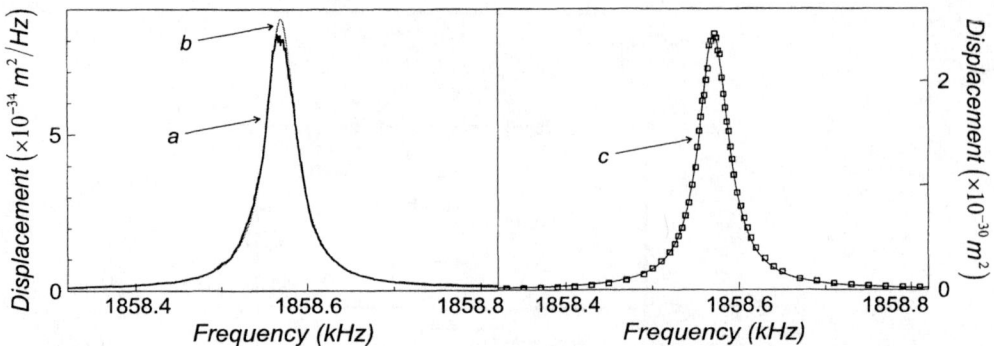

FIGURE 2. Phase noise spectra of the field reflected by the high-finesse cavity for a frequency span of 500 Hz around the fundamental resonance frequency of the mirror. Curve *a* reflects the Brownian motion of the mirror at room temperature. Curve *b* is the theoretical thermal noise deduced from the mechanical response of the mirror to the optical excitation (curve *c*) and from the fluctuation-dissipation theorem.

auxiliary laser beam reflected from the rear on the mirror (see figure 1). This beam is intensity modulated at a given frequency so that we apply a modulated radiation pressure on the mirror. Curve (2c) shows the mechanical response of the mirror observed on the phase of the field reflected by the cavity, when the modulation frequency of the auxiliary laser beam is scanned. We obtain a lorentzian mechanical resonance which corresponds to the fundamental acoustic mode of the mirror. This resonance coincides with the observed thermal peak (curve 2a). From the mechanical response of the mirror, one can derive the mechanical susceptibility $\chi\,[\Omega]$ and the expected thermal noise spectrum at room temperature (eqs. 1 to 4). The result shown in curve (2b) is in excellent agreement with the observed thermal noise.

Thermal noise of a cylindrical mirror

We have studied the thermal noise of a cylindrical mirror (1 in $\times \frac{1}{4}$ in), by replacing the plano-convex mirror by a cylindrical one. We have thus two identical mirrors in the cavity. The resulting noise spectrum is shown in figure 3 for frequencies between 0 and 500 kHz. The spectrum exhibits thermal peaks associated with the different acoustic modes of the mirrors. Each peak is actually split in two resonances corresponding to the modes of the two mirrors (insert of figure 3). The slight discrepancy between the sizes of the mirrors explains the difference between the resonance frequencies. This experimental result can be compared to a numerical calculation of the mechanical resonance frequencies. We have used the program CYPRES [4] and we have been able to identify the modes associated with each peak (modes are described by three integers in figure 3: the circumferential order n, the parity ξ, and the order m [4]). The difference between the experimental

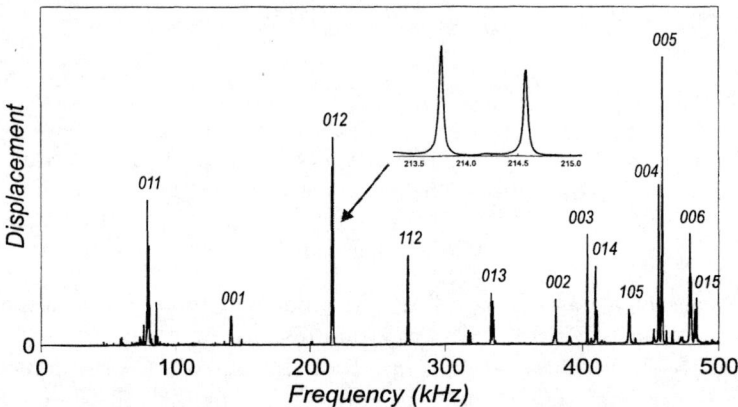

FIGURE 3. Phase noise spectrum of the field reflected by a high-finesse cavity composed of two cylindrical mirrors, for frequencies between 0 and 500 kHz. The Brownian motion appears as a collection of thermal peaks associated with the mechanical resonances of both mirrors.

and theoretical values of the resonance frequencies is less than 1%. This experimental result thus validates the theoretical model used to determine the acoustic modes and to compute the internal thermal noise of the cylindrical mirrors used in gravitational-wave interferometers.

LASER COOLING OF THE MIRROR

Since we are able to observe the thermal noise of the mirror and to change its motion by optical excitation, it is possible to freeze the mirror by feedback control [14,17]. We present in this section the basic theory of this control and the experimental results obtained in our experiment. We finally describe a possible application to gravitational-wave interferometers.

Cold damping mechanism

The principle of the control is to use the signal at the output of the homodyne detection to drive the intensity modulation of the auxiliary laser beam. For an appropriate gain of the electronics, the radiation pressure F_{rad} exerted by the auxiliary laser beam is proportional to the speed $v = -i\Omega\delta x$ of the mirror,

$$F_{rad}[\Omega] = -M\Gamma g \times (-i\Omega\delta x[\Omega]), \tag{7}$$

where g is related to the gain of the feedback loop. If we consider a single mechanical mode as in the previous section, the resulting motion is

$$\delta x[\Omega] = \chi[\Omega](F_T[\Omega] + F_{rad}[\Omega]) = \chi_{eff}[\Omega]F_T[\Omega], \tag{8}$$

313

where χ_{eff} is an effective mechanical susceptibility given by

$$\chi_{eff}[\Omega] = \frac{1}{M(\Omega_M^2 - \Omega^2 - i(1+g)\Gamma\Omega)}. \tag{9}$$

The radiation pressure F_{rad} thus corresponds to an additional viscous force which increases the effective damping by a factor $1 + g$,

$$\Gamma_{eff} = (1+g)\Gamma. \tag{10}$$

Unlike the case of a passive damping, the Langevin force F_T is not modified by the feedback. It still verifies the fluctuation-dissipation theorem (eq. 4) and it is only related to the mechanical damping Γ of the mirror. One can show that this *cold damping* mechanism leads to a cooling of the mirror [14]. More precisely, the effective temperature T_{eff} of the mirror is inversely proportional to the effective damping,

$$T_{eff} = T\frac{\Gamma}{\Gamma_{eff}} = \frac{T}{1+g}. \tag{11}$$

The cold damped mirror is therefore equivalent to a mirror with an increased damping Γ_{eff} at an effective temperature T_{eff} inversely proportional to the factor $1 + g$. The increase of the effective damping then decreases the effective temperature of the mirror.

Cooling at the mechanical resonance frequency

In the experiment, the feedback loop consists of an amplifier with variable gain and phase which drives the intensity modulator of the auxiliary laser beam from the signal provided by the homodyne detection. We also have inserted an electronic bandpass filter in order to avoid any saturation of the feedback loop. The efficiency of the loop is thus limited to a bandwidth on the order of 10 kHz around the fundamental resonance frequency of the mirror. Since this bandwidth is small compared to the resonance frequency, the frequency Ω can be considered constant over all the band. The derivation necessary to get a viscous radiation pressure $F_{rad}[\Omega]$ from the displacement $\delta x[\Omega]$ can therefore be obtained by a simple dephasing of the homodyne signal. This is practically done by an adjustment of the phase of the electronic amplifier.

Curves (b) of figure 4 show the effect of the cold damping for increasing values of the gain, and for a frequency span of 1 kHz around the fundamental resonance frequency of the mirror. The control of the mirror motion is clearly visible on those curves. The thermal peak is strongly reduced while its width, equal to the effective damping Γ_{eff}, is increased (eq. 10).

The effective temperature T_{eff} can be deduced from the variance Δx^2 of the mirror motion which is equal to the integral of the noise spectrum $S_x[\Omega]$, that is

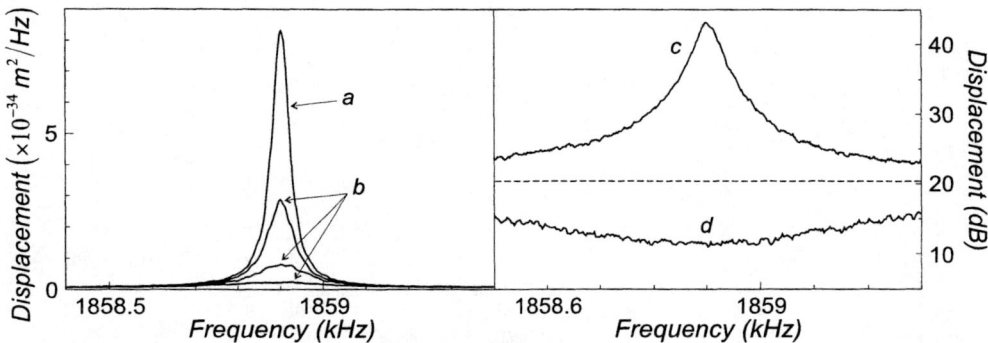

FIGURE 4. Noise spectra obtained without feedback (curves a and c) and with feedback for increasing values of the gain (curves b and d). The temperature, proportional to the area of the thermal peak, is decreased. For very large values of the gain (d), one gets a dip in the background thermal noise (dashed line) and the mirror is no longer in a thermal equilibrium. Figure on the right is in dB scale normalized to the sensitivity of the measurement.

to the area of the thermal peaks observed in figure 4. The decrease of these areas thus corresponds to a cooling of the mirror.

For very large values of the gain, it is however not possible to derive an effective temperature for the mirror since it is no longer in a thermal equilibrium. The thermal noise without feedback is actually the sum of two contributions, one corresponding to the fundamental mode which has a lorentzian shape, the other corresponding to the background thermal noise of all other modes (curve c and dashed line on the right of figure 4). For a large gain, one gets a dip in the background thermal noise (curve d). The noise spectrum still has a lorentzian shape, but its amplitude with respect to the background is reversed. This corresponds to a negative effective temperature for the fundamental mode.

This effect can be understood from the fact that the feedback does not change the background noise. One can indeed see from eqs. (8) and (9) that the feedback has no effect far from the mechanical resonance: as the mechanical susceptibility gets real, the change of its imaginary part does not modify the noise spectrum anymore. As a consequence, one can consider that only the fundamental mode is cooled, all the other acoustic modes staying at room temperature. Such a description allows to satisfactorily describe the dependence of the effective temperature with the gain of the feedback loop [14]. Curve (4d) also shows that it is possible to obtain a very significant noise reduction, larger than 30 dB. In other words, one can reduce the noise power by a factor larger than 1000 near the mechanical resonance.

Cooling at low frequency

The cold damping mechanism can reduce the thermal noise only for frequencies close to a mechanical resonance. For potential applications to gravitational-wave

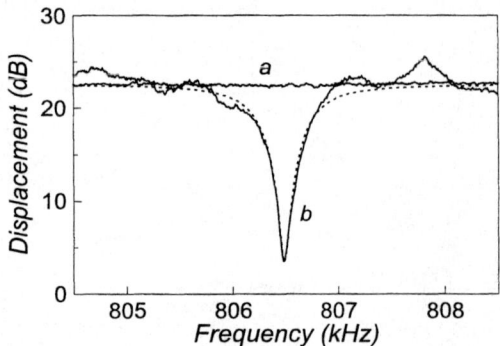

FIGURE 5. Phase noise spectra of the reflected field at low frequency. The background thermal noise (*a*) is reduced in the presence of feedback (*b*). The dashed curve is a theoretical evaluation using the experimental value of the filter bandwidth.

interferometers, it would be of interest to reduce the thermal noise at low frequency. As a matter of fact, the resonance frequencies of the mirrors used in the interferometer are larger than a few kilohertz so that the sensitivity of the interferometer is limited by their internal background noise at low frequency (between approximately 50 Hz and 900 Hz [3]). It is still possible to actively reduce the thermal noise far from the mechanical resonances, but the cooling no more corresponds to a cold damping mechanism.

For frequencies much smaller than the resonance frequency, the mechanical susceptibility becomes real ($\chi\,[\Omega \ll \Omega_M] = 1/M\Omega_M^2$, see eq. 2) and the displacement δx of the mirror is in phase with the applied force. In this quasi-static regime, the best strategy to reduce the thermal noise is to apply a radiation pressure F_{rad} proportional to the displacement of the mirror,

$$F_{rad}\,[\Omega] = -\frac{g'}{\chi\,[0]}\delta x\,[\Omega]\,, \tag{12}$$

where g' is related to the gain of the loop. In that case, the control no longer changes the effective damping, but it increases the effective spring constant of the mirror, as this can be seen from eqs. (8) and (12):

$$\chi_{eff}\,[\Omega \ll \Omega_M] = \frac{\chi\,[\Omega \ll \Omega_M]}{1 + g'}. \tag{13}$$

Since the Langevin force F_T is still unchanged, this will reduce the amplitude of the Brownian motion.

We have done the experiment with an electronic bandpass filter centered around 800 kHz (figure 5). This corresponds to a low frequency since the distance to the fundamental resonance frequency is 10^4 times larger than the width of the mechanical resonance. Curve (5a) shows the noise spectrum without feedback and

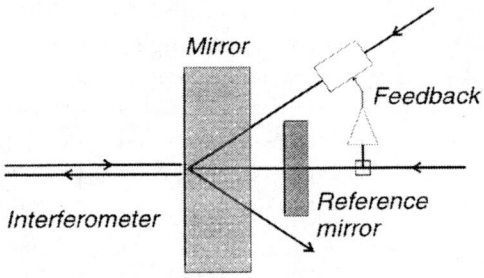

FIGURE 6. Application of the low-frequency cooling to a gravitational-wave interferometer. A short high-finesse cavity measures the thermal noise of the mirror and the signal is fed back to the mirror, without changing the response of the interferometer to a gravitational wave.

corresponds to the background thermal noise of the mirror. Curve (5b) is obtained with feedback and shows a noise reduction with a width related to the bandwidth of the electronic filter. The maximum power noise reduction is on the order of 100.

Application to gravitational-wave interferometers

Since the internal thermal noise of mirrors is one of the main limitations of gravitational-wave interferometers, several proposals have been made to control this noise, either by using cryogenic mirrors [18] or by a choice of substrates with very low dissipation [19]. Another approach consists in changing the shape of the mirrors. The effect of the internal thermal noise on the light depends on the mirror geometry and on the spatial matching between the light and the acoustic modes. For example, it has been shown theoretically that the noise induced by a plano-convex mirror can be much smaller than the one induced by a cylindrical mirror [20].

An alternative approach is to actively control the thermal noise by a feedback mechanism similar to the one demonstrated in our experiment. The main problem with such a feedback loop is that it usually does not change the signal to noise ratio: if we want to measure a displacement of the mirror, the cooling will freeze this displacement as well as the Brownian motion. This is not the case however in a gravitational-wave interferometer. One can envision to use a short high-finesse cavity to detect the thermal noise of one mirror of the interferometer and to feed this signal back to the mirror (see figure 6). Such a cavity performs a local measurement of the mirror motion and is not sensitive to a gravitational wave. As a consequence, the cooling may reduce the background thermal noise without changing the response of the interferometer to a gravitational wave [14].

To work properly, one has to ensure that the cavity only measures the thermal noise of the mirror of the interferometer. In particular, the second mirror of the cavity must be a reference mirror with a smaller thermal noise. For this purpose, one may use a cryogenic mirror by inserting both the mirror of the interferometer and the reference mirror in a cryostat. The mirror of the interferometer will then be cooled by the cryostat. However the very high light power in the interferometer, larger than 10 kW, will induce a significant heating of the mirror so that it is not expected to reduce its temperature below a few tens of Kelvin [18]. The light power in the high-finesse cavity is much weaker, on the order of 100 W. The cryostat can then be used to passively cool the reference mirror to a very low temperature, and the feedback can actively transfer this temperature to the mirror of the interferometer.

TRANSIENT EVOLUTION OF THERMAL NOISE

Another interesting feature of the cold damping mechanism presented in the previous section is the time evolution of the thermal noise when the cooling is applied or when it is stopped. We present here both a theoretical and an experimental determination of these transient regimes. We finally discuss a possible application to the cooling of the violin modes in a gravitational-wave interferometer.

Determination of the transient regime

The time evolution of the noise can be monitored on the variance $\Delta x^2(t)$ which corresponds to the integral of the noise spectrum $S_x[\Omega]$. Since we are interested in the transient feature of the cold damping around the fundamental resonance frequency of the mirror, the frequencies may be restricted to a small band around the resonance. The mirror motion can then be described as a single harmonic oscillator corresponding to the fundamental mode. Suppose now that we switch the feedback on at time $t = 0$. The equation of motion for $t > 0$ is given by

$$M\left(\ddot{x} + \Gamma\,\dot{x} + \Omega_M^2 x\right) = F_T - M\Gamma g\,\dot{x}\,. \tag{14}$$

Terms on the right represent respectively the time evolution of the Langevin force and the radiation pressure force, proportional to the speed of the mirror. To determine the variance $\Delta x^2(t)$ one has to solve this equation with initial conditions corresponding to the thermal equilibrium at temperature T. One then gets [21]

$$\Delta x^2(t) = \frac{\Delta x_T^2}{1+g}\left(1 + ge^{-\Gamma_{eff}t}\right), \tag{15}$$

where Δx_T^2 is the variance at thermal equilibrium (without feedback). The variance therefore decreases to its new value $\Delta x_T^2/(1+g)$ with a time constant Γ_{eff}^{-1}. In other words, the mirror evolves towards its damped regime with a mechanical response corresponding to its effective susceptibility χ_{eff} in presence of feedback.

One gets in a similar way the variance when the cooling is switched off at time $t = 0$:

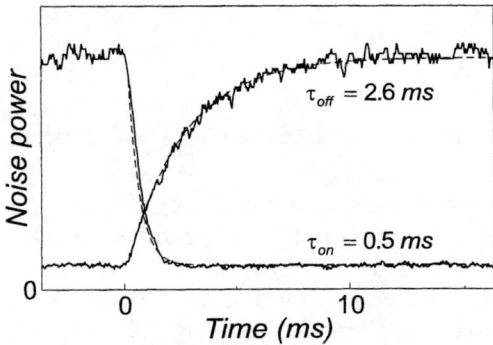

FIGURE 7. Time evolution of the displacement noise power when the cooling is applied (τ_{on}) and when it is switched off (τ_{off}). Dashed curves are theoretical fits.

$$\Delta x^2 (t) = \Delta x_T^2 \left(1 - \frac{g}{1+g} e^{-\Gamma t} \right). \tag{16}$$

The variance goes back to the thermal equilibrium Δx_T^2 with a time constant Γ^{-1} which is now related to the mechanical damping without feedback. The transient evolution to the cooled regime is then faster by a factor $1 + g$ than the relaxation towards the thermal equilibrium (see eq. 10).

Observation of time constants

To observe these transient regimes, we have inserted a fast electronic switch in the feedback loop. We can then observe on the spectrum analyzer in zero-span mode the time evolution of the thermal noise when the cooling is switched on or off. More precisely, the spectrum analyzer is centered on the resonance frequency of the mirror with a resolution bandwidth large enough to detect the whole thermal peak, but not too large in order to limit the contribution of the background thermal noise to the observed signal. Another important point is to avoid a signal filtering which would appear with a small bandwidth of the spectrum analyzer. A bandwidth of 1 kHz fulfills all these conditions, the time constant of the analyzer being on the order of 0.2 ms.

Figure 7 shows the time evolution of the signal when the cooling is applied or switched off, for a feedback gain g of 5.2. When the cooling is applied, the variance decreases to the cooled regime with a time constant $\tau_{on} \simeq 0.5$ ms, and when it is switched off, the variance goes back to the thermal equilibrium with a time constant $\tau_{off} \simeq 2.6$ ms. The experimental data are well fitted by the theoretical expressions (15) and (16) (dashed curves in figure 7). The time constants are also in reasonable agreement with the expected values. As shown theoretically, τ_{off} is related to the mechanical damping Γ which can be deduced from the thermal noise spectrum or from the mechanical response (figure 2). One gets an expected value

$\tau_{off} \simeq 3.7$ ms. The ratio $\tau_{off}/\tau_{on} \simeq 5.2$ is also in good agreement with the expected value $1 + g = 6.2$.

Application to the cooling of violin modes

The cold damping can be used to perform a cyclic cooling of thermal peaks associated with mechanical resonances of the mirrors. As we have seen experimentally, the cooling time can be much shorter than the relaxation time towards thermal equilibrium. One can then cool a mechanical resonance during short periods of time, leaving the system unaffected during much longer periods. One then gets a non stationary thermal noise since it is quickly reduced during the beginning of the cycle and it slowly increases the rest of the time. If the period of the cycle is properly chosen as compared to the relaxation time of the resonance, the noise will stay well below its level at thermal equilibrium. Furthermore, the period of time where the cooling is switched on can be made very short if the gain of the feedback is large, so that the system is free most of the time.

Such a cyclic cooling may be used to detect signals corresponding to single bursts of duration smaller than the period of the cycles. For example, one can envision to reduce the thermal noise of violin modes in a gravitational-wave interferometer. These modes are acoustic modes of the suspension wires of the mirrors and they have sharp resonances within the detection band of the interferometer [3]. Quality factors of these modes are very large so that the period of the cyclic cooling can be made much longer that the expected duration of gravitational waves. For a large feedback gain, the most probable situation is to detect a gravitational wave during the period of time where the cooling is switched off. The effect of the wave in the interferometer is then unchanged by the cyclic cooling, whereas the thermal noise is still reduced. One then gets a gain in the signal to noise ratio.

CONCLUSION

A high-finesse cavity appears to be a powerful tool to study the thermal noise of mirrors. The sensitivity reached in our experiment is on the order of 10^{-19} m/$\sqrt{\text{Hz}}$ and can even be increased by using mirrors with lower losses and a higher incident power. A sensitivity better than 10^{-20} m/$\sqrt{\text{Hz}}$ can be obtained with a cavity finesse of 10^5 and an incident power of 1 mW.

The possibility to observe the thermal noise even far on the tails of the mechanical resonances opens up the way to a quantitative study of the Brownian motion. This would allow to determine with a very high accuracy the mechanical characteristics of the mirror, such as the resonance frequencies, quality factors, spatial structures or effective masses of the various acoustic modes. The study of the spectral dependence of the background thermal noise is also of interest to determine the dissipation mechanisms in the substrates. Furthermore, a high-finesse cavity can be used to experimentally test different materials for the substrate or different geometries, in order to minimize the thermal noise.

Finally, we have demonstrated that it is possible to actively cool the mirror, either at the mechanical resonance frequencies or at low frequency. We have been able to obtain large noise reductions in both cases. We have also studied the transient regime of the cold damping mechanism and we have shown that the cooled regime is much faster to establish than the thermal equilibrium once the feedback is switched off. Such active controls may be useful to reduce the thermal noise in gravitational-wave interferometers, either by using a high-finesse cavity to measure and cool the mirrors, or by applying a cyclic cooling on the violin modes.

ACKNOWLEDGEMENTS

We gratefully thank J.M. MACKOWSKI of the *Institut de Physique Nucléaire de Lyon* for the optical coating of the mechanical resonator, F. BONDU for the CYPRES program used to compute the thermal noise of a cylindrical mirror, and R. DESALVO for helpful discussions.

REFERENCES

1. Abramovici A. *et al, Science* **256**, 325 (1992)
2. Bradaschia C. *et al, Nucl. Instrum. Methods A* **289**, 518 (1990)
3. Hello P., *Progress in Optics XXXVIII*, ed. E. Wolf, 85 (North Holland, 1998)
4. Bondu F., Vinet J.Y., *Phys. Lett. A* **198**, 74 (1995)
5. Bondu F., Hello P., Vinet J.Y., *Phys. Lett. A* **246**, 227 (1998)
6. Gillespie A., Raab F., *Phys. Rev. D* **52**, 577 (1995)
7. Levin Y., *Phys. Rev. D* **57**, 659 (1998)
8. Saulson P.R., *Phys. Rev. D* **42**, 2437 (1990)
9. Caves C.M., *Phys. Rev. D* **23**, 1693 (1981)
10. Tittonen I. *et al, Phys. Rev. A* **59**, 1038 (1999)
11. Pinard M., Hadjar Y., Heidmann A., *Eur. Phys. J. D* **7**, 107 (1999)
12. Heidmann A., Hadjar Y., Pinard M., *Appl. Phys. B* **64**, 173 (1997)
13. Hadjar Y., Cohadon P.F., Aminoff C.G., Pinard M., Heidmann A., *Europhys. Lett.* **47**, 545 (1999)
14. Cohadon P.F., Heidmann A., Pinard M., *Phys. Rev. Lett.* **83**, 3174 (1999)
15. Landau L., Lifshitz E., *Course of Theoretical Physics: Statistical Physics* (Pergamon, New York, 1958), chapt. 12
16. Wilson C.J., *J. Phys. D* **7**, 2449 (1974)
17. Mancini S., Vitali D., Tombesi P., *Phys. Rev. Lett.* **80**, 688 (1998)
18. Uchimaya T. *et al, Phys. Lett. A* **242**, 211 (1998)
19. Rowan S. *et al, Phys. Lett. A* **265**, 5 (2000)
20. Heidmann A., Cohadon P.F., Pinard M., *Phys. Lett. A* **263**, 27 (1999)
21. Pinard M., Cohadon P.F., Briant T., Heidmann A., in preparation

LASER COOLING AND DEGENERATE MATTER

^{85}Rb BEC Near a Feshbach Resonance

N. R. Claussen, S. L. Cornish, J. L. Roberts, E. A. Cornell* and
C. E. Wieman

*JILA, National Institute of Standards and Technology and the University of Colorado, and the
Department of Physics, University of Colorado, Boulder, Colorado 80309-0440*

Abstract. Bose-Einstein condensation has been achieved in a magnetically trapped
sample of ^{85}Rb atoms. Stable condensates of up to 10^4 atoms have been created by
using a magnetic-field-induced Feshbach resonance to reverse the sign of the zero-field
scattering length. These condensates provide unique opportunities for the study of
BEC physics. The variation of the scattering length near the resonance has been used
to magnetically tune the condensate self-interaction energy over a very wide range. This
range extended from very strong repulsive self-interactions to large attractive ones. The
effect of moving the condensate through the Feshbach resonance has been studied and
compared with theory. Long-lived metastable condensates with attractive interactions
have been produced near the zero of the Feshbach resonance. The transition from
repulsive to attractive interactions can lead to a "collapse" of the condensate in which
the cloud shrinks below our resolution limit, loses a significant number of atoms due
to inelastic losses, and emits a burst of high-energy atoms.

INTRODUCTION

Interactions between atoms have a strong influence on most of the properties of
Bose-Einstein condensation (BEC) in dilute alkali gases. The interactions may be
modelled by a mean-field theory with a self-interaction energy that depends only
on the density of the condensate (n) and the s-wave scattering length (a) [1]. In the
limit of strong repulsive interactions the condensate is stable and its size and shape
are fixed by the self-interaction energy. In contrast, attractive interactions ($a < 0$)
lead to a condensate state where the number of atoms is limited to a small critical
value determined by the magnitude of a [2]. Many other important properties of
BEC are determined by the scattering length, including the formation rate, the
spectrum of collective excitations, the evolution of the condensate phase, and the
coupling with the noncondensed atoms.

In most condensate experiments the scattering length is fixed by the choice of an
alkali atom. However, Tiesinga *et al.* [3] proposed a method for tuning the scattering length by utilizing the strong variation expected near a magnetic-field-induced
Feshbach resonance in ultracold ($\sim \mu$K) alkali atom collisions. Recent experiments

CP551, *Atomic Physics 17,* edited by E. Arimondo, P. DeNatale, and M. Inguscio
© 2001 American Institute of Physics 1-56396-982-3/01/$18.00

on cold ^{85}Rb and Cs atoms and Na condensates have demonstrated the variation of the scattering length via this approach [4–7]. Unfortunately, extraordinarily high inelastic losses in the Na condensates were found to severely limit the extent to which the scattering length could be varied and precluded an investigation of the interesting negative scattering length regime [8]. Nevertheless, these results inspired several authors to examine the effects of a Feshbach resonance on BEC, including the possibility of molecular formation near the resonance as a possible loss mechanism [9–11].

STABLE ^{85}RB BEC

We have successfully used a Feshbach resonance to to readily vary the self-interaction of long-lived condensates over a large range. In ^{85}Rb there exists a Feshbach resonance in collisions between two atoms in the F=2, m_f=−2 ($|2, -2\rangle$) hyperfine ground state at a magnetic field $B \sim 155$ gauss(G) [4,5]. At this field the colliding atomic state is degenerate with a quasi-bound molecular state associated with the $|3, -3\rangle + |2, -1\rangle$ potential. Near this resonance the scattering length varies dispersively as a function of magnetic field and, in principle, can have any value between $-\infty$ and $+\infty$ (see inset in Fig. 2). This has allowed us to reach novel regimes of condensate physics including the creation of strong repulsive interactions ($n_{pk}|a|^3 \simeq 10^{-2}$), where effects beyond the mean-field approximation should be readily observable. In addition, we can make transitions between repulsive and attractive interactions (or vice-versa). This now makes it possible to study condensates in the negative scattering length regime, including the anticipated "collapse" of the condensate [13], with a level of control that has not been possible in other experiments [14]. In fact, this ability to change the sign of the scattering length is essential for the existence of our ^{85}Rb condensate. Far from resonance the large negative scattering length ($\simeq -400\, a_0$ [4,15]) limits the maximum number of atoms in a condensate to ~ 80 [2]. However, we have produced condensates with up to 10^4 atoms by operating in a region of the Feshbach resonance where a is positive.

The experimental technique used to create the ^{85}Rb BEC is similar to past schemes [16]. Optical cooling is followed by transfer to a purely magnetic, Ioffe-Pritchard "baseball" trap where forced evaporative cooling is performed. Due to a strong temperature dependence in the s-wave scattering cross section (there is a zero at $E/k_B \simeq 350\,\mu$K [17]) it is favorable to use a weak trap. By avoiding the standard practice of adiabatic compression, we keep the thermal cloud temperature well below this value. ^{85}Rb also suffers from unusually high inelastic collision rates. We recently investigated these losses and observed a mixture of two- and three-body processes that varied with B [18]. The overall inelastic collision rate displayed several orders of magnitude variation across the Feshbach resonance, with a dependence on B similar to that of the elastic collision rate. However, the inelastic rate increased more rapidly than the elastic rate towards the peak of the Feshbach resonance, and was found to be significantly lower in the high field wing of the resonance than on the low field side. This knowledge of the loss rates, together with the known field dependence of the elastic cross section [4,15], has enabled us to successfully devise an evaporation path to reach BEC. Evaporation begins at a field ($B = 250\,\mathrm{G}$) far above the Feshbach resonance in a relatively weak trap ($\nu_r \simeq 15\,\mathrm{Hz}$) to minimize inelastic loss rates. The low initial elastic collision rate means that about 120 s are needed to reach $T \simeq 2\,\mu\mathrm{K}$ with the best final density, putting a stringent requirement on the trap lifetime. At this point it becomes advantageous to move to $B = 162.3\,\mathrm{G}$ where the magnitude of the scattering length is reduced but the ratio of elastic cross section to inelastic loss coefficient is improved [18]. The remainder of the evaporation is performed at this field (with a radial (axial) trap frequency of 17.5 Hz (6.8 Hz)). In contrast to field values away from the Feshbach resonance, the scattering length is positive at this field and stable condensates may therefore be produced.

The trapped atom cloud density distribution is probed using absorption imaging with a $10\,\mu\mathrm{s}$ laser pulse 1.6 ms after the rapid ($\simeq 0.2\,\mathrm{ms}$) turn-off of the magnetic trap. The shadow cast by the cloud is magnified by about a factor of 10 and imaged onto a charge-coupled device (CCD) array to determine the spatial size and the number of atoms. The BEC transition emerges at $T \simeq 15\,\mathrm{nK}$. Typically, we are able to produce "pure" condensates of up to 10^4 atoms with peak densities of $n_{pk} \simeq 1 \times 10^{12}\mathrm{cm}^{-3}$. The measured lifetime of the condensate is $\sim 10\,\mathrm{s}$ [19], which is consistent with the inelastic loss rates we measured in cold thermal clouds [18]. During the final stages of evaporation leading to the BEC transition, the number in the thermal cloud suffers a near-catastrophic decline. We approach the required BEC phase space density at $\sim$100 nK with about 10^6 atoms, but then lose a factor of about 50 in the atom number before the characteristic two-component density distribution appears. Over the final part of the trajectory, the cooling efficiency has become so low that the phase space density remains approximately constant. This is an undesirable result of our relatively weak trap. The mean free path is comparable to the cloud size while the high cloud density causes high losses. The situation improves when the number of atoms becomes sufficiently low, and we are then able to obtain a significant fraction of the atoms in a condensate. The number of atoms in the condensate after we reach this favorable low-density regime is obviously a delicate balance between the elastic (cooling) and inelastic (loss) collision processes in the cloud. Both of these are strongly field dependent near the Feshbach resonance and thus it is natural to explore the evaporation performance for different fields. Although we are able to decrease the loss rate by moving to higher field, the ratio of elastic to inelastic collisions actually decreases and it becomes harder to form condensates. For example, at $B = 164.3\,\mathrm{G}$ we can produce condensates of only a few thousand atoms. Conversely, ramping to lower fields does not help because we reach the favorable low-density cooling regime at smaller numbers of atoms. This restriction together with the increased loss rate means that at $B = 160.3\,\mathrm{G}$, for example, we are unable to form condensates.

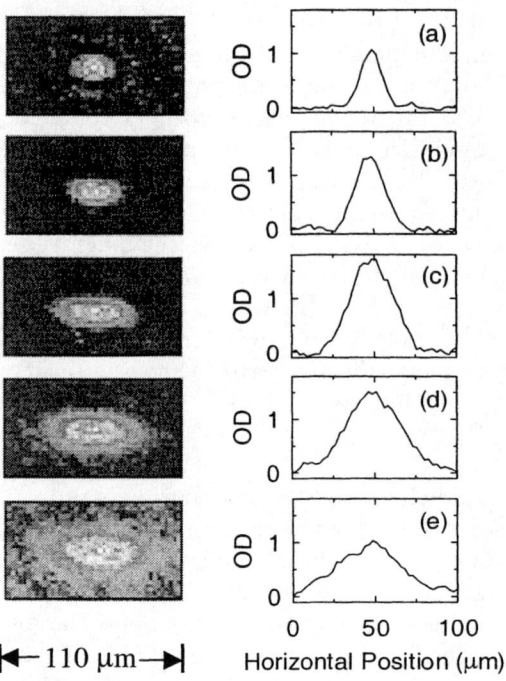

FIGURE 1. Grayscale images and horizontal cross sections of the condensate column density for various magnetic fields. The condensate number was varied to maintain an optical depth (OD) of ~ 1.5. The magnetic field values from top to bottom are: $B = 165.2\,\text{G}$, $B = 162.3\,\text{G}$, $B = 158.4\,\text{G}$, $B = 157.2\,\text{G}$, and $B = 156.4\,\text{G}$.

VARIABLE REPULSIVE INTERACTION STRENGTH

In one of our first experiments, we changed the self-interaction energy by varying the magnetic field and observed the resulting change in the condensate size and shape. By applying a slow linear ramp to the magnetic field we have varied the magnitude of a in the condensate by almost 3 orders of magnitude. The (500 ms) ramp duration was sufficiently long to ensure adiabaticity. Figure 1 shows a series of condensate images for various magnetic fields.

The sequence illustrates our ability to easily change a over a very wide range of positive values. Moving towards the Feshbach peak the condensate size increases due to the increased self-interaction energy. The density distribution approaches the parabolic distribution with an aspect ratio of $\lambda_{TF} = \omega_z/\omega_r$ expected in the TF regime [20]. Ramping in the opposite direction the condensate size quickly shrinks below the 7 micron resolution limit of our imaging system. This occurs shortly before we reach the noninteracting limit where the condensate density distribution

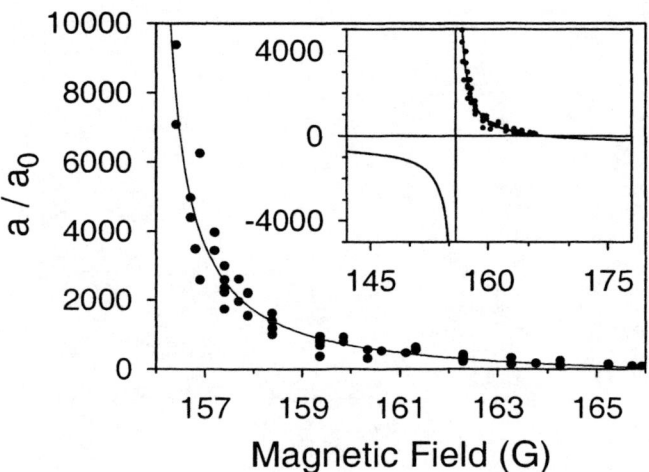

FIGURE 2. Scattering length in units of the Bohr radius (a_0) as a function of the magnetic field. The data are derived from the condensate widths. The solid line illustrates the expected shape of the Feshbach resonance using a peak position and resonance width consistent with our previous measurements [4,18]. For reference, the shape of the full resonance has been included in the inset.

becomes a Gaussian whose dimensions are set by the harmonic oscillator lengths ($l_i = (\hbar/m\omega_i)^{1/2}$ where $i = r, z$) [20]. We took condensate images similar to those shown at many field values between 156 and 166 G. From the images the full-widths at half-maximum (FWHM) of the column density distributions were determined. Then the scattering length was derived by assuming a TF column density distribution with the same FWHM ($\propto (Na)^{1/5}$). In Fig. 2 we plot the scattering length derived in this manner versus the magnetic field. It shows that these values agree with the predicted field dependence of the Feshbach resonance.

RAMPING ACROSS THE FESHBACH RESONANCE

One of the features of the high inelastic loss rates reported in the Na experiments was an anomalously high decay rate when the condensate was swept rapidly through the Feshbach resonance [8]. In light of this work, it was essential to determine to what extent the ^{85}Rb condensate was perturbed in being swept across the Feshbach peak during the trap turn-off. Such measurements also provide an additional test of coherent loss mechanisms such as those in Refs. 9-11. We applied a linear ramp to the current in the baseball coil to sweep the magnetic field experi-

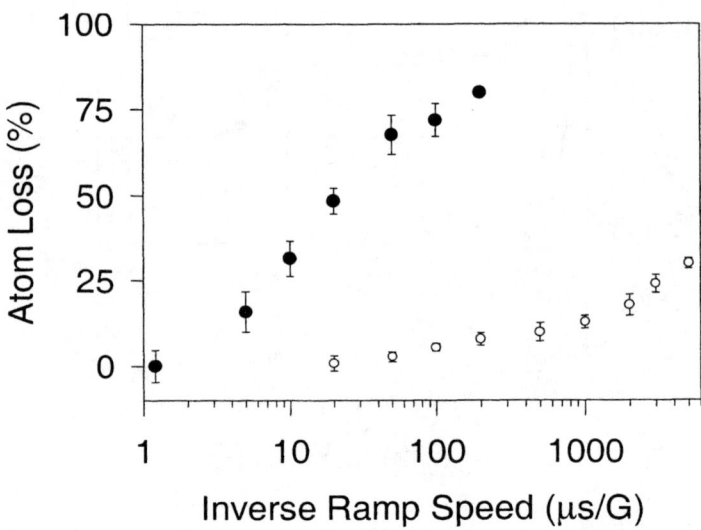

FIGURE 3. Fraction of atoms lost following a rapid sweep of the magnetic field through the peak of the Feshbach resonance as a function of the inverse speed of the field ramp. Data are shown for a condensate ($\bullet$) with a peak density of $n_{pk} = 1.0 \times 10^{12}$ cm^{-3} and for a thermal cloud ($\circ$) with a temperature $T = 430$ nK and a peak density of $n_{pk} = 4.5 \times 10^{11}$ cm^{-3}.

enced by the atoms from $B = 162.3$ G across the Feshbach peak to $B \simeq 132$ G and then immediately turned off the trap and imaged the atom cloud. From the images we determined the fraction of condensate atoms lost as a function of the inverse ramp speed (Fig. 3). The loss for the fastest ramp, which corresponds to the direct turn-off of the magnetic trap, is less than 9 %. This was determined in a separate experiment where the condensate was imaged directly in the magnetic trap both before and after the ramp. For comparison, the experiment was repeated using a cloud of thermal atoms much hotter than the BEC transition temperature. The results for the thermal atoms are consistent with the known inelastic loss rates in the vicinity of the Feshbach resonance [18]. The strong and poorly characterized temperature dependence of these known loss rates near the Feshbach peak makes it difficult to determine what fraction of the observed condensate loss can be attributed to the usual inelastic loss processes and we cannot, therefore, rule out a coherent aspect to the loss process. There have been several models of coherent loss processes put forward to explain the corresponding sodium results [9–11]. However, these calculations are based on the Timmermans theory [9] of coupled atomic and molecular Gross-Pitaevskii equations. This theory is unlikely to be applicable to the conditions of the present experiment.

Timmermans *et al.* have a novel approach to the problem of BEC near a Feshbach resonance. The authors claim that it is not sufficient to describe the condensate using the standard Gross-Pitaevski equation with an effective magnetic field-dependent scattering length: $a_{eff} = a_{bk}[1 - \Delta B/(B - B_{res})]$, where a_{bk} is the non-resonant ("background") scattering length, ΔB is the resonance width, and B_{res} is the resonant value of the magnetic field. Instead, Timmermans et $al.$ explicitly include a molecular field in the many-body formalism. They derive a pair of coupled field operator equations that can be greatly simplified by neglecting correlations, e.g., $\langle \hat{\Psi}^2 \rangle = \langle \hat{\Psi} \rangle^2 = \phi^2$. This leads to a pair of coupled equations for the expectation values (ϕ) of the atomic and molecular fields:

$$i\hbar\dot{\phi}_a = \left[-\frac{\hbar^2 \nabla^2}{2M} + \lambda_a |\phi_a|^2 + \lambda|\phi_m|^2 \right] \phi_a + \sqrt{2}\alpha\phi_m\phi_a^* \tag{1a}$$

$$i\hbar\dot{\phi}_m = \left[-\frac{\hbar^2 \nabla^2}{4M} + \delta + \lambda_m |\phi_m|^2 + \lambda|\phi_a|^2 \right] \phi_m + \frac{\alpha}{\sqrt{2}}\phi_a^2, \tag{1b}$$

where λ_a, λ_m and λ are the strengths for the atom-atom, molecule-molecule, and atom-molecule interactions, respectively; α is a measure of the width of the Feshbach resonance; and δ is the field-dependent detuning of the molecular state energy from the atomic state energy. To further simplify the problem one may neglect all of the mean-field interactions ($\lambda_a = \lambda_m = \lambda = 0$) and assume spatially homogeneous condensate fields. The coupled equations become

$$i\hbar\dot{\phi}_a = \sqrt{2}\alpha\phi_m\phi_a^* \tag{2a}$$

$$i\hbar\dot{\phi}_m = \delta\phi_m + \frac{\alpha}{\sqrt{2}}\phi_a^2. \tag{2b}$$

These nonlinear equations have a similar form to the linear equations describing a two-level system with energy splitting δ and coupling strength α. To find the energy eigenvalues of the coupled system one may look for solutions to Eq. (2) for which the probability density is stationary in time. The following solutions satisfy this criterion:

$$\phi_a = \sqrt{n_a} \exp\left[-i(\kappa t/\hbar + \theta_a)\right] \tag{3a}$$

$$\phi_m = \sqrt{n_m} \exp\left[-i(2\kappa t/\hbar + \theta_m)\right] \tag{3b}$$

where n_a, n_m are the atomic and molecular condensate densities, κ is an energy eigenvalue, and θ_a, θ_m are phase factors. Substituting Eq.(3) into Eq.(2) and using the constraint of number conservation $n_a + 2n_m = n_{tot}$ yields an equation for κ and the relative phase $\theta = 2\theta_a - \theta_m$:

$$\kappa^2 \left(\exp 2i\theta + 2\right) - \delta\kappa - \alpha^2 n_{tot} = 0. \tag{4}$$

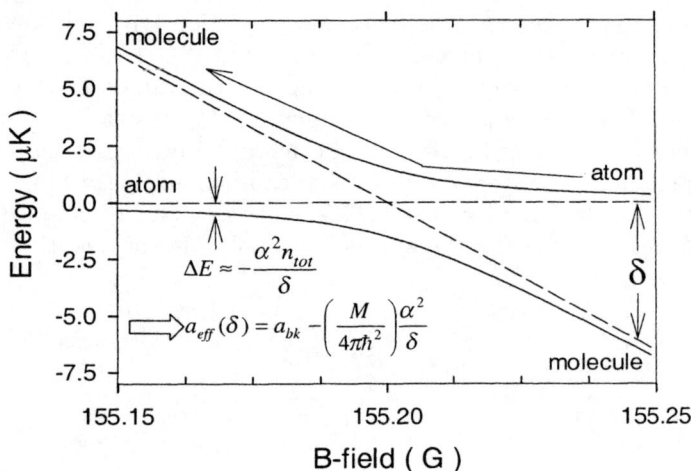

FIGURE 4. Avoided crossing of the atomic and molecular state energies at the [85]Rb Feshbach resonant field of 155.2 G. The horizontal dashed line corresponds to the incident collision threshold of $|2, -2\rangle + |2, -2\rangle$ atoms. The other dashed line shows the relative energy of the quasi-bound molecular state which belongs to the $|3, -3\rangle + |2, -1\rangle$ threshold. Magnetic field increases and δ decreases from left-to-right. Solid lines show the eigenvalues of the coupled system. The small shift ΔE between one eigenvalue of the coupled system and the uncoupled colliding atomic state energy is shown for a large, positive value of detuning δ. A possible adiabatic transition resulting from a field sweep is illustrated for the initial condition: $n_a = n_{tot}, n_m = 0$.

The imaginary part of the equation has a solution $\theta = n\pi/2, n = 0, 1, 2, \ldots$. Choosing $\theta = \pi/2$ and solving the real part of the equation for κ,

$$\kappa = \frac{\delta}{2} \pm \frac{\sqrt{\delta^2 + 4\alpha^2 n_{tot}}}{2}. \tag{5}$$

It now becomes clear that the two solutions, $\kappa_\pm$, represent the energy eigenvalues of the coupled atomic/molecular condensate system. The coupling, parametrized by $\alpha\sqrt{n_{tot}}$, breaks the degeneracy of the bare atomic and molecular states at $\delta = 0$: $\Delta\kappa_{\delta=0} = \kappa_+ - \kappa_- = 2\alpha\sqrt{n_{tot}}$. As $|\delta|$ becomes large compared to the splitting, i. e., $|\delta|/(2\alpha\sqrt{n_{tot}}) \gg 1$, the eigenvalues asymptotically approach their uncoupled values but are shifted by $\Delta E \simeq \pm\alpha^2 n_{tot}/|\delta|$ (see Fig. 4). Timmermans *et al.* show that the small energy shift ΔE leads to a shift in the effective atom-atom s-wave scattering length:

$$a_{eff} = a_{bk} - \left(\frac{M}{4\pi\hbar^2}\right)\frac{\alpha^2}{\delta} \tag{6}$$

where a_{bk} is the off-resonant value of the scattering length. Hence, far from the Feshbach resonance the Timmermans theory reproduces the well-known result of a tunable binary scattering length [3].

Our experiment of ramping the condensate from $B = 162.3\,\text{G}$ across the Feshbach peak to $B \simeq 132\,\text{G}$ corresponds to a linear ramp of the detuning through resonance. Ramping through the avoided crossing is most easily understood in terms of the Landau-Zener model [21]. For the initial condition of a pure atomic condensate ($n_a = n_{tot}$, $n_m = 0$) the probability of making an adiabatic rapid passage to the molecular state as $t \to \infty$ is given by $P = 1 - \exp(-2\pi\Gamma)$, where Γ is the ratio of the square of the coupling strength to the sweep rate: $\Gamma = \alpha^2 n_{tot}/(\hbar\dot{\delta})$.

To investigate the behavior of our ^{85}Rb condensate near the Feshbach resonance one must determine the relevant quantities appearing in the above Landau-Zener equation. The detuning dependence on magnetic field is simply calculated from knowledge of the relative magnetic moments of the colliding atom and molecular states: $\frac{1}{k_B}\frac{\Delta\delta}{\Delta B} = -132\,\mu\text{K/G}$. Our maximum magnetic field sweep rate is $-1\,\text{G}/\mu\text{s}$, so the maximum value for $\dot{\delta}/k_B$ is $132\,\text{K/s}$. The value of α is determined from our knowledge of the detuning dependence of the effective scattering length [4]. From the measured width (~ 11 G) of the Feshbach resonance we know that $a_{eff} = 0$ at a detuning $\delta_0/k_B = -132\,\mu\text{K/G} \cdot 11\,\text{G} = -1.45\,\text{mK}$. Then α may be computed using [9]

$$\alpha^2 = \delta_0 \cdot \frac{4\pi\hbar^2 a_{bk}}{M}, \tag{7}$$

which yields $\alpha/k_B = 1.5 \times 10^{-6}\,\mu\text{K cm}^{3/2}$ (here $a_{bk} \simeq -400\,a_0$). Using our typical peak density $n_{pk} \simeq 10^{12}\,\text{cm}^{-3}$ one finds that Γ can be varied over a large range by changing the sweep rate. For our maximum sweep rate of $-1\,\text{G}/\mu\text{s}$, $2\pi\Gamma = 0.014$; a factor of 100x slower ramp yields $2\pi\Gamma = 1.4$. Thus it seems possible to access both the diabatic and adiabatic transition regimes merely by varying the magnetic field ramp speed!

Unfortunately, there are several practical issues that conspire to prevent the Timmermans theory from describing our ^{85}Rb condensate system. Perhaps the most important problem is that a Gross-Pitaevski mean-field treatment is not applicable over the entire detuning sweep. The coupled Eq. (3) can be valid only under dilute conditions ($n_{pk}a_{eff}^3 \ll 1$). As we rapidly ramp over the resonance this clearly fails, because we reach values of a_{eff} for which $n_{pk}a_{eff}^3 > 1$ when the detuning is still much larger than $2\alpha\sqrt{n_{pk}}$ and hence well before the model would predict atoms are converted to molecules. At 0.2 G from the resonance, $|\delta|/(2\alpha\sqrt{n_{pk}}) \simeq 10$ and $n_{pk}a_{eff}^3 = 2.7$. In addition to this problem there are large inelastic loss rates near the Feshbach resonance. Losses could limit the minimum possible ramp speed and preclude satisfying the adiabatic condition. Finally, M. Holland's recent preprint [22] raises conceptual objections to simply equating $\langle\hat{\Psi}^2\rangle$ with ϕ^2 as in Eq. (1).

TRANSITION TO NEGATIVE SCATTERING LENGTH

One extremely exciting direction for research in ^{85}Rb is to study the effects of changing the sign of the scattering length. Our precise control over the magnetic field allows for quantitative, reproducible experiments that probe the interesting physics associated with switching from repulsive to attractive interactions. We have studied the $a \to a < 0$ transition in two ways: using slow, "adiabatic" ramps and using rapid jumps (here the most appropriate time scale for collective atomic motion is the radial breathing mode period, $1/(2\nu_r) \simeq 29$ ms).

We have used adiabatic ramps of 200 ms duration to test the well-known stability criterion for BEC with attractive interactions [2]. A harmonically trapped condensate may exist in a metastable state due to the balance between the zero-point kinetic energy of the cloud and the attractive mean-field. We have verified that we can produce such a stable ^{85}Rb BEC at a field where the scattering length is truly negative. Holding the pre-ramp number fixed and varying the ramp distance (final negative a-value) allows us to locate the B-field where a "collapse" occurs to a very high precision ($\Delta B \simeq 30$ mG). The hallmark of this collapse is a significant (up to 50%) loss of the condensate number. When we ramp past the critical field value and take an absorption image 50 ms later, we see a quasi-equilibrium remnant cloud with roughly half of the atoms in the condensate and the other half in a super-thermal distribution. We have compiled such collapse data as a function of initial condensate number and found that the dependence of the critical number, N_{crit}, on the magnitude of a is roughly consistent with theory [2]: $N_{crit}|a|/a_{ho} = k$, where k is a constant of order unity.

To study the collapse dynamics we utilize faster variations of the magnetic field. We have found it beneficial to amplify the violence of the collapse event using the Feshbach resonance. We first adiabatically increase the repulsive interactions using a slow ramp toward resonance. Then we very rapidly (~ 1 ms) jump to a scattering length well below the critical value. The condensate responds by shrinking dramatically in the radial direction for $\sim 6]$, ms before becoming too small to observe. As in the adiabatic case, a large number loss results from this compression. Nearly all of the atom loss occurs in a brief window centered 10 ms after the end of the ramp. The condensate undergoes dramatic oscillations following the number loss.

For both fast and slow ramps we have observed evidence for an explosion of relatively "hot" atoms, which is correlated in time with the condensate shrinking below our resolution limit. Apparently atoms are ejected isotropically from the condensate when it has reached a minimum size. The atoms expand outward in space and remain invisible due to their low density until the harmonic trapping potential re-focuses them onto an axis of symmetry. Thus, after one-half of an oscillation period the atoms become visible as a narrow strip of low density. In fact, the anisotropy of our magnetic trap allows us to observe these hot "plumes" extending along either the axial or the radial trap axis merely by changing the delay time before imaging. The energies measured in the two directions are consistent with one another and they range from $E/k_B = 30 \to 100\,$nK. With an initial stable condensate of ~ 6200 atoms, a typical collapse results in a remnant condensate with $N \simeq 3000$, a hot thermal cloud left over from the explosion with $N \simeq 1800$, and the remainder of atoms lost from the trap. Inelastic losses certainly play a large role in the collapse process and we are currently investigating the nature of these losses to determine the mechanism for the explosion.

ACKNOWLEDGEMENTS

We are pleased to acknowledge useful discussions with Murray Holland, Jim Burke, Josh Milstein, and Marco Prevedelli. This research has been supported by the NSF and ONR. One of us (S. L. Cornish) acknowledges the support of a Lindemann Fellowship.

REFERENCES

* Quantum Physics Division, National Institute of Standards and Technology.
1. See, for example, the review by F. Dalfovo, S. Giorgini, L. P. Pitaevskii, and S. Stringari, *Rev. Mod. Phys.* **71**, 463-512 (1999).
2. P. A. Ruprecht, M. J. Holland, K. Burnett, and M. Edwards, *Phys. Rev. A* **51**, 4704-4711 (1995).
3. W. C. Stwalley, *Phys. Rev. Lett.* **37**, 1628-1631 (1976); E. Tiesinga, B. J. Verhaar, and H. T. C. Stoof, *Phys. Rev. A* **47**, 4114-4122 (1993); E. Tiesinga, A. Moerdijk, B. J. Verhaar, and H. T. C. Stoof, *Phys. Rev. A* **46**, R1167-1170 (1992).
4. J. L. Roberts, N. R. Claussen, James P. Burke, Jr., Chris H. Greene, E. A. Cornell, and C. E. Wieman, *Phys. Rev. Lett.* **81**, 5109-5112 (1998).
5. Ph. Courteille, R. S. Freeland, D. J. Heinzen, F. A. van Abeelen, and B. J. Verhaar, *Phys. Rev. Lett.* **81**, 69-72 (1998).
6. V. Vuletić, A. J. Kerman, C. Chin and S. Chu, *Phys. Rev. Lett.* **82**, 1406-1409 (1999).
7. S. Inouye, M. R. Andrews, J. Stenger, M.-J. Miesner, D. M. Stamper-Kurn, and W. Ketterle, *Nature* **392**, 151-154 (1998).
8. J. Stenger, S. Inouye, M. R. Andrews, M.-J. Miesner, D. M. Stamper-Kurn, and W. Ketterle, *Phys. Rev. Lett.* **82**, 2422-2425 (1999).
9. E. Timmermans, P. Tommasini, M. Hussein, and A. Kerman, *Phys. Rep.* **315**, 199-230 (1999).
10. F. A. van Abeelen, and B. J. Verhaar, *Phys. Rev. Lett.* **83**, 1550-1553 (1999).
11. V. A. Yorovsky, A. Ben-Reuven, P. S. Julienne, and C. J. Williams, *Phys. Rev. A* **60**, R765-768 (1999).
12. L. Pitaevskii, and S. Stringari, *Phys. Rev. Lett.* **81**, 4541-4544 (1998); E. Braaten, and J. Pearson, *Phys. Rev. Lett.* 82, 255-258 (1999).
13. Y. Kagan, E. L. Surkov, and G. V. Shlyapnikov, *Phys. Rev. Lett.* **79**, 2604-2607 (1997); C. A. Sackett, H. T. C. Stoof, and R. G. Hulet, *Phys. Rev. Lett.* **80**, 2031-

2034 (1998); M. Ueda, and A. J. Leggett, *Phys. Rev. Lett.* **80**, 1576-1579 (1998); H. Saito, and M. Ueda, e-print cond-mat/0002393.

14. C. C. Bradley, C. A. Sackett, J. J. Tollett, and R. G. Hulet, *Phys. Rev. Lett.* **75**, 1687-1690, (1995); C. A. Sackett, J. M. Gerton, M. Welling, and R. G. Hulet, *Phys. Rev. Lett.* **82**, 876-879 (1999).

15. J. P. Burke, Jr., NIST Gaithersburg, private communication.

16. C. J. Myatt, N. R. Newbury, R. W. Ghrist, S. Loutzenhiser, and C. E. Wieman, *Opt. Lett.* **21**, 290-292 (1996).

17. J. P. Burke, Jr., J. L. Bohn, B. D. Esry, and C. H. Greene, *Phys. Rev. Lett.* **80**, 2097-3000 (1998).

18. J. L. Roberts, N. R. Claussen, S. L. Cornish, and C. E. Wieman, *Phys. Rev. Lett.* **85**, 728-731 (2000).

19. This is in dramatic contrast to the very short lifetimes obtained with Na condensates anywhere near the vicinity of a Feshbach resonance [8]. We believe that the primary difference is that we have 3 orders of magnitude lower density.

20. R. W. E. Lovelace and T. J. Tommila, *Phys. Rev. A* **35**, 3597-3606 (1987).

21. J. R. Rubbmark, M. M. Kash, M. G. Littman, and D. Kleppner, *Phys. Rev. A* **23**, 3107-3117 (1981).

22. M. Holland, J. Park, and R. Walser, "Formation of Pairing Fields in Resonantly Coupled Atomic and Molecular Bose-Einstein Condensates," e-print cond-mat/0005062.

Collective enhancement and suppression in Bose-Einstein condensates

Wolfgang Ketterle, Ananth P. Chikkatur, and Chandra Raman

Department of Physics and Research Laboratory of Electronics,
Massachusetts Institute of Technology, Cambridge, MA 02139, USA

Abstract. The coherent and collective nature of Bose-Einstein condensate can enhance or suppress physical processes. Bosonic stimulation enhances scattering in already occupied states which leads to atom amplification, and the suppression of dissipation leads to superfluidity. In this paper, we review several experiments where suppression and enhancement have been observed and discuss the common roots of and differences between these phenomena.

When a gas of bosonic atoms is cooled below the transition temperature of Bose-Einstein condensation, it profoundly changes its properties. The appearance of a macroscopically occupied quantum state leads to a variety of new phenomena which set quantum fluids apart from all other substances. Fritz London even called them the fourth state of matter [1].

Many of the key concepts in quantum fluids were derived from studying the weakly interacting Bose gas, for which rigorous theoretical treatments were possible [2,3]. In 1995, with the discovery of BEC in a dilute gas of alkali atoms [4–6], it became possible to study such a system experimentally . The theoretical framework connects the observed equilibrium and dynamic properties to the presence of long-range order and low-lying collective excitations [7].

Many special properties of Bose condensates involve the suppression or enhancement of physical processes. Our recent experiments include the suppression and enhancement of elastic collisions of impurity atoms [8], the suppression of dissipation due to superfluidity [9,10], and the suppression [11] and enhancement of light scattering [12]. Bosonically enhanced Rayleigh scattering was used to amplify either atoms [13] or light [14] in a condensate dressed by laser light. We review these experiments and discuss the properties of the Bose condensate which lead to enhancement and suppression.

CP551, *Atomic Physics 17*, edited by E. Arimondo, P. DeNatale, and M. Inguscio
© 2001 American Institute of Physics 1-56396-982-3/01/$18.00

I SCATTERING OF LIGHT AND MASSIVE PARTICLES

Before we discuss light scattering and collisions in a BEC, we want to derive some simple general expressions based on Fermi's golden rule which will be useful to see the similarities and differences between the different processes.

When a condensate scatters a photon or material particle, the scattering is described by the Hamiltonian

$$\mathcal{H}' = C \sum_{k,l,m,n} \hat{c}_l^\dagger \hat{a}_n^\dagger \hat{c}_k \hat{a}_m \delta_{l+n-k-m} \tag{1}$$

Here $\hat{c}_k$ ($\hat{c}_k^\dagger$) is the destruction (creation) operator for the scattered particles, and $\hat{a}_k$ ($\hat{a}_k^\dagger$) is the destruction (creation) operator for atomic plane waves of wavevector **k**. The strength of the coupling is parametrized by the coefficient C.

We consider the scattering process where a system with N_0 atoms in the condensate ground state and N_q quasi-particles with wavevector **q** scatters particles with incident momentum **k** into a state with momentum **k** − **q**. The initial and final states are

$$|i\rangle = |n_k, n_{k-q}; N_0, N_q\rangle$$
$$|f\rangle = |n_k - 1, n_{k-q} + 1; N_0 - 1, N_q + 1\rangle \tag{2}$$

respectively, where n_k denotes the population of scattering particles with wavevector $\mathbf{k}^1$. It should be emphasized that, due to the interatomic interactions, the quasi-particles with occupation N_q are not the plane waves created by the operator $\hat{a}_q^\dagger$, but the quanta of collective excitations with wavevector **q**.

The square of the matrix element M_1 between the initial and final state is

$$|M_1|^2 = |\langle f|\mathcal{H}'|i\rangle|^2$$
$$= |C|^2|\langle N_0 = N - 1, N_q = 1|\hat{\rho}^\dagger(\vec{q})|N_0 = N, N_q = 0\rangle|^2 (N_q + 1)(n_{k-q} + 1)n_k \tag{3}$$

where $\hat{\rho}(\mathbf{q}) = \sum_m \hat{a}_{m+q}^\dagger \hat{a}_m$ is the Fourier transform of the atomic density operator at wavevector $\vec{q}$.

The static structure factor of the condensate is $S(q) = \langle g|\hat{\rho}(\mathbf{q})\hat{\rho}^\dagger(\mathbf{q})|g\rangle/N$ where $|g\rangle = |N_0 = N, N_q = 0\rangle$ is the BEC ground state. We then obtain for the scattering matrix element M_1

$$|M_1|^2 = |C|^2 S(q)(N_q + 1)(n_{k-q} + 1)N_0 n_k \tag{4}$$

The scattering rate W_1 for the process $|n_k, n_{k-q}; N_0, N_q\rangle \rightarrow |n_k - 1, n_{k-q} + 1; N_0 - 1, N_q + 1\rangle$ follows from Fermi's golden rule as

---

[1] This choice of final states implies that we neglect scattering between quasi-particles and consider only processes involving the macroscopically occupied zero-momentum state of the condensate. Formally, we replace the Hamiltonian (Eq. 1) by $C \sum_{k,q}(\hat{c}_{k-q}^\dagger \hat{a}_q^\dagger \hat{c}_k \hat{a}_0 + \hat{c}_{k-q}^\dagger \hat{a}_0^\dagger \hat{c}_k \hat{a}_{-q})$.

$$W_1 = \frac{2\pi}{\hbar}|M_1|^2\delta(E_k - E_{k-q} - \hbar\omega_q^B) \tag{5}$$

where E_k is the energy of the incident particle with momentum $\mathbf{k}$, and $\hbar\omega_q^B$ is the energy of quasi-particles with momentum $\mathbf{q}$ (which we will later obtain from Bogoliubov theory). To obtain the net scattering rate, one has to include the reverse process $|n_k, n_{k-q}; N_0, N_q\rangle \to |n_k+1, n_{k-q}-1; N_0+1, N_q-1\rangle$ by which atoms scatter *back* into the condensate. The square of the matrix element M_2 for this process is $|C|^2 S(q)N_q n_{k-q}(N_0 + 1)(n_k + 1)$. The *net* rate W_+ of scattering atoms from the condensate into the quasi-particle mode $\mathbf{q}$ is the difference of the two partial rates $W_+ = W_1 - W_2$. Assuming $N_0 \gg 1$ (i.e. $N_0 + 1 \approx N_0$), we obtain for the net rate

$$W_+ = \frac{2\pi}{\hbar}|C|^2 S(q)N_0[n_k(N_q + n_{k-q} + 1) - N_q n_{k-q}]\delta(E_k - E_{k-q} - \hbar\omega_q^B) \tag{6}$$

For large n_k (e.g. a laser beam illuminating the condensate) the dominant bosonic stimulation term $(N_q + n_{k-q} + 1)$ is approximately $(\tilde{N} + 1)$ with $\tilde{N} = \max(N_q, n_{k-q})$. This illustrates that there is no bosonic stimulation of the net rate by the *least* populated final state. With the dynamic structure factor $S(\mathbf{q}, \omega) = S(q)\delta(\omega - \omega_q^B)$ Eq. 6 simplifies to

$$W_+/N_0 = \frac{2\pi}{\hbar^2}|C|^2 n_k\, S(\mathbf{q}, \omega)(N_q + n_{k-q} + 1) \tag{7}$$

The rate W_+ in Eq. 6 is the rate for the Stokes process where $E_k > E_{k-q}$. Momentum transfer $\mathbf{q}$ to the condensate is also possible as an anti-Stokes process where a quasi-particle with momentum $-\mathbf{q}$ is scattered into the condensate, and the scattered particle *gains* energy. The net rate W_- for this process is obtained in an analogous way as

$$W_- = \frac{2\pi}{\hbar}|C|^2 S(q)N_0[n_k(N_{-q} - n_{k-q}) - (N_{-q} + 1)n_{k-q}]\delta(E_k - E_{k-q} + \hbar\omega_q^B). \tag{8}$$

The net scattering rates in Eqs. 6 and 8 are the product of three terms.

- The static structure factor $S(q)$ represents the squared matrix element for the condensate to absorb momentum $\mathbf{q}$.

- The delta function denotes the density of final states.

- The bosonic stimulation term represents stimulation by the occupation in the final state either of the scattering particles or the condensate.

The interplay of these three terms is responsible for the suppression and enhancement of physical processes in a condensate. The properties of the condensate as an intriguing many-body system are reflected in the structure factor. In section II, we discuss its measurement through stimulated light scattering. The density of states is responsible for superfluidity because it vanishes for initial velocities of the incident particles which are smaller than the Landau critical velocity (Sections III and IV). Finally, bosonic stimulation by the occupancy N_q of final states was responsible for superradiance, matter wave amplification and optical amplification in a condensate (Section V).

II   DETERMINATION OF THE STATIC STRUCTURE FACTOR BY BRAGG SPECTROSCOPY

The BEC matrix element or the structure factor in Eqs. 6 and 8 can be directly determined experimentally by stimulated light scattering. The density of states (the δ function in Eq. 6) does not restrict the scattering since the photon energy is much higher than the quasi-particle energy. As a result, photons can be scattered into the full solid angle and provide the necessary recoil energy of the atom by a small change ($\approx 10^{-9}$) in the frequency of the scattered photon.

When the condensate is illuminated by two strong laser beams with wavevectors $\mathbf{k}$ and $\mathbf{k} - \mathbf{q}$ and a difference frequency ω, the rate W of transferring photons from one beam to the other is

$$W/N_0 = (W_+ + W_-)/N_0 = 2\pi\omega_R^2 S(q)(\delta(\omega - \omega_q^B) - \delta(\omega + \omega_q^B)) \qquad (9)$$

where the two-photon Rabi frequency ω_R is given by $\omega_R^2 = |C|^2 n_k n_{k-q}/\hbar^2$. When ω is scanned the "spectrum" of a condensate consists of two peaks at $\pm\omega_q^B$ (Fig. 1). The strength (or integral) of each peak corresponds to $S(q)$. We refer to this method as Bragg spectroscopy since the basic process is Bragg scattering of atoms from an optical standing wave [15].

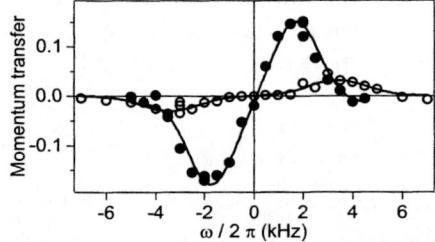

FIGURE 1. Bragg scattering of phonons and of free particles. Momentum transfer per particle, in units of $\hbar q$, is shown vs. the frequency difference $\omega/2\pi$ between the two Bragg beams. The two spectra were taken at different atomic densities. The open symbols represent the phonon excitation spectrum for a trapped condensate (at a chemical potential $\mu/h = 9.2$ kHz, much larger than the free recoil shift of ≈ 1.4 kHz). Closed symbols show the free-particle response of a twenty-three times more dilute (ballistically expanded) cloud. Lines are fits to the difference of two Gaussian line shapes representing excitation in the forward and backward directions. See Refs. [11,16] for more details. Figure is taken from Ref. [11].

We observed that the scattering rate was strongly suppressed when the momentum transfer $\hbar\mathbf{q}$ became smaller than the speed of sound c in the condensate, i.e., when the light scattering excited a phonon and not a free particle. These observations are in agreement with the Bogoliubov theory, which obtains the elementary excitations or Bogoliubov quasiparticles as eigenstates of the Hamiltonian (Eq.

1). A quasi-particle with wavevector q is annihilated by the Bogoliubov operator $\hat{b}_q = u_q \hat{a}_q^\dagger + v_q \hat{a}_{-q}$, where $u_q = \cosh \phi_q$, $v_q = \sinh \phi_q$ and $\tanh 2\phi_q = \mu/(\hbar \omega_q^0 + \mu)$. Its energy is given by

$$\hbar \omega_q^B = \sqrt{\hbar \omega_q^0 (\hbar \omega_q^0 + 2\mu)}, \tag{10}$$

where $\hbar \omega_q^0 = \hbar^2 q^2 / 2m$ and μ is condensate's chemical potential. Expressing $\hat{\rho}(\mathbf{q})$ by Bogoliubov operators leads to an expression for the static structure factor $S(q) = (u_q - v_q)^2$. The static structure factor tends to $S(q) \rightarrow \hbar q / 2mc$ and vanishes in the long wavelength limit, as required of a zero–temperature system with finite compressibility [17].

This reveals a remarkable many-body effect. For free particles, the matrix element for momentum transfer is always 1, which reflects the fact that the operator $e^{i\mathbf{q}\cdot\mathbf{r}}$ connects an initial state with momentum $\mathbf{p}$ to a state with momentum $\mathbf{p} + \hbar \mathbf{q}$ with unity overlap. For an interacting Bose-Einstein condensate, this overlap vanishes in the long-wavelength limit. As we have discussed in previous publications [11,16], this suppression of momentum transfer is due to an destructive interference between the two pathways for a condensate to absorb momentum $\mathbf{q}$ and create a quasi-particle: one pathway annihilates an admixture with momentum $-\mathbf{q}$, the other creates an admixture at momentum $+\mathbf{q}$.

The leading terms in stimulated light scattering do not depend on temperature. The rates W_+ and W_- (Eqs. 6 and 8) are independent of the thermally excited population of quasiparticles N_q and N_{-q} in the limit of large n_k, $n_{k-q} \gg N_{\pm q}$, i.e. when the scattering is induced by a second laser beam. For spontaneous scattering $(n_{k-q} = 0)$, one obtains for the total scattering rate W instead of Eq. 9

$$W/N_0 = \frac{2\pi}{\hbar^2} |C|^2 S(q) n_k \left[(1 + N_q)(\delta(\omega - \omega_q^B) + N_{-q} \delta(\omega + \omega_q^B)) \right]. \tag{11}$$

Absorption of and bosonic stimulation by thermally excited quasi-particles become important when the temperature is comparable to or larger than the quasi-particle energy $\hbar \omega_q^B$. Eq. 11 is proportional to the temperature dependent dynamic structure factor $S_T(\mathbf{q}, \omega)$ [18].

III   SUPPRESSION OF IMPURITY COLLISIONS

The key difference between the scattering of light and massive particles (or impurities) is their energy-momentum dispersion relation. The dispersion relation for impurities is $E_k = (\hbar k)^2 / 2M$, whereas for light, $E_k = \hbar k c_l$ with c_l denoting the speed of light. This difference is responsible for the complete suppression of impurity scattering at low velocities. For simplicity, we assume equal mass M for the impurity and condensate atoms.

Energy-momentum conservation, the delta function in Eq. 6, requires $E_k - E_{k-q} = \hbar \omega_q^B$. For impurity particles, the l.h.s. is always less than $v \hbar q$, where $v = \hbar k / M$ is the initial velocity of the impurities. Thus, collisions with the condensate are only possible, if this maximum energy transfer is sufficient to excite a quasi-particle, i.e., $v > \min(\omega_q^B/q) = v_L$, where v_L is the Landau critical velocity [19] for superfluidity below which no dissipation occurs because the density of final states vanishes.

For the excitation spectrum of the condensate (Eq. 10), the Landau velocity is the Bogoliubov speed of sound $v_L = c = \sqrt{\mu/M}$. Impurity particles moving below this speed cannot dissipate energy in collisions. In contrast, for photons, $dE_k/d(\hbar k) = c_l \gg c$, and scattering is always possible.

In the perturbative limit (no stimulation by the final occupation), the total rate of scattering Γ is given by integrating Eq. 6 over all possible momentum transfers.

$$
\begin{aligned}
\Gamma &= \frac{2\pi|C|^2}{\hbar^2} N_0 \sum_{\mathbf{q}} S(q)\, \delta\left(\frac{\hbar \mathbf{k}\cdot\mathbf{q}}{M} - \frac{\hbar q^2}{2M} - \omega_q^B\right) \\
&= (N_0/V)\left(\frac{2\hbar a}{M}\right)^2 \int dq d\Omega\, q^2 S(q)\, \delta\left(\frac{\hbar kq\cos\theta}{M} - \frac{\hbar q^2}{2M} - \omega_q^B\right) \\
&= 2\pi(N_0/V)\left(\frac{2\hbar a}{M}\right)^2 \frac{1}{v}\int_0^Q dq\, qS(q) \\
&= (N_0/V)\,\sigma(\eta)\,v,
\end{aligned}
\tag{12}
$$

where $\hbar Q = Mv(1 - 1/\eta^2)$ is the maximum possible momentum transfer, and $\eta = v/c$ must be larger than 1. We have used $C = 4\pi\hbar^2 a/MV$ for s-wave scattering between condensate and the impurity particles where a denotes the scattering length and V the condensate volume. The collision cross section is $\sigma(\eta) = \sigma_0 F(\eta)$ where $\sigma_0 = 4\pi a^2$. For $\eta < 1$, $F(\eta) = 0$ and for $\eta > 1$, $F(\eta) = 1 - 1/\eta^4 - \log(\eta^4)/\eta^2$.

The suppression factor $F(\eta)$ is determined by two factors: the phase space restriction due to the delta function in Eq. 12, and additional suppression at low momentum transfers by the structure factor of the condensate. For decreasing velocity, the possible scattering angles become restricted to a forward scattering cone, which shrinks to zero solid angle at the Landau critical velocity. A graph of the suppression factor as function of impurity velocity is shown in Fig. 2.

Experimentally, the Landau critical velocity can usually only be observed by moving *microscopic* particles through the superfluid which do not create a macroscopic flow pattern. Studies of superfluidity with microscopic objects were pursued in liquid ^{4}He by dragging negative ions through pressurized ^{4}He [20,21], and by scattering ^{3}He atoms off superfluid ^{4}He droplets [22].

To study the effects of impurities interacting with the condensate, we created microscopic impurity atoms using a stimulated Raman process which transferred a small fraction of the condensate atoms in the $|F = 1, m_F = -1\rangle$ hyperfine state into an untrapped hyperfine state $|F = 1, m_F = 0\rangle$ with a well-defined initial velocity [23]. The initial velocity could be adjusted between zero and two recoil velocities by varying the angle between the two Raman beams. As these impurities traversed the condensate, they collided with the stationary condensate, resulting

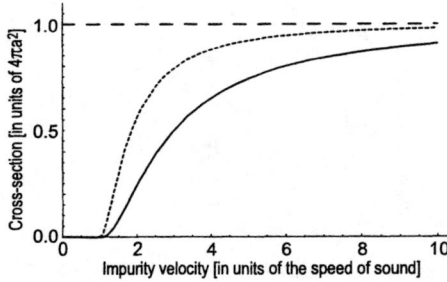

FIGURE 2. Suppression of collisions. Shown is the suppression factor as a function of the impurity velocity (normalized by the speed of sound c, solid line). The dotted line represents the suppression due to phase-space restriction alone (i.e., setting the structure factor $S(q) = 1$).

in a redistribution of the impurity momenta which was detected by a time-of-flight analysis [8] (see Fig. 3).

To probe for the suppression of collisions, one has to vary the impurity velocity around the speed of sound. For that, we produced impurity atoms at low velocities (7 mm/s) and varied the speed of sound by changing the condensate density. The small axial velocity imparted by Raman scattering allowed us to identify products of elastic collisions in time-of-flight images (Fig. 3b, c) since collisions with the stationary condensate redistributed the impurity atoms toward lower axial velocities. However, the impurity velocity was predominantly determined by the gravitational acceleration g, which imparted an average velocity of $v_g = \sqrt{2gz_c}$ where z_c is the Thomas-Fermi radius of the condensate in the z-direction. Thus, the effect of su-

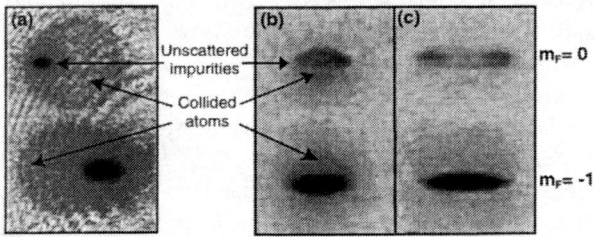

FIGURE 3. Observation of elastic collisions between the condensate and impurity atoms. (a) Impurities traveling at 6 cm/s along the radial axis (to the left in images) were scattered into a s-wave halo. Absorption image after 50 ms of time-of-flight shows the velocity distribution after collisions between the condensate (bottom) and the outcoupled $m_F = 0$ atoms (top), spatially separated by a Stern-Gerlach type magnetic field gradient. The collisional products are distributed over a sphere in momentum space. The image is 4.5 × 7.2 mm. (b) Similar image as (a) shows the collisional products (arrow) for impurity atoms (top) traveling at 7 mm/s along the condensate axis (upward in image). For this image, $v_g/c = 2.7$ (see text). Collisions are visible below the unscattered impurities. (c) Similar image as (b) with $v_g/c = 1.6$. Collisions are suppressed. The outcoupled atoms were distorted by mean-field repulsion. The images are 2.0 × 4.0 mm.

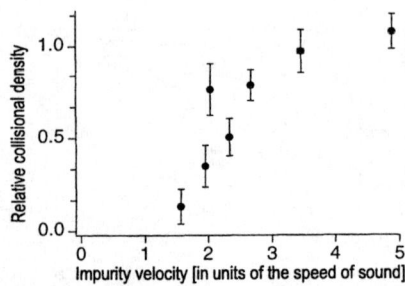

FIGURE 4. Onset of superfluid suppression of collisions. Shown is the observed collisional density normalized to the predicted one in the limit of high velocities as a function of $\bar{\eta} = v_g/c$, which is a measure of the impurity velocity in units of the condensate's speed of sound. The error bars represent the statistical uncertainty. Data is taken from Ref. [8]

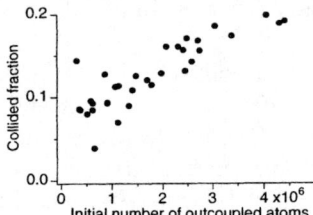

FIGURE 5. Collective amplified elastic scattering in a Bose-Einstein condensate. Shown is the fraction of collided atoms vs. the number of outcoupled atoms. For this data, $v_g/c=4.9$ and the chemical potential was 1.8 kHz. Figure is taken from Ref. [8]

perfluidity on impurity scattering depends primarily on the parameter $\bar{\eta} = v_g/c$ which is the ratio of the typical impurity velocity v_g to the speed of sound c at the center of the condensate.

A time-of-flight analysis of impurity scattering for the case of a low-density condensate (small c) and large condensate radius (large v_g) is shown in Fig. 3b. The effect of collisions is clearly visible with about 20% of the atoms scattered to lower axial velocities (below the unscattered impurities in the image). In contrast, in the case of tight confinement, the condensate density is higher (larger c) and its radius is smaller (smaller v_g), and the collision probability is greatly suppressed due to superfluidity (Figs. 3c and 4).

The bosonic stimulation factor in Eq. 6 is relevant if the final states are populated, either by scattering or thermally. We observed that the fraction of collided atoms increased with the number of outcoupled impurities (Fig. 5). For a large outcoupled fraction, population n_{k-q} and N_q is built up in the final states and stimulates further scattering. This collisional amplification is not directional, and is similar to the recently observed optical omnidirectional superfluoresence [24].

Gain of momentum and thus transfer of energy from the condensate to the impurity atoms is impossible at zero temperature, but may happen at finite temperature

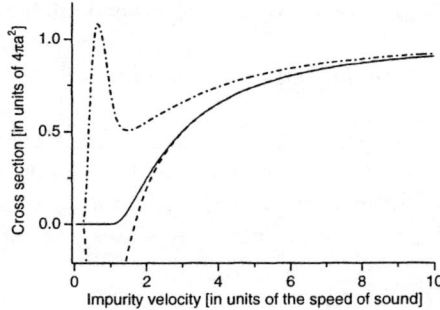

FIGURE 6. Temperature dependent cross-section vs. impurity velocity. Shown is the cross-section at zero temperature (solid line) and at a finite temperature $kT = \mu$ which is typical for our experimental conditions (dash-dotted line). The finite temperature cross-section includes collisions involving thermally occupied quasi-particles where the impurities lose or gain energy. In the experiment, we measured the number of impurities which lost its energy minus the number which gained energy. Thus, the experimental measured cross-sections (Fig. 4) should be compared to $\sigma_{coll,loss} - \sigma_{coll,gain}$ (dashed line).

due to the presence of thermal excitations (the N_{-q} term in Eq. 8)[2]. Thus finite temperature enhances the elastic cross section by two effects: Absorption of quasi-particles (anti-Stokes process) and stimulation of momentum transfer by the final state population (Stokes process).

Fig. 6 shows the dramatic variation of the elastic scattering cross section with temperature. However, the finite temperature did not affect our data in a major way: Due to gravitational acceleration we couldn't probe the velocity regime well below the Landau critical velocity where only thermally assisted collisions are possible. Furthermore, when we counted the number of collided atoms we had to use a background subtraction method where we subtracted the small signal of the energy gain collisions from the energy loss collisions (see Ref. [8] for details), thus cancelling most of the finite-temperature effects.

IV SUPPRESSION OF DISSIPATION FOR A MOVING MACROSCOPIC OBJECT

So far, we have discussed the suppression and enhancement of microscopic processes (light scattering and impurity collisions). The suppression of dissipation is even more dramatic on the macroscopic scale. The flow of liquid ^{4}He and the motion of macroscopic objects through it are frictionless below a critical velocity [25]. Recently, we have explored such frictionless flow in a gaseous BEC [9,10].

The microscopic and macroscopic cases bear many parallels. The onset of scattering or dissipation has two requirements: one needs final states which conserve en-

---

[2] We are grateful to S. Stringari for pointing out the importance of finite-temperature effects.

ergy and momentum, and an overlap matrix element which populates these states. In the case of macroscopic flow, the first requirement leads to a critical velocity for vortex creation and the second requirement addresses the nucleation process of vortices.

The Landau criterion for superfluidity shows that excitations with momentum p and energy $E(p)$ are only possible when the relative velocity between the fluid and the walls or a macroscopic object exceeds the Landau critical velocity v_L which is given by $v_L = \min(E(p)/p)$ (see e.g. [2,25]. A similar criterion applied to vortex formation yields

$$v_c = \frac{E_{\text{vortex}}}{I_{\text{vortex}}} \sim \frac{\hbar}{Md} \ln\left(\frac{d}{\xi}\right) \tag{13}$$

where $I = \int P d^3 r$ is the integrated momentum of the vortex ring or line pair, E is its total energy, d is the dimension of the container, and ξ the core radius of a vortex which in the case of dilute gases is the healing length $\xi = 1/\sqrt{8\pi na}$. Ref. [25] derived Eq. 13 for vortex rings with a maximum radius d, Ref. [26] looked at pairs of line vortices at distance d. Feynman [27] found a similar result for superflow through a channel of diameter d.

A similar result is obtained for a Bose condensed system placed under uniform rotation with angular velocity Ω. A vortex becomes energetically allowed when its energy E' in the rotating frame drops to zero,

$$E' = E - \Omega L = 0 \tag{14}$$

where E and L are the energy and angular momentum in the laboratory frame. This defines a critical *angular* velocity below which a vortex cannot be sustained due to conservation of angular momentum and energy [28]:

$$\Omega_c = \frac{E_{\text{vortex}}}{L_{\text{vortex}}} \sim \frac{\hbar}{Md^2} \ln\left(\frac{d}{\xi}\right). \tag{15}$$

The critical velocity at the wall of the rotating container, $v_c = d\Omega_c$, agrees with Eq. 13.

However, Eqs. 13 and 15 only reflect the energy and momentum required to generate vortices, and do not take into account the nucleation process. If the scattering particle is macroscopic in size, the coupling is between the ground state and a state containing a vortex. Populating such a state requires nucleation of the vortex by the perturbing potential, which usually does not occur until higher velocities are reached than those predicted by Eqs. 13 and 15. The other option, the formation of the vortex by macroscopic quantum tunneling between the two states is an extremely slow process. In recent experiments in which a Bose condensate was placed in a rotating potential, the critical angular velocity for the formation of a single vortex was observed to be 1.6 times higher than the value given by Eq.

15 [29]. This discrepancy may be due to a nucleation barrier associated with the excitation of surface modes, as some authors have recently suggested [30,31].

To study frictionless flow in a Bose-condensate, we focused a blue-detuned 514 nm Argon laser beam onto the sodium condensate, which repelled atoms from the focus. The laser beam was scanned back and forth along the axial direction of the condensate, creating a moving "hole" that simulated a macroscopic object. Rapid sequence phase-contrast imaging allowed us to directly measure the flow pattern of the superfluid around the moving laser beam.

For a weakly interacting Bose-condensed gas at density $n(\mathbf{r})$ and chemical potential μ, pressure is identical to the mean-field energy density $P = \mu(\mathbf{r})n(\mathbf{r})/2$ [7]. A drag force arises due to the pressure difference across the moving object. The chemical potential is given by $\mu(\mathbf{r}, t) = gn(\mathbf{r}, t)$, where $g = 4\pi\hbar^2 a/M$ is the strength of two-body interactions. The drag force F is given by

$$F \simeq gSn\Delta n = S\mu\Delta\mu/g \tag{16}$$

where Δn and $\Delta\mu$ are the differences in density and chemical potential across the stirring object, and S the surface area the macroscopic object presents to the condensate.

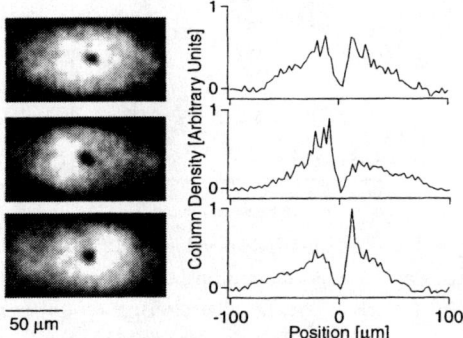

FIGURE 7. Pressure difference across a laser beam moving through a condensate. On the left side *in situ* phase contrast images of the condensate are shown, strobed at each stirring half period: beam at rest (top); beam moving to the left (middle) and to the right (bottom). The profiles on the right are horizontal cuts through the center of the images. The stirring velocity and the maximum sound velocity were 3.0 mm/s and 6.5 mm/s respectively. Figure is taken from Ref. [10].

If the laser beam is stationary, or moves slowly enough to preserve the superfluid state of the condensate, there will be no gradient in the chemical potential across the laser focus, and therefore zero force according to Eq. 16. A drag force between the moving beam and the condensate is indicated by an instantaneous density distribution $n(\mathbf{r}, \mathbf{t})$ that is distorted asymmetrically with respect to the laser beam. Fig. 7 shows phase contrast images strobed at half the stirring period (where the laser beam is in the center of the condensate). A bow wave and stern wave form in front of and behind the moving laser beam, respectively. We define the asymmetry A as the relative difference between the peak column densities in front $(\tilde{n}^f)$ and behind $(\tilde{n}^b)$ the laser beam $A = 2(\tilde{n}^f - \tilde{n}^b)/(\tilde{n}^f + \tilde{n}^b)$. The asymmetry A is proportional to the drag force F.

In Fig. 8 we show measurements of the asymmetry for two maximum densities n_0 of 9×10^{13} and 1.9×10^{14} cm^{-3}. In each data set there is a threshold velocity v_c below which the drag force is negligible, and this threshold increases at higher density. Its value is close to $0.1\ c$ for both data sets, where c is the sound velocity. Above this critical velocity, the drag force increases monotonically, with a larger slope at low density. In addition, the heating rate due to friction against the laser beam was directly measured through time of flight absorption imaging [9,10,32] and found to be in good agreement with the value $\mathbf{F} \cdot \mathbf{v}$, with $\mathbf{F}$ estimated from Eq. 16.

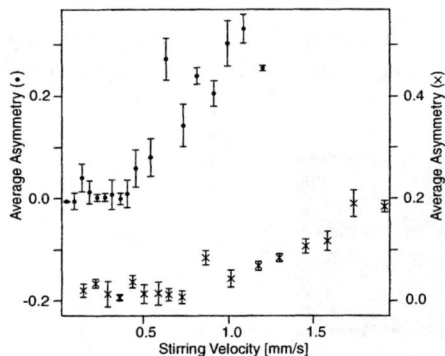

FIGURE 8. Density dependence of the critical velocity. The onset of the drag force is shown for two different condensate densities, corresponding to maximum sound velocities of 4.8 mm/s ($\bullet$, left axis) and 7.0 mm/s ($\times$, right axis). The stirring amplitudes are 29 μm and 58 μm, respectively. The two vertical axes are offset for clarity. The bars represent statistical errors. Figure is taken from Ref. [10].

The observed critical velocity may be related to the formation of vortices. An estimate based on Eq. 13 for typical experimental parameters in sodium ($d = 10\ \mu$m, peak density $n_0 = 1.5 \times 10^{14}$ cm^{-3}, $a = 2.75$ nm) yields $v_c \simeq 1.0$ mm/s, close to the experimental observations. However, Eq. 13 depends only weakly on the speed of sound, through the logarithmic dependence on the healing length ξ. In contrast, our measurements show an approximate proportionality to the sound velocity [10], suggesting that vortex nucleation, determines the onset of dissipation.

Time-dependent simulations of the Gross-Pitaevskii equation show the formation of vortex line pairs, above a critical velocity which is close to the observed value [33]. Several authors have emphasized the role of locally supersonic flow around the laser beam in the nucleation of vortices. [33–35]. In one theoretical model [34], the

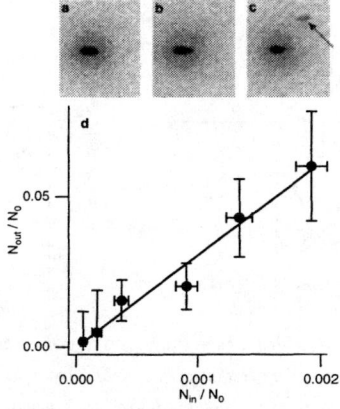

FIGURE 9. Input–output characteristic of the matter-wave amplifier. (**a-c**) Typical time-of-flight absorption images demonstrating matter wave amplification. The output of the seeded amplifier (**c**) is clearly visible, whereas no recoiling atoms are discernible in the case without amplification (**a**) or amplification without the input (**b**). The size of the images is 2.8 mm × 2.3 mm. (**d**) Output of the amplifier as a function of the number of atoms at the input. A straight line fit shows a number gain of 30. Reprinted by permission from Nature, Ref. [13], copyright 1999 Macmillan Magazines Ltd.

vortices are emitted periodically at a rate that increases with velocity, and reduce the pressure gradient across the object. The predicted heating rate [34,36] is in rough agreement with the data. Moreover, this model also predicts that the slope of the asymmetry should increase at lower density, in accord with our observations.

V   AMPLIFICATION OF LIGHT AND ATOMS

Spontaneous light scattering can be stimulated when the atomic recoil state is already populated (the N_q term in Eq. 6). We have explored this process in our studies of superradiance [12], phase-coherent atom amplification [13] and optical amplification [14].

In all these experiments, the condensate was illuminated with a laser beam (mode k, also called "dressing beam"). A condensate atom scatters a photon from the laser beam into another mode and receives the corresponding recoil momentum and energy. Injection of atoms or light turns this *spontaneous* process into a *stimulated* process and realizes an amplifier for either atoms or light. If atoms are injected, they form a matter wave grating (an interference pattern with the condensate at rest) and this grating diffracts light. The diffraction transfers recoil momentum and energy to the atoms, which results in a growth of the grating and therefore the number of atoms in the recoil mode — this is the intuitive picture for atom gain.

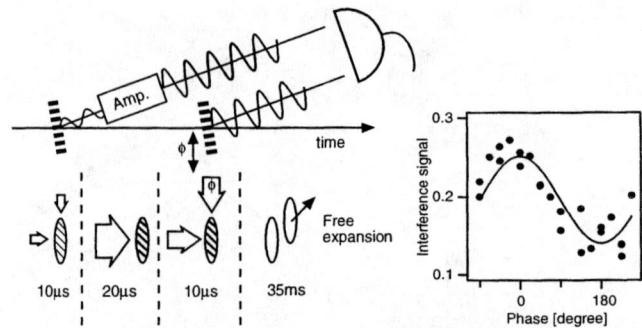

FIGURE 10. Experimental scheme for observing phase coherent matter wave amplification. A small-amplitude matter wave was split off the condensate by applying a pulse of two off-resonant laser beams (Bragg pulse). This input matter wave was amplified by passing it through the condensate pumped by a laser beam. The coherence of the amplified wave was verified by observing its interference with a reference matter wave, which was produced by applying a second (reference) Bragg pulse to the condensate. The interference signal was observed after 35 ms of ballistic expansion. The fringes on the right side show the interference between the amplified input and the reference matter wave. Reprinted by permission from Nature, Ref. [13], copyright 1999 Macmillan Magazines Ltd.

Eq. 6 describes gain for atoms. In the limit of an empty mode for the scattered light ($n_{k-q} = 0$), one obtains

$$W_+ = \frac{2\pi}{\hbar}|C|^2 N_0 n_k (N_q + 1)\delta(E_k - E_{k-q} - \hbar\omega_q^B) \tag{17}$$

For the high momentum transfers considered here (on the order of the photon recoil momentum), $S(q) = 1$. Each scattering event which transfers momentum $\mathbf{q}$ to the condensate, generates a recoiling atom—therefore the scattering rate W_+ integrated over all final states for the scattered photon gives the growth rate N_q for the recoiling atoms:

$$\dot{N}_q = (G_q - \Gamma_{2,q})N_q \tag{18}$$

with the gain coefficient

$$G_q = R N_0 \frac{\sin^2\theta_q}{8\pi/3}\Omega_q. \tag{19}$$

Here R is the rate for single-atom Rayleigh scattering which is proportional to the pump light intensity, N_0 the number of atoms in the condensate at rest, θ_q the angle between the polarization of the incident light and the direction of the scattered light, and Ω_q the phase-matching solid angle for scattering into mode q.

In addition, a loss term $\Gamma_{2,q}$ was included which describes the decoherence rate of the matter-wave grating and determines the threshold for exponential growth. It represents the linewidth of the two-photon process generating recoil atoms in mode q.

Fig. 9 shows the input-output characteristics of the amplifier. The number of input atoms was below the detection limit of our absorption imaging (Fig. 9a). The amplification pulse alone, although above the threshold for superradiance [12], did not generate a discernible signal of atoms in the recoil mode (Fig. 9b). When the weak input matter wave was added, the amplified signal was clearly visible (Fig. 9c). The gain was controlled by the intensity of the pump pulse (see Eq. 19) and typically varied between 10 and 100. Fig. 9d shows the observed linear relationship between the atom numbers in the input and the amplified output with a number gain of 30.

The phase of the amplified matter wave was determined with an interferometric technique. For this, a reference matter wave was split off the condensate in the same way as the first (input) wave (see Fig. 10). The phase of the reference matter wave was scanned by shifting the phase of the radio-frequency signal that drove the acousto-optic modulator generating the axial Bragg beam. We then observed the interference between the reference and the amplified matter waves by measuring the number of atoms in the recoil mode.

The atom amplification is described by the Hamiltonian (Eq. 1) as a four-wave mixing process between two electromagnetic fields and two Schrödinger fields. The symmetry between light and atoms indicates that a dressed condensate should not only amplify injected atoms, but also injected light. In Ref. [16] we have discussed that matter wave gain and optical gain emerge as two limiting cases, depending on whether the atomic population N_q or occupation of the optical mode n_{k-q} dominates the bosonic stimulation term in Eq. 7. In our experiments, the recoiling atoms move ten orders of magnitude more slowly than the scattered light. Therefore, we always have $N_q \gg n_{k-q}$ and $n_{k-q} < 1$ and optical stimulation should play no role. Still, as we want to discuss now, the dressed condensate can act as an amplifier for light.

The physical picture behind the optical gain is as follows: If a very weak probe beam is injected into the dressed condensate, it acts together with the dressing beam as a pair of Bragg beams and creates recoiling atoms. Those recoiling atoms move out of the condensate (or decohere) on a time scale Γ_2^{-1} which is the inverse of the linewidth of the Bragg transition. In steady state, the number of recoiling atoms N_q in the volume of the condensate is proportional to the intensity of the probe light. Those recoiling atoms interfere with the condensate at rest and form a diffraction grating which diffracts the dressing beam into the path of the probe light resulting in amplification of the probe light (Fig. 11).

An expression of the gain can be derived in analogy to a fully inverted two-level system with dipole coupling which would have a gain cross section of $6\pi\lambda^2$ for radiation with wavelength $\lambda(= 2\pi\lambda)$. For the Raman-type system in Fig. 11b, the gain is reduced by the excited state fraction, R/Γ_1 (where R is the Rayleigh scattering rate for the dressing beam and Γ_1 is the linewidth of the single-photon

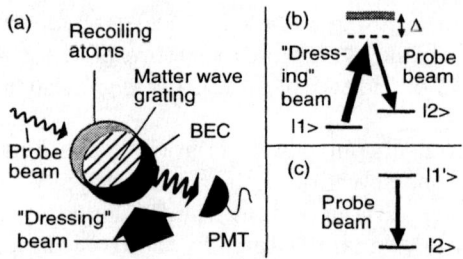

FIGURE 11. Amplification of light and atoms by off-resonant light scattering. (a) The fundamental process is the absorption of a photon from the "dressing" beam by an atom in the condensate (state $|1\rangle$), which is transferred to a recoil state (state $|2\rangle$) by emitting a photon into the probe field. The intensity in the probe light field was monitored by a photomultiplier. (b) The two-photon Raman-type transition between two motional states ($|1\rangle, |2\rangle$) gives rise to a narrow resonance. (c) The dressed condensate is the upper state ($|1'\rangle$) of a two-level system, and decays to the lower state (recoil state of atoms, $|2\rangle$) by emitting a photon. Figure is taken from Ref. [14]

atomic resonance) and increased by Γ_1/Γ_2, the ratio of the linewidths of the single-photon and two-photon Bragg resonances. Thus the expected cross-section for gain is

$$\sigma_{\text{gain}} = 6\pi\lambda^2 \frac{R}{\Gamma_2}. \tag{20}$$

We observed quasi-steady state gain by a factor of up to three. This optical gain has a narrow bandwidth due to the long coherence time of a condensate. The gain represents the imaginary part of the complex index of refraction. A sharp peak in the gain implies a steep dispersive shape for the real part of the index of refraction $n(\omega)$. This resulted in an extremely slow group velocity for the amplified light. Fig. 12 shows that light pulses were delayed by about 20 μs across the 20 μm wide condensate corresponding to a group velocity of 1 m/s. This is one order of magnitude slower than any value reported previously (see Ref. [37] and references therein).

The (amplitude) gain g for the probe light is related to the matter wave gain G

$$g = 1 + \frac{n_0\sigma_{\text{gain}}l}{2} = 1 + \frac{G}{\Gamma_2} \tag{21}$$

where n_0 is the condensate density and l its length. However, when the gain G is above the threshold for superradiance, $G > \Gamma_2$ (Eq. 18) the optical gain should diverge: a single recoiling atom created by the probe light and dressing light is exponentially amplified and creates a huge matter wave grating which will diffract the dressing light into the probe light path, thus amplifying the probe light by a large factor. In order to describe this non-linear feedback, one has to use coupled

352

equations for optical gain and atom gain [14]. As a result, one has to replace the loss rate Γ_2 in Eq. 21 by the dynamic rate $\Gamma_2 - G$ (see Eq. 18)

$$g = 1 + \frac{G}{\Gamma_2 - G} = 1 + \frac{n_0 \sigma_{\text{gain}} l}{2} \frac{\Gamma_2}{\Gamma_2 - G} \tag{22}$$

which agrees with Eq. (21) in the low-intensity limit. By raising the gain over the threshold we could map out the transition from single-atom gain to collective gain [14]. The expansion $g = 1 + (G/\Gamma_2) + (G/\Gamma_2)^2 + \dots$ shows the transition from (linear) single-atom gain to (non-linear) collective gain and illustrates that the dressed condensate is a clean model system for discussing linear and non-linear behavior, optical and atom-optical properties and their interplay.

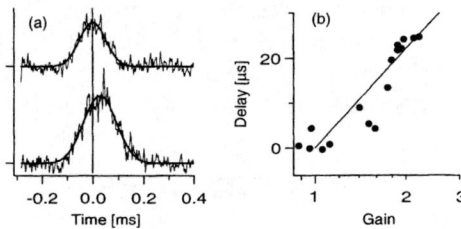

FIGURE 12. Pulse delay due to light amplification. (a) About 20 μs delay was observed when a Gaussian pulse of about 140 μs width and 0.11 mW/cm^2 peak intensity was sent through the dressed condensate (bottom trace). The top trace is a reference taken without the dressed condensate. Solid curves are Gaussian fits to guide the eyes. (b) The observed delay was proportional to the logarithm of the observed gain. Figure is taken from Ref. [14]

VI   DISCUSSION

This paper has summarized our recent experiments on Bose-Einstein condensation with the unifying theme of suppression and enhancement. Suppression of scattering or dissipation can arise for two different reasons. The phonon and vortex nature of the collective excitations together with energy and momentum conservation allow dissipation only above a critical velocity. In addition, one has to consider the dynamics of the excitation process. For microscopic particles, this is reflected in the matrix element $S(q)$ which characterizes how easily can the condensate absorb momentum. For macroscopic motion, it is reflected in a critical velocity for vortex nucleation. Scattering processes are also enhanced by the population in the final states (bosonic stimulation). Optical stimulation was used in Bragg scattering, and matter wave stimulation led to superradiance and matter wave amplification.

In closing, let us note that there are some subtleties which go beyond the simple picture using rate equations and occupation numbers. A condensate in its ground state is in a coherent superposition state of the zero-momentum state with correlated pairs with momenta $\pm\mathbf{q}$ (the quantum depletion). The population in the quantum depletion can cause bosonic stimulation of spontaneous emission [38]. However, for a scattering situation, there are two bosonically enhanced pathways which destructively interfere (causing $S(q) < 1$, Sect. II). Therefore, the concept of bosonic stimulation can be applied to the quantum depletion, but with caution [38].

Finally, the matter wave amplification described in Sect. V can be regarded as the self-amplification of a density modulation (caused by matter wave interference between the condensate and the amplified recoiling atoms). A similar amplification can happen for a density modulation in a fermionic gas [39], but, the coherence time of such a density modulation is generally much shorter than in a condensate. This example shows that a more general description of stimulation and amplification has to address the symmetry and coherence of the prepared state. Bosonic quantum-degeneracy is sufficient, but not necessary for stimulated scattering [39].

We are grateful to Axel Görlitz for helpful discussions. This work was supported by NSF, ONR, ARO, NASA, and the David and Lucile Packard Foundation.

REFERENCES

1. F. London, *Superfluids, Vol.II* (Dover, New York, 1964).
2. K. Huang, *Statistical Mechanics* (Wiley, New York, 1987).
3. N. N. Bogoliubov, J. Phys. (USSR) **11**, 23 (1947).
4. M. H. Anderson, J. R. Ensher, M. R. Matthews, C. E. Wieman, and E. A. Cornell, Science **269**, 198 (1995).
5. K. B. Davis, M.-O. Mewes, M. A. Joffe, M. R. Andrews, and W. Ketterle, Phys. Rev. Lett. **74**, 5202 (1995).
6. C. C. Bradley, C. A. Sackett, J. J. Tollet, and R. G. Hulet, Phys. Rev. Lett. **75**, 1687 (1995).
7. F. Dalfovo, S. Giorgini, L. P. Pitaevskii, and S. Stringari, Rev. Mod. Phys. **71**, 463 (1999).
8. A. P. Chikkatur, A. Görlitz, D. M. Stamper-Kurn, S. Inouye, S. Gupta, and W. Ketterle, Phys. Rev. Lett. **85**, 483 (2000).
9. C. Raman, M. Köhl, R. Onofrio, D. S. Durfee, C. E. Kuklewicz, Z. Hadzibabic, and W. Ketterle, Phys. Rev. Lett. **83**, 2502 (1999).
10. R. Onofrio, C. Raman, J. M. Vogels, J. Abo-Shaeer, A. P. Chikkatur, and W. Ketterle, Phys. Rev. Lett. **85**, 2228 (2000).
11. D. M. Stamper-Kurn, A. P. Chikkatur, A. Görlitz, S. Inouye, S. Gupta, D. E. Pritchard, and W. Ketterle, Phys. Rev. Lett. **83**, 2876 (1999).
12. S. Inouye, A. P. Chikkatur, D. M. Stamper-Kurn, J. Stenger, D. E. Pritchard, and W. Ketterle, Science **285**, 571 (1999).
13. S. Inouye, T. Pfau, S. Gupta, A. P. Chikkatur, A. Görlitz, D. E. Pritchard, and W. Ketterle, Nature **402**, 641 (1999).
14. S. Inouye, R.F.Löw, S. Gupta, T. Pfau, A. Görlitz, T. L. Gustavson, D. E. Pritchard, and W. Ketterle, cond-mat/0006455; Phys. Rev. Lett., in print.

15. J. Stenger, S. Inouye, A. P. Chikkatur, D. M. Stamper-Kurn, D. E. Pritchard, and W. Ketterle, Phys. Rev. Lett. **82**, 4569 (1999).
16. D. Stamper-Kurn and W. Ketterle, proceedings of the Les Houches Summer School on Bose-Einstein Condensation in 1999, in print; cond-mat/0005001.
17. P. J. Price, Phys. Rev. **94**, 257 (1954).
18. A. Griffin, *Excitations in a Bose-condensed liquid* (Cambridge University Press, Cambridge, 1993).
19. L. D. Landau, J. Phys. (USSR) **5**, 71 (1941).
20. L. Meyer and F. Reif, Phys. Rev. **123**, 727 (1961).
21. D. R. Allum, P. V. E. McClintock, A. Phillips, and R. M. Bowley, Phil. Trans. R. Soc. A **284**, 179 (1977).
22. J. Harms and J. P. Toennies, Phys. Rev. Lett. **83**, 344 (1999).
23. E. W. Hagley, L. Deng, M. Kozuma, J. Wen, K. Helmerson, S. L. Rolston, and W. D. Phillips, Science **283**, 1706 (1999).
24. A. I. Lvovsky and S. R. Hartmann, Phys. Rev. Lett. **82**, 4420 (1999).
25. P. Nozières and D. Pines, *The Theory of Quantum Liquids* (Addison-Wesley, Redwood City, CA, 1990).
26. M. Crescimanno, C. G. Koay, R. Peterson, and R. Walsworth, preprint cond-mat/0001163; Phys. Rev. A, in print.
27. R. P. Feynman, in *Progress in Low Temperature Physics*, edited by C. Gorter (North-Holland, Amsterdam, 1955), Vol. 1, p. 17.
28. E. Lundh, C. J. Pethick, and H. Smith, Phys. Rev. A **55**, 2126 (1997).
29. K. W. Madison, F. Chevy, W. Wohlleben, and J. Dalibard, Phys. Rev. Lett. **84**, 806 (2000).
30. D. L. Feder, C. W. Clark, and B. I. Schneider, Phys. Rev. A **61**, 011601(R) (1999).
31. T. Isoshima and K. Machida, Phys. Rev. A **60**, 3313 (1999).
32. C. Raman, R. Onofrio, J. M. Vogels, J. R. Abo-Shaeer, and W. Ketterle, preprint cond-mat/0008423.
33. B. Jackson, J. F. McCann, and C. S. Adams, Phys. Rev. A **61**, 051603(R) (2000).
34. T. Frisch, Y. Pomeau, and S. Rica, Phys. Rev. Lett. **69**, 1644 (1992).
35. B. Jackson, J. F. McCann, and C. S. Adams, Phys. Rev. Lett. **80**, 3903 (1998).
36. T. Winiecki, J. F. McCann, and C. S. Adams, Phys. Rev. Lett. **82**, 5186 (1999).
37. L. V. Hau, S. E. Harris, Z. Dutton, and C. H. Behroozi, Nature **397**, 594 (1999).
38. A. Görlitz, A. P. Chikkatur, and W. Ketterle, preprint, cond-mat/0008067.
39. W. Ketterle and S. Inouye, preprint, cond-mat/0008232.

Laser cooling: Beyond optical molasses and beyond closed transitions

Vladan Vuletić, Andrew J. Kerman, Cheng Chin, and Steven Chu

Department of Physics, Stanford University, Stanford, CA 94305-4060

Abstract. We present a simple and general optical cooling method based on 3D degenerate Raman sideband cooling with adiabatic release that goes significantly beyond the density and temperature limitations of optical molasses. In 10 ms we cool a sample of 3×10^8 cesium atoms to a temperature of 330 nK at a density of 1.1×10^{11} cm^{-3}, which corresponds to a phase-space density $n\lambda_{dB}^3 = 1/500$.

We further propose to cool atoms or molecules inside an optical cavity that enhances the scattering of blue-detuned photons and show that the dissipative mechanism can be viewed as a cavity-induced generalized Doppler cooling. Since the cooling depends on the atom's internal level structure only through the photon scattering rate, cavity Doppler cooling is applicable to particles that do not possess a closed optical transition, which may allow one to extend laser cooling to a greater class of atoms or molecules. Large samples are cooled at the same rate as single atoms if the effect of one atom on the cavity resonance frequency is small. We also show how to achieve 3D cooling with a single optical cavity.

BEYOND OPTICAL MOLASSES: 3D DEGENERATE RAMAN SIDEBAND COOLING

Optical molasses has long been the best simple optical cooling method available for atoms in free space and it remains the most widely used technique to prepare large samples of cold atoms for various experiments. Some applications, such as atomic fountain clocks, mainly require a simple and fast method that produces the lowest possible temperatures with the largest number of atoms, while other experiments, among them those where the atoms are loaded into magnetic or optical traps, require moderately low temperatures at high atomic densities. A simple and fast method that can cool large atomic samples to high phase-space density is particularly useful for evaporative cooling to Bose-Einstein condensation, since it allows one to reduce the atom loss during evaporation, and to produce significantly larger condensates in a shorter time.

Here we present a simple and general optical cooling method that requires only one weak laser in addition to the lasers used to create a magneto-optical trap (MOT), while it increases the phase-space density in 10 ms by almost three orders

CP551, *Atomic Physics 17,* edited by E. Arimondo, P. DeNatale, and M. Inguscio
© 2001 American Institute of Physics 1-56396-982-3/01/$18.00

of magnitude over the value attainable with optical molasses for all 3×10^8 atoms prepared in our cesium MOT. In particular, this technique may allow an even stronger relative improvement for atoms where the excited-state hyperfine structure is not resolved and therefore sub-Doppler cooling is not effective in optical molasses. Important atoms belonging to this class are lithium and potassium, that have both bosonic and fermionic isotopes.

The new method is based on Raman sideband cooling [1–3] inside a 3D optical lattice in combination with adiabatic release [4,5]. The 3D lattice is formed by two counterpropagating beams along the x-axis and two running waves along the y and z-axes. This four-beam setup does not require interferometric stabilization of the relative time phases of different beams, since variations of the time phases correspond merely to a translation of the lattice in space. As these variations occur at a rate that is much lower than the typical lattice vibration frequency of $\omega/2\pi = 30$ kHz, the atoms that are trapped at the lattice sites follow the motion of the lattice adiabatically and consequently the cooling performance is not affected. We choose e^{-2} beam waists of 1 mm at the position of the atoms, such that the lattice is only slightly larger than the trapped cloud in the MOT containing up to 5×10^8 atoms, which reduces the power requirement for the lattice laser to a total of 20 mW for all four beams. The polarizations of the four beams are linear and both the polarization directions and the relative beam intensities are independently adjusted to optimize the cooling. A typical detuning for the lattice laser is ± 10 GHz relative to the $6S_{1/2}, F = 3 \rightarrow 6P_{3/2}, F'$ transitions. As the lattice geometry differs significantly for red and for blue detuning of the lattice laser, different beam intensity ratios and polarization angles are found to yield the lowest temperatures in these two cases.

As discussed in Refs. [3,5], degenerate Raman sideband cooling in the lower hyperfine manifold starts with atoms optically pumped into the lowest-energy magnetic sublevel $F = 3, m_F = 3$ in a trap where the vibration frequency exceeds the recoil energy (Lamb-Dicke regime). A small magnetic field on the order of 50 mG is applied that produces a Zeeman splitting between the neighboring magnetic sublevels $m_F = 3$ and $m_F = 2$ that equals the vibration energy. The atoms are transferred from $m_F = 3$ to $m_F = 2$ by means of an energy selective Raman transition with two photons of equal frequency but different polarizations, such that the vibrational quantum number is lowered by one unit. Optical pumping back to the sublevel $m_F = 3$ preferentially maintains the vibrational state, and the atoms are cooled by one vibration energy in the cycle. The cooling continues until the atoms accumulate in the vibrational ground state of the $m_F = 3$ sublevel that is dark to both the optical pumping light and to the Raman transition. Since the $P_{3/2}$ state of cesium has a sufficiently large hyperfine splitting , our optical pumping is performed on the $F = 3 \rightarrow F = 2$ transition of the D_2 line. For atoms where the hyperfine structure of the excited P state is not resolved optical pumping with circularly polarized light on the D_1 line that always provides a dark state should be used for Raman sideband cooling.

To attain fast cooling in our 3D lattice where the trapping potential at each site is neither isotropic nor harmonic, we adjust the polarizations of the lattice beams to achieve a large Raman coupling that is comparable to the trap vibration frequency. Consequently, and in contrast to the best cooling in the 1D case where the red cooling sideband and its second harmonic were clearly resolved [3], here our cooling spectrum exhibits only a broad minimum for the final temperature as a function of magnetic field, as shown in Figure 1.

Although the large Raman coupling leads to a non-negligible off-resonant rate out of the nominally 'dark' vibrational ground state of the $m_F = 3$ sublevel, we still achieve a very good cooling performance. Figure 2 shows the phase-space density $n\lambda_{dB}^3$ as a function of density for all 3×10^8 atoms prepared in our MOT [5]. The dotted and the dashed line indicate the results for standard red-detuned polarization gradient cooling (bright molasses) and for blue-detuned polarization gradient cooling ('grey' molasses) for cesium from Ref. [6], respectively. The phase-space density obtained with our cooling scheme exceeds that of optical molasses by almost three orders of magnitude. This improvement results from a combination of three factors: First, the low-density temperature limit of our scheme of 290 nK is lower than both for bright optical molasses (3μK) and for dark optical molasses (1.1μK). More important is the fact that for our method the final temperature increases very slowly with density. We measure a linear density dependence of only 8nK$/10^{10}$cm^{-3}, which is 75 times lower than the coefficient of 600nK$/10^{10}$cm^{-3} measured for dark optical molasses [6], while standard red-detuned polarization gradient cooling performs even worse. The observed low value of density dependent heating is probably partly due to the fact that in the lattice the atoms quickly bind to the lattice sites and are then isolated from each other, which leads to a suppression of radiative collisions, and partly due to the fact that in the Lamb-Dicke regime the relevant recoil heating (expressed in terms of change of quantum state per scattering event) is smaller than in free space. The weak dependence of the heating on density allows us to significantly increase the density before the

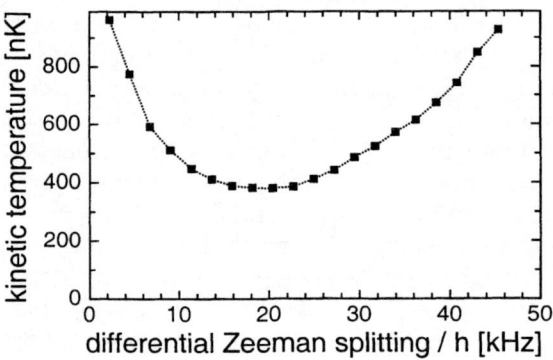

FIGURE 1. The final temperature after adiabatic release as a function of Zeeman energy splitting between two neighboring magnetic sublevels for 3D cooling with large Raman coupling.

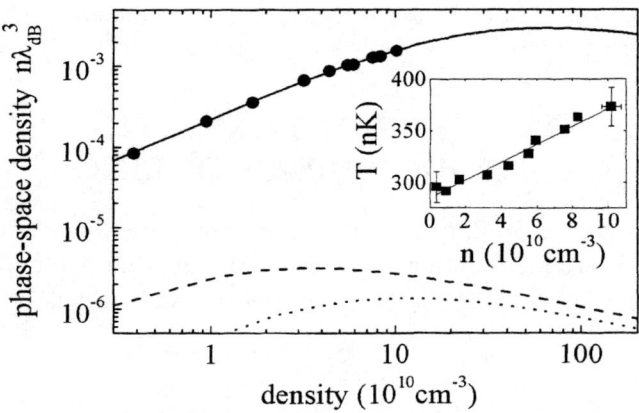

FIGURE 2. The phase space density as a function of atomic density for red detuned polarization gradient cooling (dotted line), blue detuned polarization gradient cooling (dashed line) and for 3D Raman sideband cooling with adiabatic release (solid line).

temperature increase eventually overcomes the effect of the density increase and limits the attainable phase-space density. Finally, our method simultaneously spin-polarizes the sample into one magnetic sublevel, while optical molasses leaves the atoms largely unpolarized with respect to an external quantization axis.

The highest phase-space density achieved with all 3×10^8 atoms prepared in our MOT is $n\lambda_{dB}^3 = 1/500$ at a density of $1.1 \times 10^{11} \text{cm}^{-3}$ and a temperature of 330nK. In this experiment the cooling was optimized with the constraint that it should not induce a loss of atoms and the measured transfer from the MOT to the cold sample was better than 95%. Higher values for the phase-space density are possible if one allows for atom loss during the cooling or if one uses smaller samples. Under such conditions a further improvement of the phase-space density by more than one order of magnitude has been recently demonstrated in Ref. [7], and we also observe higher phase-space density in combination with significant loss due to radiative collisions at densities above 10^{12}cm^{-3}. The critical value for Bose-Einstein condensation of $n\lambda_{dB}^3 = 2.6$ after release into free space would be obtained if one could prepare one atom at each lattice site and cool each atom to the local 3D vibrational ground state. However, this will be difficult to achieve since two or more atoms at one lattice site are lost by a radiative collision [8,4] long before they can be optically pumped, while a Poisson distribution of occupation numbers permits only a maximum single-atom occupancy of the fraction e^{-1} of sites. Even if it should not be possible to go beyond unity phase-space density using Raman sideband cooling alone, a simple cooling technique that provides significant improvement is important for various applications and precision experiments with cold atoms, while the phase-space density attained for a given atom number in a given cooling time represents a good measure for the brightness of the cold atom source.

BEYOND CLOSED TRANSITIONS: CAVITY DOPPLER COOLING

When an atom of mass m illuminated by monochromatic light of wavevector $\vec{k_i}$ scatters a photon into a mode with wavevector $\vec{k_s}$ (Figure 3), the atom's momentum changes from $\vec{p} = m\vec{v}$ to $\vec{p'} = \vec{p} + \hbar\vec{k_i} - \hbar\vec{k_s}$, and its kinetic energy from $W = \vec{p}^2/2m$ to

$$W' = W + \hbar(\vec{k_i} - \vec{k_s}) \cdot \vec{v} + \frac{\hbar^2(\vec{k_i} - \vec{k_s})^2}{2m}. \tag{1}$$

Energy conservation then requires that the angular frequency of the scattered photon differs from that of the incident photon by an amount

$$\Delta = ck_s - ck_i = \frac{W - W'}{\hbar} = -(\vec{k_i} - \vec{k_s}) \cdot \vec{v} - \frac{\hbar(\vec{k_i} - \vec{k_s})^2}{2m}. \tag{2}$$

This formula shows that the frequency of the scattered light is determined by the Doppler effect along the two-photon wavevector $\vec{k_i} - \vec{k_s}$, while the last term describes recoil heating that is independent of the atomic velocity. If $\Delta > 0$, i.e. if the scattered photon is blue-detuned relative to the incident photon, the atom's kinetic energy will be reduced in the scattering event. An atomic gas is cooled if on average $\Delta > 0$, i.e. if on average the two-photon Doppler effect $\langle(\vec{k_i} - \vec{k_s}) \cdot \vec{v}\rangle$ is negative and exceeds the two-photon recoil heating $\langle\hbar(\vec{k_i} - \vec{k_s})^2/2m\rangle$ in magnitude.

In conventional Doppler cooling, the direction of the scattered photon is random and therefore $\langle\vec{k_s} \cdot \vec{v}\rangle = 0$, while the desired negative Doppler effect for the incident beam $\langle\vec{k_i} \cdot \vec{v}\rangle < 0$ is achieved by tuning the laser frequency below the atomic

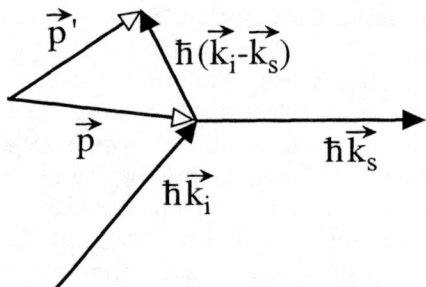

FIGURE 3. The atom scatters a photon from a mode with wavevector $\mathbf{k}_i$ into a mode with wavevector $\mathbf{k}_s$ and changes its momentum from $\mathbf{p}$ to $\mathbf{p'}=\mathbf{p}+\hbar(\mathbf{k}_i-\mathbf{k}_s)$.

resonance, leading to a preferential absorption of photons from a beam opposing the atomic velocity. Conventional Doppler cooling requires a closed two-level system and a small detuning relative to the atomic resonance that is comparable to the Doppler effect of the moving atom. If the two-level system is not closed, the atom will be optically pumped to a different internal state where the detuning is different and typically large, and the cooling will stop.

To design a cooling scheme that works at arbitrary detuning from atomic transitions and that therefore can be applied to multilevel atoms, one can use an optical cavity to enhance scattering events where the two-photon Doppler effect $\langle (\vec{k_i} - \vec{k_s}) \cdot \vec{v} \rangle$, rather than the incident-photon Doppler effect $\langle \vec{k_i} \cdot \vec{v} \rangle$, is negative. In a quantum mechanical description the cavity changes the density of electromagnetic modes in a frequency-dependent manner [9,10] which according to Fermi's Golden Rule leads to a scattering rate that depends strongly on the frequency $ck_s = ck_i + \Delta$ of the scattered photon. If the cavity is blue-detuned relative to the incident light, it will enhance scattering events with $\Delta > 0$, i.e. those events where the scattered photon carries away a larger energy than that of the incident photon. Energy conservation then requires that the atom's energy is reduced by $\hbar\Delta$ in such a process.

In a classical description, the cavity acts to direct the scattered field back onto the atom with a phase that depends on the frequency of the scattered light. The field that has been reflected by the cavity mirrors interferes with the field scattered by the atom at a later time, "stimulating" further scattering through the addition of field amplitudes. This frequency-dependent feedback provided by the cavity leads to a scattering rate into the cavity mode that is proportional to the classical intensity enhancement function L of the cavity, in agreement with the quantum mechanical result. This cavity response function $L(\delta)$ not too far from resonance is given by the Lorentzian form [11,12]

$$L(\delta) = \frac{2E}{1 + \delta^2/\gamma_c^2}.$$ (3)

Here E is the cavity intensity enhancement factor that is related to the cavity finesse F by $E = F/\pi$, while δ is the detuning of the scattered light from cavity resonance and γ_c is the cavity decay rate constant for the field amplitude.

With respect to atomic properties, cavity-induced Doppler cooling relies on the fact that at low saturation of the atomic transition and at a detuning that exceeds the atomic linewidth, essentially all the scattered power is concentrated in the coherent Rayleigh scattering peak [13]. The phase and frequency of the scattered light are then completely determined by the incident light in combination with the atomic position and velocity. (For an atom fixed in space, the frequency spectrum of this classically scattered light is a delta function at the frequency of the incident light, while the amplitude depends on the real part of the atomic polarizability [13].) Since at low saturation the frequency of the scattered light is independent of the atomic level structure, temperatures below the atomic Doppler limit can be reached with cavity Doppler cooling. Furthermore in contrast to conventional Doppler cooling the atomic properties and in particular the detuning between the incident light and the atomic transitions enter only through the free-space photon scattering rate. It follows that cavity Doppler cooling is possible at arbitrary detuning from atomic resonances, provided sufficient photon scattering rates can be achieved with an intense laser beam. This should enable one to cool simultaneously different atoms or isotopes in the same light field, and to cool atoms with a complicated level structure or even molecules that possess a sufficiently large polarizability in the optical frequency domain.

To calculate the cooling force, let us consider an atom scattering photons at a rate Γ_1, where the index indicates that the atom is illuminated by a single linearly polarized running plane wave. The scattering rate Γ_w along one direction $\vec{k_s}$ of a Gaussian mode with a polarization that is parallel to that of the incident beam and that has a waist w centered on the atom (Figure 4) is then given by $\Gamma_w = (3/k_s^2 w^2)\Gamma_1$. (This formula can be obtained by considering the dipole force that an atom experiences if an external light field is applied in the same mode. As the dipole force arises from the interference of the field that is already present in a particular mode with the scattered field, momentum conservation requires that the force on the atom equals the rate at which net momentum is carried away by the total (incident and scattered) electromagnetic field. The known expression for the dipole force [14] can then be used to calculate the scattered field.) The factor 3 takes into account that such a mode with parallel polarization is coupled three times stronger to the incident light field than the spatial average of the dipole pattern $\langle \cos^2 \theta \rangle = 1/3$, while the denominator $k_s^2 w^2$ defines the solid angle subtended by the scattering mode.

If a linear cavity is placed around the scattering mode, the modified scattering rate is obtained by multiplying the free-space scattering rate Γ_w into mode $\vec{k_s}$ with the frequency-dependent cavity response function $L(\delta)$. For an atom moving at velocity $\vec{v}$, the scattered light will be shifted in frequency by the Doppler and

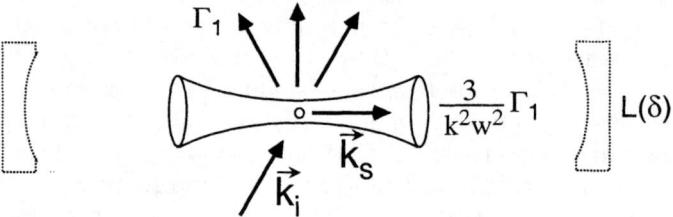

FIGURE 4. The atom with a total scattering rate Γ_1 scatters a fraction $3/k_s^2 w^2$ of the light along one direction of a mode with waist w. The polarization of the scattering mode is parallel to the polarization of the incident light.

recoil effects as described by Eq. (2). It follows that the photons scattered along the two directions of the cavity $\pm\vec{k}_s$ will be detuned by $\delta_\pm = \delta_l - (\vec{k}_i \mp \vec{k}_s) \cdot \vec{v} - \hbar(\vec{k}_i \mp \vec{k}_s)^2/2m$ relative to the cavity resonance. Here δ_l denotes the detuning of the incident light relative to the cavity resonance. The cooling force $\vec{f}$ is simply given by the sum of the two possible scattering rates into the cavity mode $\Gamma_w L(\delta_\pm)$, multiplied by the corresponding two-photon momenta $\hbar(\vec{k}_i \mp \vec{k}_s)$:

$$\vec{f} = \hbar(\vec{k}_i - \vec{k}_s)\Gamma_w L(\delta_+) + \hbar(\vec{k}_i + \vec{k}_s)\Gamma_w L(\delta_-). \tag{4}$$

Note the close resemblance of this formula with the expression for conventional Doppler cooling, where the absorption momentum $\pm\vec{k}_i$ has been replaced by the two-photon momentum $\vec{k}_i \mp \vec{k}_s$, the total free-space scattering rate Γ_1 has been replaced by the free-space scattering rate Γ_w into the cavity mode, and the atomic lineshape has been replaced by the cavity line $L(\delta)$ in the Lorentzian approximation. Eq. (4) clearly shows that cooling by means of an optical cavity at low saturation of the atomic transition can be viewed as a generalized Doppler cooling technique.

In the special case that the incident light is a standing wave that circulates inside the cavity or encloses a very small angle with the cavity mode, one finds the following result for the 1D cooling force [15]:

$$f = 2\hbar k \Gamma_{sc} \frac{12E}{k^2 w^2} \frac{2xy}{[1 + (2x + y)^2][1 + (2x - y)^2]} \tag{5}$$

Here $\Gamma_{sc} = 2\Gamma_1$ is the average free-space scattering rate in the standing wave, while $x = kv/\gamma_c$ and $y = (\delta_l - 4(\hbar k)^2/2m)/\gamma_c$ are the Doppler shift and recoil-shifted incident-light detuning relative to the cavity resonance, respectively, both normalized to the cavity linewidth. The cooling force is proportional to the free-space scattering rate in the incident beam of a given intensity, but otherwise independent of atomic properties. The force is cooling ($f \cdot v < 0$) if $y < 0$, i.e. if the incident light is red detuned *relative to the cavity resonance*, independent of the detuning relative to the atomic resonance. The last factor in Eq. (5) shows that the velocity dependence of the force, and therefore the cooling limit, are determined by the cavity parameters alone. The minimum attainable temperature is given by the usual Doppler limit $kT = \hbar\gamma_c$ for the Lorentzian cavity line in the case that the enhanced scattering into the cavity at a rate $(6E/k^2 w^2)\Gamma_{sc}$ exceeds scattering into free space that proceeds at the rate Γ_{sc} [15].

We point out that in the limiting case of low saturation of the atomic transition, the sole purpose of the cavity is to enhance the scattering into electromagnetic modes with a frequency which exceeds that of the incident light by the two-photon Doppler effect $-(\vec{k}_i - \vec{k}_s) \cdot \vec{v}$. Therefore it is by no means necessary that the incident light itself circulate inside the cavity, a feature which is less obvious in the time domain treatment [16,15] of the problem.

In the time domain description the atom acts as a position-dependent refractive index, tuning the cavity in and out of resonance as the atom moves along the standing wave, while the intracavity field changes with a time delay that is determined by the resonator linewidth. The delayed circulating light field then acts back on the atom via the light shift, which under appropriate conditions gives rise to a Sisyphus-like cooling force [16]. This dissipative force is proportional to the square of the real part of the atomic polarizability α, since both the detuning of the cavity by the atom and the ac light shift exerted by the intracavity field on the atom are proportional to $\mathrm{Re}(\alpha)$. It follows that in the classical limit the cooling force is proportional to the photon scattering rate as given by the imaginary part of the atomic polarizability [15].

To understand in the time domain picture why it is not necessary that the incident field circulate in the cavity, we note that the refractive index has its origin in the interference between the incident field and the field radiated in quadrature by the atom, which in lowest order leads to a phase shift of the total field. Even if the incident field encloses a nonzero spatial angle with the scattered field circulating inside the cavity, the interference between incident and scattered field still produces a delayed time-dependent light shift that can give rise to a cooling force.

The fact that the incident field need not circulate inside the cavity can be used to achieve cooling in all three spatial dimensions with a single optical resonator. Since the cooling mechanism depends on the two-photon Doppler shift $(\vec{k_i} - \vec{k_s}) \cdot \vec{v}$, for 3D cooling it is sufficient to illuminate the atom with four running waves along $\pm x$ perpendicular to the cavity axis that is oriented along z (Figure 5). Each incident running wave $\vec{k_i}$ in combination with scattering along one direction $\pm \vec{k_s}$ of the cavity leads to a dissipative force along one diagonal direction $\vec{k_i} \mp \vec{k_s}$. (It is also possible to use a symmetric arrangement of three incident beams in the xy-plane.) The force is cooling if the cavity is blue detuned relative to the incident field. This

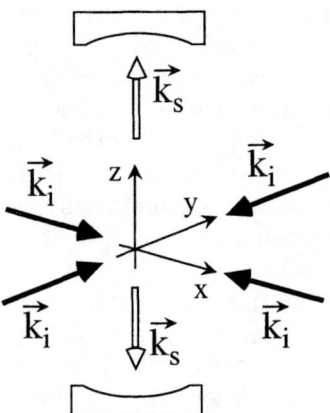

FIGURE 5. Setup for three-dimensional cooling using only one optical cavity. Since the cooling force is directed along the two-photon wavevector $\mathbf{k}_i$-$\mathbf{k}_s$, three-dimensional cooling can e.g. be achieved with the cavity oriented along the z axis and four incident beams along the $\pm x$ and $\pm y$ axes.

simple setup that requires only one high-finesse cavity should greatly facilitate 3D cavity Doppler cooling of an atomic or molecular gas.

When the effect of a single atom on the cavity resonance is small, i.e. when the atom inside the cavity changes the cavity resonance frequency by much less than a cavity linewidth, then a sample containing more than one atom will be cooled at the same rate as a single atom: If a sample of N uncorrelated atoms is placed inside the cavity, the effects of most of the atoms on the cavity will cancel each other. However, fluctuations from equilibrium will lead to an effective coupling parameter β_N between sample and cavity mode that is given by $\beta_N = \sqrt{N}\beta$, where $\beta = (k_s/\pi w^2)Re(\alpha)/\epsilon_0$ is the coupling between a single atom and the cavity [15]. As long as $\beta_N \ll 1$, i.e. as long as the coupling of the sample to the cavity is below saturation, the scattering rate into the cavity mode will be proportional to β_N^2, and therefore to the atom number N. Therefore the sample will act as if it consisted of N independent scatterers, and it follows that the cooling time will be comparable to that of a single atom. (Note that the same line of argument applies to an optically thin sample of N independent scatterers in free space, where the total scattered power is simply N times the power scattered by a single atom, even when the scattering of the incident monochromatic radiation is coherent and the correct treatment is to add up the electric fields from all scatterers.) Therefore cavity Doppler cooling of a sample for $\beta_N \ll 1$ resembles Doppler cooling of an optically thin sample in free space, where the atoms act effectively as independent scatterers. This makes cavity Doppler cooling of large samples significantly more favorable than stochastic cooling [17], where the cooling slows down as the sample size increases. The scaling will be different and less favorable in the case of critical coupling $\beta \approx 1$ between the atom and the cavity [18], that although leading to a larger cooling force, is difficult to achieve with atoms at large detuning from atomic transitions and out of reach with molecules.

When the scattering rate of photons into the cavity by a single atom or by an atomic sample exceeds the cavity linewidth, the number of photons in steady state inside the resonator will exceed one, and the coherent coupling between atoms and scattered light becomes important. Although this situation will give rise to complicated dynamics of the particle motion and of the intracavity field, energy conservation requires that the atom's motion is cooled at the rate at which the blue-detuned photons leaking out of the cavity carry away the excess energy $\hbar\Delta = \hbar(ck_s - ck_i)$. Therefore even if the cavity contains more than one photon, the cooling force will be given by the simple formula Eq. (4) or (5) derived above.

In conclusion, cavity Doppler cooling relies on the classical scattering of light by particles and therefore may allow one to extend laser cooling to atomic or even molecular species with a complicated internal level structure. Simultaneous cooling of different species by the same light field is possible. The application of cavity Doppler cooling at large detuning from atomic transitions may open a new way to cool high-density samples. The free-space cooling considered here can be easily generalized to trapped particles (cavity sideband cooling) and to internal degrees of freedom, such as rovibrational states of molecules, or possibly even to transitions in liquids or solids. The basic idea is always that the cavity is used to enhance the scattering of higher-energy and suppress the scattering of lower energy photons. The irreversibility of the coherent scattering process is ensured by the coupling of the intracavity field to the vacuum outside the cavity.

ACKNOWLEDGMENTS

This work was supported in parts by grants from the AFOSR and the NSF. V.V. acknowledges support from the Humboldt foundation.

REFERENCES

1. Hamann S.E., Haycock D.L., Klose G., Pax, P.H., Deutsch I.H. and Jessen P.S., *Phys Rev. Lett.* **80**, 4149 (1998).
2. Perrin H., Kuhn A., Bouchole I., and Salomon C., *Europhys Lett.* **42**, 395 (1998).
3. Vuletić V., Chin C., Kerman A.J., and Chu S., *Phys Rev. Lett.* **81**, 5768 (1998).
4. DePue M.T., McCormick C., Winoto S.L., Oliver S., and Weiss D., *Phys Rev. Lett* **82**, 2262 (1999).
5. Kerman A.J., Vuletić V., Chin C., and Chu S., *Phys Rev. Lett.* **84**, 439 (2000).
6. Boiron D., Michaud A., Lemonde P., Castin Y., Salomon C., Weyers S., Szymaniec K., Cognet L., and Clairon A., *Phys Rev. A* **53**, R3734 (1996).
7. Han D.-J., Wolf S., Oliver S., McCormick C., DePue M.T., and Weiss D., *Phys Rev. Lett* **85**, 724 (2000).
8. Burnett K., Julienne P.S., and Suominen K.-A., *Phys Rev. Lett.* **77**, 1416 (1996).
9. Purcell E.M., *Phys Rev.* **69**, 681 (1946).
10. Kleppner D., *Phys Rev. Lett.* **47**, 233 (1981).
11. Heinzen D. and Feld M.S., *Phys Rev. Lett.* **59**, 2623 (1987).
12. Mossberg T.W., Lewenstein M., and Gauthier D.J., *Phys Rev. Lett.* **67**, 1723 (1991).
13. Mollow, B.R., *Phys Rev.* **188**, 1969 (1969).
14. Letokhov V.S., Minogin V.G., and Pavlik B.D., *Opt. Commun.* **19**, 72 (1976).
15. Vuletić V. and Chu S., *Phys Rev. Lett.* **84**, 3787 (2000).
16. Horak P., Hechenblaikner G., Gheri K.M., Stecher H., and Ritsch H., *Phys Rev. Lett.* **79**, 4974 (1997).
17. Raizen M.G., Koga J., Sundaram B., Kishimoto Y., Takuma H., and Tajima T., *Phys Rev. A* **58**, 4757 (1998).
18. Gangl M. and Ritsch H., *Phys Rev. A* **61**, 011402(R) (1999); *ibid.* **61** 043405 (2000); Gangl M. and Ritsch H., *Eur. J. Phys. D* **8**, 29 (2000).

Atom Trap Trace Analysis

Zheng-Tian Lu[1*], Kevin Bailey[1], Chun-Yen Chen[1], Xu Du[1,3], Yi-Min Li[1],
Thomas P. O'Connor[1], Linda Young[2]

[1]Physics Division, [2]Chemistry Division, Argonne National Laboratory, Argonne, IL 60439, USA
[3]Department of Physics and Astronomy, Northwestern University, Evanston, IL 60208, USA

Abstract. A new method of ultrasensitive trace-isotope analysis has been developed based upon the technique of laser manipulation of neutral atoms. It has been used to count individual ^{85}Kr and ^{81}Kr atoms present in a natural krypton sample with isotopic abundances in the range of 10^{-11} and 10^{-13}, respectively. The atom counts are free of contamination from other isotopes, elements, or molecules. The method is applicable to other trace-isotopes that can be efficiently captured with a magneto-optical trap, and has a broad range of potential applications.

INTRODUCTION

Much can be learned from the concentrations of the ubiquitous long-lived radioactive isotopes. W. Libby and coworkers first demonstrated in 1949 that trace analysis of ^{14}C ($t_{1/2}$ = 5.7 kyr, isotopic abundance = 1×10^{-12}) can be used for archaeological dating [1]. Since then, two well established methods, low-level counting [2] and accelerator mass spectrometry [3], have been used to analyze many other trace-isotopes at about the parts-per-trillion level and to extract valuable information encoded in the production, transport, and decay processes of these isotopes. The impact of ultrasensitive trace-isotope analysis has reached a wide range of scientific and technological fields.

We have recently developed a new method, Atom Trap Trace Analysis (ATTA) [4], and utilized it to analyze two rare krypton isotopes, ^{81}Kr and ^{85}Kr, with isotopic abundances near the parts-per-trillion level. This new method promises to enhance the capabilities and expand the applications of ultrasensitive trace-isotope analysis.

In this paper, we will first describe the motivation of analyzing ^{81}Kr and ^{85}Kr. We will then survey the existing techniques, describe ATTA, and discuss some of the potential applications of ATTA.

RARE KRYPTON ISOTOPES

Krypton gas constitutes 1 ppm of the earth's atmosphere in fractional volume. It has six stable isotopes, ^{78}Kr(isotopic abundance = 0.35%), ^{80}Kr(2.25%), ^{82}Kr(11.6%),

---

* Email: lu@anl.gov   URL: www-mep.phy.anl.gov/atta/

CP551, *Atomic Physics 17*, edited by E. Arimondo, P. DeNatale, and M. Inguscio
2001 American Institute of Physics 1-56396-982-3

^{83}Kr(11.5%), ^{84}Kr(57%), ^{86}Kr(17.3%), and two long-lived radioactive isotopes, ^{81}Kr and ^{85}Kr (Table 1). There are about 2×10^4 ^{81}Kr atoms and 3×10^5 ^{85}Kr atoms in 1 liter STP of air. There are roughly 10^3 ^{81}Kr atoms in one kilogram of modern water or ice.

^{81}Kr is produced in the upper atmosphere by cosmic-ray induced spallation and neutron activation of stable krypton isotopes [5]. As a result of its long lifetime in the atmosphere, ^{81}Kr is well mixed and evenly distributed over the earth with a homogeneous isotopic abundance. Human activities with nuclear fission have had a negligible effect on the ^{81}Kr concentration, largely because the stable ^{81}Br shields ^{81}Kr from the neutron-rich isotopes that are produced in nuclear fission [6]. These physical and chemical properties make ^{81}Kr an ideal tracer for dating ice and groundwater that are older than 100,000 years [7], which is beyond the range of ^{14}C-dating. The ages of ancient ice in this range are currently determined with the less reliable glaciological models [8].

^{85}Kr is a fission product of ^{235}U and ^{239}Pu. Its present-day concentration in the environment has been released primarily by nuclear fuel reprocessing plants. As a result, its abundance in the atmosphere has increased by six orders of magnitude since the 1950s. It has been used as a general tracer to study air and ocean currents [9], date shallow groundwater, and monitor nuclear-fuel reprocessing activities [10]. Due to its fast mobility, it may be used as a leak sensor to check the seals of nuclear fuel cells and nuclear waste containers.

Noble gas tracers in general have the advantages that they can be chemically separated from large amounts of raw samples, and that their transport processes in environment are easy to understand.

Table 1. The properties of Kr-81 and Kr-85.

Isotope	Half-life (year)	Atmospheric Isotopic abundance	Applications
Kr-81	2.3×10^5	$(5.9\pm0.6)\times10^{-13}$, LLC [5] $(4.5\pm0.3)\times10^{-13}$, LLC [11] $(5.3\pm1.2)\times10^{-13}$, AMS [6]	Geological dating of polar ice and groundwater; Detecting solar neutrinos via ^{81}Br(v_e, e)^{81}Kr.
Kr-85	10.8	$\sim10^{-11}$, LLC [9]	Monitor nuclear-fuel reprocessing activities; Short-term tracer for environmental studies.

EXISTING TECHNIQUES

Here we briefly review the existing techniques with an emphasis on the analysis of the rare krypton isotopes.

Low-Level Decay Counting (LLC)

Long-lived trace-isotopes decay in various ways including α-decay, β-decay, or e-capture. A single decay event releases energy in the range of 10^4–10^7 eV, and can be readily detected with a scintillation counter or a proportional counter with high efficiency (>50%). The overall detection efficiency is usually limited by the short counting time (t_c) compared with the half-life of isotopes ($t_{1/2}$) since the fraction of nuclei decayed during the counting time, $f \approx \ln2 \times (t_c / t_{1/2})$ when $t_c \ll t_{1/2}$. For example, in one week of counting, factors of approximately 10^{-3} of ^{85}Kr, 10^{-6} of ^{14}C, or 10^{-7} of ^{81}Kr in the initial sample would decay. The shorter the half-life, the more efficient this method is.

LLC is often carried out in a specially designed underground laboratory in order to avoid background due to cosmic-rays and the radioactivity present in common materials. Environmental samples often contain other radioactive isotopes, which can be reduced by chemical purification or, in the case of short-lived impurities such as ^{222}Rn ($t_{1/2} = 3.8$ days), by waiting.

LLC is used to analyze ^{85}Kr [9]. It was also used in the first observation of atmospheric ^{81}Kr [5], but this is no longer possible because in today's atmosphere the decay activity of ^{85}Kr is 10^5 times that of ^{81}Kr. Pre-nuclear-age samples are also affected, and are now extremely difficult to analyze with LLC due to the inevitable small contamination of modern krypton during the sampling and preparation stages.

Mass spectrometry (MS)

Atom counting has a number of advantages over decay counting. The efficiency and speed of atom counting is not fundamentally limited by the long half-lives of isotopes, nor is it affected by radioactive backgrounds in the environment or in samples.

The most popular atom-counting method is mass spectrometry, which separates and detects individual ions of a chosen mass. However, the selectivity of this method is limited by interference from isobars, i.e., atoms of other elements or molecules with the same mass number, which are present in trace amounts even after the most careful chemical purification. Conventional mass spectrometry in general has a detection limit at an isotopic abundance of 10^{-9} [12], and is not suitable for analyses at the PPT level.

Accelerator mass spectrometry (AMS)

Isobar contamination can be eliminated in some cases by performing mass spectrometry with a high energy (~MeV) beam from an accelerator [13-15]. First, molecular isobars can be eliminated by passing the accelerated beam through a thin foil where molecules disintegrate. Second, some atomic isobars can be eliminated by exploiting the stability property of negative ions that are used in the first acceleration stage of a tandem accelerator. For example, ^{14}N$^-$, the only abundant isobar of ^{14}C$^-$, is not stable, and consequently not accelerated.

The advantages of atom counting are indeed realized with AMS, which has replaced LLC as the standard method of ^{14}C-dating. Furthermore, AMS has opened up new applications with other trace-isotopes whose half-lives are too long to be counted with LLC. The commonly utilized isotopes in AMS are ^{10}Be, ^{14}C, ^{26}Al, ^{36}Cl, ^{41}Ca, and ^{129}I [3]. The AMS community has grown steadily since the late 1970s, even at a cost of several million U.S. dollars for a typical AMS setup. As of 1998, there were about 40 dedicated AMS facilities around the world, and more facilities where AMS is performed on a routine basis [16].

369

Krypton isotopes cannot be analyzed at a standard AMS facility that uses a tandem accelerator because krypton negative ions are unstable. A new approach has been developed by P. Collon et. al. [6], in which an ECR source is used to produce positive krypton ions, and a GeV-scale cyclotron (K1200, MSU) is used to produce a fully stripped krypton ion beam. Once fully stripped, $^{81}Kr^{+36}$ can be cleanly separated from its abundant isobar $^{81}Br^{+35}$ (Z=35). They have thus realized radiokrypton dating and determined the ages of groundwater, ranging from 200 to 400 kyr, at several sites in the Great Artesian Basin in Australia [17]. In a typical run, 16 tons of groundwater were processed to extract 0.4 cm^3 STP of krypton gas, and resulted in 60-100 ^{81}Kr counts with a detection efficiency of $\sim 1 \times 10^{-5}$. Because old ice is much more difficult to extract than old groundwater in similar quantities, an improved efficiency, at 10^{-3} or higher, is needed before dating ice can be realized.

Resonance ionization mass spectrometry (RIMS)

An alternative method of reducing isobar contamination is to use resonant photons to selectively ionize the element of choice [18, 19]. Figure 1a shows a three-step ionization scheme where the first two steps are resonant excitations that select an element and, in some cases, an isotope, followed by a third non-resonant ionization step [20]. Additional steps of resonant excitation can be added to enhance the selectivity. The combination of isobar selection by resonance ionization and isotope selection by mass spectrometry would, in principle, enable RIMS to reach a selectivity well below the PPT level. In practice, however, complications such as thermal or collisional ionization limit both sensitivity and selectivity.

G.S. Hurst and coworkers have counted ^{81}Kr atoms using the excitation scheme illustrated in figure 1a [20]. Krypton is a difficult case because the first excitation requires a laser of 116.5 nm wavelength. In their work, ^{81}Kr had to be pre-enriched three times with a mass spectrometer in order to reach an isotopic abundance of $\sim 10^{-3}$ before atom-counting was performed. The whole process, including the enrichment cycles, has a total detection efficiency of >50% [20]. This scheme, although extremely efficient, involves a multi-step operation that is difficult to implement in practical applications.

RIMS is successfully implemented in cases where cw narrow-bandwidth lasers are available to excite atoms. For example, K. Wendt et. al. have analyzed ^{90}Sr, a nuclear fission product, in dust particles collected in Munich after the Chernobyl accident [21]. Applying collinear laser spectroscopy to a fast atomic beam, they have reached an isotopic selectivity (Sr/^{90}Sr) of $\sim 10^{11}$ with a detection limit of 3×10^6 atoms. A simplified version [22], in which laser spectroscopy is applied to a thermal atomic beam, has demonstrated an isotopic selectivity (Sr/^{90}Sr) of 3×10^9 with a detection limit of 1×10^4 atoms. Work aimed at analyzing ^{41}Ca at a higher selectivity is in progress [23, 24].

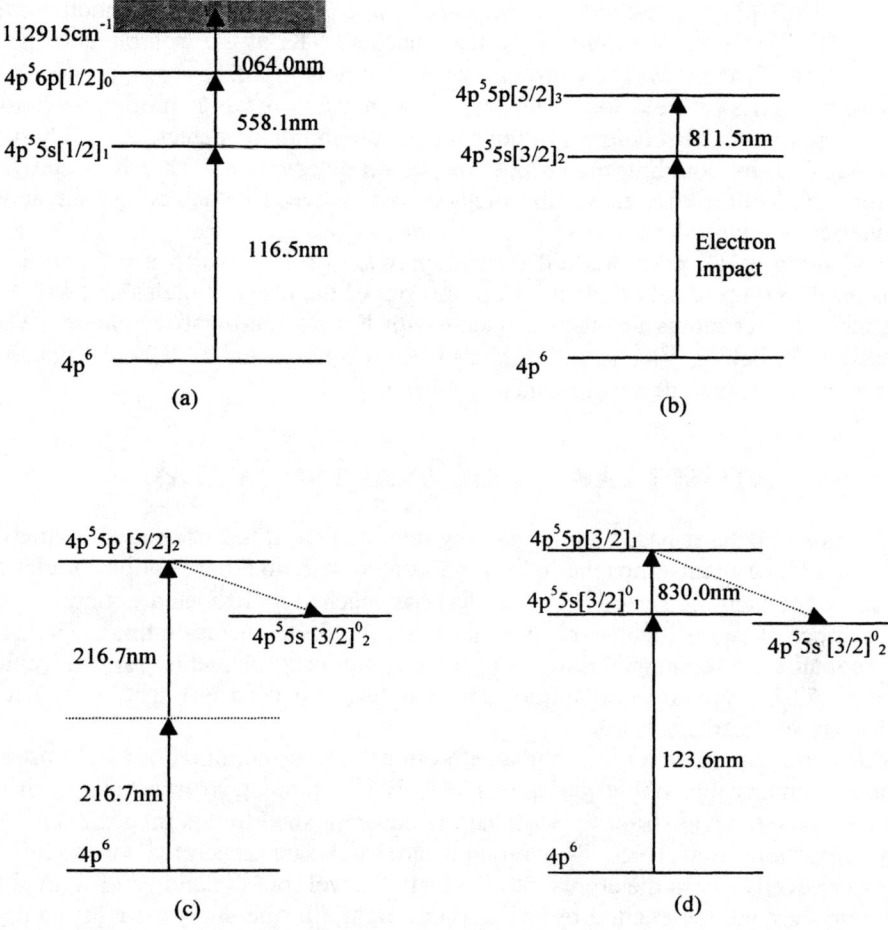

Figure 1. Krypton energy diagrams. (a) Excitation scheme used in the resonance ionization spectroscopy of krypton; (b) Excitation scheme used in the laser trapping of krypton; (c) Populating the metastable level via a non-resonant UV+UV excitation; (d) Populating the metastable level via a resonant VUV+IR excitation.

Photon-burst mass spectrometry (PBMS)

A single atom can also be detected by observing its fluorescence burst in a resonant laser beam [25, 26]. By detecting multiple photons in coincidence during the short transit time, both the detector dark counts and the noise photon-counts due to light scattered off walls can be suppressed. Furthermore, multiple photon detection also enhances isotopic selectivity.

Using PBMS [27], a method that combines mass spectrometry and photon-burst detection, W. M. Fairbank Jr. and coworkers detected ^{85}Kr at the isotopic abundance of 6×10^{-9} [28]. In their work, metastable krypton atoms in a fast beam, produced by neutralizing a mass-selected ion beam, are counted when passing through a photon-burst detection region that consists of ten avalanche-photodiode detectors. They have pointed out that, by doubling the number of photon detectors, and thereby improving the photon collection efficiency, this method may succeed in detecting ^{85}Kr at the atmospheric abundance level ($\sim 10^{-11}$).

B.D. Cannon et. al. proposed a different approach [29], in which a resonant laser beam is used to transversely deflect ^{85}Kr atoms out of the primary metastable krypton. The separated ^{85}Kr atoms are then detected with the photon-burst technique. They succeeded in enriching ^{85}Kr in the deflected beam by a factor of 1.2×10^{4}, which is limited by non-resonant deflection due to collisions [30].

ATOM TRAP TRACE ANALYSIS (ATTA)

ATTA is a new laser-based atom-counting method [4]. It has been used to analyze both ^{81}Kr and ^{85}Kr in an atmospheric krypton sample with no other isotope enrichment processes. The isotopic selectivity (Kr/^{81}Kr) has reached 1×10^{13}, and is only limited by the number of atoms it can sort through during the finite operation time. The atom counts contain no contamination from other isotopes, elements, or molecules. Therefore, ATTA can tolerate impure gas samples, and does not require a special operation environment.

Our design is based on a type of magneto-optical trap system that had been used to trap various metastable noble gas atoms [31, 32]. Trapping krypton atoms in the $5s[^{3}/_{2}]_{2}$ metastable level (lifetime ≈ 40 sec) is accomplished by exciting the $5s[^{3}/_{2}]_{2}$ - $5p[^{5}/_{2}]_{3}$ transition (Fig. 1b). Two repump sidebands are generated via additional AOMs to optically pump the atoms into the F=13/2 level for ^{85}Kr and F=11/2 level for ^{81}Kr where they can be excited by the trapping light. In the analysis, a krypton gas sample is injected into the system through a discharge region, where about 1×10^{-4} of the atoms are excited into the $5s[^{3}/_{2}]_{2}$ level via electron impact excitation. The thermal (300°C) atoms are then transversely cooled, decelerated with the Zeeman slowing technique [33, 34], and loaded into a magneto-optical trap (MOT) [35]. Atoms remain trapped for an average of 1.8 sec as the vacuum is maintained at 2×10^{-8} Torr. This trap system can capture the abundant ^{83}Kr atoms at the rate of 2×10^{8} sec^{-1}. The ratio of the capture rate to the injection rate gives a total capture efficiency of 1×10^{-7}.

With expected capture rates between 10^{-3} sec^{-1} and 10^{-2} sec^{-1} for the rare krypton isotopes, the system must be able to detect a single atom in the trap [36]. In the trap, a single atom scatters resonant photons at a rate of $\sim 10^{7}$ sec^{-1}, of which 1% are collected, spatially filtered to reduce background light, and then focused onto an avalanche photodiode with a photon counting efficiency of 25%. In order to achieve a high capture efficiency and a clean single-atom signal, the setup is switched at 2 Hz between the different parameters optimized for capture and for atom counting. The resulting fluorescence signal of a single atom is 16 kcps (kilo-counts per second) while the background level is 3.4 kcps (Fig. 2).

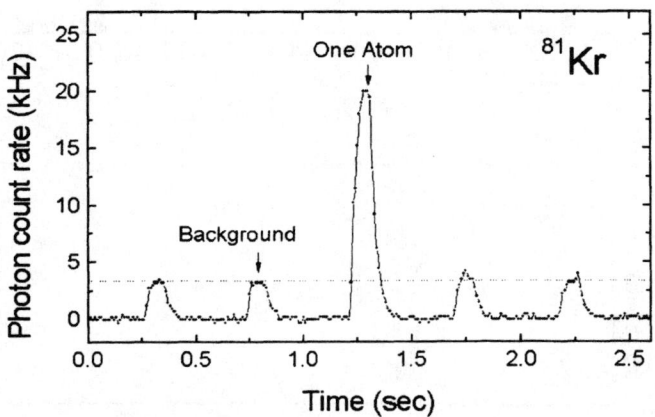

Figure 2. Signal of a single trapped ^{81}Kr atom. The photon counter is only open during the detection phase. Single atom signal ≈ 1600 photon counts, background ≈ 340 photon counts.

We have trapped and counted ^{85}Kr and ^{81}Kr atoms from a natural krypton gas sample. The frequency settings of the trapping laser and the two sidebands are in good agreement with previous spectroscopic measurements obtained using enriched ^{85}Kr gas and enriched ^{81}Kr gas [37]. We have also mapped the atom capture rates versus laser frequency (Fig. 3). Furthermore, repeated tests were performed under conditions in which a ^{85}Kr (^{81}Kr) trap should not work, such as turning off repump sidebands and tuning the laser frequency above resonance. These tests always yielded zero atom counts, and showed that the recorded counts are solely due to laser-trapped ^{85}Kr (^{81}Kr) atoms.

Previous efforts to develop a laser-based technique have encountered serious problems as a result of contamination from nearby abundant isotopes or isobars. ATTA is immune from the contamination for several reasons: fluorescence is only collected in a small region (φ0.5 mm) around the trap center; a trapped atom is cooled to a speed below 1 m/s so that its laser induced fluorescence is virtually Doppler-free; the long observation time (>100 ms) allows the atom to be unambiguously identified (S/N ≈ 40); and trapping allows the temporal separation of capture and detection so that both capture efficiency and detection sensitivity can be optimized. Our design also provides additional features, such as chopping off the atomic beam before detecting the trapped atom

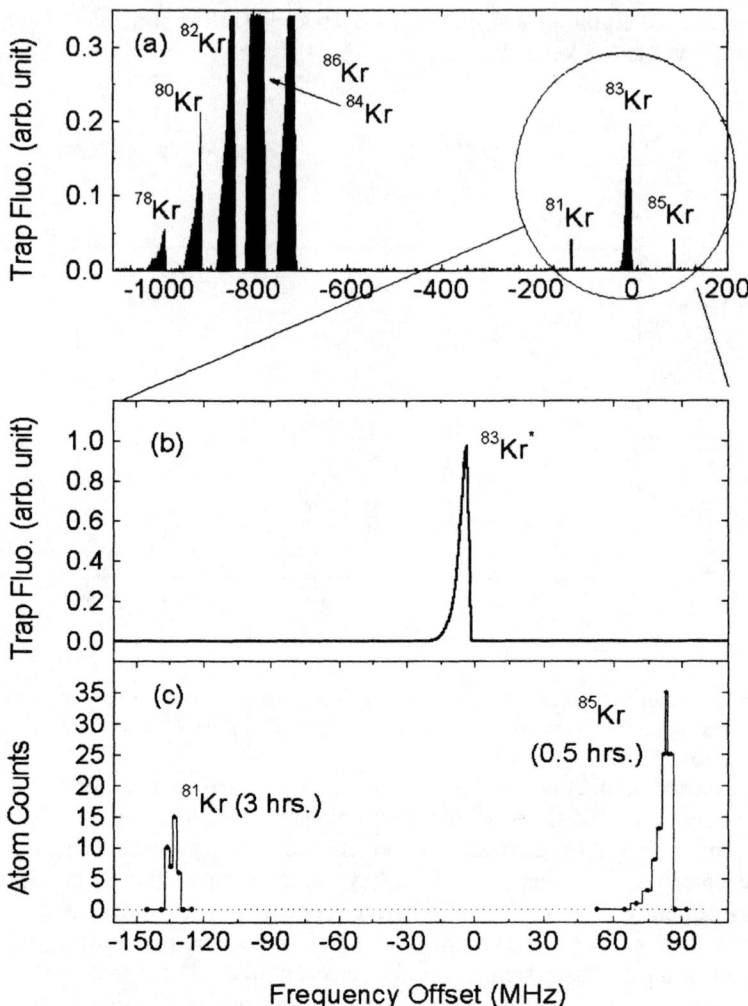

Figure 3. (a) Fluorescence of trapped krypton atoms. Dark bands are the signal of stable isotopes measured with a low-gain photo-diode detector. Line markers mark the positions of the two rare isotopes. (b) Fluorescence of trapped ^{83}Kr atoms versus laser frequency. (c) Number of ^{81}Kr and ^{85}Kr atoms counted versus laser frequency. Each data point represents the number of ^{81}Kr atoms counted in 3 hours, and ^{85}Kr atoms counted in 0.5 hours.

The capture rate of our system depends on the discharge current, laser power, as well as optical alignment. At one particular setting, we measured capture rates of

^{83}Kr, ^{85}Kr, and ^{81}Kr, which were $(1.5\pm0.3)\times10^8$ sec^{-1}, $(1.9\pm0.3)\times10^{-2}$ sec^{-1}, and $(1.3\pm0.4)\times10^{-3}$ sec^{-1} respectively. If we assume the same detection efficiency for all three isotopes, then we get isotopic abundances of $(1.5\pm0.4)\times10^{-11}$ for ^{85}Kr and $(1.0\pm0.4)\times10^{-12}$ for ^{81}Kr, which are in good agreement with previous measurements performed using other methods [5, 6, 11]. The capture efficiencies can be calibrated with enriched samples of known isotopic abundance to correct for any isotope-dependent effects and measure isotopic ratios in unknown samples. For example, in ^{81}Kr-dating, a known amount of ^{85}Kr can be mixed into the sample, thus allowing the ^{81}Kr abundance be extracted by measuring the ratio of ^{81}Kr/^{85}Kr.

Our system has achieved an overall efficiency of 1×10^{-7}. Use of this system to measure the abundance of ^{85}Kr to within 10% would require 2 hours and a krypton sample of 3 cm^3 STP while measurement of ^{81}Kr to within 10% would require 2 days and a sample of 60 cm^3 STP. This limits the current system to atmospheric applications where large samples of gas are available. Improvements, such as a liquid-nitrogen cooled discharge source and recirculation of krypton gas [38], are presently under investigation.

The metastable level of krypton can also be populated via photon excitations (Fig. 1c, 1d). With a suitable laser or lamp, the excitation efficiency could be much higher than the $\sim10^{-4}$ currently achieved with a discharge. Furthermore, without the constraint on gas pressure imposed by a discharge, atoms can be well collimated and cooled, thus further reducing the inefficiencies. It should be noted that the laser excitation discussed here is different from what was used in RIMS . In ATTA, laser excitation is desirable only to boost efficiency, with selection being mainly accomplished by the atom-manipulation process.

Trace Analysis of Cesium Isotopes

By combining a mass separator and a MOT, Dave Vieira and coworkers have recently analyzed two long-lived radioactive isotopes, ^{135}Cs and ^{137}Cs [42]. Applications of this analysis are in environmental science [39] and nonproliferation monitoring. In this work, cesium is ionized, accelerated, mass selected, and implanted into a foil located inside a glass cell. Neutral atoms released from the foil are then captured by a MOT via the vapor-cell loading technique [40] and detected by observing their fluorescence in the trap. With a sample where the isotopic abundances of ^{135}Cs and ^{137}Cs were within the range of 10^{-6}-10^{-4} [41], $\sim10^4$ atoms of each isotope have been trapped and an ion-current normalized isotopic ratio of ^{135}Cs / ^{137}Cs = 1.21 $\pm 0.10^{stat} \pm 0.30^{sys}$ has been measured (a nominal value of 1 is expected) [42]. Since the time of this measurement they have improved the overall efficiency of their system to the 10^{-3} level [41].

Periodic table of long-lived radioisotopes:

Legend: Element → Kr, Mass number 81(5) 85(1), Half-life: 10^i years. ■ Laser manipulation demonstrated.

H 3(1)																	He
Li	Be 10(6)											B	C 14(3)	N	O	F	Ne
Na 22(0)	Mg											Al 26(5)	Si 32(2)	P	S	Cl 36(5)	Ar 39(2) 42(1)
K 40(9)	Ca 41(5)	Sc	Ti 44(1)	V 50(17)	Cr	Mn 53(6)	Fe 55(0) 60(6)	Co 60(0)	Ni 59(4) 63(2)	Cu	Zn	Ga	Ge	As	Se 79(5)	Br	Kr 81(5) 85(1)
Rb 87(10)	Sr 90(2)	Y	Zr 93(6)	Nb 91(2) 92(7) 93(1) 94(4)	Mo 93(3)	Tc 97(6) 98(6) 99(5)	Ru 106(0)	Rh 101(0) 102(0)	Pd 107(6)	Ag 108(2)	Cd 109(0) 113(1)	In 115(14)	Sn 121(1) 126(5)	Sb 125(0)	Te 130(21)	I 129(7)	Xe
Cs 135(6) 137(1)	Ba 133(1)	La 138(11)	Hf 172(0) 174(15) 178(1) 182(6)	Ta 179(0) 180(15)	W	Re 186(5) 187(10)	Os 186(15) 194(0)	Ir 192(1)	Pt 190(11) 193(1)	Au	Hg 194(2)	Tl 204(0)	Pb 202(4) 205(7) 210(1)	Bi 207(1) 208(5) 210(6)	Po 209(2)	At	Rn
Fr	Ra 226(3) 228(0)	Ac 227(1)	Rf	Db	Sg	Bh	Hs	Mt	Uun	Uuu	Uub		Uuq				

Lanthanides / Actinides:

Ce	Pr	Nd 144(15)	Pm 145(1) 146(0) 147(0)	Sm 146(8) 147(11) 148(15) 151(1)	Eu 150(1) 152(1) 154(0) 155(0)	Gd 148(1) 150(6) 152(12)	Tb 157(2) 158(2)	Dy 154(6)	Ho 163(3) 166(3)	Er	Tm 171(0)	Yb	Lu 173(0) 174(0)
Th 228(0) 229(3) 230(4) 232(10)	Pa 231(4)	U *	Np 235(0) 236(5) 237(6)	Pu *	Am 241(2) 242(3) 243(3)	Cm *	Bk 247(3) 248(0)	Cf 248(2) 250(1) 251(2) 252(0)	Es 252(0)	Fm	Md	No	Lr

* U: 232(1), 233(5), 234(5), 235(8), 236(7), 237(6), 238(9).
* Pu: 236(0), 238(1), 239(4), 240(3), 241(1), 242(5), 244(7).
* Cm: 243(1), 244(1), 245(3), 246(3), 247(7), 248(5), 250(3).

Figure 4. A table of long-lived ($t_{1/2} > 1$ year) radioisotopes [72]. The mass numbers and the exponents of the half-lives in years are tabulated. Elements upon which laser manipulation has been experimentally demonstrated are marked.

OTHER POTENTIAL APPLICATIONS

Trace-isotope analysis has become an essential tool in modern science. In basic research, isotope tracers are used to detect solar neutrinos, study cosmic rays [43, 44], study rare $\beta\beta$ decays [45], and search for exotic particles [46]. In environmental sciences, isotope tracers are used to track atmospheric, oceanic, and groundwater currents [9], and to help understand the earth climate. In archaeology [47] and geology [48], various long-lived isotopes are used to determine ages and to help understand the causality of historical events. Isotope tracing is also widely used in biology and medicine [49]. Furthermore, fission isotopes are monitored to assess the contamination of the environment either by the regular operation of a nuclear facility or by a nuclear accident. It is also a means to verify compliance with nuclear nonproliferation treaties.

Laser manipulation of neutral atoms has been demonstrated on an increasing number of elements Based on these demonstrated cases, there are already ~10 long-lived radioisotopes (Fig. 4) that can be analyzed with ATTA.

In the following text, some examples are discussed in more detail.

Solar neutrinos

Solar neutrinos were first detected with a radioisotope tracer [50]. In this experiment, ^{37}Ar ($t_{1/2}$ = 35 days) atoms were produced at the rate of 0.5 atoms/day in a sealed tank containing 615 tons of perchlorethylene via the ^{37}Cl(v_e, e)^{37}Ar reaction (threshold = 0.814 MeV). These ^{37}Ar atoms were recovered with over 90% efficiency and counted in a proportional counter. LLC works very well here due to the short half-life and because the source material is artificially maintained in a clean environment. The measured neutrino flux disagrees with the prediction of the Standard Solar Model [51]. Indeed the measured neutrino flux is 2.56±0.22 SNU while the theoretical prediction is 7.7±1.2 SNU, thus establishing the so-called solar neutrino puzzle. Later experiments as well as theoretical work continue to support this disagreement [52].

An analogous detector that counts ^{81}Kr produced in the ^{81}Br(v_e, e)^{81}Kr reaction (threshold = 0.471 MeV) has been proposed [53, 51]. Due to a lower reaction threshold, this detector would be more sensitive to the ^{7}Be-branch neutrinos than the chlorine detector, and would help separate the contributions from the ^{8}B and ^{7}Be branches. For a detector containing 1000 tons of a Br-chemical, ^{81}Kr is produced at a rate of a few atoms per day. Thus an atom-counter of ~10% efficiency is required.

Besides of measuring the present-day solar neutrino flux, radioisotopes can play a unique role in measuring the long-term average of the solar neutrino flux, which would test the long-term constancy of the solar neutrino flux and the thermal stability of the solar core [54, 55]. In such a geochemical experiment, tracer atoms (for example, ^{98}Tc) are chemically extracted from an ancient mineral deposit (for example, molybdenite, MoS$_2$), and then counted. The number of tracer atoms present in the target mineral is a measure of the solar neutrino flux integrated over the lifetime of the tracer. Suitable geological deposits have been located for the ^{98}Tc and ^{205}Pb experiments. However, an efficient (10^{-3}) and unambiguous method of counting these tracer atoms has to be developed before such experiments can be carried out. MS [56], AMS [57], and RIMS [58] have been attempted, but none has succeeded so far.

Table 2. Proposed radioisotope experiments to measure time-integrated solar neutrino fluxes.

Tracer	Half-life (yr)	Target	Threshold (MeV)	Reference
Ca-41	1.0E5	K-41	2.36	[59]
Kr-81	2.3E5	Br-81	0.471	[53]
Tc-98	4.2E6	Mo-98	1.68	[60]
Pb-205	1.5E7	Tl-205	0.062	[61]

Radiocalcium dating

^{41}Ca ($t_{1/2}$ = 100 kyr) tracer could be used to date bones ranging from 50 thousand to 1 million years old [62, 63]. This period covers an important stage of early human development, and is beyond the reach of ^{14}C-dating. ^{41}Ca is produced by cosmic rays on the ground within 3 meters of depth from surface, therefore its isotopic abundance varies depending on the erosion rate of the ground. Early AMS work indicated that ^{41}Ca/Ca in different bone samples vary in the range of 10^{-15}-10^{-14} [64]. However, the detection limit of AMS, a few×10^{-15}, is so close to the measured level that certain unaccounted systematic errors may contribute to some of the observed variation. Work aimed at improving the detection limit of AMS is in progress [65]. If ATTA can achieve a detection limit of 1×10^{-15} or lower, then dedicated setups can be used to investigate in detail on the feasibility of ^{41}Ca-dating. Furthermore, ^{41}Ca/Ca can be used to determine exposure age of geological samples.

^{39}Ar/^{40}Ar dating

^{40}K/^{40}Ar dating is one of the most commonly used of all radiometric techniques [47]. In a variation of this method, the task of measuring ^{40}K/^{40}Ar is converted to a much easier and more reliable one of measuring ^{39}Ar/^{40}Ar. Here ^{39}Ar is artificially produced by irradiating the sample with neutrons via ^{39}K(n,p)^{39}Ar, and its resulting abundance can be linked to the abundance of ^{40}K with the known reaction rates and the ^{39}K/^{40}K ratio. In this case the isotopic abundance of each argon isotope is quite high (>10%) and mass spectrometry, with a detection limit of ~10^7 atoms, is the standard analysis tool. ATTA may be useful here due to its immunity to contamination [66]. With ATTA, steps of gas purification could be reduced or eliminated, thereby minimizing the contamination by environmental argon. By employing optical excitation to populate argon atoms into the metastable level, ATTA could lower the detection limit significantly and open up new applications of ^{39}Ar / ^{40}Ar dating.

Medical diagnostics

Radioisotope tracers, ^{14}C and ^{3}H in particular, are widely used in biomedical research. With the advancements in AMS, other long-lived isotopes, such as ^{41}Ca and ^{26}Al, are also becoming available to the biomedical field. It is advantageous to use long-lived isotopes as tracers on human subjects due to the low radioactivity of these isotopes.

One particularly interesting proposal that is currently under investigation with AMS is to use ^{41}Ca-tracing to diagnose osteoporosis, a disease commonly found in women after menopause [67, 68]. A patient with osteoporosis loses bone mass, and hence calcium, at an excessive rate. The current standard medical practice is to monitor the bone density with X-ray imaging. Measuring the bone loss rate directly could be more sensitive to the patient's condition, and provide a faster feedback in assessing treatments. In this proposal, a subject of the high-risk group would take a ^{41}Ca pill.

^{41}Ca in tissues would be depleted within a few weeks, but ^{41}Ca in bones would last a lifetime. ^{41}Ca in urine samples, with an isotopic abundance in the range of 10^{-13}-10^{-9}, can then be measured on a regular basis as a monitor of the bone metabolism. Note that AMS has met the technical needs of this proposal. The advantage of using ATTA lies in a significant reduction of cost (greater than a factor of 10).

Other biomedical applications are nutrition studies using isotopes such as ^{22}Na, ^{40}K and ^{41}Ca, and toxicology studies using ^{10}Be and ^{26}Al [49].

Mapping ocean currents

Earth climate depends closely on the global ocean currents, which are mapped and modeled with the help of radioisotope tracers [69]. Radioisotopes are used to determine the "age" of water, defined as the time since the water was near the ocean surface and exchanged gas with the atmosphere. While the analyses of ^{3}H and ^{14}C have been successfully implemented, ^{39}Ar-tracing is of strong interest because it can be used to map ocean currents in the range of 10^{2}-10^{3} years and fill the gap between the dating ranges of ^{3}H and ^{14}C. Not to be confused with ^{39}Ar/^{40}Ar-dating discussed earlier, the ^{39}Ar discussed here is produced by cosmic rays, and its isotopic abundance in the atmosphere is 8.1×10^{-16}. Counting ^{39}Ar atoms in water is difficult because 1 liter of surface water contains only $\sim 1\times10^{4}$ ^{39}Ar atoms. At present, LLC is used to count ^{39}Ar, but requires a large amount ($\sim$1000 liters) of water and several weeks of counting time [70]. AMS has succeeded in detecting ^{39}Ar at the atmospheric level, but its efficiency needs to be improved [71]. A quick and efficient ($\sim$1%) atom-counting method would make a global mapping possible.

ACKNOWLEDGMENTS

This work is supported by the U.S. Department of Energy, Nuclear Physics Division. L.Y. was supported by the Office of Basic Energy Sciences, Division of Chemical Sciences (contract W-31-109-ENG-38). We thank Carl Wieman and Walter Kutschera for stimulating discussions that helped initiate this project. We thank the following people for help us understand the other trace analysis techniques: Philippe Collon (AMS), Bernhard Lehmann (LLC, RIMS), Michael Paul (AMS), and Klaus Wendt (RIMS). We thank the following people for help us understand some potential applications of ATTA: Wick Haxton and Kurt Wolfsberg (solar neutrino); Paul Renne (^{39}Ar/^{40}Ar dating); Stewart Freeman (^{41}Ca-tracing for osteoporosis); Kenny Gross (nuclear waste monitor); Philippe Collon and Roland Hohman (^{39}Ar-dating in oceanography). We thank Gerhard Winkler for many thoughtful suggestions on this paper.

REFERENCES

1) J.R. Arnold and W.F. Libby. Age determinations by radiocarbon content: checks with samples of known age. Science **110**, 678 (1949).
2) Edited by A.J.T. Jull *et al*. *Methods of low-level counting and spectrometry*. (IAEA, Vienna, 1981).
3) *Proc. of the 7th International Conference on AMS*. Nucl. Instr. Meth. **B123** (1997).
4) C.Y. Chen *et al*. Ultrasensitive isotope trace analyses with a magneto-optical trap. Science **286**, 1139 (1999).
5) H.H. Loosli and H. Oeschger. 37Ar and 81Kr in the atmosphere. Earth Planet. Sci. Lett. **7**, 67 (1969).
6) P. Collon *et al*. Measurement of 81Kr in the atmosphere. Nucl. Instr. Meth. **B123**, 122 (1997).
7) H. Oeschger. Accelerator mass spectrometry and ice core research. Nucl. Instr. Meth. **B29**, 196 (1987).
8) J.R. Petit *et al*. Climate and atmospheric history of the past 420,000 years from the Vostok ice core, Antarctica. Nature **399**, 429 (1999).
9) *Isotope of Noble Gases as Tracers in Environmental Studies*. International Atomic Energy Agency (IAEA, Vienna, 1992).
10) F. von Hippel *et al*. Stopping the production of fissile materials for weapons. Sci. Am. **253**, 40 (September, 1985).
11) V.V. Kuzminov *et al*. New measurement of the 81Kr atmospheric abundance. Radiocarbon **22**, 311 (1980).
12) K.G. Heumann *et al*. Recent developments in thermal ionization mass spectrometric techniques for isotope analysis. Analyst **120**, 1291 (1995).
13) R.A. Muller. Radioisotope dating with a cyclotron. Science **196**, 489 (1977).
14) D.E. Nelson *et al*. Carbon-14: direct detection at natural concentrations. Science **198**, 507 (1977).
15) C.L. Bennett et. al. Radiocarbon dating using electrostatic accelerators: negative ions provide the key. Science **198** 508 (1977).
16) P. Collon. A new method to detect cosmogenic 81Kr. Ph.D. Thesis (University of Vienna, Austria 1999).
17) P. Collon *et al*. 81Kr in the Great Artesian Basin, Australia: A new method for dating very old groundwater. To be published in Earth Planet. Sci. Lett. (2000).
18) V.S. Letokhov. *Laser Photoionization Spectroscopy* (Academic Press, Orlando, 1987).
19) G.S. Hurst, M.G. Payne. *Principles and Applications of Resonance Ionisation Spectroscopy* (Adam Hilger, 1988).
20) N. Thonnard *et al*. Resonance ionization spectroscopy and the detection of 81Kr. Nucl. Inst. Meth. **B29**, 398 (1987).
21) K. Wendt *et al*. Rapid trace determination of 89,90Sr in environmental samples by collinear laser resonance ionization mass spectrometry. Radiochimica Acta **79**, 183 (1997).
22) B.A. Bushaw and B.D. Cannon. Diode laser based resonance ionization mass spectrometric measurement of strontium-90. Spectrochimica. Acta. **B52**, 1839 (1997).
23) K. Wendt *et al*. Recent developments in and applications of resonance ionization mass spectrometry. Fresenius J. Anal. Chem. **364**, 471 (1999).
24) B.A. Bushaw *et al*. Diode-laser-based resonance ionization mass spectrometry of the long-lived radionuclide 41Ca with abundance sensitivity <10E-12. Submitted to J. Radioanal. Nucl. Chem.
25) G.W. Greenlees *et al*. High resolution laser spectroscopy with minute samples. Opt. Commun. 23, 236 (1977).
26) V.I. Balykin *et al*. Laser detection of single atom fluorescence. JETP lett. 26, 357 (1977).
27) W.M. Fairbank Jr. Photon burst mass spectrometry. Nucl. Instr. Meth. **B29**, 407 (1987).
28) W.M. Fairbank *et al*. Photon burst mass spectrometry for the measurement of 85Kr at ambient levels. SPIE **3270**, 174 (1998).
29) B.D. Cannon and T.J. Whitaker. A new laser concept for isotopically selective analysis of noble gases. Appl. Phys. **B38**, 57 (1985).
30) G. R. Janik *et al*. Isotopically selective optical deflection of a krypton atomic beam. J. Opt. Soc. Am. **B6**, 1617 (1989).
31) Fujio Shimizu *et al*. A high intensity metastable neon trap. Chem. Phys. **145**, 327 (1990).
32) M. Walhout *et al*. Magneto-optical trapping of metastable xenon: Isotope-shift measurements. Phys. Rev. **A48**, R879 (1993).
33) W. Phillips and H. Metcalf. Laser deceleration of an atomic beam. Phys. Rev. Lett. **48**, 596 (1982).
34) T.E. Thomas *et al*. Slowing atoms with σ⁻ polarized light. *ibid*. **67**, 3483 (1991).
35) E.L. Raab *et al*. Trapping of neutral sodium atoms with radiation pressure. *ibid*. **59**, 2631 (1987).
36) Z. Hu and H.J. Kimble. Observation of a single atom in a magneto-optical trap. Opt. Lett. **19**, 1888 (1994).
37) B.D. Cannon. Hyperfine spectra of the radioactive isotopes 81Kr and 85Kr. Phys. Rev. **A47**, 1148 (1993).
38) B.E. Lehman *et al*. An isotope separator for small noble gas samples. Nucl. Instr. Meth. **B28**, 571 (1987).

39) T. Lee *et al.* First detection of fallout Cs-135 and potential applications of 137Cs/135Cs ratios. Geochimica et Cosmochimica Acta **57**, 3493 (1993).

40) M. Stephens and C. Wieman. High collection efficiency in a laser trap. Phys. Rev. Lett. **72**, 3787 (1994).

41) D.J. Vieira and X. Zhao, Los Alamos National Laboratory. Private communications.

42) D.J. Vieira *et al.* Trapping radioactive atoms for basic and applied research. To be published in Hyperfine Int.

43) S. Baumgartner *et al.* Geomagnetic modulation of the 36Cl flux in the GRIP ice core, Greenland. Nature 279, 1330 (1998).

44) K. Knie *et al.* Indication for supernova produced 60Fe activity on earth. Phys. Rev. Lett. **83**, 18 (1999).

45) T. Kirsten *et al.* Rejection of evidence for nonzero neutrino rest mass from double beta decay. Phys. Rev. Lett. 50, 474 (1983)

46) R.N. Cahn and S.L. Glashow. Chemical signatures for superheavy elementary particles. Science 213, 607 (1981).

47) Edited by R.E. Taylor and M.J. Aitken. *Chronometric dating in archaeology* (Plenum Press, New York, 1997).

48) R. Bowen. *Isotopes in the earth sciences* (Elsevier Applied Science, London, 1988).

49) J.S. Vogel *et al.* Elements in biological AMS. Nucl. Instr. Meth. B123, 241 (1997).

50) R. Davis Jr. *et al.* Solar neutrinos. Annu. Rev. Nucl. Part. Sci. **39**, 467 (1989).

51) J.N. Bahcall and R.K. Ulrich. Solar models, neutrino experiments, and helioseismology, Rev. Mod. Phys. **60**, 297 (1988).

52) W.C. Haxton. The solar neutrino problem. Annu. Rev. Astrophys. **33**, 459 (1995).

53) G.S. Hurst *et al.* Counting the atoms. Phys. Today **33**, 24 (September, 1980).

54) W.A. Fowler. What cooks with solar neutrinos? Nature **238**, 24 (1972).

55) A. Cumming and W.C. Haxton. 3He transport in the sun and the solar neutrino problem. Phys. Rev. Lett. **77**, 4286 (1996).

56) D.J. Rokop *et al.* Mass spectrometry of technetium at the subpicogram level. Anal. Chem. **62**, 1271 (1990).

57) W. Henning *et al.* in *Solar Neutrino and Neutrino Astronomy.* Edited by M.L. Cherry (AIP, New York, 1985).

58) B. Trautmann. Ultratrace analysis of technetium. Radiochimica Acta **63**, 37 (1993).

59) W.C. Haxton and G.A. Cowan. Solar neutrino production of long-lived isotopes and secular variations in the sun. Science **210**, 897 (1980).

60) G.A. Cowan and W.C. Haxton. Solar neutrino production of technetium-97 and technetium-98. Science **216**, 51 (1982).

61) M.S. Freedman *et al.* Solar Neutrino: Proposal for a New Test. Science **193**, 1117 (1976).

62) W. Henning *et al.* Calcium-41 concentration in terrestrial materials: prospects for dating of Pleistocene samples. Science, **236**, 725 (1987).

63) R.E. Taylor. Dating techniques in archaeology and paleoanthropology. Anal. Chem. **59**, 317A (1987).

64) D. Fink, J. Klein and R. Middleton. 41Ca: past, present and future. Nucl. Inst. Meth. **B52**, 572 (1990).

65) M. Paul. Racah Institute of Physics, Hebrew University. Private communication.

66) P. Renne. Berkeley Geochronology Center, University of California at Berkeley. Private communication.

67) D. Elmore *et al.* Calcium-41 as a long-term biological tracer for bone resorption. Nucl. Inst. Meth. **B52**, 531 (1990).

68) S.P.H.T. Freeman *et al.* Human calcium metabolism including bone resorption measured with 41Ca tracer. Nucl. Instr. Meth. B123, 266 (1997).

69) W.S. Broecker. Chaotic climate. Sci. Am. 62 (November, 1995).

70) H.H. Loosli. A dating method with 39Ar. Earth Planet. Sci. Lett. **63**, 51 (1983).

71) W. Kutschera *et al.* Long-lived noble gas radionuclides. Nucl. Inst. Meth. **B92**, 241 (1994).

72) *Nuclides and Isotopes.* (General Electric Co. and KAPL, Inc., San Jose, 1996).

Laser cooling of strontium atoms toward quantum degeneracy

Hidetoshi Katori[*,†], Tetsuya Ido[*], Yoshitomo Isoya[*],
and Makoto Kuwata-Gonokami[*]

[*] *Cooperative Excitation Project, ERATO, Japan Science and Technology Corporation,*
KSP D-842, 3-2-1 Sakado Takatsu-ku, Kawasaki 213-0012, Japan
[†] *Engineering Research Institute, School of Engineering, the University of Tokyo,*
2-11-16 Yayoi, Bunkyo-ku, Tokyo 113-8656, Japan.
E-mail: katori@amo.t.u-tokyo.ac.jp

Abstract. We report on a narrow-line laser cooling and trapping of strontium atoms near quantum degeneracy. Employing a magneto-optical trap (MOT) on the spin-forbidden transition $^1S_0 - {}^3P_1$ at 689 nm, we have laser-cooled an atomic sample down to the photon recoil temperature of 400 nK with a phase space density of 10^{-2}. The atoms were then compressed into a new type of far-off resonance optical dipole trap that was designed to allow simultaneous Doppler cooling, resulting in a phase space density of 10% to that required for quantum degeneracy. It is shown that this value is finally limited by light-assisted collisions occurring in the optical cooling. To reduce these inelastic losses, we applied evaporative cooling and demonstrated a creation of two-dimensional atomic gases.

INTRODUCTION

The attainment of quantum degeneracy in atomic gases has been one of the strong driving forces promoting development of laser cooling. Subrecoil cooling schemes have been successfully demonstrated for dilute atomic gases [1,2], clearly showing that no intrinsic lower limits existed in laser cooling temperatures. However, in higher atom density that is required to attain quantum degeneracy, laser cooling is not completely successful as many experiments have witnessed so far [3,4], because the condition requires atom spacing to be on the order of the laser wavelength λ even for ultracold atoms cooled to the photon recoil momentum of h/λ. In these atom densities, strong atom-atom interactions mediated by the near resonant photons, such as the radiation trapping [3] and the light-assisted collisions [4], take place and drastically disturb the cooling dynamics. Therefore the development of new laser cooling schemes is awaited in these regimes.

Because of these difficulties, Bose-Einstein condensation (BEC) has been first achieved by using evaporative cooling in magnetic traps [5,6]. Nevertheless, optical

CP551, *Atomic Physics 17,* edited by E. Arimondo, P. DeNatale, and M. Inguscio

cooling approach attracts strong interest due to its better handling and quick production of the condensates, which would greatly benefit their future applications. Moreover, optical schemes will provide an alternative or unique access to some atom species toward quantum degeneracy; polarized Fermions [7], some kinds of bosons with unfavorable scattering length [8], or spinless particles, to which thermalization due to elastic collisions, or magnetic trapping cannot be applied. Among them, alkaline earth atoms are of special interest as a new source for BEC and collision studies [9,10] because of their importance in metrology.

Laser cooling of alkaline earth atoms

As a prime candidate for an optical frequency standard in neutral atoms [11], laser cooling of alkaline earth atoms have been studied in Mg, Ca, and Sr [12–15]. Cold atomic samples with temperatures of a few mK are routinely generated by laser cooling on the $^1S_0 - {}^1P_1$ transition, which reduced the Doppler shifts and extended the observation time, allowing a high resolution spectroscopy on the $^1S_0 - {}^3P_1$ clock transition with a line Q-factor above 10^{12} [13–15]. The configuration of two outer electrons introduces a unique complexity in their energy structure, specifically, the spin-singlet ground state and the triplet excited states where radiative decays to the ground state are spin-forbidden, as shown in Fig. 1 for strontium atoms. These spin states favor experiments such as the scalar atom-wave interferometry [16] free from magnetic field perturbations and optical frequency standards.

The following reasons, which may break the existing limitations so far experienced in alkali atoms, make the laser cooling of alkaline earth atoms attractive. 1) A very narrow spin-forbidden transition $^1S_0 - {}^3P_1$ enables Doppler cooling in μK to nK regime [17–20], while reducing radiation trapping effects; 2) Fewer inelastic collision channels are expected in the $^1S_0$ ground state, which is crucial to achieve high-density trapping as well as efficient evaporation; 3) Abundant fermionic isotopes can be laser-cooled as well to explore the physics in degenerate Fermi-systems. An ideal transition linewidth γ for cooling and trapping can be found in strontium; the Doppler cooling limit, $\sim \hbar\gamma$ [17,21], is close to the photon recoil energy $E_R = (\hbar k)^2/2m$, while the acceleration due to photon recoils $\sim \hbar k\gamma/2m$ is an order of magnitude larger than that of gravity, allowing one to hold atoms stable, where m is the atomic mass and $k = 2\pi/\lambda$.

In this paper, we review our experiments on laser cooling of strontium atoms, with the emphasis on how we cope with the two major difficulties known in radiative cooling of atoms near quantum degeneracy. First we demonstrate a narrow-line magneto-optical cooling as an effective way to suppress the radiation trapping occurring at high atom densities [18]. By quickly compressing the atoms into the newly developed optical trap [22], we obtained a nearly degenerate atomic sample with a phase space density of 10 % to that required for quantum degeneracy [23]. We have then shown that the attainable phase space density is stringently limited by the near-resonant-light assisted cold collisions [4]. To reduce these processes, we

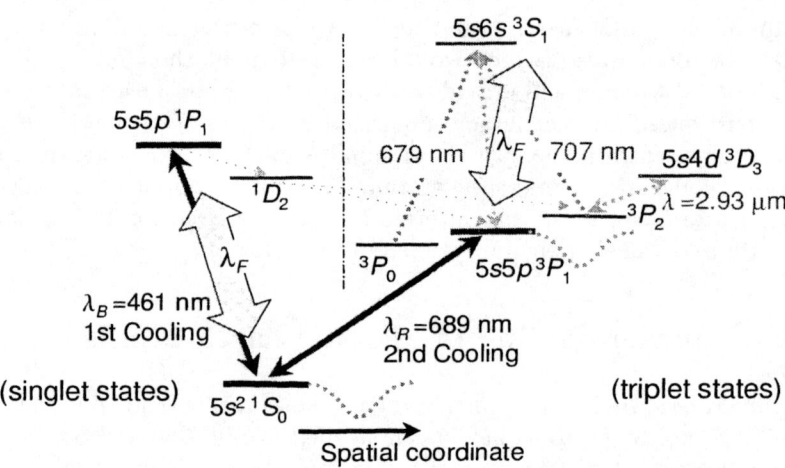

FIGURE 1. Strontium energy levels. The dipole-allowed transition $5s^2 \, ^1S_0 - 5s5p \, ^1P_1$ at $\lambda_B = 461$ nm is used for the preliminary cooling, and the spin-forbidden transition $5s^2 \, ^1S_0 - 5s5p \, ^3P_1$ at $\lambda_R = 689$ nm for the secondary cooling down to the photon recoil temperature. In the final stage, the dipole-trapping laser at $\lambda_F \sim 800$ nm couples the $^1S_0$ and $^3P_1$ states to the upper singlet and triplet states, respectively, to produce same amount of light-shift potentials as shown by the dotted curves. The $^3P_0$ and $^3P_2$ states are the long-lived metastable states.

applied evaporative cooling in an optical lattice and obtained 2D atomic gases. In the last section we discuss experiments with long-lived metastable states.

NARROW-LINE MAGNETO-OPTICAL COOLING AND TRAPPING

Laser cooling essentially relies on the spontaneous emissions for energy dissipations. Thus the reabsorption of these spontaneously emitted photons, which prevents energy dissipation and causes repulsive forces between trapped atoms [3], raises serious obstacles in obtaining cold and dense atomic samples [24]. As a result, in conventional magneto-optical traps (MOTs') [25] or polarization gradient molasses [26] on broad transitions $\gamma \, (\gg \hbar k^2/m)$, the phase space density is typically less than 10^{-5}. One of the strategies against these limitations is to minimize the excitation of atoms once they are cooled and trapped. A dark-SPOT has been proposed to achieve the density close to 10^{12} cm^{-3} with $T \sim 1$ mK [27]. Later, Raman cooling was applied for sodium and cesium atoms [28,29] trapped in far off resonant optical dipole traps (FORTs') to increase the phase space density up to 10^{-3}. Similar value has been obtained using gray molasses [30]. Recently phase space density over 10^{-2} has been demonstrated by using Raman sideband cooling, etc [31]. In these schemes, however, they cannot suppress the reabsorption of the spontaneous photons emitted in the course of the laser cooling, because of their broad cooling transitions. In contrast to these approaches, our strategy is to use a spin-forbidden transition with a linewidth even narrower than the single photon recoil shift $\gamma < \hbar k^2/m$, to substantially reduce the reabsorption process itself.

Experiment on narrow-line MOT

We used two cooling transitions as shown in Fig. 1 to realize narrow-line cooling. Thermal strontium atoms were precooled and trapped by the broad $5s^2\ {}^1S_0 - 5s5p\ {}^1P_1$ transition at $\lambda_B = 461$ nm with the linewidth $\gamma_B/2\pi = 32$ MHz. They were then further cooled down to the photon recoil temperature by the spin-forbidden $5s^2\ {}^1S_0 - 5s5p\ {}^3P_1$ transition at $\lambda_R = 2\pi/k_R = 689$ nm with $\gamma_R/2\pi = 7.6$ kHz.

Apparatus. The precooling light was produced by frequency doubling a diode laser operated at 922 nm. Fundamental laser power of 240 mW was coupled into a power buildup cavity with a $KNbO_3$ crystal to generate 150 mW of blue light. By this laser detuned $8\gamma_B$ below the resonance, strontium atoms from an oven were decelerated in a 30-cm-long tapered solenoid. The slowed atoms were fed into a MOT formed by a pair of anti-Helmholtz coils and three pairs of counter-propagating red-detuned laser beams with opposite circular polarization [25]. This blue-MOT collected 8×10^7 atoms at a temperature of a few mK in a typical loading time of 50 ms. We will address an issue on the optical pumping loss to the ${}^3P_2$ state via the ${}^1D_2$ state [12,19] in the latter section.

A 2nd stage red-MOT was applied for further cooling the atoms, by switching over the laser to excite the intercombination transition at λ_R. This cooling laser was generated by diode lasers: An extended cavity loaded diode laser, electronically stabilized to a high-finesse reference cavity, was used to injection-lock a slave laser to amplify laser power up to 10 mW. The linewidth as well as the frequency stability was estimated to be less than 10 kHz.

Narrow-line MOT. Since the velocity capture range $(v_c \sim \gamma_R/k_R)$ of this narrow transition is even less than the photon recoil velocity of $\hbar k_R/m$, we manipulated the laser linewidth so as to cover the whole Doppler shift $k_R v_0$ ($\sim 2\pi \times 3$ MHz) of the precooled atoms. The laser frequency tuned 1.6 MHz below the ${}^1S_0 - {}^3P_1$ resonance was sinusoidally modulated at 50 kHz with a maximum frequency shift of 1.5 MHz. The timing chart for the cooling is depicted in Fig. 2(a). Figure 2(b) shows the change of ${}^1S_0 - {}^3P_1$ fluorescence intensity from the trap, which is proportional to the number of atoms inside the 3 mm diameter observation region. The ratio of the intensity just after the loading ($t = -90$ ms) and at the end of broadband cooling ($t = 0$ ms) indicates the transfer efficiency ($> 30\%$) of precooled atoms into the red-MOT. In the final stage ($t > 0$ ms), the frequency modulation was turned off to operate at a single frequency: The laser detuning δ and the total laser intensity I for the MOT were adjusted to maximize the final phase space density.

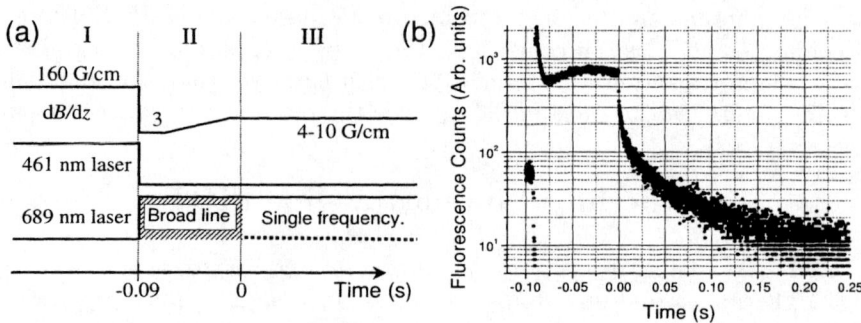

FIGURE 2. (a) Sequence of the experiment; a timing chart for the magnetic field gradient for the MOT, $^1S_0 - {}^1P_1$ cooling laser, and $^1S_0 - {}^3P_1$ cooling laser, respectively. The period I, II, and III correspond to a precooling on broad transition, broadband cooling on narrow transition, and narrow-line cooling, respectively. (b) Change of $^1S_0 - {}^3P_1$ fluorescence from the red-MOT.

Characteristics of narrow-line MOT

Figure 3 shows the temperature of atoms, which was measured at $t = 10$ ms (see Fig. 2(b)) by Time of Flight (TOF) technique. By reducing the laser intensity, the temperature was decreased down to 400 nK. This is due to the reduced power broadening in the cooling transition. However, for intensity $I < 10 \ \mu W/cm^2$, the radiation force could no longer hold atoms against the gravity, so that the number of trapped atoms decreased rapidly. As shown in the inset of Fig. 3, the temperatures were found to be rather insensitive to the laser detunings δ. This flat dependence is explained by the radiation pressure that depends on the Doppler-shift and the position-dependent Zeeman shift [32]. The effective detuning $\delta'_{\pm}$ for atoms moving along a particular axis in a quadrupole field $B(z)$ is written as $\delta'_{\pm} = \delta \pm (k_R v + \mu_R B(z))$, with $\beta = dB(z)/dz = 4 \sim 10$ G/cm, $\mu_R/2\pi = 2.1$ MHz/G, and $\pm$ corresponding to circularly polarized light from right and left with opposite helicities, respectively. Large detuning $|\delta| \gg \gamma_R$, as used in the experiment, simplifies the discussion. Atoms moving toward their turning points rebound at $z_0 \sim \pm\delta/\mu_R\beta$ with nearly zero velocity. They are then accelerated toward the origin until their Doppler shift exceed the transition linewidth, i.e., $k_R v = \gamma_R$, which indicates the MOT temperature $k_B T$ on the order of $m(\gamma_R/k_R)^2 = (\hbar\gamma_R)^2/2E_R \sim k_B \times 300$ nK, in reasonable agreement to our minimum temperature.

The phase space density just before the TOF measurement is determined by $\rho = (N/V)\lambda_{dB}^3$ with $\lambda_{dB} = h/(2\pi m k_B T)^{1/2}$. Obtained maximum phase space density was $\rho = 10^{-2}$ with $T = 550$ nK and $N/V = 6.5 \times 10^{11}$ cm^{-3}. We measured the density-dependent atom temperature of $dT/dn \sim 0.4 \ \mu K/(10^{12}$ cm$^{-3})$ in the range $n = (1 \sim 5) \times 10^{11}$ cm^{-3}. This heating rate may be attributed to the radiation trapping as is discussed in Ref [30]. Our value is an order of magnitude smaller than the best value reported for gray molasses [30], demonstrating the efficient

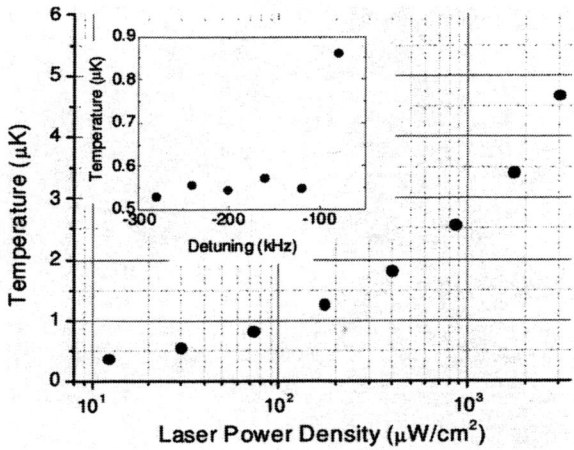

FIGURE 3. The change of the atomic temperatures as a function of total laser intensity I with laser frequency detuned 120 kHz below $^1S_0 - {}^3P_1$ resonance. Inset shows the frequency dependence of the temperature with $I = 30 \ \mu\text{W/cm}^2$.

suppression of the radiation trapping due to the narrow linewidth γ_R.

AN OPTICAL TRAP COMPATIBLE WITH NARROW-LINE COOLING

In the spontaneous-force-trap as described above, light-assisted collisions limit the atom density, as inferred from the rapid non-exponential decay seen in Fig. 2(b) at $t \sim 0$ ms. These inelastic collisions are attributed to the excitations of the attractive quasi-molecular potential by the red-detuned cooling photon, followed by the state-changing collisions or radiative escape process [4]. Therefore it is necessary to transfer the atoms into a conservative atom trap as soon as the cooling process is finished, preferably with an increase of phase space density. The latter can be realized by designing a conservative trap to be compatible with dissipative forces.

Far-off-resonance optical dipole traps (FORTs') [33,34], where the dipole trapping laser is detuned so far from the atomic resonance to reduce photon scattering, have been successfully used to form nearly conservative potentials with tight confinement. In red-detuned dipole traps, since the attractive light-shift potential in the ground state is accompanied by the repulsive light shift in the upper cooling state, the trap potential spatially modulates the atomic resonance frequency. Therefore Doppler cooling was believed to be incompatible with the dipole trapping, as discussed earlier [35]. To overcome this difficulty, alternating the Doppler cooling and dipole-trapping phases by chopping the dipole-trapping laser, which realizes a time-averaged trap potential and dissipation, has often been used to load

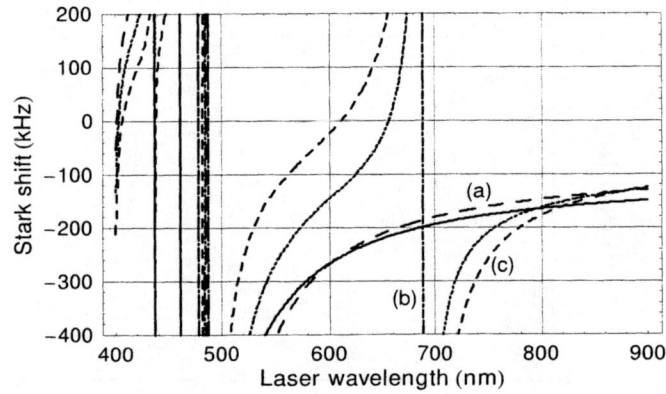

FIGURE 4. Light shifts for $^1S_0$ and $^3P_1$ states are shown by the solid and dashed lines, respectively, as a function of the laser wavelength λ_F, for a linearly polarized light with $I_F = 100$ mW$/\pi(17\ \mu\text{m})^2$. The dashed lines correspond to the light shifts depending on the choice of quantization axis and the magnetic sublevels; the light shift of $m_J = 0$ sublevel by the π polarized light (a), $m_J = \pm 1$ by the $\sigma^+ + \sigma^-$ light (b), and $m_J = \pm 1$ by the π light and $m_J = 0$ by the $\sigma^+ + \sigma^-$ light (c).

atoms into the optical dipole traps [33,36].

Different situation arises if a spin-forbidden transition is used as the cooling transition as in our case. Since the dipole moment of the cooling transition $^1S_0 - {}^3P_1$ is much less than that of allowed transitions, these two states can be independently coupled to the respective singlet and triplet states by applying the light field at λ_F as shown in Fig. 1. Figure 4 shows the Stark shifts [22] for the $5s^2\ ^1S_0$ and the $5s5p\ ^3P_1$ states, plotted by the solid and dashed lines, respectively, as a function of the laser wavelength λ_F for the peak power density of $I_F = 100$ mW$/(\pi a^2)$ with $a = 17\ \mu$m. At $\lambda_F \sim 800$ nm, the Stark shifts for the $^1S_0$ and $^3P_1$ states cross with the negative sign, where we expect the dipole trap to operate in the $^1S_0$ ground state while the atomic resonance frequency remains the same. These Stark shifts allow the Doppler cooling or magneto-optical trapping to work as in the free space. In addition, as λ_F is more than 300 nm away from the $^1S_0 - {}^1P_1$ transition, the photon-scattering rate for atoms in $^1S_0$ is only 10^{-1} s^{-1} for the above parameters, which guarantees a long trap lifetime with negligible heating.

Experiment on 1D-FORT

Ultracold strontium atoms with a few recoil temperatures were prepared by the narrow-line magneto-optical cooling and trapping [18]. Simultaneously, FORT laser beam at $\lambda_F \sim 800$ nm was focused into the MOT (see Fig. 5(a)) to start loading atoms. The atom transfer into the FORT continued for 35 ms in the presence of the

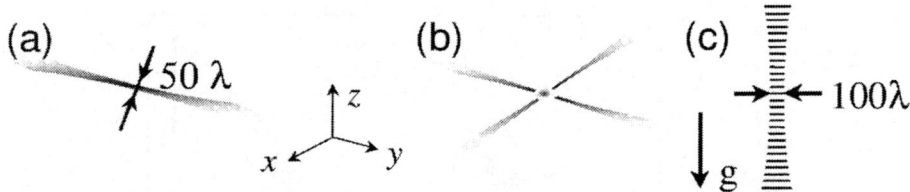

FIGURE 5. Three types of FORTs used in the experiment. (a) 1D-FORT with single tight focused laser beam, (b) crossed FORT, and (c) Lattice-FORT formed by a single standing wave along the vertical direction.

MOT laser. The trapped atoms in the FORT were observed using the absorption imaging technique [5] with a weak 20-μs-long probe laser near-resonant to the $^1S_0 - {}^1P_1$ transition. For a deep potential depth ($\sim 15\mu K$), the loading efficiency of atoms into the FORT was measured to be as high as 80%.

Narrow-line Doppler cooling. An important feature of this novel FORT design, long confinement time without affecting the atomic resonance frequency, can be best demonstrated in testing the Doppler cooling theory on narrow transitions. For $\gamma \leq \hbar k^2/m$, the optimal cooling parameters are predicted to be scaled by the photon recoils; minimum temperature of $k_B T = E_R/2$ is obtained for the laser detuning of $\delta = -1.7\hbar k^2/m$ [21]. This is in contrast with the familiar Doppler cooling theory in the broad line limit, $\gamma \gg \hbar k^2/m$, where they are scaled solely by a linewidth of the transition [37].

We applied a cooling laser on $^1S_0 - {}^3P_1$ for atoms trapped in the 1D-FORT at a temperature of ~ 1.2 μK. After 30-ms-long irradiation, the atom temperature was measured by the TOF image by turning off the trap. The obtained velocity distribution was found to be Gaussian, and was attributed temperatures as in Fig. 6. Although Castin *et al.* [21] predicted a non Gaussian feature of atomic velocity distribution in narrow-line cooling, our results were not the case. This is attributed to our FORT confining potential of $\nu \sim 400$ Hz, by which momentum diffusion is moderated. One can notice the remarkable difference in frequency dependence in Fig. 3 inset and in Fig. 6. This supports the previous discussion on magneto-optical cooling: In the MOT, atoms seek their optimal laser detuning by their motion in the inhomogeneous magnetic field, thus recoil-limited atom temperature can be easily achieved without requiring fine tuning of laser frequency.

Compressing atoms into a crossed FORT

A crossed dipole trap formed by two horizontal beams propagating along the x- and y- axes as depicted in Fig. 5(b) was applied to increase the phase space density. The shape of the trap potential, determined by the combination of the Stark shift potentials and the gravitational potential mgz, is written as

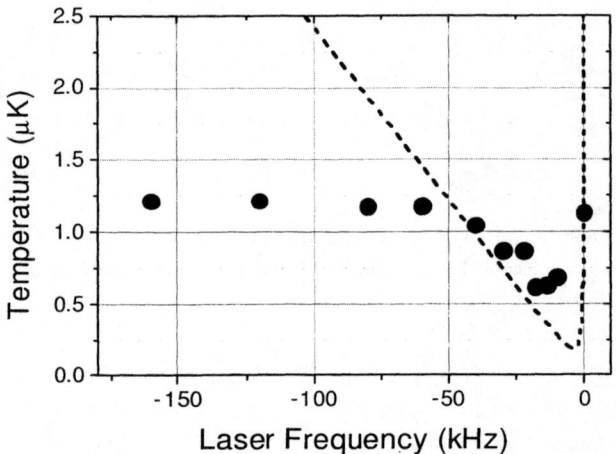

FIGURE 6. Narrow-line Doppler cooling in the FORT. Temperature was measured as a function laser detunings from the $^1S_0 - {}^3P_1$ resonance. Minimum temperature on the order of E_R was obtained for the laser detuning of ~17 kHz in agreement to the theory. Dashed line show the prediction of broad line limit.

$$U(x, y, z) = -U_0\{e^{-(x^2+z^2)/a^2} + e^{-(y^2+z^2)/a^2}\} + mgz, \tag{1}$$

where a is the e^{-1} radius of the laser beam and U_0 is the Stark shift potential given by a single beam. The effective trap depth is determined by the lowest potential barrier in either direction. $u_{xy} = U_0$ provides the potential depth in the xy-plane, and $u_z = U(0, 0, z)$ the depth along the z-axis depending on the competition between gravity and the optical dipole force. For high laser intensity, where the Stark shift potential dominates over the gravitational one and $u_{xy} = U_0 < u_z < 2U_0$ holds, energetic atoms leak along each beam axis in the horizontal plane. Whereas for lower intensity with $u_z < u_{xy} = U_0$, the atoms leak along the z direction.

In order to form the crossed FORT, we applied two laser beams perpendicularly crossed one another at their waists almost in the center of the MOT. These beams had orthogonal linear polarization to avoid interference. Figure 7(a),(b) show the atom transfer process from the narrow-line MOT to this crossed FORT. As shown in the figure, thousand-fold compression in volume has been achieved, while the atom temperature remained constant. However we observed that the atom density was strongly limited by cooling-light assisted collisions as discussed below.

Cold-collision-limited phase space density. The phase space density obtained in the crossed FORT is summarized in Fig. 7(c) as a function of FORT potential depth. The asymmetric error bars shown in the figure is mainly due to the uncertainties in the volume measurements as discussed in Ref. [23]. The phase space density exceeding 0.1 was obtained with slightly increasing tendency

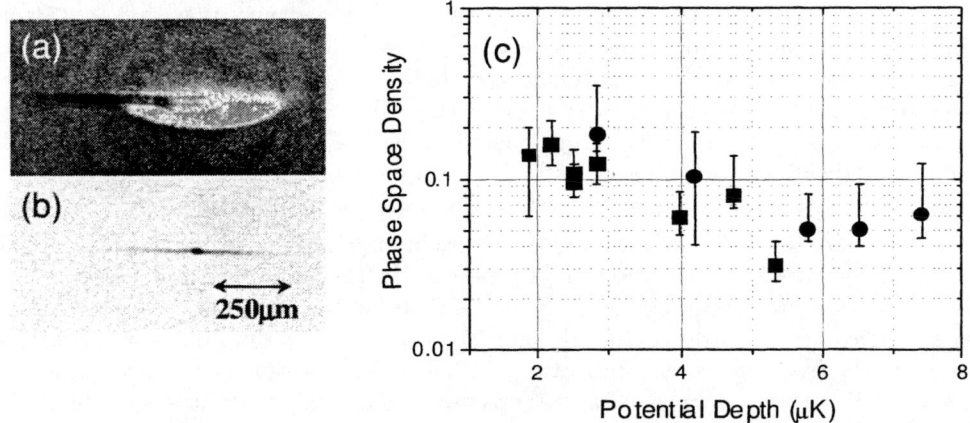

FIGURE 7. Absorption image of atoms transferred from a MOT(a) into a crossed FORT(b), resulting in a factor of thousand compression in volume. The typical e^{-1} radius of the compressed atom cloud was $\sigma_v = 6\mu m$ vertically and $\sigma_h = 12\mu m$ horizontally. (c) Phase space density of atoms trapped in the crossed FORT as a function of the effective FORT barrier height. A phase space density exceeding 0.1 was obtained.

as the FORT depth decreases, and was mainly limited by the light-induced inelastic collisions occurring during atom transfer into the FORT. To examine the influence, we measured the binary loss rate β from the fluorescence decay of the MOT as shown in Fig. 2(b) for $t > 0$. The light-induced collision loss rate was $5 \times 10^{-12} < \beta/(\text{cm}^3\text{s}^{-1}) < 1.5 \times 10^{-11}$, rather insensitive to the MOT laser parameters that were used in the FORT loading experiment.

With the loading flux ϕ into the FORT volume V, the change of atom density n obeys the rate equation, $dn/dt = \phi/V - \beta n^2$. Assuming a constant ϕ in the loading period, the rising time constant is defined as $\tau = 1.5(\beta\phi/V)^{-1/2} = 1.5\,(\beta n_f)^{-1}$, in which the density reaches 90% of the equilibrium density n_f. The measured rising time $\tau = 15$ ms and the loss rate $\beta = 10^{-11}$ cm^3/s, yielded the attainable density $n_f = 10^{13}$ cm^{-3}, which is close to the maximum density we observed in the FORT experiment. Since n_f is proportional to $\sqrt{\phi/(\beta V)}$, this indicates that two orders of enhancement in atom flux is necessary to reach quantum degeneracy.

EVAPORATIVE COOLING IN A LATTICE-FORT

To reduce these inelastic collision losses, we employed an evaporative cooling in a FORT [38] with 1D lattice configuration [8] formed by a vertical standing wave as depicted in Fig 5 (c). This Lattice-FORT configuration produces highly asymmetric trap potential while providing nearly isotropic potential barrier even in the presence of gravity: The axial confinement of atoms along the gravity is within $\lambda/2$, thus the potential energy difference between upper and lower barriers is as small as $mg\lambda/2 \sim k_B \times 40$ nK, which is much less than the typical potential height of $\sim \mu$K. These isotropic potential barriers allow three-dimensional efficient evaporation, in contrast to the crossed FORT discussed above. The strong anisotropy of this lattice potential was exploited to estimate a thermalization time constant of atoms confined in the trap, to design the following evaporation experiment.

Generation of 2D atomic gas. By reducing the FORT depth in a typical time constant of 0.5 s, we observed an increase in the phase space density as well as the drop in atom temperature. Using TOF technique, we have measured highly anisotropic temperatures of 100 nK and 270 nK for the radial and axial direction respectively (see Fig. 8(a)), indicating the vibrational ground-state occupation in the axial direction. In which case, we typically had a total of 10^5 atoms in about 10^2 pancake potentials. The atom temperature in the axial direction was determined by the strong confinement of $\lambda/2$, resulting in the photon recoil energy of E_R, while in the radial direction many vibrational levels were still occupied, as the level spacing was thousand times less than that of the axial direction. Preliminary results are shown in Fig. 8. The quantitative analysis for this process is underway and will be discussed elsewhere. These 2D samples are of interest in view of Kosterlitz-Thouless transition and the control of mean-field interactions [39], etc.

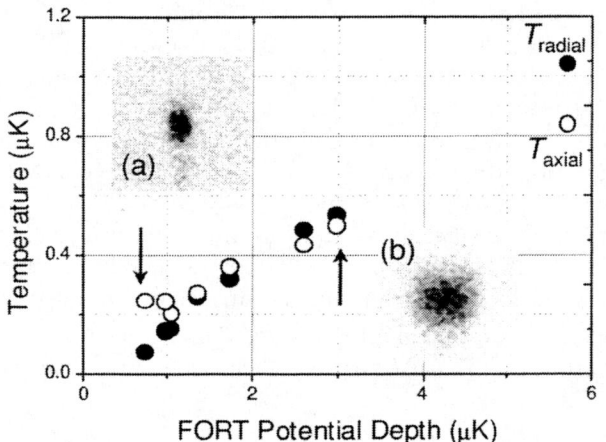

FIGURE 8. Evaporative cooling of atoms in a Lattice-FORT. After loading atoms into the FORT, the FORT laser intensity was gradually decreased in 0.5 s. The atom temperature was measured as a function of the FORT potential depth u_f at the end of the evaporation. Inset shows the atom image after 20-ms-long expansion for $u_f = 0.7~\mu$K (a) and $u_f = 3~\mu$K (b), respectively. For the lower potential depth, the axial temperature reached a constant value, showing the clear anisotropy in temperature.

EXPERIMENTS WITH METASTABLE STATES

Existence of long-lived metastable states is one of the fascinating features in alkaline earth atoms. These states permit ultra-high resolution spectroscopy well below the Hz level, making alkaline earth atoms promising candidates for the future optical standards. Furthermore, because of the predicted very long lifetime [40], $5s5p\ ^3P_0$ and $^3P_2$ states can be used as a lower state for cooling, or a reservoir for atoms as described below.

Shielding atoms from resonant light. Currently the number of trapped atoms is severely limited by the radiation trapping occurring in the broad transition of $^1S_0 - ^1P_1$. To reduce these light-assisted trap losses, we have demonstrated a shielding of atoms from the cooling light by magnetically trapping them in the $5s5p\ ^3P_2$ state. During cooling and trapping on the $^1S_0 - ^1P_1$ transition, precooled atoms were optically pumped to the $5s5p\ ^3P_2$ state via the $5s4d\ ^1D_2$ state in a typical time constant of 20 ms [12]. A part of the atoms pumped into the $^3P_2$ state could be trapped by a quadrupole magnetic field with a field gradient of 100 G/cm that was applied to from the MOT.

The number of these magnetically trapped atoms was estimated by optical pumping them to the $^1S_0$ ground state, by introducing two lasers [19] exciting $^3P_0 - ^3S_1$ and $^3P_2 - ^3S_1$ at 679 nm and 707 nm, respectively. Figure 9 monitors the change of fluorescence from a MOT on the $^1S_0 - ^1P_1$ transition. After operating the MOT for one second, we introduced the pump lasers at $t = 0$ ms. Which increased the fluorescence intensity or the atom number by a factor of 5. The atoms then rapidly

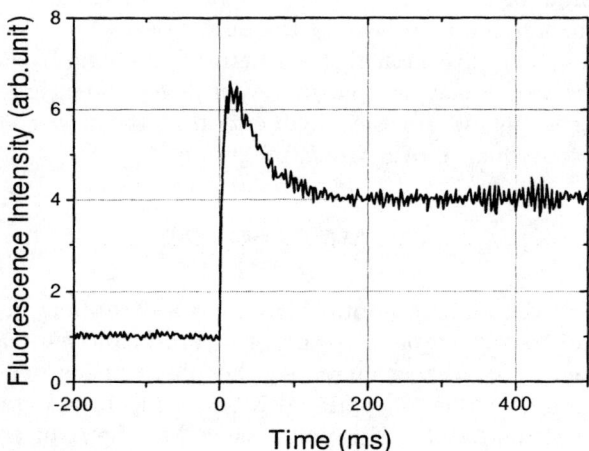

FIGURE 9. Magnetic trapping of precooled atoms in the $5s5p\ ^3P_2$ metastable state. Fluorescence intensity monitored the number of atoms in the MOT on $^1S_0 - ^1P_1$ transition. At $t = 0$ the atoms magnetically trapped in the $^3P_2$ state were pump backed into the $^1S_0$ state, enabling a transient increase of the atom number.

decreased in 100 ms, indicating a strong resonant-light-assisted collisions.

Continuous cold atom generation by state transfer. The above experiment raises the possibility of realizing continuous ultracold atom source. In the existing pulsed atom-laser scheme, several different cooling stages are accomplished successively in time domain to obtain cold atoms. In contrast, strontium allows continuous cooling by transferring the population between metastable states. After 1D deceleration and MOT by the blue transition, cold atoms with a few mK populate the $^3P_2$ metastable state, where a Sisyphus cooling may be applied on the cyclic transition of $^3P_2 - {}^3D_3$ at $\lambda = 2.93$ μm to cool them down to a few recoil energies of ~ 10 nK. Cooled atoms can be coupled to another $5s5p\ ^3P_0$ metastable state by stimulated Raman transition via the $^3S_1$ state, thus scalar atom wave can be continuously extracted into a free space.

Spectroscopy in the optical trap. In neutral atoms, precise spectroscopic measurements are usually accomplished during the free flight or in atomic fountains, which limit the observation time less than 1 second due to gravity. As is clearly demonstrated in the narrow-line Doppler cooling experiment, the Stark shift differences δf in the "clock transition" $^1S_0 - {}^3P_1$ can be adjusted within $\delta f / \delta \lambda_L \sim$ 300 Hz/nm at $\lambda_F \sim 800$ nm because of our FORT design as shown in Fig. 4. This suggests that the spectroscopic measurement in the FORT should permit frequency accuracy well below 1 Hz level, by adjusting the FORT laser wavelength within 10^{-3} nm. Furthermore, low photon scattering rate of 10^{-1} s^{-1} as well as the Lamb-Dicke confinement can be achieved in the lattice configuration. Therefore we can realize a situation similar to the single ion spectroscopy in an ion trap. Marked difference in these two systems is the number of particles. Because of very weak interactions, number of quantum absorbers can be increased for neutral atoms, allowing the significant improvement in the shot noise that is determined by the number of particles. The extension of this scheme to the long lived $^3P_0$ or $^3P_2$ states will be straightforward. Spectroscopic survey of degenerate Fermi-systems will be another intriguing target, where the modification of the upper state lifetime may be expected as a consequence of Fermi suppression [41].

CONCLUSION

We demonstrated the magneto-optical cooling and trapping of strontium atoms down to the photon recoil temperature using a spin-forbidden transition. The use of narrow transition $\gamma < \hbar k^2/m$ increased the phase space density more than 3 orders of magnitude over the conventional MOT using broad transitions, because of the suppression of the radiation trapping as well as the reduced Doppler cooling limit. By compressing these atoms into an optical trap, we attained a phase space density of 10% of that required for quantum degeneracy, which is, to the authors' knowledge, the highest ever attained by purely optical means. Owing to the rapid radiative-cooling process, the total cooling time required to obtain such near degenerate atomic gas was as short as a few hundred milli-second. This short production time will open up possibilities in applying the gas sample for precision measurements. We have shown, however, that light-assisted inelastic collisions in the optical cooling finally set a stringent limit on the phase space density. Therefore, in order to increase the density toward the quantum degeneracy, the reduction of the inelastic collisions is crucial. Application of the developed cooling scheme to fermionic ^{87}Sr isotope is straightforward. Assuming experimentally realized trap conditions for ^{88}Sr atoms in the crossed FORT, 10^6 atoms in the harmonic potential with $\nu \sim 400$ Hz, the Fermi temperature is expected at 1.5 μK, which is sufficiently higher than the recoil-limited Doppler-cooling temperature of 400 nK. Thus the Doppler-cooling temperature depending on the particles' quantum statistics could be explored.

ACKNOWLEDGEMENT

We would like to thank M. Daimon for his interest in this work, and T. Mukaiyama, Y. Li and M. Yasuda for their experimental assistance and discussions.

REFERENCES

1. Aspect, A., Arimondo, E., Kaiser, R., Vansteenkiste, N., and Cohen-Tannoudji, C., *Phys. Rev. Lett.* **61**, 826-829 (1988).
2. Kasevich, M. and Chu, S., *Phys. Rev. Lett.* **69**, 1741-1744 (1992).
3. Walker, T., Sesko, D., and Wieman, C., *Phys. Rev. Lett.* **64**, 408-411 (1990).
4. Gallagher, A., and Pritchard, D. E., *Phys. Rev. Lett.* **63**, 957-960 (1989); Weiner, J., Bagnato, V. S., Zilio, S., and Julienne, P. S., *Rev. Mod. Phys.* **71**, 1-85 (1999).
5. Anderson, M., *et al.*, *Science* **269**, 198-201 (1995); Davis, K. B., *et al.*, *Phys. Rev. Lett.* **75**, 3969-3972 (1995); Bradley, C. C., Sackett, C. A., and Hulet, R. G., *Phys. Rev. Lett.* **78**, 985-988 (1997).
6. Fried, D.G., Killian, T. C., Willmann, L., Landhuis, D., Moss, S.C., Kleppner, D. and Greytak, T. J., *Phys. Rev. Lett.* **81**, 3811-3814 (1998).
7. DeMarco, B. and Jin, D. S., *Science*, **285**, 1703-1706 (1999).
8. Vuletić, V., Kerman, A. J., Chin, C., and Chu, S., *Phys. Rev. Lett.* **82**, 1406-1409 (1999).
9. Dinneen, T. P., Vogel, K. R., Arimondo, E., Hall, J. L., and Gallagher, A., *Phys. Rev. A* **59**, 1216-1222 (1999).
10. Machholm, M., Julienne, P. S. and Suominen, K.-A., *Phys. Rev. A* **59**, R4113-R4116 (1999).
11. Hall, J. L., Zhu, M., and Buch, P., *J. Opt. Soc. Am. B* **11**, 2194-2205 (1989).
12. Kurosu, T., and Shimizu, F., *Jpn. J. Appl. Phys.* **29**, L2127-L2130 (1990).
13. Schnatz, H., Lipphardt, B., Helmcke, J., Riehle, F., and Zinner, G., *Phys. Rev. Lett.* **76**, 18-21 (1996).

14. Ruschewitz, F. Peng, J., Hinderthür, H., Schaffrath, N. Sengstock, K., and Ertmer, W., *Phys. Rev. lett.* **80**, 3173-3176 (1998).

15. Zibrov, A. S., Fox, R. W., Ellingsen, R., Weimer, C. S., Velichansky, V. L., Tino, G. M., and Hollberg, L., *Appl. Phys. B* **59**, 327-331 (1994).

16. Riehle, F. *et al.*, *Phys. Rev. Lett.* **67**, 177-180 (1991).

17. Wallis H. and Ertmer W., *J. Opt. Soc. Am. B* **11**, 2211-2219 (1989).

18. Katori, H., Ido, T., Isoya, Y., and K.-Gonokami, M., *Phys. Rev. Lett.* **82**, 1116-1119 (1999).

19. Vogel, K. R., Dinneen, T. P., Gallagher, A., and Hall, J. L., *Proc. SPIE Int. Soc. Opt. Eng.* **3270**, 77-80 (1998).

20. Kuwamoto, T., Honda, K., Takahashi, Y., and Yabuzaki, T., *Phys. Rev. A* **60**, R745-R748 (1999).

21. Castin, Y., Wallis, H., and Dalibard, J., *J. Opt. Soc. Am. B* **11**, 2046 (1989).

22. Katori, H., Ido, T., and K.-Gonokami, M., *J. Phys. Soc. Jpn.* **68**, 2479-2482 (1999).

23. Ido, T., Isoya, Y., and Katori, H., *Phys. Rev. A*, **61**, 061403(R)- 061406(R) (2000).

24. Castin, Y., Cirac, J. I., and Lewenstein, M., *Phys. Rev. Lett.* **80**, 5305-5308 (1998).

25. Raab, E. L., Prentiss, M., Cable, A., Chu, S., and Pritchard, D., *Phys. Rev. Lett.* **59**, 2631-2634 (1987).

26. Drewsen, M., *et al.*, *Appl. Phys. B* **59**, 283-298 (1994).

27. Ketterle, W., Davis, K. B., Joffe, M. A., Martin, A., and Pritchard, D. E., *Phys. Rev. Lett.* **70**, 2253-2256 (1993).

28. Lee, H. J., Adams, C. S., Kasevich, M., and Chu, S., *Phys. Rev. Lett.* **76**, 2658-2661 (1996).

29. Reichel, J., Bardou, F., Ben Dahan, M., Peik, E., Rand, S., Salomon, C., and Cohen-Tannoudji, C., *Phys. Rev. Lett.* **75**, 4575-4578 (1995).

30. Boiron, D., Michaud, A., Fournier, J. M., Simard, L., Sprenger, M., Grynberg, G., and Salomon, C., *Phys. Rev. A* **57**, R4106-R4109 (1998).

31. Hamann, S. E., Haycock, D. L., Klose, G., Pax, P. H., Deutsch, I. H., and Jessen, P. S., *Phys. Rev. Lett.* **80**, 4149-4152 (1998); DePue, M. T., McCormick, C., Winoto, S. L., Oliver, S., and Weiss, D. S., *Phys. Rev. Lett.* **82**, 2262-2265 (1999); Vuletić, V., Chin, C., Kerman, A. J., and Chu, S., *Phys. Rev. Lett.* **81**, 5768-5771 (1998).

32. Metcalf, H., *J. Opt. Soc. Am. B* **11**, 2206-2210 (1989).

33. Miller, J. D., Cline, R. A., and Heinzen, D. J., *Phys. Rev. A***47**, R4567-R4570 (1993).

34. Takekoshi, T., Yeh, J. R., and Knize, R. J., *Opt. Commun.* **114**, 421-424 (1995).

35. Gordon, J. P. and Ashkin A., *Phys. Rev. A* **21**, 1606-1617 (1980).

36. Chu, S. , Bjorkholm, J. E., Ashkin, A., and Cable, A., *Phys. Rev. Lett.* **57**, 314-317 (1986).

37. Stenholm, S., *Rev. Mod. Phys.* **58**, 699-739 (1986).

38. Adams, C. S., Lee, H. J., Davidson, N., Kasevich, M., and Chu, S., *Phys. Rev. Lett.* **74**, 3577-3580 (1995).

39. Petrov, D. S., Holtzmann, M., and Shlyapnikov, G. V., *Phys. Rev. Lett.* **84**, 2551-2554 (2000).

40. Garstang, R. H., *The Astrophysical Journal* **148**, 579-584 (1967).

41. Busch, Th., Anglin, J. R., Cirac, J. I., and Zoller, P., *Europhys. Lett.* **44**, 755-760 (1998).

Bouncing Atoms: A Coherent Nano-probe

N. Westbrook, V. Savalli, V. Josse, L. Cognet, P. Featonby, D. Stevens,
C.I. Westbrook and A. Aspect

Laboratoire Charles Fabry de l'Institut d'Optique
B.P. 147, 91403 Orsay CEDEX France

Abstract. We describe experiments with atoms reflecting from atomic mirrors. We show that the Zeeman structure of the atoms permits the realization of an nano scale interferometer which can be used to probe the van der Waals interaction between an atom an a dielectric surface. We also discuss preliminary results on an experiment to observe sub-angstrom surface roughness using stimulated Raman transitions to achieve velocity selection of 0.3 times the single photon recoil velocity. This resolution corresponds to a 1.5 mrad maximum deviation in the effective reflecting surface.

INTRODUCTION

For several years now, many workers have been studying atom mirrors. Experimental geometries have included both flat and curved surfaces, fiber-like waveguides as well as more complex structures [1]. The evanescent wave mirror, based on the force between a laser field and an induced atomic dipole, is in principle quite simple, but the behavior of real atoms in real evanescent waves has yielded several surprises. The same can be said of the magnetic mirror, which is based on the interaction of a permanent magnetic moment with a periodically magnetized structure. Our group in Orsay has spent considerable effort in elucidating this behavior and we will summarize some of our recent results here.

We will describe two types of experiment. In the first we use the Zeeman substructure of the atoms to create the analog of a Michelson atom interferometer which is entirely confined within the decay length of an evanescent wave, less than 100 nm. In the second experiment we discuss measurements of the effect of the mirror's roughness on the observed behavior of the atoms. The common thread in these experiments is that the atoms are sensitive to features on the order of nanometers. In the first experiment we will show that we are sensitive to the effects of the van der Waals interaction, in the second the atoms are sensitive to surface roughnesses of order 0.1 nm. A second common feature of these experiments is that they both rely on stimulated Raman transitions.

A NANO-SCALE ATOM INTERFEROMETER

This idea and the experiment were first described in Refs. [2] and [3]. In that work we emphasized the relation of our work to the problem of atom diffraction from an

CP551, *Atomic Physics 17*, edited by E. Arimondo, P. DeNatale, and M. Inguscio
© 2001 American Institute of Physics 1-56396-982-3/01/$18.00

evanescent wave mirror at grazing incidence. Here we will discuss the experiment from a different point of view, that of a stimulated Raman transition between two Zeeman sub-levels in an atom.

To understand the effect, consider an evanescent wave resulting from a TM polarized incident optical field. We choose the quantization axis to be the direction of the magnetic field of the optical field (the y-axis in Fig. 1). A straightforward calculation [3,4] shows that the polarization in the evanescent wave is predominately σ^- polarized. We will neglect the σ^+ component of the light polarization. Next, consider an atom with a non-zero angular momentum F within the evanescent wave, and assume that the laser is detuned to the blue of the atomic transition. The circularly polarized light will produce light shifts that increase with decreasing m_F. Thus the atom follows different trajectories depending on its internal state.

If one adds a weak TE polarized wave (corresponding to π polarization), one can consider the potential curves unchanged to first order in the TE electric field and treat the effect of the TE wave as a weak coupling between states with different m_F. This coupling arises from a stimulated Raman transition involving absorption of a σ^- polarized photon and the stimulated emission of a π polarized photon. The coupling is resonant only if the difference frequency $\Delta\omega$ of the two laser beams is equal to the energy difference (the light shift plus van der Waals interaction) between the two states.

The process can be pictured as in Fig. 2. We label the two states 1 and 2, and we have shifted one with respect to the other by the amount $\hbar\Delta\omega$, corresponding to the difference in the two dressed states that are connected by the Raman transition. The curves cross at a distance z_C at which the Raman resonance condition is satisfied. The atom approaches on one of the two potentials and is "split in two" at the crossing. Its

SCHEMATIC DIAGRAM OF THE INTERFEROMETER EXPERIMENT

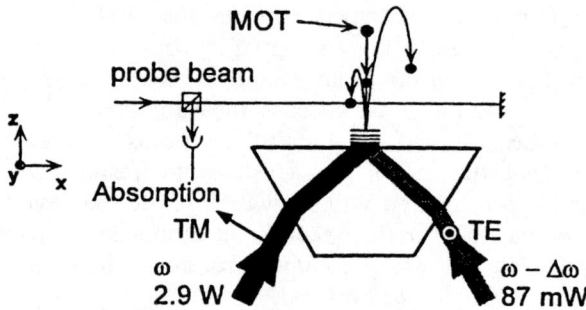

FIGURE 1. Atoms from a MOT fall on an evanescent wave mirror formed by a strong TM polarized beam. The TE polarized beam is much weaker and shifted in frequency, and served to stimulate Raman transitions between states of different m_F. Different internal state change is accompanied by a kinetic energy change on reflection allowing the probe beam absorption signal to distinguish the different internal states.

momentum is then reversed by the potential and it reencounters the crossing which again splits probability amplitude. Which potential curve the atom ends up on, depends on the splitting amplitude and on the difference in the phases accumulated on the two possible paths. Thus fringes, or Stueckelberg oscillations appear in the transition probability either as a function of the incident energy or of the position of the crossing.

In the experiment, we used ^{87}Rb atoms, captured and laser cooled in a magneto-optical trap (MOT). Immediately after turning off the MOT, the atoms were prepared in the $F = 2$, $m_F = 2$ state by an optical pumping pulse and allowed to drop 18 mm onto the mirror. The evanescent wave was produced by a Ti:S laser tuned 1.3 GHz to the blue of the D2 Rb resonance line (780 nm). The TE beam was derived from the same laser using two double-pass acousto-optic modulators to give a tunable Raman difference frequency $\Delta\omega$ between 6 and 22 MHz. We chose an intensity of the TE beam to be about 3% of the TM intensity so that it was sufficiently weak to cause only a single Raman transition ($\Delta m_F = \pm 1$). This permits comparison with a simple two level analysis of the process [2]. Transitions were observed by recording the absorption from a near resonant probe laser beam as shown in Fig. 1.

Figure 3 shows a plot of the measured transition probability as a function of the Raman detuning $\Delta\omega$. Clear oscillations are present. We can compare the data to a model based on the analogy with a Michelson interferometer. At the level crossings, we can get a good estimate of the splitting amplitudes by using the Landau-Zener

POTENTIAL CURVES

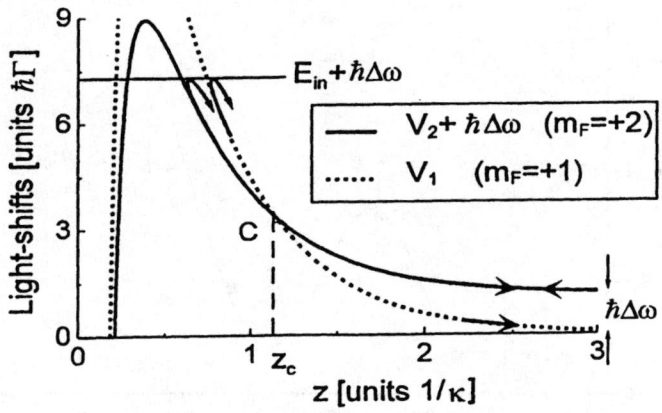

FIGURE 2. Plotted are the potential curves, including the van der Waals interaction experienced by the atoms during the reflection. V_2 (V_1) corresponds to the $m_F = 2$ ($m_F = 1$) state. The horizontal and vertical scales correspond to our experimental parameters. The atoms arrive on the V_2 potential, pass through the curve crossing twice, and can end up on either V_2 or V_1. Two paths are possible and can interfere producing fringes as a function of the location of the crossing.

formula. The phase accumulated along each path is calculated using a JWKB treatment. The fact that the laser beam is Gaussian means that the effect of the van der Waals interaction is not the same on every point of the mirror. Therefore we must perform an average over the mirror to accurately calculate the fringe pattern. This causes a slight loss of contrast. Figure 3 shows the effect of entirely neglecting the presence of the van der Waals interaction. Although the qualitative shape of the fringes does not change, their phase changes dramatically.

In the above analysis we assumed that the van der Waals interaction was described by a simple z^{-3} law. It is of course natural to ask whether the experiment is sensitive to deviations from this law as one expects for distances of order or larger than 100 nm, at which retardation effects should become important. We have analyzed this question in detail in Ref. [5] and conclude that indeed there are slight differences in the expected fringes if retardation effects are taken into account. Our present signal-to-noise however, is not sufficient to resolve them. Retardation effects may become more visible for incident atomic momenta much lower than used here (about 100 recoil velocities). In Ref. [5] we find that atoms with a velocity close to the recoil velocity may permit a measurement of the effect of retardation in an evanescent wave mirror. A related question has also been discussed in Ref. [6]

INTERFERENCE IN AN EVANESCENT WAVE

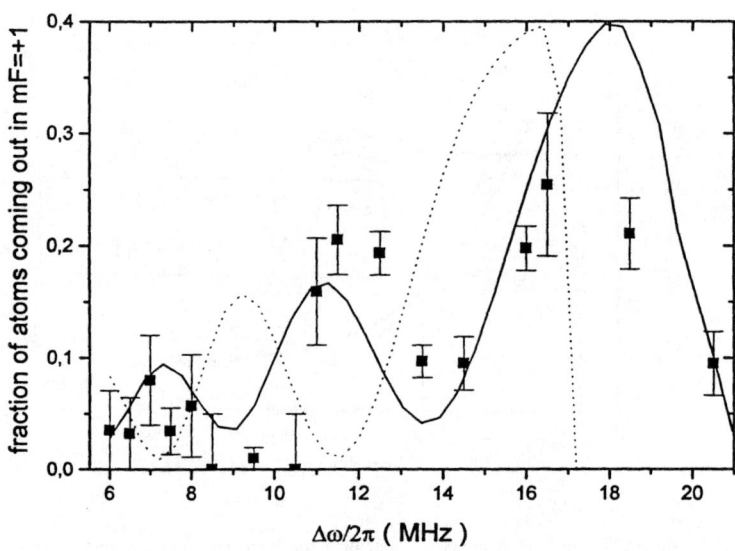

FIGURE 3. Transition probability vs the Raman detuning $\Delta\omega$, showing oscillations. The solid curve is a prediction based on the Landau-Zener model including the van der Waals potential. The dotted line is the same calculation but neglecting the van der Waals potential.

ARE ATOM MIRRORS FLAT?

First Generation Experiments

Our group has also done extensive work on studying the flatness of the mirrors. The first generation of these experiments was quite simple, employing an imaging technique, in which we collimated the incident atoms by reducing the size of the mirror and checking to see whether the width of the transverse velocity distribution increased after reflection (see Fig. 4). To our surprise, the initial experiments using evanescent wave mirrors [7] indicated a substantial increase despite a very high quality optical prism.

It was natural for us to extend our measurements to magnetic mirrors[8]. This we did in collaboration with a group from Harvard University, which had been developing lithographic techniques to fabricate current carrying magnetic mirrors[9]. An important innovation in this experiment was the use of extra wires to compensate for the residual magnetic fields due to the finite size of the mirror[10,11]. We were able to demonstrate that these compensation wires did indeed improve the flatness of the magnetic mirror [12]. The lowest angular deviation observed was 13 mrad. Several other groups have performed related measurements of the roughness of various types of magnetic mirrors [11,13]. It appears to us that there is still room for a lot of progress. Roughness can be reduced using more elaborate wire designs[14] and potential barrier heights as well as useful surface areas can be increased.

ATOMIC MIRROR ROUGHNESS EXPERIMENT

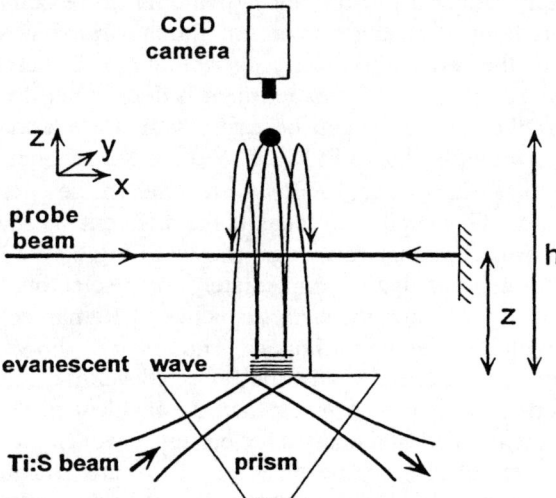

FIGURE 4. First generation technique to observe diffuse reflection from an evanescent wave mirror. The MOT size is about 1 mm as is the Ti:S laser spot on the prism. The velocity resolution is about 1.5 v_{RECOIL} rms.

Recent Experiments

In the case of the evanescent wave mirror, the cause of the diffuse reflection is quite interesting. Diffuse reflection appeared to be related to the surface roughness of the prism. This result was surprising because the measured rms surface roughness of these prisms was of order 0.1 nm, while the distance of closest approach of the atoms was of order 50 nm. How do the atoms know about the roughness when they are so far away from the surface? The result was interpreted in terms of the interaction of the atoms with the small amount of light scattered by the surface [15]. The scattered light interferes with the unscattered evanescent wave producing an rough optical dipole potential even at the height of 50 nm. In addition, the fact that the atom's wavelength is so small (of order 10 nm), means that they are more sensitive to small roughnesses than are optical waves. An important prediction of this understanding is that, for sufficiently small roughness, the velocity distribution of the reflected atoms should contain a perfectly reflected specular peak and a pedestal whose width is a few times the recoil velocity due to the surface roughness.

In order to test this prediction it is therefore necessary to achieve a velocity resolution better than the recoil velocity. The resolution of the first generation experiments were limited by the size of the magneto-optical trap to a few times the single photon recoil velocity v_{RECOIL}. No matter how small one makes the effective mirror surface, the width of the velocity distribution cannot be narrower than the size of the MOT divided by the fall time. One way around this problem was demonstrated in Ref. [13]. In that experiment atoms bounced several times from a curved mirror, and successive increases in the velocity width added up.

We have adopted a different approach based on using velocity selective Raman transitions to directly select a narrow slice of atoms in velocity space [16]. After selection, the atoms bounce on the mirror and Raman transitions are again used to analyze the width of the reflected velocity distribution. The method is quite flexible because the velocity resolution of the experiment is determined by the duration of the Raman pulse. Thus the resolution can be easily varied electronically. A schematic diagram of our experiment is shown in Fig. 5. We use the magnetic field independent transition between the $F=2, m_F=0$ and $F=3$, $m_F=0$ states in the ^{85}Rb atom. A magnetic guiding field of about 1 G is applied to separate the different m_F states.

An example of our preliminary results are shown in Fig 6. We begin by measuring our velocity resolution by analyzing immediately after selecting. A typical result is shown on the left of Fig. 6. Then the same sequence of Raman selection and analysis pulses is repeated with an intervening bounce. The result is shown on the right of Fig 6. The data indicate that at least part of the reflected velocity distribution is indeed contained within a peak which is no wider than the incident peak. The data also may be consistent with a wider pedestal and we are currently working to improve the signal to noise ratio in order to observe the pedestal. If we are able to observe both the narrow peak and the pedestal we will be able to quantify the 'coherent reflectivity' of an atom mirror. As in ordinary optics, the reflectivity of a mirror refers to the quantity of light which is perfectly specularly reflected, while the diffusely reflected part is considered as loss. Thus the reflectivity is simply the area under the coherent peak divided by the sum of the areas of the specular and diffuse contributions.

RAMAN VELOCITY SELECTION PROCEDURE

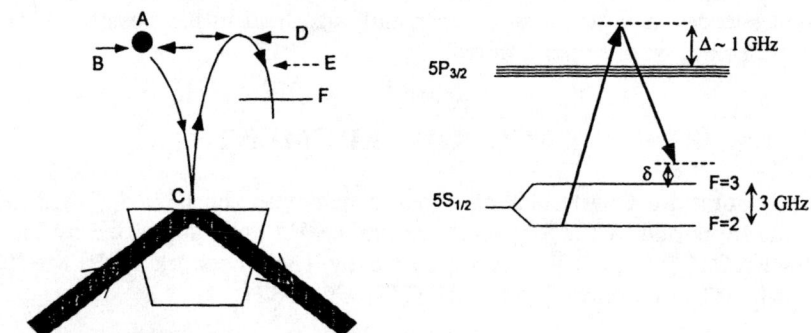

A   Atoms prepared in F=2 dropped from trap
B   Raman pulse for velocity selection
C   Atoms in F=3 reflect, atoms in F=2 stick
D   Raman pulse with variable detuning δ for velocity probing
E   Atoms in F=3 are pushed
F   Remaining atoms detected

FIGURE 5. Left: The temporal sequence used for the velocity selection. Right: A schematic diagram of the ^{85}Rb structure showing the two Raman lasers and their detunings.

PRELIMINARY MEASUREMENT OF THE REFLECTED VELOCITY DISTRIBUTION

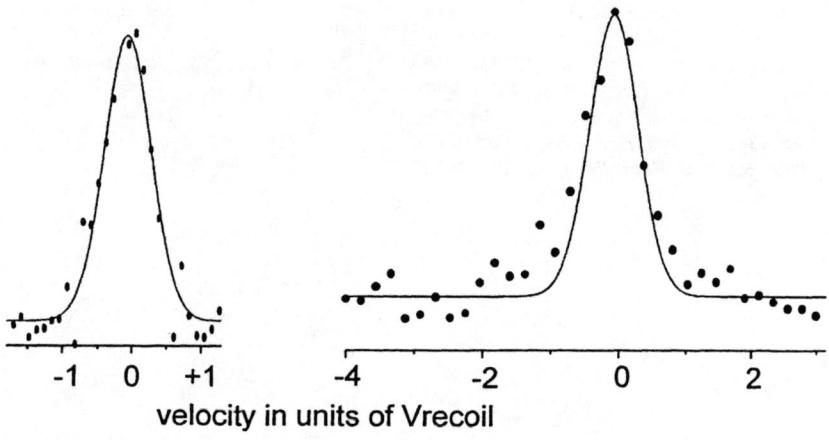

velocity in units of Vrecoil

FIGURE 6. Left: Velocity distribution after Raman selection, but before reflection from a mirror. The fitted rms is 0.34 v_{RECOIL}. Right: Velocity distribution after reflection from an mirror evanescent wave mirror. For comparison, a Gaussian with rms 0.34 v_{RECOIL} (solid curve) has been superimposed.

In future work we hope to be able to use an analysis of the diffusely reflected peak to characterize the power spectrum of the surface roughness. Experiments using coherent sources such as an atom laser may also lead to the possibility of observing 'atomic speckle' from an atom mirror.

ACKNOWLEDGMENTS

The Laboratoire Charles Fabry is associated with the CNRS (UMR 8501). This work was supported by the Région Ile de France. P.F. was supported by TMR network ERBFMRXCT960002. D.S. was supported by TMR network ERBFMRXCT960002 and by Marie Curie Fellowship HPMF-CT-1999-00321.

REFERENCES

1. For recent reviews, see J. Dowling and J. Gea-Banaloche, *Adv. At. Mol. and Opt. Phys.* **37**, 1 (1996), N. Westbrook, et al., *Phys. Scripta* **T78**, 7 (1998), V. Balykin, *Adv. At. Mol.Opt. Phys.* **4**, 181 (1999), and E. Hinds and I. Hughes, *J. Phys. D: Appl. Phys.* **32**, 119 (1999).
2. C. Henkel *et al. Phys. Rev.* A **56**, 9 (1997).
3. L. Cognet *et al.*, *Phys. Rev. Lett.* **81**, 5044 (1998).
4. M. Born, and E. Wolf, *Principles of Optics 6th ed.*, Oxford: Pergamon, 1980, pp. 40, and L. Cognet, Ph. D. Thesis, Université de Paris XI, 1999.
5. R. Marani, *et al.*, *Phys. Rev.* A **61**, 053402 (2000).
6. B. Segev, R. Coté and M. Raizen, *Phys. Rev.* A **56**, 3350 (1997).
7. A. Landragin et al., *Opt. Lett.* **21**, 1591 (1996).
8. V. Vladimiriski, *JETP Lett.* **12**, 740 (1960), and G. Opat, S. Wark and A. Cimmino, *Appl. Phys. B* **54**, 396 (1992).
9. M. Drndic et al., *Appl. Phys. Lett.* **65**, 2503-2504 (1994), K. Johnson *et al.*, *Phys. Rev. Lett.* **81**, 1137 (1998).
10. G. Zabow *et al.*, *Eur. Phys. J. D* **7**, 351 (1999).
11. D. Lau *et al.*, *Eur. Phys. J. D* **5**, 193 (1999).
12. L. Cognet *et al.*, *Europhys. Lett.* **47**, 538 (1999), M. Drndic *et al.*, *Phys. Rev. A.* **60**, 4012 (1999).
13. C. Saba *et al.*, *Phys. Rev. Lett.* **82**, 468 (1999).
14. G. Zabow *et al.*, *Eur. Phys. J. D* **7**, 351 (1999).
15. C. Henkel *et al. Phys. Rev. A* **55**, 1160 (1997).
16. M. Kasevich *et al. Phys. Rev. Lett.* **60**, 2297 (1991).

Miniature Guides For Neutral Atoms

E. A. Hinds

Sussex Centre for Optical and Atomic Physics, University of Sussex, Brighton BN1 9QH, U.K.

Abstract. This article discusses the manipulation of cold atoms by miniature magnetic structures. In particular we describe some recent advances in atomic de Broglie waveguides, a magnetic conveyor belt for transporting atoms across a surface, and steps towards making an array of atoms for integrated atom optics.

INTRODUCTION

There is widespread interest in guiding cold atoms [1, 2]. For atom lithography, the goal is to dispense precise quantities of atoms onto specific regions of a surface [3]. In atom optics, the guide can be a "hosepipe", delivering large quantities of atoms to an inaccessible region, or it can operate in a single transverse mode, permitting coherent propagation of de Broglie waves for applications such as interferometry [4] and integrated atom optics [5, 6, 7]. In statistical physics, a quantum guide for atoms permits the study of degenerate quantum gases in 1D. Here one might realize a Tonks gas of Bosons whose elementary excitations obey Fermi statistics or observe Luttinger liquid behaviour with correlation functions that decay algebraically [8].

In several previous experiments [9], cold atoms have been guided inside hollow optical fibres where optical dipole forces confined them on axis. With this approach it is a challenging problem to avoid heating the atoms through intensity fluctuations and spontaneous emission. Confinement by static magnetic fields provides an alternative. On the macroscopic scale, magnetic guiding has been demonstrated using arrangements of permanent magnets [10] and along the side of a current-carrying wire [11]. Here we report on two miniature magnetic guides being developed in our laboratory. The first is single guide based on current-carrying wires supported by an optical fibre, in which we observe an atom cloud of 100 μm radius propagating for several centimetres. The second is an array of guides made by recording a pattern on videotape. The tape can serve as a corrugated reflector for atoms and a conveyor belt. It can also be used to form a microscopic pattern of magneto-optical traps spaced 100 μm apart in lines of 10 μm radius.

CP551, *Atomic Physics 17*, edited by E. Arimondo, P. DeNatale, and M. Inguscio
© 2001 American Institute of Physics 1-56396-982-3/01/$18.00

WIRE GUIDES SUPPORTED BY OPTICAL FIBRES

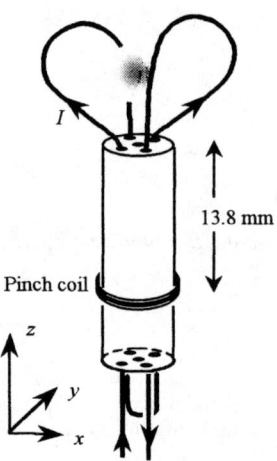

FIGURE 1. Schematic view of the quadrupole guide. A cloud of cold atoms falls into the central hole where it is guided down to the pinch coil, reflected, and guided back up to the top. The diameter of the guide is greatly exaggerated.

Our first guide is illustrated in figure 1. The supporting structure is a 25 mm-long silica tube fabricated by the Optoelectronics Research Centre at Southampton. Five parallel holes, each of radius $R = 261(5)\,\mu m$, run through the length of the tube parallel to the vertical z-axis. Four of them intersect the (x, y) plane on the corners of a square at $(\pm a, \pm a)$, with $a = 522(10)\,\mu m$. These contain copper wires which carry a current I to produce the guiding magnetic field. The fifth hole is on the tube axis where the atoms are guided. To a good approximation the magnetic field vector is a purely quadrupole one: $\mathbf{B} = (B'x, -B'y)$ with $B' = \mu_0 I / \pi a^2 = 1.46(6)I\ \mathrm{T/m}$. An atom at radius ρ interacting with this field has Zeeman energy $-B'\mu_\zeta \rho$, where μ_ζ is the projection of its magnetic moment onto the local magnetic field direction. For the modest field strengths of interest here, the gyromagnetic ratio is independent of field strength, so provided the spin does not flip, μ_ζ is a constant of the motion and the potential is linear in ρ. For $\mu_\zeta < 0$ the atom then experiences a constant force directed toward the axis. Non-adiabatic behaviour is suppressed at the centre by adding ~300 mG magnetic field along the z-direction to maintain an adequate splitting of the magnetic sublevels [12]. This causes the potential to become harmonic at distances less than a few μm from the axis. At the top of the guide the wires spread out (figure 1) to form a magnetic funnel of apex angle ~90°, centred on the axis. A "pinch coil" wound on the outside of the guide, 13.8 mm below the top, allows us to add a field in the z-direction to close off the guide.

The guide is mounted in a vacuum chamber (10^{-9} Torr) with its entrance 10 mm below a magneto optical trap (MOT) of standard design [2]. The MOT collects ~5×10^6 ^{85}Rb atoms from the background vapour (filling lifetime ~5 s). These are then cooled in optical molasses to form a $T = 25\,\mu K$ cloud with rms radius $\sigma = 0.72$ mm, which we optically pump into the $(F = 3, m_F = 3)$ ground-state sublevel relative to a uniform magnetic field $B_x \hat{x}$. As the atoms fall in the dark, B_x is reduced to zero and the guide current is adiabatically turned on. With a field of 40 G at the centre of the pinch coil, atoms travelling down the guide should be reflected to re-emerge from the entrance aperture. These are detected by a 2 ms pulse of laser light (10 mW/cm^2, -10 MHz molasses), which produces enough fluorescence for a CCD camera to record the distribution of the atoms without appreciably moving them under radiation pressure. Figure 2 shows the evolution of the cloud viewed in this way at 10 ms intervals. In the

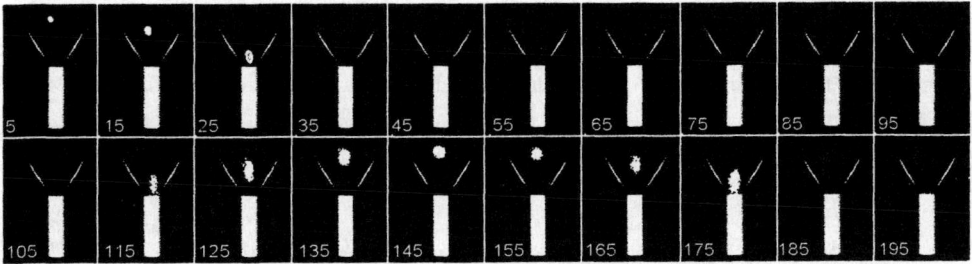

FIGURE 2. Images of the atom cloud. Time after release is shown in milliseconds in the corner of each frame. Atoms released from the MOT at t = 0 propagate down the guide and back up again, rising up almost to the original height after 139 ms.

first three frames we simply image the cloud falling into the guide. After 65 ms, we take CCD pictures with the pinch coil on and off and the images shown are the on-off differences as a function of the propagation time. The cloud clearly re-emerges from the fibre and rises to its original height, 139 ms after being released, demonstrating that the atoms are indeed being guided in the fibre. Since the total drop height is $h = 2.38\,\text{cm}$, this round-trip time indicates that the vertical component of the motion is close to free-fall. The velocity of atoms along the guide is therefore approximately 45 cm/s near the top and 65 cm/s near the bottom.

In order to measure the transverse distribution of atoms in the guide, we adiabatically add a transverse magnetic field B_x to the guide while the atoms are propagating in it. This translates the magnetic potential to the side, lowering the energy required for atoms to escape onto the wall by $\mu_\zeta B_x$. Figure 3 shows how the number of atoms returning from the guide decreases. Fitting these points to the integral of a Gaussian distribution,

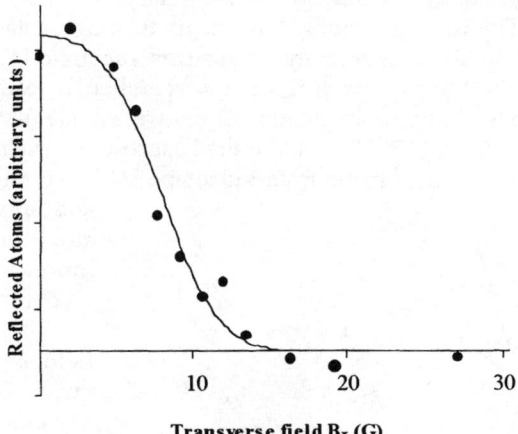

FIGURE 3. When a transverse magnetic field is applied to the guide, the number of guided atoms decreases because of loss to the walls. Dots: experiment. Line: simple theory corresponding to a transverse temperature of 670 μK.

shown by the solid line, we find a mean transverse a energy of $670\,\mu K$ with a standard deviation of $270\,\mu K$, in excellent agreement with a Monte Carlo simulation of the experiment. This temperature is much larger than the initial $25\,\mu K$ of the cloud because the vertical velocity due to falling under gravity is partly converted into transverse motion by the funnel. Noting that the mean potential energy is 2/3 of this total (the virial theorem for a linear potential) we deduce from the known field gradient that the radius of the atom cloud in the guide $98\,\mu m$.

The quantum mechanical eigenmodes of this guide [12] are characterized by an energy $(\hbar \mu_B B' / \sqrt{m})^{2/3}$, which scales with current and wire spacing as $I^{2/3} / a^{4/3}$. For our experimental parameters, this energy divided by Boltzmann's constant is $\sim 1\,\mu K$, indicating that many transverse modes of the de Broglie waves are occupied in our present experiment. However, we have already demonstrated that similar currents can be passed through structures that are 20 times smaller, in which the mode spacings are $\sim 50\,\mu K$. In order to achieve good coupling to such a waveguide it will be important to introduce some dissipation into the funnel or alternatively to use an atom source of smaller phase-space volume, such as a Bose-Einstein condensate. In conclusion, we have shown that cold atoms can be efficiently coupled into a miniature magnetic quadrupole waveguide where they propagate within a 100 μm radius without appreciable loss other than background gas collisions. This is a prototype which shows the feasibility of miniature guides as a tool in the new field of integrated atom optics, leading to single-mode propagation of de Broglie waves and the possible preparation of 1D atom clouds.

MAGNETIC TAPE: AN ATOM CONVEYOR BELT

In recent years our laboratory has shown that a magnetic mirror can be used to retro-reflect cold atoms. The basic principle of the magnetic mirror relates to the forces acting on atoms in the region above a suitably magnetised substrate [13]. For a sinusoidally magnetised surface the Stern-Gerlach force is perpendicular to the surface and can be used to repel the atoms. Previous magnetic mirrors were made first from audio-tape [14], and then from floppy disks, [15, 16, 17], but the best results have recently been obtained using a third generation mirror made from videotape [18]. Further details of the theory and experimental results for atom reflection with magnetic mirrors can be found in a recent review [2].

When a uniform magnetic field is superimposed on the magnetic mirror, the contours of constant magnetic field strength are no longer planes parallel to the surface but take on the more complicated form illustrated in Figure 4. Since the magnetic moment of a cold

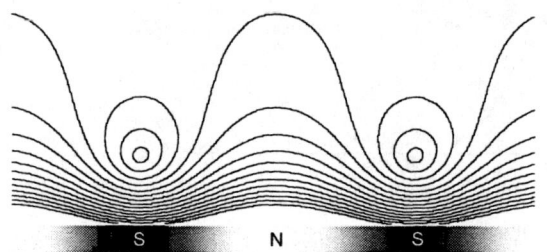

FIGURE 4. Contours of constant B above a magnetic mirror when a constant magnetic field is applied. In the upper half of the figure we see closed contours indicating the presence of microscopic guides centred on $B = 0$. Closer to the mirror, the equipotential surfaces have sinusoidal corrugations.

408

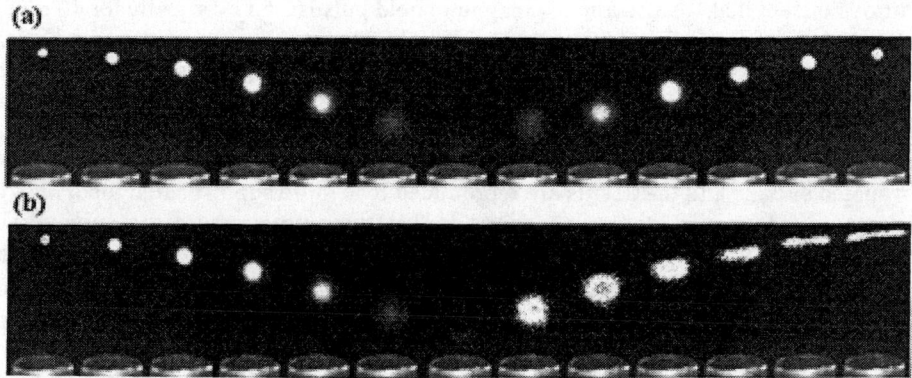

FIGURE 5. Atom cloud images taken from the side at 10 ms intervals. The reflector is seen at the bottom of the picture. Near the mirror the atoms are shadowed from the imaging beam. (**a.**) The cloud expands thermally as it falls, with the centre of mass following a parabolic trajectory. It bounces on a smooth mirror and is subsequently refocused at its initial height. (**b.**) The mirror is corrugated and the reflected cloud splits into two peaks at the refocusing height h=27 mm.

atom can follow the field adiabatically, these contours are also the equipotentials for the interaction of the atom with the magnetic field. We see two interesting features as a result of the applied bias field. First, some of the contours become closed circles, corresponding to magnetic guiding tubes, very similar to the quadrupole guide discussed above. Second, the reflecting potential becomes corrugated.

We have investigated both these features experimentally using atom mirrors made from 12 mm-wide Ampex 398 Betacam SP videotape. To study the corrugations [19], we recorded a sinusoidal pattern on the tape with a wavelength of 25.4 µm, which gave a field at the surface of $667\,G$. This tape is stuck by vacuum epoxy to a 12 mm-diameter ring and then deformed to make a concave reflector with a radius of curvature $R = 54$ mm. A MOT collects a cloud of approximately 5×10^6 ^{85}Rb atoms to be reflected from the mirror. These are cooled to 24 µK in optical molasses and then optically pumped into the ($F = 3$, $m_F = +3$) state. The atoms are dropped onto the mirror and after an adjustable delay a CCD camera records the distribution of atoms using laser-induced fluorescence. We are careful to ensure that the detection process itself does not blur or displace the image of the cloud by virtue of the radiation pressure exerted on the atoms.

For the first set of experiments, shown in Figure 5a, the atoms are dropped from a height of $R/2 = 27$ mm, chosen so that the initial cloud will be reconstructed after a single bounce. The first 6 images show the cloud falling freely under gravity with its radius expanding thermally at 4.8 cm/s The corrugating external magnetic field is turned off so that the atoms interact with a smooth mirror potential and therefore we see in the last 6 images that the cloud is indeed refocused in the plane of the initial MOT. (A small holding field is applied in the z direction, perpendicular to the plane of Figure 4, in order to prevent any spin flips as the atoms move through the magnetic field). The images are viewed from an angle that makes the magnetic reflector visible at the bottom of each frame. The sequence of images in Figure 5b is taken under identical experimental conditions except that we also apply a magnetic field pulse (0.5 G vertically for 40 ms) so that the atoms reflect from a corrugated potential. The motion is unchanged when the atoms are falling downward, but after reflection from the corrugated surface, we see the refocused cloud split into a double-peaked structure, with each peak located symmetrically about the initial cloud position prior to free fall. The main features of this distribution can be understood by considering the simple classical reflection of balls from a corrugated surface and the details are reproduced by a more sophisticated theory [19].

An exciting feature of the corrugations is that they move across the surface with velocity $f\lambda$ when the bias field is rotated in the x-y plane at frequency f [20]. This offers the prospect of a magnetic conveyor belt for transporting cold atoms across the surface. To see the effect we rotated the bias field, first clockwise and then anticlockwise, and measured the shift 2Δ in the centre of mass of the cloud at the peak of the bounce. Division of Δ by the 74 ms flight time gives the velocity v_{cloud} imparted to the cloud. Although the atoms interact only briefly with the grating (~50 μs) they nevertheless acquire an appreciable centre-of-mass velocity. This implies that they are carried along by the motion of the grating during the time they interact with the field of the mirror. For our magnetic field strength (~1 G) the atoms are carried much less than one recording wavelength by the moving wave, but it is nevertheless the "surfing" of the atom on the moving grating that produces v_{cloud}. Figure 6 shows our experimental results for v_{cloud} at two different magnetic field strengths and for a range of frequencies [21]. The lines show that there is good agreement with the cloud velocity calculated by a Monte Carlo program. The inset graph shows the difference between cloud profiles for the two directions of a 6 kHz, 1.1 G grating. There is little information in the centre of the graph because the cloud profile is rather flat there and consequently it is not very sensitive to small displacements. By contrast, the wings of the profile are steep and it is their displacement that causes the peaks at each side of the graph. The solid line calculated by numerical integration fits the data well. To a first approximation the curve is very similar to a simple derivative of the cloud profile, although it differs in detail because different parts of the cloud acquire slightly different velocities from the movement of the grating.

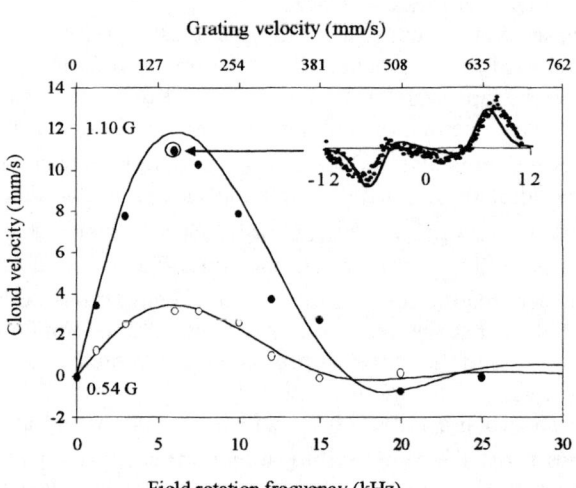

FIGURE 6. Centre-of-mass velocity transferred to the cloud from a travelling grating. Two abscissae show the frequency of the field and the corresponding grating velocity. Circles: experimental results. Lines: calculation using full equations of motion. Inset: difference between cloud profiles with left and right moving gratings at 6 kHz and 1.1 G.

MAGNETIC TAPE: A MINIATURE ARRAY OF ATOM GUIDES

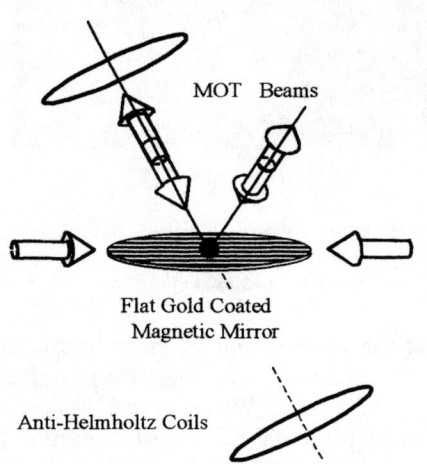

FIGURE 7. A mirror MOT collects 2×10^7 rubidium atoms in a small cloud approximately 1.5 mm above the gold-coated magnetic tape.

We turn now to the circles above the magnetic tape in figure 4, which represent cylindrical surfaces of constant magnetic field strength. The field above the tape is given by $B_1 \exp(-ky)(-\cos(kx)\hat{\mathbf{x}} + \sin(kx)\hat{\mathbf{y}})$, with $B_1 = 227\,\mathrm{G}$ being the expected field strength at the surface and k being the wavevector of the recorded pattern on the tape. For an applied field $B_0\hat{\mathbf{x}}$, the tubes lie at positions where $kx = n\pi$ and $ky = \ln(B_1/B_0)$. Therefore they form an array spaced by the wavelength of the recording and parallel to the magnetic surface. The height is comparable with the recording wavelength and governed by the strength of the applied field. If the applied field is rotated in the x-y plane, the tubes are translated across the surface. In the vicinity of a tube, the gradient of the field, which determines the trapping force on atoms, is a constant kB_0.

We are currently investigating these tubes in our laboratory using the same videotape as above, but with a recording of 100 μm wavelength [22]. A 1 cm square piece of the tape is glued flat onto a glass slide. This assembly is covered with a 400 nm-thick coating of gold to reflect the 780 nm trapping light, allowing us to collect an initial cloud of 2×10^7 rubidium atoms in a mirror MOT [23], as illustrated in figure 7. The anti-Helmholtz coils of the mirror MOT are then switched off, allowing the cloud to fall down onto the magnetic mirror. Now a bias field is applied parallel to the surface of the magnetic mirror to create the quadrupole tubes. By choosing the correct direction for the applied field we ensure that the magnetic field lines of the tubes are correlated with the circular polarisations of the four reflected MOT beams so as to make each tube work as a cylindrical MOT. There is no trapping force along the axis of the cylinder. Figure 8a shows the fluorescence viewed almost at grazing incidence to the mirror. The upper line is the array of tubes, one behind the other, whilst the lower line is a reflection in the gold surface. Figure 8b shows the same array seen from above, with the tubes spaced by the 100 μm period of the magnetic recording on the tape. In this view, the tubes are partially obscured by a haze due to atoms distributed in a cloud above the mirror.

For our bias field of 0.7 G, the calculated height of the tubes above the mirror is $\ln(B_1/B_0)/k = 92\,\mu\mathrm{m}$. This is completely consistent with the height of 97 μm deduced from the separation between the atoms and their reflection in figure 8a. We estimate from the image in figure 8b that the radius of the MOT tubes is only 10 μm, placing these among the smallest MOTs to date (see also reference [24]). The field gradient is 4 T/m, approximately 40 times larger than the field gradient of the mirror MOT, while the size is

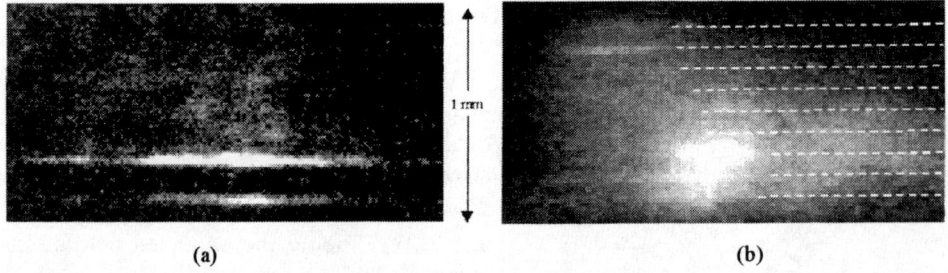

(a)   (b)

FIGURE 8. (a): Upper line: the array of atom tubes viewed from near grazing incidence. Lower line: a reflection in the gold surface. (b) Array of atom tubes viewed from above. The tubes are partially obscured by the cloud of atoms above the mirror: dashed lines are drawn in as an aid to the eye.

approximately 50 times less. This suggests that the temperature of the microscopic MOTs is not significantly higher than that of the mirror MOT. The rate at which these microscopic guides decay depends on the detuning of the laser. We observe a lifetime of ~30 ms when the detuning is −10 MHz, increasing to 150 ms for a −30 MHz detuning. We interpret this as being due to the rate at which atoms propagate along the guides, through the optical molasses formed by the two light beams parallel to the surface.

ACKNOWLEDGEMENTS

This work was supported by grants and fellowships from EPSRC (UK) and the European Union.

REFERENCES

1. J. P. Dowling and J. Gea-Banacloche *Adv. At. Mol. Opt. Phys.* **37**, 1 (1996); V. Balykin *Adv. At. Mol. Opt. Phys.* **41**, 181 (1999).
2. E. A. Hinds and I. G. Hughes *J. Phys. D* **32**, R119 (1999).
3. Jabez J. McClelland, in Handbook of Nanostructured Materials and Nanotechnology, ed. H. S. Nalwa (Academic Press San Diego 1999).
4. *Atom Interferometry*, ed. P. R. Berman (Academic Press, Boston 1997).
5. E. A. Hinds, M. G. Boshier and I. G. Hughes *Phys. Rev. Lett.* **80**, 645 (1998).
6. J. Schmiedmayer *European Physical Journal D* **4**, 57 (1998).
7. J. H. Thywissen, R. M. Westervelt, and M. Prentiss, physics/9909020.
8. M. Olshanii, *Phys. Rev. Lett.* **81**, 938 (1998); H. Monien, M. Linn, and N. Elstner *Phys. Rev. A* **58** R3395 (1998).
9. M. J. Renn *et al.*, *Phys. Rev. Lett.* **75**, 3253 (1995); *Phys. Rev. A* **53**, R648 (1996); *Phys. Rev. A* **55** 3684 (1997); H. Ito *et al.*, *Appl. Phys. Lett.* **70**, 2496 (1997).
10. C. J. Myatt *et al.*, *Opt. Lett.* **21**, 290 (1996); A. Goepfert *et al.*, *Appl. Phys. B* **69**, 217 (1999).
11. J. Schmiedmayer, *Phys. Rev. A* **52**, R13 (1995); J. Denschlag, D. Cassettari and J. Schmiedmayer *Phys. Rev. Lett.* **82**, 2014 (1999).
12. E. A. Hinds and C. C. Eberlein, *Phys Rev. A* **61**, 33614 (2000).
13. G. I. Opat, S. J. Wark and A. Cimmino, *Appl. Phys. B* **54**, 396 (1992).
14. T. M. Roach, H. Abele, M. G. Boshier, H. L. Grossman, K. P. Zetie and E. A. Hinds, *Phys. Rev. Lett.* **75**, 629 (1995).
15. I. G. Hughes, P. A. Barton, T. M. Roach, M. G. Boshier and E. A. Hinds, *J. Phys. B* **30**, 647 (1997).

16. I. G. Hughes, P. A. Barton, T. M. Roach and E. A. Hinds, *J. Phys. B* **30**, 2119 (1997).
17. I. G. Hughes, P. A. Barton, M. G. Boshier and E. A. Hinds, *Atom Optics*, SPIE Proceedings **2995**, 182 (1997).
18. C. V. Saba, P. A. Barton, M. G. Boshier, I. G. Hughes, P. Rosenbusch, B. E. Sauer and E. A. Hinds, *Phys. Rev. Lett.* **82**, 468 (1999).
19. P. Rosenbusch, B. V. Hall, I. G. Hughes, C. V. Saba and E. A. Hinds, *Phys. Rev. A* **61**, 31404 (2000).
20. E. A.Hinds, *Phil. Trans Roy. Soc. Lond.* A **357**, 1409 (1999).
21. P. Rosenbusch, B. V. Hall, I. G. Hughes, C. V. Saba and E. A. Hinds, *Appl. Phys. B* **70**, 709 (2000).
22. J Retter, B. V. Hall, P. Rosenbusch, B. E. Sauer, and E. A. Hinds, *to be published*.
23. J. Reichel, W. Hansel and T. W. Hänsch, *Phys. Rev. Lett.* **83**, p. 3398 (1999).
24. B. Ueberholz, S. Kuhr, D. Frese, D. Meschede, and V. Gomer, *J. Phys B* **33**, L135 (2000).

413

Exploring a Quantum Degenerate Fermi Gas

D. S. Jin*, B. DeMarco, S. Papp

JILA, NIST and the University of Colorado
*NIST Quantum Physics Division

Abstract. We discuss the production and study of a quantum degenerate Fermi gas of atoms. The fundamental difficulty in cooling a gas of fermionic atoms is the lack of rethermalizing collisions in a spin-polarized sample. We overcome this difficulty using simultaneous forced evaporation of a two spin-state cloud of ^{40}K atoms held in a magnetic trap. In the quantum degenerate regime we explore the effects of the Fermi statistics on the thermodynamics of the ultracold gas. Initial studies of the dynamics of a two-component Fermi gas are also discussed.

QUANTUM DEGENERATE GASES

In quantum degenerate matter the quantum statistics of the constituent particles govern the behavior of the system as a whole. Quantum degeneracy occurs in a regime of temperature, T, and density, n, where the thermal deBroglie wavelength of the particles (atoms in a gas, for example) is comparable to the mean spacing between those particles. Under these conditions the deBroglie waves of neighboring particles overlap and the ensemble is clearly in a quantum regime. Quantum degenerate matter is readily found in high-density condensed matter systems but is difficult to realize in a low-density system such as a gas. However the recent advent of powerful cooling techniques in atomic physics now makes it possible to cool a trapped gas of atoms to 100 nK temperatures. At these temperatures the deBroglie wavelength of an atom becomes on the order of a micron and it is possible to reach conditions of quantum degeneracy in a dilute gas[1-3].

Because quantum degeneracy is reached in a regime of low density and ultralow temperature, a gas of neutral atoms has a number of unique properties that facilitate the study of quantum effects. The low density combined with the relatively short-range interactions between neutral atoms yields a weakly interacting system. In addition the atom-atom interactions are accessible theoretically from a microscopic level, which is not possible in more complicated condensed matter systems. Furthermore, the trapped atomic gas can be manipulated and probed with fine control over both the internal atomic states and the external motional states. These factors make the quantum degenerate gas system opportune for detailed comparisons between theory and experiment.

There are two classes of quantum particles found in nature – bosons and fermions. Since atoms, which are composite particles, can be either bosons or fermions, the

CP551, *Atomic Physics 17*, edited by E. Arimondo, P. DeNatale, and M. Inguscio
© 2001 American Institute of Physics 1-56396-982-3/01/$18.00

possibility exists for studying either type of quantum statistics in the atomic gas system. While initial experiments on quantum degenerate gases used bosonic atoms, this paper describes the first quantum degenerate Fermi gas of atoms[4]. The very different behaviors of bosonic and fermionic atom gases arise from the distribution of the particles in the available energy states of the system. The gas is typically confined by a harmonic potential and the relevant energy states are the motional states of this external potential. For bosonic atoms, as the temperature of the gas approaches absolute zero, large numbers of atoms will occupy the lowest energy state of the harmonic potential. This macroscopic occupation of a single quantum state is Bose-Einstein condensation (BEC). Fermionic atoms, on the other hand, obey the Pauli exclusion principle and at T=0 occupy the lowest energy states by stacking one per state in an arrangement called the Fermi sea.

Even though quantum degeneracy requires the same combination of temperature and density for either bosons or fermions, their very different behaviors makes it convenient to give different labels to the temperature that marks the crossover between classical and quantum regimes. For bosons, BEC is a phase transition and the critical temperature for this phase transition marks the quantum degenerate regime. For fermions, there is no phase transition and the Fermi sea arrangement emerges gradually as the temperature is lowered. In this case the Fermi temperature T_F, which corresponds to the energy of the highest occupied state at T=0, marks the crossover to the quantum regime.

COOLING FERMIONIC ATOMS

Apparatus

In our experiment the overall strategy used to cool fermionic atoms is based on techniques developed to cool bosonic atoms to BEC. We use two distinct stages of cooling: laser cooling and trapping followed by magnetic trapping and forced evaporative cooling. To probe the ultracold gas we release the atoms from the trap and then take a picture of the cloud using resonant absorption imaging. A schematic of the apparatus is shown in Fig. 1.[4] The fermionic atoms, ^{40}K, are initially collected in a vapor cell magneto-optical trap (MOT). The atoms are captured out of a room temperature background gas provided by an atom source that is enriched to approximately 5% in the fermionic isotope ^{40}K.[5] Using a series of light pulses the atoms are transferred to a second MOT in a part of the vacuum chamber that is maintained at lower vacuum pressure. At this second MOT the atoms are optically pumped and then loaded into a Ioffe-Pritchard type magnetic trap. In the magnetic trap, the gas has a measured exponential lifetime of 300 seconds, limited by collisions with residual gas in the vacuum chamber. A gas of several times 10^8 ^{40}K atoms at a temperature of approximately 150 μK is loaded into the magnetic trap.

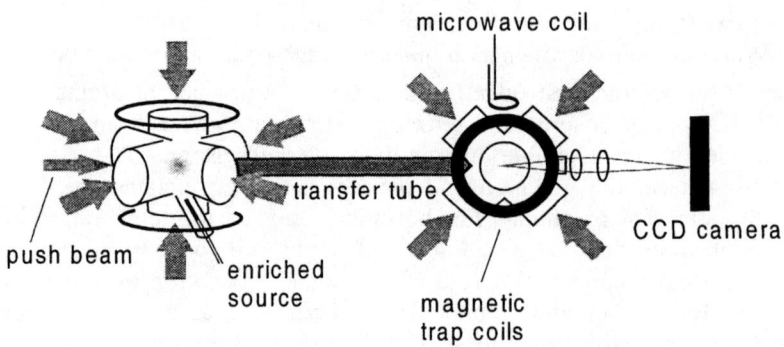

microwave coil

push beam

transfer tube

enriched
source

magnetic
trap coils

CCD camera

FIGURE 1. Schematic of apparatus, reprinted with permission from B. DeMarco and D.S. Jin, *Science* **285**, 1703-6 (1990), Copyright 1999 American Association for the Advancement of Science. Atoms are captured by the first MOT (in the glass cell on the left) and then pushed with light pulses to the second MOT (right) in a lower pressure portion of the vacuum chamber. Coils around the second glass cell provide a magnetic trap in which the atoms are evaporatively cooled using an applied microwave field. Images of the cold clouds are taken with a resonant probe pulse (incident along the axis between the two MOT's) and captured on a CCD camera.

Cold Collisions

Once in the magnetic trap, the atomic gas is ready to be evaporatively cooled. However, a fundamental difficulty in the evaporative cooling of fermionic atoms arises from the nature of cold collisions between fermions. Because the atoms are held in a magnetic trap, experiments typically employ a gas of spin-polarized atoms. Unfortunately collisions in a low temperature gas of spin-polarized fermionic atoms are strongly suppressed. This thwarts successful forced evaporative cooling which requires a reasonably large elastic collision rate to rethermalize the gas after removal of the highest energy atoms.

This effect can be seen in the results of a cold collision study that we performed using magnetically trapped ^{40}K atoms[6]. Fig. 2 shows that for spin-polarized fermionic atoms (solid circles) the elastic collision cross-section plummets as the temperature decreases. This strong temperature dependence comes from the quantum statistics of fermions in atom-atom collisions. The wavefunction describing a pair of identical, colliding fermions must be antisymmetric with respect to exchange of the particles (this is a statement of the Pauli exclusion principle). For spin-polarized atoms the spin part of the two-particle wavefunction is of course symmetric, therefore the spatial part of the wavefunction must be antisymmetric. Since *p*-waves are the lowest order partial waves that are antisymmetric with respect to exchange, collisions between spin-polarized atoms will be *p*-wave in character. However, *p*-wave collisions have one quanta of angular momentum between the two colliding atoms and therefore require a collision energy that is above the centrifugal barrier. This barrier is roughly 200 μK for ^{40}K atoms, and at temperatures below this threshold the elastic collision cross section in Fig. 2 decreases as T^2, as is expected from the Wigner threshold law for *p*-wave collisions between neutral scatterers.

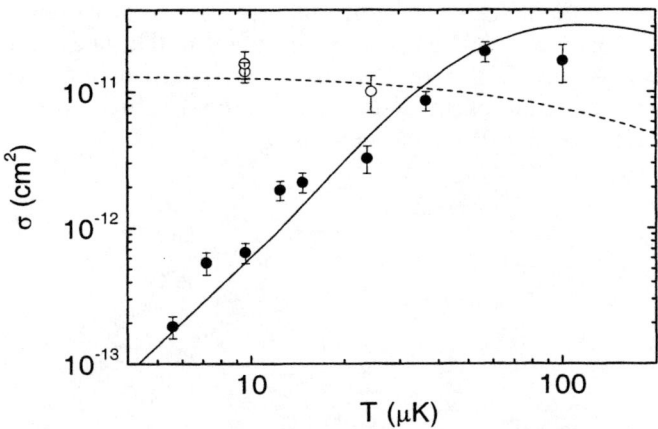

FIGURE 2. Cold collisions of [40]K atoms, reproduced from Ref. 6. The elastic collision cross section σ, measured via cross-dimensional relaxation, is plotted for a spin-polarized gas (solid circles) in the |9/2, 9/2> spin state and for a mixture of two spin-states (open circles) |9/2, 9/2> and |9/2, 7/2>. Lines represent the best fit to the data yielding a triplet s-wave scattering length a=157±20 Bohr radii for [40]K.

Because of this suppression of collisions, it is exceedingly difficult to evaporatively cool a spin-polarized Fermi gas to quantum degeneracy. A solution to this difficulty is to use a mixture of fermionic atoms in two or more spin states. In this mixture, s-wave collisions are allowed between atoms in the two different internal states since the spin part of the two-particle wavefunction can be antisymmetric with respect to exchange. For s-wave collisions, which have zero angular momentum between the colliding atoms, there is no centrifugal barrier and therefore the elastic collision cross section is, in general, constant at low temperature (open circles in Fig. 2). Because of its temperature-independent, and relatively large, elastic collision cross section, a mixture of [40]K atoms in two spin states can be used for efficient forced evaporative cooling of fermionic atoms.

Two-Species Evaporation

Holding atoms in two different spin states in a magnetic trap is possible using [40]K atoms because of the large number of spin states. The ground state energy levels of [40]K are shown in Fig. 3. The fermionic alkali isotope [40]K has a nuclear spin of 4 giving rise to two hyperfine ground states with total atomic spin f=9/2 and f=7/2. A magnetic field, for instance the bias field of the magnetic trap, lifts the degeneracy among states with different magnetic quantum number m_f. Half of the 18 spin states of [40]K can in principle be confined by the standard magnetic traps used in cold atom experiments. However, a gas of atoms in a mixture of spin states is subject to spin relaxation; inelastic spin-exchange collisions generally lead to rapid loss from the magnetic trap. Therefore we chose to trap ^{40}K atoms in the states $|f, m_f> = |9/2, 9/2>$ and $|9/2, 7/2>$. At low T this mixture is stable against spin-exchange collisions, which preserve the sum of m_f. In addition Fermi-Dirac statistics suppresses collisions between atoms in the same spin state.

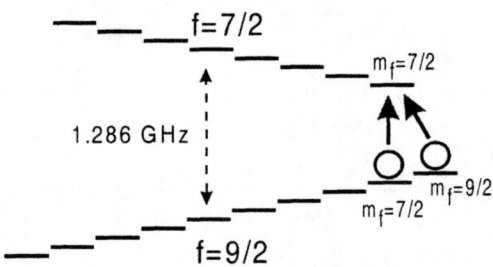

FIGURE 3. Ground-state energy levels of ^{40}K[7], shown with exaggerated Zeeman splitting. Atoms in the states $|9/2, 9/2>$ and $|9/2, 7/2>$ are simultaneously confined in the magnetic trap. Solid arrows indicate the hyperfine-changing transitions used for forced evaporation.

We execute a simultaneous forced evaporation of the two component gases, keeping the system in a roughly 50/50 spin mixture; this is required to maintain a large elastic collision rate for efficient cooling. For evaporative removal of atoms from the trap, we apply a frequency-swept microwave field around 1.3 GHz to drive transitions to an untrapped spin state, $|7/2, 7/2>$, in the upper hyperfine manifold (see Fig. 3). The two-species evaporation successfully cools the ^{40}K gas to the quantum degenerate regime. Typical evaporation starts with 2×10^8 atoms at $T/T_F = 200$ in the magnetic trap and reaches $T/T_F = 1$ with $N = 10^6$ atoms. Once in the quantum degenerate regime we find that evaporation becomes more difficult[8] and the most quantum degenerate temperature we have reached is $T/T_F = 0.2$.

QUANTUM DEGENERATE FERMIONS

Thermodynamic Measurement

In the quantum degenerate regime we have explored the effects of Fermi-Dirac statistics with thermodynamic measurements of the trapped ultracold ^{40}K gas.[4] To simplify interpretation of the data these measurements were performed on a single-component Fermi gas. Atoms in the second spin state, $|9/2, 7/2>$, were removed using a frequency-swept microwave field to drive hyperfine-changing transitions. These atoms were removed slowly, compared to the collision rate in the gas, in order to ensure that the remaining spin-polarized cloud stayed in thermal equilibrium; in fact this removal provides the last stage of evaporative cooling. The momentum distribution of the remaining single-component Fermi gas was measured using a time-of-flight imaging technique. The magnetic trap is turned off quickly and the atom gas expands ballistically for 10 to 20 ms. Then resonant absorption imaging is used to take a picture of the expanded cloud. Interaction effects are negligible in the expansion since even in the original two-component Fermi gas the interaction energy[9,10] is always less than 0.3% of the Fermi energy. Since the cloud expands due to its kinetic energy, the momentum distribution in the magnetic trap can be extracted from these absorption images.

Momentum Distribution

Classically the momentum distribution is always gaussian; however, Fermi-Dirac statistics distorts this gaussian distribution at low T/T_F. Measured expanded cloud profiles at three different temperatures, $T/T_F = 0.9$, 0.5, and 0.2, are shown in Fig. 4. The Fermi temperature T_F is determined using[11]

$$T_F = \frac{\hbar\bar{\omega}}{k_b}(6N)^{1/3} \tag{1}$$

where k_b is Boltzmann's constant, $\hbar$ is Planck's constant divided by 2π, $\bar{\omega}$ is the geometric mean of the angular trap frequencies in three orthogonal directions, and N is the total number of atoms determined from the total absorption. The temperature T is determined from a fit to the expanded cloud images as discussed below. There is an overall systematic uncertainty in T/T_F of 20%, coming primarily from uncertainty in obtaining N from absorption images.

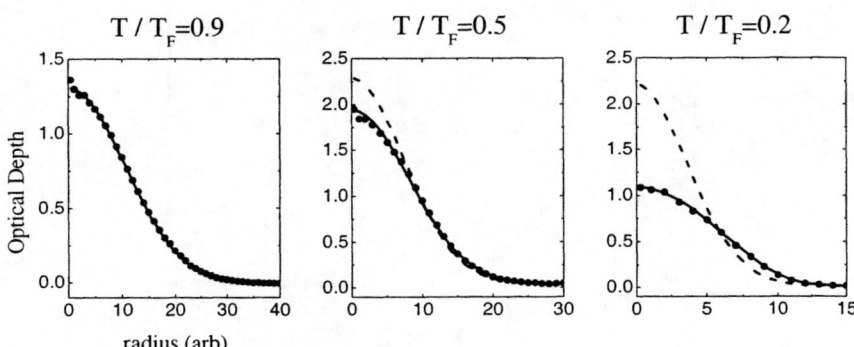

FIGURE 4. Expanded cloud profiles of the Fermi gas. The optical depth, which is proportional to density, is plotted vs. the distance from the center of the cloud. These expanded cloud profiles correspond to the momentum distribution of the gas. Solid points represent angularly averaged data, the dashed line shows the classical prediction, and the solid line is a fit to the expected form for fermions. The clouds, from left to right, have $N=2\times10^6$, 1.7×10^6, and 4×10^5 atoms, T=700 nK, 350 nK, and 90 nK, and $T/T_F=0.9$, 0.5, and 0.2 respectively.

Also plotted Fig. 4 is the classical gaussian distribution predicted for the same N and T (dashed line) as well as fits to a Thomas-Fermi form[11] for an ideal Fermi gas (solid line). While high T/T_F data agree with the classical prediction, the low T/T_F clouds show fewer atoms at low momentum (near the center of the expanded cloud image) and a wider distribution. This agrees with the expected behavior for fermions which are forced to occupy progressively higher energy states as the lowest states become filled.

The results of fitting the expanded cloud shapes to the Thomas-Fermi profile[12] are summarized in Fig. 5 where we plot the fugacity vs. T/T_F. Thermodynamically the fugacity, defined by $\chi = \exp(\mu/k_b T)$ where μ is the chemical potential, is an analytic function of T/T_F (solid line in Fig. 5). However when fitting the expanded cloud images we do not impose this constraint on the fitting function but rather leave both χ and T as free parameters. The fugacity χ gauges the overall deviation of the expanded cloud shape from the classical gaussian, with higher χ corresponding to a larger deviation. The fitting function asymptotes to a gaussian at large radii (distance from the center of the cloud), and T is proportional to this gaussian's width squared. The fact that fits to the data give the χ that thermodynamically corresponds to T/T_F (determined from T, N, and $\bar{\omega}$ as described above) demonstrates that the momentum distribution of the ultracold gas of ^{40}K atoms follows the expected form for an ideal Fermi gas.

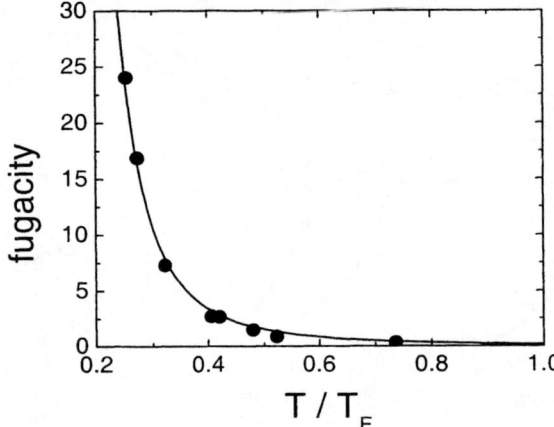

FIGURE 5. Results of fitting the expanded cloud images to a Thomas-Fermi profile for an ideal Fermi gas. The fugacity vs. temperature T/T_F, determined from fits to the expanded cloud images (solid points), agrees well with the thermodynamic prediction (solid line).

Energy

We can also determine the expansion energy by measuring the second moment of the expanded cloud images[13] and then infer the total energy in the original trapped gas. In Fig. 6 we plot the ratio of the measured cloud energy per particle U to the classical prediction $U_{cl} = 3\ k_b\ T$ as a function of T/T_F. For high T/T_F the measured energy in the ^{40}K atom gas agrees with the classical prediction, while at T/T_F below 1 we observe more energy in the gas than would be expected classically. The data agree with the thermodynamic prediction for an ideal Fermi gas (solid line in Fig 6). The "excess" energy in the Fermi gas comes from the formation of a Fermi sea of atoms; as atoms start to fill the lowest energy states with an occupancy near unity they are forced by Fermi-Dirac statistics to fill higher energy states of the external potential.

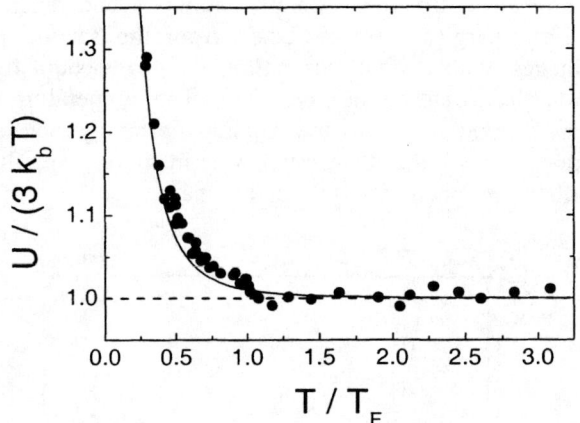

FIGURE 6. Energy of the ultracold Fermi gas. The ratio of the measured energy to the classical prediction (solid points) is plotted as a function of T/T_F. The data agree with the classical limit (dashed line) at high temperature but increase above the classical value at low T/T_F as predicted thermodynamically for an ideal Fermi gas (solid line).

TWO-COMPONENT FERMI GAS

Energy

The fact that our evaporative cooling strategy employs two spin-states makes it straightforward to extend these studies to a two-component Fermi gas. The two-component Fermi gas is interesting because of the possibility that interactions, in addition to the quantum statistics, could become important in determining the behavior of the system. Initial studies discussed here will be relevant to future investigations of an interacting Fermi gas and the possibility of Cooper pairing of atoms and fermionic condensates.

To examine the two-component Fermi gas we apply a Stern-Gerlach magnetic field, with a gradient of 70 gauss/cm at the atoms, during the expansion in order to separate the m_f=9/2 and 7/2 components for imaging. The two components are well overlapped in the magnetic trap and are only separated after the gas is released from the magnetic trap. We can then study the thermodynamics of the two-component Fermi gas by analyzing the expanded cloud images as discussed in the previous section.

Fig. 7 shows the ratio of the average energy per particle for the m_f=9/2 and m_f=7/2 clouds as a function of the T/T_F for the more quantum degenerate m_f=9/2 cloud, T/T_F 9/2. For this experiment the fractional number in the m_f=9/2 cloud was 92±3% where the uncertainty represents the variation in the data shown in Fig. 7. The number N in each cloud was controlled through evaporative removal and determined from the total absorption. The energies were extracted from the second moment of the expanded cloud images, with a small correction (4%) to account for the measured effect of the Stern-Gerlach field on the expansion. The temperature was determined by fitting the m_f=9/2 cloud image to the Thomas-Fermi distribution for an ideal Fermi gas. In order to check that the system was in thermal equilibrium we have varied the rethermalization time before taking the image.

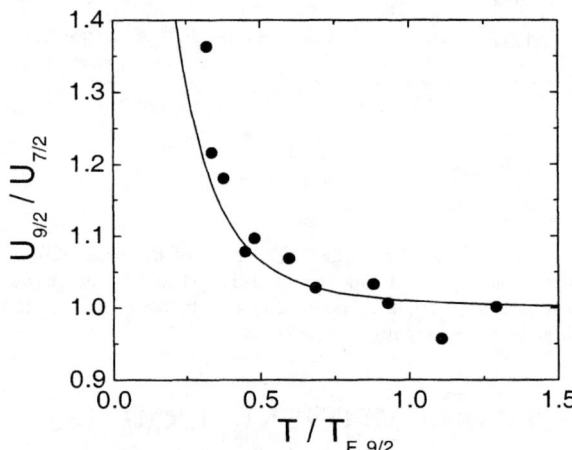

FIGURE 7. Energy of a two-component Fermi gas. The ratio of the average energy per particle in the two components is plotted as a function of T/T_F for the more abundant m_f=9/2 cloud. Because of the difference in the number of atoms in each component, T/T_F for the m_f=7/2 cloud is 2.5 times higher than that for the m_f=9/2 cloud. The solid line is the thermodynamic prediction for ideal Fermi gases.

Fig. 7 is reminiscent of the single-component Fermi gas data shown in Fig. 6. While classically the average energy per atom is the same for the two components, the Fermi-Dirac statistics results in a higher energy per particle for the more quantum degenerate cloud. Because of the difference in the number of atoms in each component, the $m_f=9/2$ cloud is more quantum degenerate; T/T_F for the $m_f=7/2$ cloud is 2.5 times higher than $T/T_{F\,9/2}$. One can view this data as a demonstration of a new kind of thermometry. With the single-component gas we are forced to extract T and U from the same data. However with the two-component gas we can use a second component with low N as a thermometer that is independent of the quantum gas. In the limit of very low N the second component behaves classically with the temperature simply proportional to its expansion energy. This is a thermometer in the usual sense of a well-understood system put in thermal contact with the quantum gas.

Pauli Blocking

Fig. 7 also shows that we can now explore a new system – a two-component Fermi gas. The results of this experiment reveal an imbalance of the energy between the two components that is surprising from a microscopic picture of collisional interactions between the atoms. Classically collisions, averaged over all possible initial and final states, would tend to equalize the energy per particle in the two components. So why doesn't the $m_f=9/2$ cloud transfer its excess energy to the $m_f=7/2$ cloud? The answer comes from Pauli blocking[14,15], which is a dynamical effect of the quantum statistics on collisions. A collision that would transfer energy from a $m_f=9/2$ atom to a $m_f=7/2$ atom can be "blocked" by the quantum degeneracy of the $m_f=9/2$ cloud. At low T/T_F there is a high probability for low energy states of the harmonic potential to be occupied and this, combined with the Pauli exclusion principle, reduces the density of final states for collisions. This Pauli blocking, which gives rise to an additional suppression factor that must be included in the average over all possible collisions, allows the $m_f=9/2$ atoms to have higher energy on average than the less quantum degenerate $m_f=7/2$ atoms.

We have explored this effect theoretically in preparation for a future experiment where we intend to look for the Pauli blocking directly in a collision measurement. Using quantum kinetic theory[8,16] we calculate the thermal relaxation of a two-component Fermi gas that is initially out of thermal equilibrium with too little energy in the $m_f=7/2$ component. In the calculations we assume a Fermi-Dirac distribution for each component and we include the small difference in the trap frequency due to the different magnetic moments of the two spin states. To model the time dependence of the relaxation, we calculate the time rate of change of the energy of the $m_f=7/2$ cloud by weighting each collision by the energy transferred to the $m_f=7/2$ cloud and then averaging over all possible collisions.

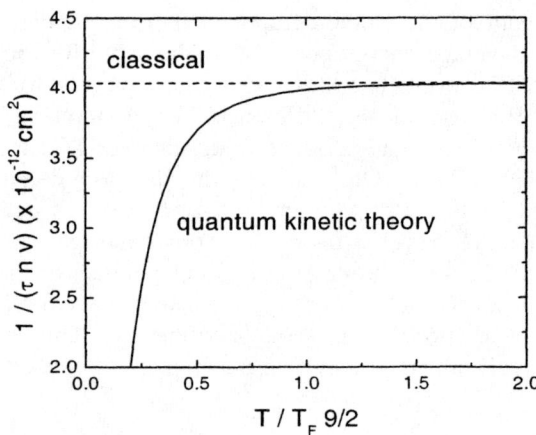

FIGURE 8. Calculated thermal relaxation for a two-component Fermi gas. Classically the elastic collision cross section is proportional to the relaxation rate $1/\tau$, divided by the density n and relative speed v. However Pauli blocking in the collisional dynamics of the Fermi gas causes the relaxation rate to decrease at low T/T_F. The calculation was done for an ideal gas with 90% of the atoms in the m_f=9/2 state.

We find that the energy of the m_f=7/2 cloud can be well approximated by an exponential in time.[17] We calculate the exponential time constant, τ, for the thermal relaxation of a gas with 90% of the atoms in the m_f=9/2 spin state. For the calculations, we assume a harmonic trap with an axial frequency of 137 Hz and a radial frequency of 19.5 Hz for atoms in the m_f=9/2 spin state. The results of the calculation are shown in Fig. 8 where we plot $\dfrac{1}{\tau n v}$ as a function of $T/T_{F\,9/2}$; here n is the relevant density-weighted density $n = \dfrac{1}{N_{7/2}} \int n_{7/2}(\vec{r}) \cdot n_{9/2}(\vec{r})\, d^3\vec{r}$ and v is the average relative speed between colliding atoms. Both n and v are calculated assuming a Thomas-Fermi profile. The quantity plotted in Fig. 8 has units of cross section and indeed classically would be proportional to the elastic collision cross sections in the gas. However, because of Pauli blocking we find that $\dfrac{1}{\tau n v}$ is not constant but instead decreases as the m_f=9/2 cloud becomes quantum degenerate. Thus an experimental study of the thermal relaxation rate in the two-component Fermi gas should provide a direct probe of the Pauli blocking.

ACKNOWLEDGMENTS

This work is funded by the National Science Foundation, the National Institute of Standards and Technology, and the Office of Naval Research.

REFERENCES

1. M. H. Anderson, J. R. Ensher, M. R. Matthews, C. E. Wieman and E. A. Cornell, *Science* **269**, 198-201 (1995).
2. K. B. Davis, M. O. Mewes, M. R. Andrews, N. J. van Druten, D. S. Durfee, D. M. Kurn and W. Ketterle, *Phys. Rev. Lett.* **75**, 3969-73 (1995).
3. C. C. Bradley, C. A. Sackett and R. G. Hulet, *Phys. Rev. Lett.* **78**, 985-9 (1997).
4. B. DeMarco and D. S. Jin, *Science* **285**, 1703-6 (1999).
5. B. DeMarco, H. Rohner and D. S. Jin, *Rev. Sci. Instrum.* **70**, 1967 (1999).
6. B. DeMarco, J. L. Bohn, J. P. Burke, Jr., M. Holland and D. S. Jin, *Phys. Rev. Lett.* **82**, 4208-11 (1999).
7. E. Arimondo, M. Inguscio, P. Violino, L. Vichi and S. Stringari, *Reviews of Modern Physics* **49**, 31-75 (1977).
8. M. J. Holland, B. DeMarco and D. S. Jin, *Phys. Rev. A* **61**, 053610/1-6 (2000).
9. L. Vichi and S. Stringari, *Phys. Rev. A* **60**, 4734-7 (1999).
10. M. Amoruso, A. Minguzzi, S. Stringari, M. P. Tosi and L. Vichi, *Euro. Phys. D* **4**, 261-5 (1998).
11. D. A. Butts and D. S. Rokhsar, *Phys. Rev. A* **55**, 4346-50 (1997).
12. The fitting form includes the effect of integrating along the line-of-sight of the probe beam.
13. The second moment of the images is determined using a gaussian fit that minimizes the root mean square deviation in energy.
14. J. M. V. A. Koelman, H. T. C. Stoof, B. J. Verhaar and J. T. M. Walraven, *Phys. Rev. Lett.* **59**, 676-9 (1987).
15. G. Ferrari, *Phys. Rev. A* **59**, R4125-8 (1999).
16. O. J. Luiten, M. W. Reynolds and J. T. M. Walraven, *Phys. Rev. A* **53**, 381-9 (1996).
17. For the conditions of the calculation, where 27% of the total energy in the system is transferred between the two components during rethermalization, the ratio of $dU_{7/2}/dt$ to the difference $U_{7/2}(t)-U_{7/2}(t \rightarrow \infty)$ is constant in time t to within 10%.

Manipulation and Phase Engineering of Bose-Einstein Condensates

K. Bongs, S. Burger, K. Sengstock and W. Ertmer

Institut für Quantenoptik
Universität Hannover Welfengarten 1
D-30167 Hannover, Germany

Abstract. We present experiments investigating different aspects concerning the coherent manipulation of a Bose-Einstein condensate wavefunction with far blue detuned dipole potentials. In a first set of experiments we study a waveguide for matter waves as an important atom optical element allowing to hold and guide the atoms for long observation times. We present a flexible realization of this waveguide based on a tube shaped Laguerre Gaussian laser beam and investigate the dynamics of a Bose condensed ensemble transferred into this guide. In a second set of experiments we study the dynamics of dark solitons created by the manipulation of the phase of the condensate wave function by temporally applied dipole potentials. This method of phase imprinting represents a powerful tool for the design of matter waves and the controlled creation of various excitations. The dynamics of dark solitons in matter waves is investigated in detail by experimentally and numerically studying the temporal evolution of the condensate density as well as by mapping the velocity field of the excited condensate wavefunction. In these measurements we directly detect the flux of the wave function associated with the dark soliton using spatially resolved Bragg scattering. Experiments investigating the creation and interaction of several dark solitons conclude this chapter.

INTRODUCTION

The experimental realization of Bose-Einstein condensation (BEC) in weakly interacting atomic gases [1–4] provides access to quantum mechanical systems which are unique in several respects. The dilute nature in combination with the low temperature and the purity of these systems allows to observe and to compare phenomena known from other fields, like superfluidity [5,6], vortices [7,8], spin domains [9] or nonlinear optics [10], and their theoretical treatment from first principles [11]. Moreover, the macroscopic size of a Bose-Einstein condensate opens the exciting possibility to manipulate the quantum mechanical wave function itself. This aspect is also of great importance for one of the most interesting future prospects for Bose-Einstein condensates, their application as a source of coherent matter waves [12–15], e.g. in atom optics and atom interferometry. The nonlinear dynamics of

CP551, *Atomic Physics 17*, edited by E. Arimondo, P. DeNatale, and M. Inguscio
© 2001 American Institute of Physics 1-56396-982-3/01/$18.00

BEC leads to a rich variety of phenomena, e.g. self interference structures after free fall and reflection from a mirror [16].

The nonlinear aspects can be controlled in atom optical applications, e.g. by using matter wave guides in geometries similar to fibers in light optics. Single mode operation can be realized by using Bose-Einstein condensates to load the lowest mode of matter waveguides. Confinement in such a guide opens the perspective of long observation times and thus high sensitivity when used in interferometric sensors. In addition, coherent waveguides for matter waves are the basis for the development of integrated atom optics.

We present experiments on the non-adiabatic transfer of Bose condensed ensembles into a harmonic waveguide with observation times up to 400 ms.

Besides coherent confinement, an important aspect for future applications is the controlled preparation of matter waves in a desired state. A powerful tool in this respect is the method of phase manipulation of the condensate wave function by a local application of pulsed dipole potentials which facilitates the application of specifically tailored wave functions. This method is also well suited for the controlled creation of fundamental excitations and we present its use to create dark solitons in Bose-Einstein condensates.

We will concentrate on the case of far detuned light fields where spontaneous emission processes can totally be neglected during the atom-light interaction. This allows to treat the corresponding dipole potentials as conservative and thus coherent potentials.

LOADING BEC'S INTO A DE BROGLIE WAVEGUIDE

Optical fibers for the guiding of photons represent a robust tool for many applications ranging from high speed data communication to fundamental physics. Correspondingly, waveguides for atomic matter waves promise to be extremely useful in various atom optical applications (for a review see, e.g. [17]). Guiding of atoms has been demonstrated, e.g. in experiments using hollow fibers [18–21] or freely propagating light beams [22,24]. An additional purpose of waveguides for matter waves is to "hold" atomic ensembles against gravity thus allowing long interaction and observation times for earth based systems. Of particular interest are new interferometer designs with an unprecedented sensitivity, e.g. long arm Sagnac interferometers using guided atoms.

In this respect the availability of coherent matter waves increases the demand for coherent guides. Configurations similar to those well established in linear and nonlinear light optics with laser light in fibers can be a basis for the new regimes of coherent linear and nonlinear matter wave optics.

The operation of single mode guiding, in reach within the nearest future, will allow for fascinating aspects, like pure 1D configurations for coherent matter waves, controlled collisions in strongly confined 1D systems, as well as the study of fundamental excitations like solitons in 1D or quasi 1D regimes.

As a class of waveguides, the dipole potentials of tube shaped light fields are well suited for gaining insight into different confinement regimes by changing the beam parameters. In addition they have the advantage of being roughness free manipulation tools, well suited for flexible geometries and for fast timing. A further advantage of dipole traps, e.g. compared to magnetic traps, is the possibility of trapping atoms independently of their magnetic substate [9,23]. We use the repulsive dipole potential of blue detuned Laguerre-Gaussian ("doughnut") laser beams of different orders leading to waveguides with different transverse power law potentials. In earlier experiments, the loading of a Bose-Einstein condensate into a TEM_{05}^* waveguide was studied [25]. In this chapter, we report on experiments using a Laguerre-Gaussian beam of first order, TEM_{01}^*, to hold and guide the atoms. The radially symmetric intensity distribution, $I(r)$, for this doughnut mode is given by

$$I(r) = \frac{4Pr^2}{\pi r_0^4} \cdot e^{-2\frac{r^2}{r_0^2}}, \tag{1}$$

where P and r_0 are the laser power and the beam waist, respectively. With a power of $P = 1\,\text{W}$ at 532 nm and a beam waist of $r_0 \approx 10\,\mu\text{m}$, the resulting dipole potential at the focal plane has a maximum value of $\sim 120\,\mu\text{K}\cdot k_B$ with a transverse oscillation frequency of $\sim 6\,\text{kHz}$ for Rb atoms.

As guiding of the atoms is realized in the low intensity region of the light field, light scattering which leads to decoherence is suppressed for atoms in the transverse ground state by more than three orders of magnitude in comparison to the scattering rate in the intensity maximum. In the experiment described here the

FIGURE 1. Schematic setup for the guiding of Bose-Einstein condensates within a doughnut mode light field.

Bose-Einstein condensates were (non-adiabatically) transferred into the waveguide potential with a loading rate of up to 100%. The waveguide was aligned with the long condensate axis (see Fig. 1) with a slight tilt allowing for gravitational acceleration. We studied the ensemble for observation times as long as 400 ms inside the waveguide after which the atomic cloud became too dilute to be monitored due to axial expansion (Fig. 2). These results show the high potential of this type of waveguide even for non-adiabatic loading.

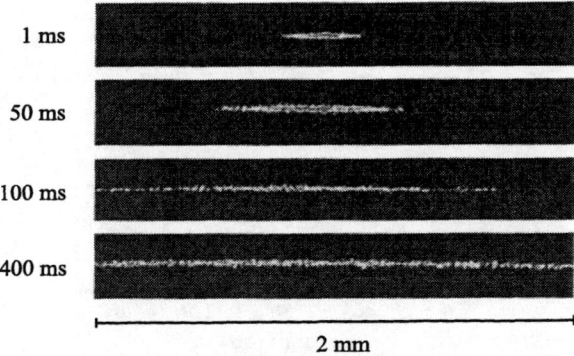

1 ms	
50 ms	
100 ms	
400 ms	

2 mm

FIGURE 2. Atoms from a BEC loaded to a doughnut waveguide with different evolution times inside the waveguide. The atomic ensemble is accelerated to the left by gravity due to a slight tilt of the light field with respect to the horizontal plane. Broadening of the cloud occurs due to heating effects and a release of mean-field energy.

We expect the axial expansion of the ensemble to be governed by mean field energy as well as by the alignment-dependent potential energy in the waveguide after the non-adiabatic transfer. The mean field energy leads to an immediate (accelerated) expansion after the axial confinement is turned off. The potential energy can be distributed among the different degrees of freedom under certain conditions due to rethermalization processes and is thus expected to show up in the axial expansion after some time delay. The axial expansion rate then corresponds to the temperature of the ensemble and can be used as indicator for the alignment. Fig. 3 shows results on the broadening of the atomic density distribution within the waveguide for two different radial alignments of the waveguide. The data gives evidence of two time scales in the axial expansion due to a transfer of the initial radial energy to axial energy which occurs with a time delay on the order of 30 to 50 ms. From the experimental data axial temperatures of approximately $1.5\,\mu$K and $20\,\mu$K can be deduced from the constant expansion corresponding to energies much larger than the mean-field energy of the original condensate ($\sim 40\,$nK).

The studies of non-adiabatic loading of a BEC into a matter waveguide give insight into the dynamics during and after the transfer and can help to optimize the alignment. Currently we study an improved scheme by adiabatically transferring the condensate into the waveguide leading to a coherent single mode operation. Also a direct condensation into the dipole potential trap in combination with a magnetic trap is currently under investigation.

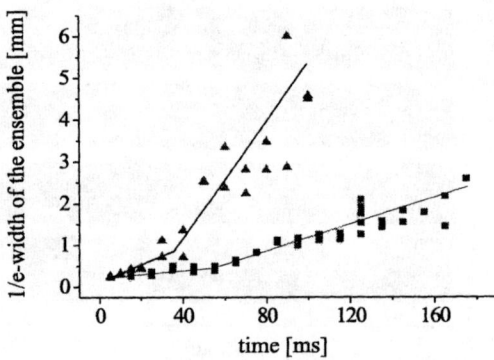

FIGURE 3. Longitudinal expansion of Bose-Einstein condensates loaded into a doughnut waveguide with two different radial alignments. The squares (triangles) are data points for a slight (relatively large) misalignment between the center of the doughnut mode and the BEC in the magnetic trap. (The lines are added to guide the eye.)

CREATION AND INVESTIGATION OF DARK SOLITONS IN BOSE-EINSTEIN CONDENSATES

In contrast to "classical" atom optical elements directly influencing the density distribution of the atomic ensemble, recently developed holographic methods open the exciting possibility to locally effect the phase of an atom or the collective phase of a Bose condensed ensemble. In this section we describe the application of the technique of *"phase imprinting"*, which allows for engineering of a variety of excited states in coherent matter waves, e.g. to convert the ground state of a Bose-Einstein condensate into the state of a dark soliton [26].

Solitons as shape preserving waves are typical for nonlinear media. *Dark solitons* as an important class of macroscopically excited Bose condensed states clearly demonstrate the intrinsic nonlinearity of these systems. In this respect they are of particular interest within the new field of *nonlinear* atom optics to explore nonlinear properties of matter waves.

Solitonlike solutions of the Gross-Pitaevskii equation are closely related to similar solutions in nonlinear optics describing the propagation of light pulses in optical fibers. Here, bright soliton solutions correspond to short pulses where the dispersion of the pulse is compensated by the self-phase modulation, i.e. the shape of the pulse does not change. Similarly, optical dark solitons correspond to intensity minima within a broad light pulse [27].

In the case of nonlinear matter waves, bright solitons are expected only for an attractive interparticle interaction (*s*-wave scattering length $a < 0$) [28], whereas dark solitons are expected to exist for repulsive interactions ($a > 0$). In these systems the repulsive interaction (trying to reduce the size of density minima) counteracts the kinetic energy (minimized by small density gradients and thus widespread density minima) leading to a self stabilizing dark soliton solution. Recent theoretical studies discuss the dynamics and stability of dark solitons [29–32], as well as concepts for their creation [26,33,34].

Conceptually, solitons provide a link of BEC physics to other fields like fluid mechanics and nonlinear optics.

Dark solitons in matter waves are characterized by a local density minimum and a sharp phase gradient of the wave function at the position of the minimum (Fig. 4(a)).

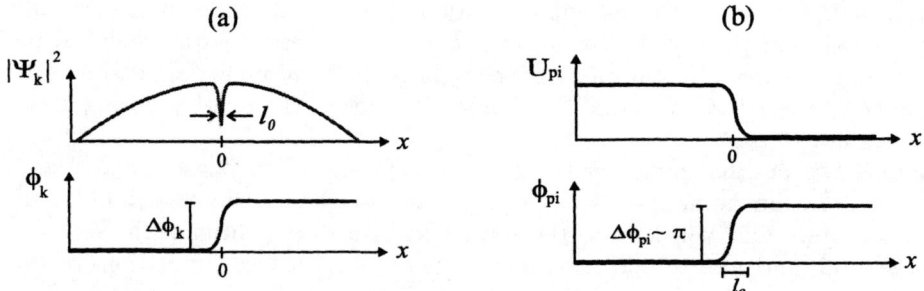

FIGURE 4. (a) Density and phase distribution of a dark soliton state with $\Delta\phi_k \sim \pi$. The density minimum has a width of $\sim l_0$. (b) Phase imprinting potential, U_{pi}, and associated phase distribution.

A dark soliton in a homogeneous 1D BEC of density n_0 is described by the wave function (see [32], and references therein)

$$\Psi_k(x) = \sqrt{n_0} \left(i \cdot \frac{v_k}{c_s} + \sqrt{1 - \frac{v_k^2}{c_s^2}} \cdot \tanh\left[\frac{x - x_k}{l_0}\sqrt{1 - \frac{v_k^2}{c_s^2}}\right]\right), \qquad (2)$$

with the position x_k and velocity v_k of the dark soliton, the correlation length $l_0 = (4\pi a n_0)^{-1/2}$, and the speed of sound $c_s = \sqrt{4\pi a n_0}\hbar/m$, where m is the mass of the atom. The corresponding density distribution shows a minimum at the position of the dark soliton with a width of the order of the (local) correlation length.

For $T = 0$ in 1D, dark solitons are stable. In this case, only solitons with zero velocity in the trap center do not move, otherwise they oscillate along the trap axis [30]. However, in 3D at finite temperature, dark solitons exhibit thermodynamic and dynamical instabilities. The interaction of the soliton with the thermal cloud causes dissipation which accelerates the soliton. Ultimately, it reaches the speed of sound and disappears [32]. The dynamical instability originates from the transfer of the (axial) soliton energy to the radial degrees of freedom and leads to the undulation of the minimum density plane, and ultimately to the destruction of the soliton. This instability is essentially suppressed for solitons in cigar-shaped traps with a strong radial confinement [31], such as in our experiment [35].

As can be seen from equation (2), the local phase of the dark soliton wave function varies only in the vicinity of the density minimum, $x \approx x_k$, and is constant in the outer regions, with a phase difference $\Delta\phi_k$ between the parts left and right to the minimum (see Fig. 4(a)).

To generate dark solitons experimentally, we apply the method of phase imprinting [26], which also allows for the excitation of vortices and other textures in Bose-Einstein condensates. A homogeneous potential, U_{pi}, generated again by the dipole potential of a far detuned laser beam is applied to one half of the condensate wave function (Fig. 4(b)). The potential is pulsed on for a time t_p, such that the wave function locally aquires an additional phase factor $e^{-i\Delta\phi_{pi}}$, with $\Delta\phi_{pi} = U_{pi} \cdot t_p/\hbar \sim \pi$. The pulse duration t_p is chosen to be short compared to the correlation time of the condensate, $t_c = \hbar/\mu$, where μ is the chemical potential. This ensures that the effect of the light pulse corresponds approximately to a change of the phase of the BEC, whereas changes of the density during this time can be neglected.

Note, however, that at larger times due to the imprinted phase $\Delta\phi_{pi}$ (Fig. 4(b)) one expects an adjustment of the phase and of the density distribution in the condensate. This will lead to the formation of a dark soliton with $\Delta\phi_k \neq \Delta\phi_{pi}$ in general, and also to additional structures. Fig. 5 shows simulations of the 1D

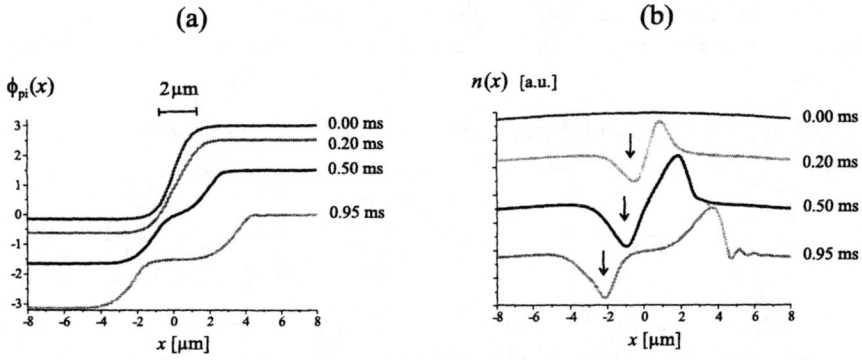

FIGURE 5. Phase distribution (a) and density distribution (b) of the condensate wave function after the phase imprinting obtained numerically from 1D-simulations of the Gross-Piteavskii equation for different evolution times in the magnetic trap. The dark soliton is indicated by an arrow.

Gross-Pitaevskii equation for the dynamics of the condensate wave function after the phase imprinting at $t = 0$ ms. After 0.2 ms the imprinted phase step leads to the formation of a density minimum and a maximum both travelling in opposite directions. These two structures "take away" part of the initial phase gradient, $\Delta\phi_{pi}$. Due to matter wave dispersion and the repulsive interparticle interaction the density maximum moves with the speed of sound and broadens, whereas the density minimum travels with smaller velocity and preserves its shape. This density minimum thus corresponds to a moving dark soliton.

In these experiments we produce Bose-Einstein condensates of ^{87}Rb every 20 s containing typically 1.5×10^5 atoms in the $(F = 2, m_F = +2)$ state, with less than 10% of the atoms being in the thermal cloud. The fundamental frequencies of our static magnetic trap are $\omega_x = 2\pi \times 14$ Hz and $\omega_\perp = 2\pi \times 425$ Hz along the axial and radial directions, respectively. The condensates are cigar-shaped with the long axis (x-axis) oriented horizontally.

For the phase imprinting potential, U_{pi}, we use a blue detuned, far off-resonant laser field ($\lambda = 532$ nm) of intensity $I \approx 50$ W/mm^2 pulsed for a time $t_p = 20\,\mu$s resulting in a phase shift $\Delta\phi_{pi}$ on the order of π [36]. Spontaneous processes can be totally neglected.

A high quality optical system is used to image an intensity profile onto the BEC, nearly corresponding to a step function with a width of the edge, l_e, smaller than $3\,\mu$m. The corresponding potential gradient leads to a force transferring momentum locally to the wave function and supporting the creation of a density minimum for the dark soliton. Note that the velocity of the soliton also depends on l_e.

After applying the dipole potential we let the atoms evolve within the magnetic trap for a variable time t_{ev}. We then release the BEC from the trap (switched off within 200 μs) and take an absorption image of the density distribution after a time of flight $t_{TOF} = 4$ ms (reducing the density in order to get a good signal-to-noise ratio in the images).

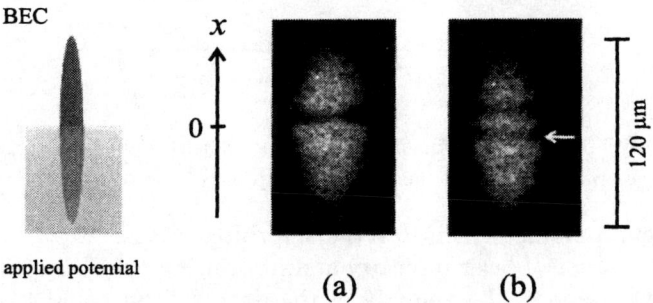

FIGURE 6. Absorption images of Bose-Einstein condensates with kink-wise structures propagating in the direction of the long condensate axis for two different evolution times, t_{ev}, in the magnetic trap: (a) $t_{ev} = 200\,\mu$s and (b) $t_{ev} = 3$ ms. The dark soliton is marked by an arrow. ($\Delta\phi \sim \pi$, $N \sim 1.5 \times 10^5$, and $t_{TOF} = 4$ ms).

In a series of measurements we have studied the creation and successive dynamics of dark solitons as a function of the evolution time and the imprinted phase. Figures 6(a) and (b) show density profiles of the atomic clouds for two different evolution times in the magnetic trap, t_{ev}. The phase imprinting potential, U_{pi}, has been applied to the lower part of the condensate with $x < 0$ and the potential strength in this measurement corresponded to a phase shift of $\sim \pi$.

For short evolution times (Fig. 6(a)) the density profile of the condensate shows a pronounced minimum (contrast about 40%). After a time of typically $t_{ev} \sim 1.5\,\mathrm{ms}$, a second minimum appears. Both minima (contrast about 20% each) travel in opposite directions and in general with different velocities along the long condensate axis (Fig. 6(b)).

One of the most important results is that both structures move with velocities which are *smaller* than the speed of sound ($c_s \approx 3.7\,\mathrm{mm/s}$ for our experimental parameters) and depend on the applied phase shift $\Delta\phi_{pi}$. We identify the minimum moving slowly in the negative x direction as a dark soliton. Figure 7 shows the experimentally observed evolution of this dark soliton. The other minimum can be attributed to a density wave which "consumes" part of the imprinted phase $\Delta\phi_{pi}$ and which travels in the positive x direction with a velocity close to c_s. After opening the trap, a complex dynamics results in the appearance of a second minimum behind the density wave [37].

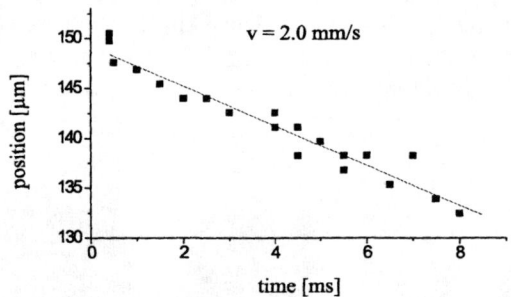

FIGURE 7. Position of the experimentally observed density minimum corresponding to the dark soliton versus evolution time in the magnetic trap.

With different parameter sets for the imprinted phase, $\Delta\phi_{pi}$, which is determined by the product of laser intensity and imprinting time, and the width of the imprinted phase step, l_e, the velocity of the dark soliton could be varied experimentally between $v_k = 2.0\,\mathrm{mm/s}$ (Fig. 7) and $v_k = 3.0\,\mathrm{mm/s}$.

Our experimental observations agree very well with theoretical investigations and numerical simulations of the 3D Gross-Pitaevskii equation performed by A. Sanpera, M. Lewenstein, and G.V. Shlyapnikov [37].

The movement of the dark soliton is connected to a matter wave flux in the opposite direction. This flux is associated with the phase gradient across the soliton as follows if one writes the condensate wave function as

$$\Phi(\vec{r}, t) = \sqrt{n(\vec{r}, t)}e^{iS(\vec{r}, t)} \tag{3}$$

with the local density $n(\vec{r}, t)$ and local phase $S(\vec{r}, t)$. Via

$$n(\vec{r}, t)\vec{v}(\vec{r}, t) = \frac{\hbar}{2im}(\Phi^* \nabla \Phi - \Phi \nabla \Phi^*) \tag{4}$$

the resulting velocity field is given by the gradient of the phase of the matter wave

$$\vec{v}(\vec{r}, t) = \frac{\hbar}{m} \nabla S(\vec{r}, t). \tag{5}$$

In order to measure the matter wave flux associated with the dark soliton we performed velocity selective Bragg spectroscopy [38,39] to the condensate wave function as shown schematically in Fig. 8. For these experiments the magnetic

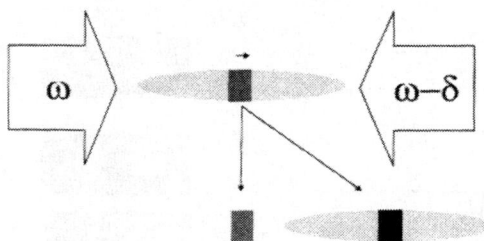

FIGURE 8. Schematic setup for matter wave flux measurements by velocity selective Bragg spectroscopy.

trapping fields were left on for a variable evolution time after phase imprinting, allowing the dark soliton to form. Then 2 ms time of flight followed before the Bragg pulse. The images were taken after another 13 ms of time of flight, allowing the Bragg deflected atoms to seperate. The velocity sensitivity of the π-Bragg pulse was $\sim 1.5\,\mathrm{mm/s}$.

The matter wave velocity for which the resonance condition was satisfied was varied by changing the frequency difference of the Bragg lasers allowing to measure the velocity distribution of the wave function.

We compared our results with numerical simulations of the 1D Gross Pitaevskii equation for our experimental parameters. The left part of Fig. 9 shows schematically the density distribution and the velocity distribution prior to the Bragg pulses as deduced from these simulations. From the numerical simulation it is clear, that the density wave as well as the dark soliton are associated with a matter flux in the same direction and with similar velocities. These regions with relatively high velocities are well localized inside the condensate wavefunction which is essentially at rest. When scanning the Bragg frequency the resonance condition can be tuned over the full range of velocities occurring inside the excited matter wave. In this way the different velocity components can be selectively scattered by the Bragg grating.

This is shown in Fig. 9. For the upper absorption image no Bragg pulse was applied and we observe the condensate with two minima corresponding to the dark soliton and the density wave as discussed above. The second image corresponds to Bragg scattering with a Bragg laser detuning resonant for the zero velocity part of the wavefunction. In this case most of the wavefunction is scattered and appears in the right part of the image. Only the two high velocity peaks corresponding to the dark soliton and the density wave remain unscattered. The third case shown in the lowest image corresponds to the opposite case of Bragg scattering of the high velocity peaks associated with the density wave and the dark soliton while the resting part of the wavefunction remains unscattered.

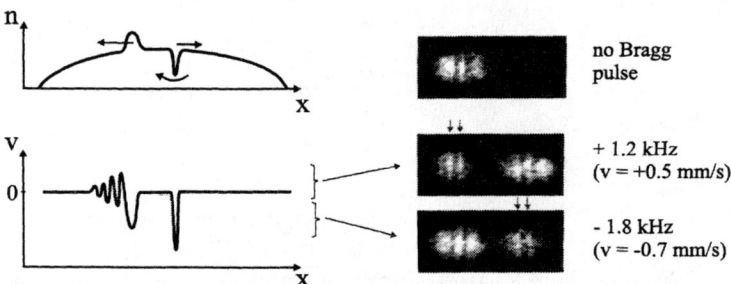

FIGURE 9. Density and velocity distribution of a 1D dark soliton state as obtained from numerical simulations of the 1D Gross Pitaevskii equation (left side). The corresponding density distribution of the condensate and for condensates after Bragg spectroscopy for different Bragg detunings are shown in absorption images on the right.

The experimental results also clearly show the signature of the presence of *dissipation* in the dynamics originating from the interaction of the dark soliton with the thermal cloud. We observe a decrease of the contrast of the dark soliton by $\sim 50\%$ on a time scale of 10 ms. This is in contradiction with a nondissipative dynamics, where the contrast should even increase for a dark soliton moving away from the trap center [32]. The decrease of the contrast can therefore only be explained by the presence of dissipation decreasing the energy of the dark soliton. As the lifetime of the dark soliton is sensitive to the ensemble temperature, the studies of dissipative dynamics of dark solitons will offer a possibility for thermometry of Bose-Einstein condensates in the condition where the thermal cloud is not discernible.

We would like to point out that parallel to our work, dark solitons in nearly spherical Bose-Einstein condensates were studied at NIST/Gaithersburg [40].

INTERACTION OF SEVERAL SOLITONS IN BOSE-EINSTEIN CONDENSATES

The study of dark solitons as *particle-like* objects can be extended to several solitons within a Bose-Einstein condensate. The characteristics of their interaction can provide a direct link of the physics of nonlinear matter waves to other fields like nonlinear optics.

Recent theoretical studies discuss the dynamics of *two* dark solitons in Bose-Einstein condensed ensembles [41–43]. In this case, two different interaction regimes depending on the relative orientation of the two phase gradients can be studied (see Fig. 10).

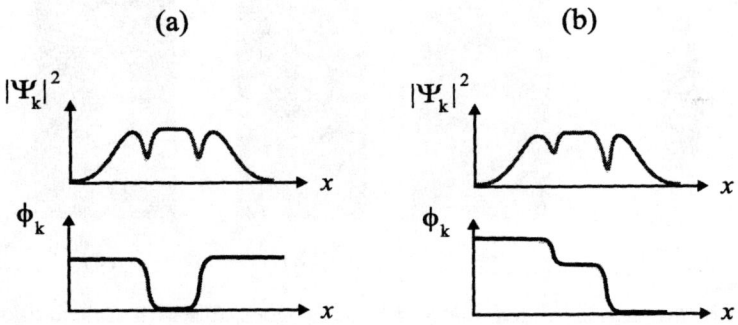

FIGURE 10. Density and phase distribution of two dark solitons in a BEC (a) with opposite phase gradients (*"kink"*- *"anti-kink"*) and (b) with the same phase gradient (*"kink"*- *"kink"*).

Two "kinks", i.e. two dark solitons with the same phase gradient, lead to an effective interaction which is attractive, whereas the interaction between a "kink" and an "anti-kink" with opposite phase gradients is effectively repulsive [41].

We present here first experimental results on the interaction of two counter-propagating dark solitons within a Bose-Einstein condensate. To generate them experimentally, we apply again the method of phase imprinting as discussed in the previous section . The homogeneous potential, U_{pi}, of the far detuned laser beam ($\lambda = 532\,\mathrm{nm}$) is now applied to the middle of the condensate wave function (Fig. 12) generating two dark solitons with opposite phase gradients (i.e. a "kink" and an "anti-kink"). After a variable evolution time, t_{ev}, within the magnetic trap we release the condensate and take an absorption image of the density distribution after a time of flight $t_{TOF} = 4\,\mathrm{ms}$.

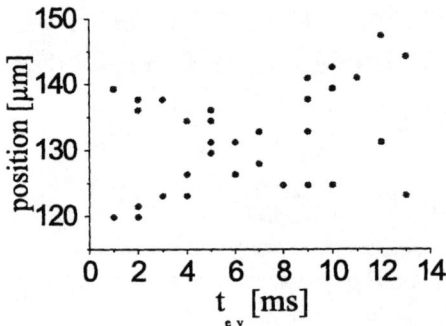

FIGURE 11. Position of the experimentally observed density minima corresponding to the two dark solitons versus evolution time in the magnetic trap.

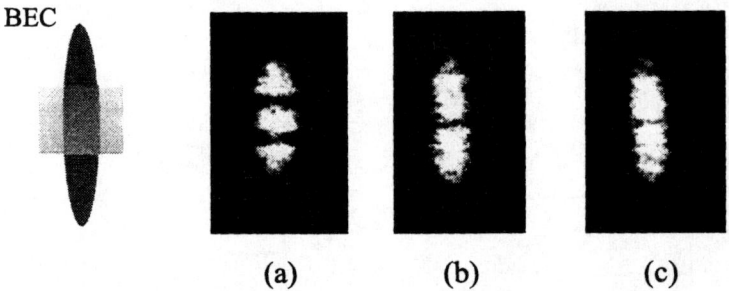

(a)   (b)   (c)

FIGURE 12. Absorption images of Bose-Einstein condensates with kink/anti-kink structures propagating in the direction of the long condensate axis for different evolution times, t_{ev}, in the magnetic trap: (a) $t_{ev} = 100\,\mu s$, (b) $t_{ev} = 5\,ms$, and (c) $t_{ev} = 12\,ms$. ($\Delta\phi \sim \pi, N \sim 1.5 \times 10^5$, and $t_{TOF} = 4\,ms$).

Figure 11 shows the evolution of two dark solitons, i.e. the experimentally observed positions of the two density minima during the first few ms in the magnetic trap after the phase imprinting. The dark solitons approach each other with approximately constant velocity and after about 5 to 6 ms they "overlap" in the central region of the condensate (see also Fig. 12(a) and (b)).

For longer evolution times up to about 20 ms, we observed quite different situations. The absorption image in Fig. 12(c) shows one single minimum of the density distribution after 12 ms still at the same position as it was after only 5 ms of evolution time within the magnetic trap. This gives clear indication for an interaction of the solitons.

The fact that our experimental results fluctuate reproducibly indicates a very critical dependence of the soliton interaction on the specific experimental parameters. We currently investigate in detail the successive dynamics of a crossing or reflection of dark solitons as well as possible "bound" states of two solitons in the presence of dissipation.

CONCLUSIONS

We have discussed the coherent manipulation of matter waves by far blue detuned dipole potentials. We presented a waveguide as "classical" element as well as phase manipulation methods realized with pulsed dipole potentials.

Bose-Einstein condensates of ^{87}Rb have been loaded to a linear waveguide for atomic de Broglie waves. The waveguide is created by the optical dipole force of a far off-resonant, blue detuned Laguerre-Gaussian laser beam of high order and the transport and broadening of the atomic cloud inside the waveguide has been studied.

The method of phase imprinting has been used to create dark soliton states in Bose-Einstein condensates as fundamental excitations in nonlinear matter waves. Investigations of the dynamics of dark solitons in Bose-Einstein condensates were presented monitoring the density as well as the velocity distribution. First experiments concerning the interaction of two dark solitons were discussed.

Acknowledgements
Part of the work presented here has been done in fruitful and stimulating co-operation with Ł. Dobrek, M. Gajda, M. Lewenstein, K. Rzążewski, A. Sanpera, and G.V. Shlyapnikov. We also acknowledge important support in different stages of the experiment by G. Birkl, S. Dettmer, D. Hellweg, M. Kottke, M. Kovacev, and T. Rinkleff.
This work is supported by SFB 407 of the *Deutsche Forschungsgemeinschaft*.

REFERENCES

1. Anderson, M. J., Ensher, J. R. , Matthews, M. R., Wieman, C. E., and Cornell, E. A., *Science* **269**, 198 (1995).
2. Davis, K. B., Mewes, M.-O., Andrews, M. R., van Druten, N. J., Durfee, D. S., Kurn, D. M., and Ketterle, W., *Phys. Rev. Lett.* **75**, 3969 (1995).
3. Bradley, C. C., Sackett, C. A., Tollett, J. J., and Hulet, R. G., *Phys. Rev. Lett.* **75**, 1687 (1995).
   Bradley, C. C.,*et al.*, *Phys. Rev. Lett.* **78**, 987 (1997).
4. Fried, D. G., Killian, T. C., Willmann, L. , Landhuis, D., Moss, S. C., Kleppner, D., and Greytak, T. J., *Phys. Rev. Lett.* **81**, 3811 (1998).
5. Raman, C., Köhl, M., Onofrio, R., Durfee, D. S., Kuklewicz, C. E., Hadzibabic, Z., and Ketterle, W., *Phys. Rev. Lett.* **83**, 2502 (1999).

6. Chikkatur, A. P., Görlitz, A., Stamper-Kurn, D. M., Inouye, S., Gupta, S., and Ketterle, W., cond-mat/0003387.

7. Matthews, M. R., Anderson, B. P., Haljan, P. C., Hall, D. S., Wieman, C. E., and Cornell, E. A., *Phys. Rev. Lett.* **83**, 2498 (1999).

8. Madison, K. W., Chevy, F., Wohlleben, W. and Dalibard, J., *Phys. Rev. Lett.* **84**, 806 (2000).

9. Stenger, J., Inouye, S., Stamper-Kurn, D. M., Miesner, H.-J., Chikkatur, A. P., and Ketterle, W., *Nature* **396**, 345 (1998).

10. PhillipsNL Deng, L., Hagley, E. W., Wen, J., Trippenbach, M., Band, Y., Julienne, P. S., Simsarian, J. E., Helmerson, K., Rolston, S. L., Phillips, W. D., *Nature* **398**, 218 (1999).

11. Dalfovo, F., Giorgini, S., Pitaevskii, L. P., and Stringari, S., *Rev. Mod. Phys.* **71**, 463 (1999).

12. Mewes, M.-O., Andrews, M. R. , Kurn, D. M., Durfee, D. S., Townsend, C. G., and Ketterle, W., *Phys. Rev. Lett.* **78**, 582 (1997).

13. Anderson, B. P. and Kasevich, M. A., *Science* **282**, 1686 (1998).

14. Hagley, E. W., Deng, L., Kozuma, M., Wen, J., Helmerson, K., Rolston, S. L., and Phillips, W. D., *Science* **283**, 1706 (1999).

15. Bloch, I., Hänsch, T. W., and Esslinger, T., *Phys. Rev. Lett.* **82**, 3008 (1999).

16. Bongs, K., Burger, S., Birkl, G., Sengstock, K., Ertmer, W., Rzążewski, K., Sanpera, A., and Lewenstein, M., *Phys. Rev. Lett.* **83**, 3755 (1999).

17. Dowling, J. P. and Gea-Banacloche, J., *Adv. At. Mol. Opt. Phys.* **37**, 1 (1997).

18. Renn, M. J., Montgomery, D., Vdovin, O., Anderson, D. Z., Wieman, C. E. and Cornell, E. A., *Phys. Rev. Lett.* **75**, 3253 (1995).

19. Renn, M. J., Donley, E. A., Cornell, E. A., Wieman, C. E., and Anderson, D. Z., *Phys. Rev. A*, **53**, R648 (1996).

20. Ito, H., Nakata, T., Sakaki, K., Ohtsu, M., Lee, K. I., and Jhe, W., *Phys. Rev. Lett.*, **76**, 4500 (1996).

21. Wokurka, G. , Keupp, J. , Sengstock, K., and Ertmer, W., *Procs. EQEC*, QFG 5, 235 (1998).

22. Schiffer, M., Rauner, M., Kuppens, S., Zinner, M., Sengstock, K., and W. Ertmer, W., *Appl. Phys. B* **67**, 705 (1998).

23. Miesner, H.-J., Stamper-Kurn, D. M., Stenger, J., Chikkatur, A. P., and Ketterle, W., *Phys. Rev. Lett.* **82**, 2228 (1999)

24. Kuppens, S., Rauner, M., Schiffer, M., Sengstock, K., and Ertmer, W., *Phys. Rev. A* **58**(4), 3068 (1998).

25. Burger, S., Bongs, K., Sengstock, K. and Ertmer, W., in Martellucci, S., *editor*, 2000, *Int. School of Quant. Electr.*, 27th *Course, Erice, Sicily (1999), Kluwer Academic Publishers, in print.*

26. Dobrek, L., Gajda, M., Lewenstein, M., Sengstock, K., Birkl, G., and Ertmer, W., *Phys. Rev. A* **60**, R3381 (1999).

27. Kivshar, Y. S., and Luther-Davies, B., *Physics Reports* **298**, 81 (1998).

28. Ruprecht, P. A., Holland, M. J., Burnett, K., and Edwards, M., *Phys. Rev. A* **51**, 4704 (1995).

29. Zhang, W., Walls, D. F., and Sanders, B. C., *Phys. Rev. Lett.* **72**, 60 (1994); Rein-

hardt, W. P., and Clark, C. W., 1997, *J. Phys. B* **30**, L785 (1997); Jackson, A. D., Kavoulakis, G. M., and Pethick, C. J., *Phys. Rev. A* **58**, 2417 (1998).

30. Busch, T., and Anglin, J., cond-mat/9809408.

31. Muryshev, A. E., van Linden van den Heuvell, H. B., and Shlyapnikov, G. V., *Phys. Rev. A* **60**, R2665 (1999).

32. Fedichev, P. O., Muryshev, A. E., and Shlyapnikov, G. V., *Phys. Rev. A* **60**, 3220 (1999).

33. Dum, R., Cirac, J. I., Lewenstein, M., and Zoller, P., *Phys. Rev. Lett.* **80**, 2972 (1998).

34. Scott, T. F., Ballagh, R. J., and Burnett, K., *J. Phys. B* **31**, L329 (1998).

35. In our case, the dark solitons move with a velocity of about $0.5 \cdot c_s$, and they are already dynamically stable at a smaller radial confinement than in the case of standing dark solitons.

36. The imprinted phase is estimated from measured laser parameters and tables of atomic data. In addition, the onset (contrast $> 10\%$) for dark soliton creation was measured to correspond to $\Delta\phi \sim 0.5\pi$, in agreement with the theoretical results.

37. Burger, S., Bongs, K., Dettmer, S., Ertmer, W., Sengstock, K., Sanpera, A., Shlyapnikov, G. V., and Lewenstein, M., *Phys. Rev. Lett.* **83**, 5198 (1999).

38. Kozuma, M., Deng, L., Hagley, E. W., Wen, J., Lutwak, R., Helmerson, K., Rolston, S. L., and Phillips, W. D., *Phys. Rev. Lett.* **82**, 871 (1999)

39. Stenger, J., Inouye, S., Chikkatur, A. P., Stamper-Kurn, D. M., Pritchard, D. E. and Ketterle, W., *Phys. Rev. Lett.* **82**, 4569 (1999)

40. Denschlag, J., Simsarian, J. E., Feder, D. L., Clark, C. W., Collins, L. A., Cubizolles, J., Deng, L. Hagley, E. W., Helmerson, K., Reinhardt, W. P., Rolston, S. L., Schneider, B. I. and Phillips, W. D., *Science*, **287**, 97 (2000).

41. Sanpera, A., *et al.*, *to be published*

42. Busch, T., *private communication.*

43. Shlyapnikov, G.V., *private communication.*

Quantized vortices in a gaseous Bose-Einstein condensate

K. Madison, F. Chevy, W. Wohlleben, and J. Dalibard

Laboratoire Kastler Brossel, 24 rue Lhomond, 75005 Paris, France.

Abstract. Using a focused laser beam we stir a Bose-Einstein condensate confined in a magnetic trap. When the stirring frequency reaches a critical value, we observe the formation of a vortex at the center of the condensate. Using the quadrupolar excitation of the condensate we measure the angular momentum of the condensate with the vortex, and we find that it is $\sim \hbar$ per particle, as expected. For larger stirring frequencies, regular arrays of vortices are observed.

I INTRODUCTION

The discovery of Bose-Einstein condensation of atomic gases [1–4] has led to a new impulse in the physics of quantum fluids. Among the several questions that can be studied in these systems, superfluidity is one of the most intriguing and fascinating. This problem has received several experimental answers during the last year, all showing evidence for superfluid behaviour. A first class of study has been performed at MIT and consists in measuring the energy deposited in the condensate by a moving "object" (*i.e.* the hole created by a blue detuned laser) [5]. The second class of study is based on specific oscillation patterns, such as the *scissors mode*. This was adressed theoretically by D. Guéry-Odelin and S. Stringari [6] and investigated experimentally by the Oxford group [7]. Finally another clue for superfluidity is related to the behaviour of vortices. Using a "phase printing method", the Boulder group has created a vortex in a double component condensate with one component standing still at the center of a magnetic trap and the other component in quantized rotation around the first one [8]. In our work in Paris, we have succeeded in nucleating one or more vortices in a single component condensate, which is stirred by a laser beam [9–11].

Our experiment is a direct transposition to atomic gases of the famous "rotating bucket experiment" performed with liquid helium. In § II we briefly recall some essential results obtained with this experimental scheme for superfluid He. We then turn in § III to the description of our experimental setup. The observation of single and multiple vortices is presented in § IV where we also discuss recent data concerning the measurement of the angular momentum of the condensate, with

CP551, *Atomic Physics 17,* edited by E. Arimondo, P. DeNatale, and M. Inguscio
© 2001 American Institute of Physics 1-56396-982-3/01/$18.00

and without vortices. These data show in particular that the nucleation of the first vortex is associated with a jump of the angular momentum per particle from 0 to $\hbar$.

II THE ROTATING BUCKET EXPERIMENT

Consider a superfluid placed in a bucket rotating at angular frequency Ω (fig. 1). If Ω is small enough no motion of the superfluid occurs. This is a direct manifestation of superfluidity and analogous to the existence of a critical linear flow velocity below which the condensate exhibits viscous free behaviour: here the slow motion of the rough walls of the bucket does not perturb the superfluid. As shown by Landau, superfluidity is a direct consequence of repulsive interactions which gives rise to a phonon-like dispersion relation for the lowest lying excitations [12].

When Ω is increased beyond the critical frequency Ω_c the superfluid is set into motion. As pointed out by Onsager [13] and Feynman [14] the corresponding velocity field is subject to very strong constraints due to its quantum nature. Consider indeed the macroscopic wave function $\psi(\mathbf{r})$ describing the state of the superfluid. It can be written:

$$\psi(\mathbf{r}) = \sqrt{\rho(\mathbf{r})}\, e^{i\theta(\mathbf{r})} \ , \tag{1}$$

where $\rho(\mathbf{r})$ is the superfluid density. The velocity field is given by:

$$\mathbf{v}(\mathbf{r}) = \frac{\hbar}{m}\nabla\theta(\mathbf{r}) \ . \tag{2}$$

Therefore the circulation of the velocity along any closed contour is quantized as a multiple of h/m:

$$\oint \mathbf{v} \cdot d\mathbf{r} = n\frac{h}{m} \qquad \text{where } n \text{ is an integer.} \tag{3}$$

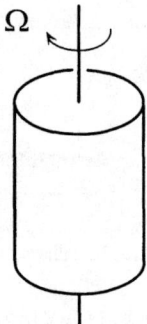

FIGURE 1. The rotating bucket experiment.

Just above Ω_c, the superfluid wave function involves a singular line or *vortex line*, along which the density is zero. On any closed path going around this line the phase of the wave function varies continuously from 0 to 2π.

For a rotating frequency Ω notably larger than Ω_c, several vortex lines can be generated, and they form a regular triangular lattice (Abrikosov lattice). These lines have been imaged by R. Packard and his team by trapping electrons at the core of each vortex and then accelerating them to a phosphorus screen (for a review see [15]).

It is worth emphazing that these vortices are universal structures associated with circulating quantum flow. Besides superfluid liquid helium, other large quantum systems such as neutron stars and superconductors support quantized vortices. In the latter case the rotation vector is induced by an applied magnetic field which modifies the motion of the charges, and the quantization of the circulation of the velocity results in flux quantization.

III   THE EXPERIMENTAL SETUP

Our experiment is performed with a ^{87}Rb gaseous condensate (fig 2). This condensate is prepared from a sample of 10^9 atoms confined in a magneto-optic trap. The atoms are transferred into a Ioffe-Pritchard magnetic trap, and they are evaporatively cooled down through the Bose-Einstein transition, which occurs at a temperature of the order of 500 nK. For the experiment reported here, we continue the evaporative cooling to a temperature below 80 nK, and the number N of atoms in the condensate is varied between 10^5 and $4\,10^5$.

The magnetic trap is axisymetric and it creates a harmonic potential:

$$U(\mathbf{r}) = \frac{1}{2}m\omega_t^2(x^2 + y^2) + \frac{1}{2}m\omega_z^2 z^2 \,, \qquad (4)$$

with a transverse frequency $\omega_t/2\pi \sim 200$ Hz and an axial frequency $\omega_z/2\pi \sim 10$ Hz. The condensate is cigar-shaped, with a length ~ 120 μm and a diameter of 6 μm for $N = 2 \times 10^5$.

The stirring of the condensate is provided by a far detuned laser, whose motion is controlled by two acousto-optic modulators. This laser propagates along the axis

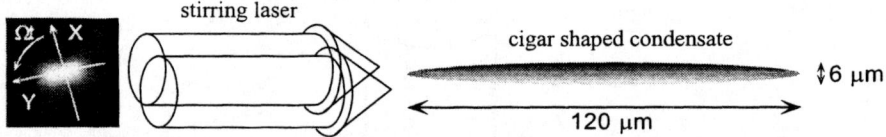

FIGURE 2. The cigar-shaped condensate is strirred using the dipole potential created by a laser "spoon". The laser waist is 20 μm; its axis is toggled between two symmetric positions about the trap axis, separated by 16 μm. The laser intensity profile averaged after this toggling is displayed on the left of the figure. The resulting anisotropy is then rotated at the stirring frequency Ω.

of the cigar, and it toggles back and forth very rapidly between two positions which are symmetric with respect to this axis (toggling frequency 100 kHz $\gg \omega_z, \omega_t$). It creates for the atoms an average dipole potential which is anisotropic in the xy plane (fig. 2). This potential can be written:

$$\delta U(\mathbf{r}) = \frac{1}{2} m \omega_t^2 (\epsilon_X X^2 + \epsilon_Y Y^2) \tag{5}$$

where $\epsilon_X = 0.09$ and $\epsilon_Y = 0.03$. Using the acousto-optic modulators, the XY basis is rotated at constant angular frequency Ω in the xy plane, and the turning anisotropy plays the role of the rough walls of the material bucket used with liquid He. The stirring frequency Ω is chosen in the interval $(0, \omega_t)$. At the upper value of this interval, the centrifugal force equals the transverse restoring force of the trap, and a dynamical instability of the trap occurs.

We now address the question of the vortex visibility. In a fluid of density ρ, the radius of the vortex core is determined by the healing length $\xi = (8\pi\rho a)^{-1/2}$, where a is the scattering length characterizing the two-body interactions in the ultra-low temperature regime [12]. For our experimental conditions, $\xi \sim 0.2\ \mu m$, which is too small to be observed *in situ* by optical means. Fortunately this size can be expanded using a time-of-flight technique [16–18]. When we release the atomic cloud from the magnetic trap and let it expand for a duration T, the transverse dimensions of the condensate and of the vortex core are increased by a factor $(1 + \omega_t^2 T^2)^{1/2} \sim 35$ for $T = 27$ ms. The cloud is then imaged using resonant absorption imaging, along the z axis.

IV   SINGLE AND MULTIPLE VORTICES

We now discuss the results of this experiment. When the stirring frequency is below a critical frequency $\Omega_c \simeq 0.65\ \omega_t$, no modification of the condensate is observed. Just above this critical frequency (within 1 or 2 Hz), a density dip appears at the center of the cloud, with a reduction in optical thickness at this location which reaches 50% (figure 3).

When we stir the condensate at a higher frequency more vortices are nucleated (fig. 4). The maximal number of vortices that can be observed in this experiment depends on the oscillation frequency ω_t. For a relatively tight trap ($\omega_t/2\pi = 225$ Hz), we have observed up to 4 vortices [9]. When we used a less confining trap ($\omega_t/2\pi = 170$ Hz), we obtained configurations with up to 11 vortices [10]. In all cases, when Ω is increased past the range of stability for the multiple vortex configuration, the density profile takes on a "turbulent" structure. When the stirring frequency is close to ω_t the centrifugal force destabilizes the cloud and the condensate is no longer confined.

After this qualitative observation of vortex nucleation, we turn to a more quantitative measurement, whose goal is to measure the angular momentum of the rotating condensate. For this purpose we use a theoretical result derived by F. Zambelli

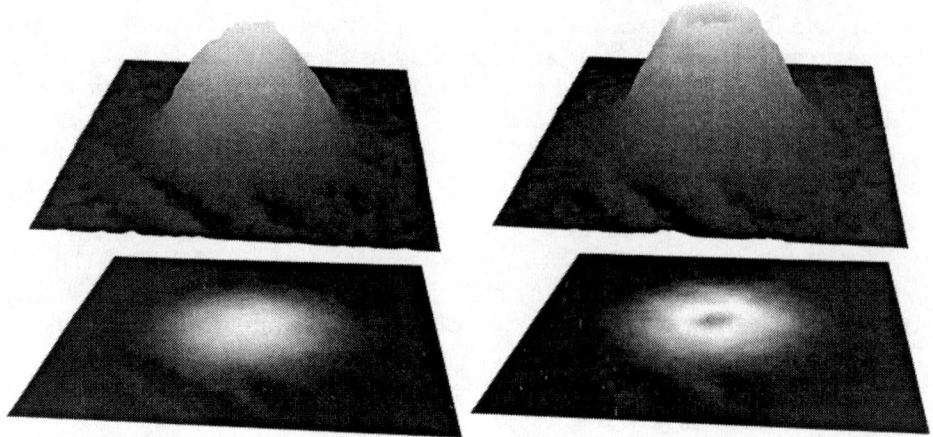

FIGURE 3. Density profile after a time-of-flight expansion of a Bose-Einstein condensate stirred below (left) and above (right) the critical frequency for nucleating a vortex. For these data, $\Omega_c/2\pi = 147$ Hz, $\Omega/2\pi = 145$ Hz and 152 Hz for the left and the right column respectively. The number of atoms is $N = 1.4 \times 10^5$ and the temperature is below 80 nK.

and S. Stringari [19]. These authors have studied the two transverse modes of a cigar shape condensate, corresponding respectively to an excitation with angular momentum $m = 2$ and $m = -2$. Because of symmetry for a condensate at rest these two modes have the same frequency. However, for a condensate in rotation, the degeneracy is lifted, and the frequency difference between the two modes is related to the average angular momentum $\langle L_z \rangle$ and to the transverse size of the condensate $\langle r_t^2 \rangle$:

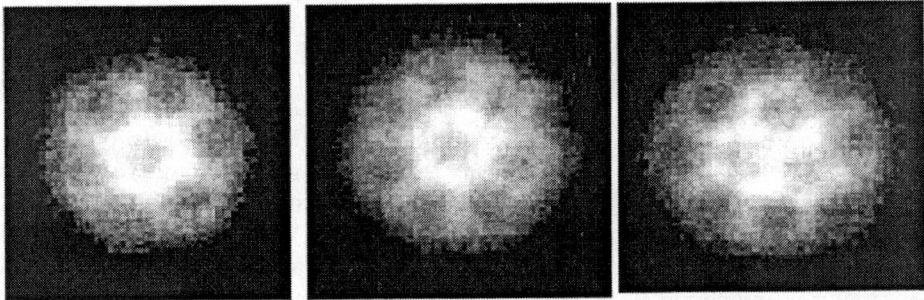

FIGURE 4. Arrays of vortices obtained in a magnetic trap with $\omega_t/2\pi = 169$ Hz and $\Omega/2\pi = 135$ Hz.

$$\omega_+ - \omega_- = \frac{2\langle L_z \rangle}{m\langle r_t^2 \rangle} \tag{6}$$

The measurement of the difference $\omega_+ - \omega_-$ is performed by looking at the transverse quadrupolar oscillation of the condensate which is a superposition of both the $m = 2$ and $m = -2$ excitations [11,20]. In the absence of rotation $\omega_+ = \omega_-$ and the oscillation occurs along fixed axes. However, if ω_+ and ω_- differ, the axes of the quadrupolar oscillation precess, and the precession frequency is $\dot{\theta} = (\omega_+ - \omega_-)/4$.

To study the angular momentum $\langle L_z \rangle$ of the condensate, we first stir it with the rotating laser. We then excite the transverse quadrupolar oscillation using again the dipole potential created by the stirring laser but now on a fixed basis $(x, y = X, Y)$. This potential is applied to the atoms for a duration of 0.3 ms, which is short compared to the quadrupole oscillation period ω_Q ($\omega_Q = \sqrt{2}\omega_t$). We then let the atomic cloud oscillate freely in the magnetic trap for an adjustable period τ, between 0 and 8 ms, after which we perform the time-of-flight + absorption imaging sequence. A typical result is shown in fig. 5. It displays three images taken after 1, 3 and 5 ms of quadrupolar excitation of a condensate with a vortex present. The precession of the quadrupolar axes is clearly visible and has an angular velocity $\dot{\theta} = 5.9$ degrees/ms. For this set of pictures the *in situ* transverse size of the condensate (inferred from the size measured after time-of-flight and from the expansion factor $(1 + \omega_t^2 T^2)^{1/2}$) is $\langle r_t^2 \rangle^{1/2} = 2.0$ μm. From $\dot{\theta}$ and $\langle r_t^2 \rangle$ we deduce the value $L_z = 1.2$ (± 0.1)$\hbar$ from the data shown in fig. 5. This measurement is analogous to the experiment performed with superfluid liquid helium by Vinen in which he detected a single quantum of circulation in rotating He II by studying two opposing circular vibrational modes of a thin wire placed at the center of the rotating fluid [21].

We have repeated this angular momentum measurement for various stirring frequencies, and the results are shown in fig. 6. For $\Omega < \Omega_c$, the angular momentum is measured to be zero. Right at Ω_c, i.e. when the first density dip appears in the absorption image of the condensate, the measured angular momentum jumps

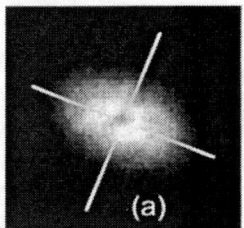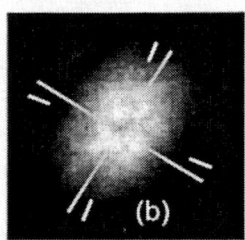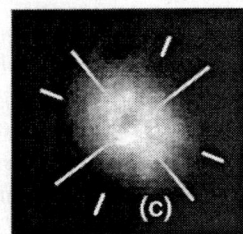

FIGURE 5. Transverse oscillation of a condensate with $N \sim 3.7 \times 10^5$ atoms and $\omega_t/2\pi = 171$ Hz. The stirring frequency is 120 Hz, slightly above the vortex nucleation threshold $\Omega_c/2\pi = 115$ Hz. For a,b,c: $\tau = 1, 3, 5$ ms respectively. The fixed axes indicate the excitation basis and the rotating ones the condensate axes. A single vortex is visible at the center of the condensate.

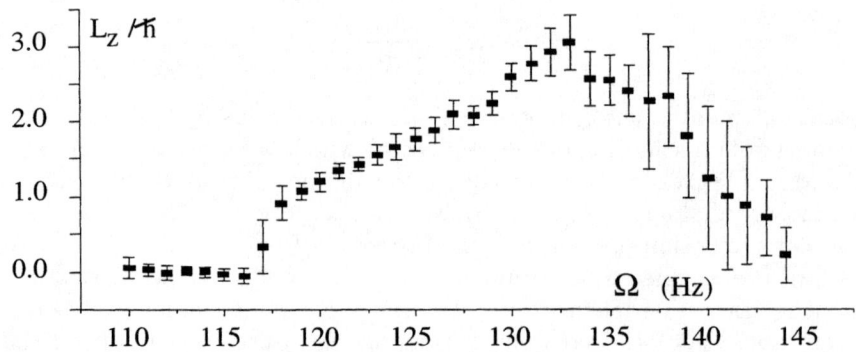

FIGURE 6. Variations of the angular momentum deduced from (6) as a function of the stirring frequency Ω for $\omega_t/2\pi = 175$ Hz and $N = 2.5 \times 10^5$ atoms.

from 0 to $\hbar$. Then it increases lineraly with Ω until the "turbulent" region is reached in which the rate of precession decreases with increasing Ω. Finally, when $\Omega/2\pi \geq 145$ Hz there is no precession at all and the condensate appears to be at rest. This regime requires a more detailed investigation to determine if the condensate is indeed not rotating. Another possibility is that the cloud is rotating but that its structure is more complex than a normal Bose-Einstein condensate and this might invalidate the prediction (6).

In conclusion we report on a rotating bucket experiment performed with a gazeous condensate and show that the results are in full agreement with what is expected for a superfluid: (i) that there exists a critical frequency Ω_c below which no motion of the superfluid occurs; (ii) that at Ω_c a vortex line appears as evidenced by the density dip at the center of the cloud, and that it is associated with an angular momentum of $\hbar$ per particle; (iii) that for a condensate stirred at a frequency larger than Ω_c, multiple vortices can be nucleated which arrange themselves in regular patterns.

Still much work is needed to achieve quantitative agreement between the experimental findings and theoretical predictions. For instance, the value found experimentally for Ω_c is notably higher than what is predicted from thermodynamic arguments [22–28]. This discrepancy may be due to "anomalous" decay modes of the vortex line which are suppressed only at a high rotation frequency, or it may be due to the existence of an energy barrier to nucleation which is overcome only for large stirring frequencies [29–33]. Also we want to pursue a detailed investigation of the decay of vortices which is relevant to the physics of rotating neutron stars [34,35]. Finally the interaction of a rotating condensate with light may allow for the study of a new class of phenomena in which the vortex core behaves as an optical black hole as a consequence of the very slow group velocity of light in a condensate [36].

REFERENCES

1. M. H. Anderson, J. Ensher, M. Matthews, C. Wieman, and E. Cornell, Science **269**, 198 (1995).
2. C. C. Bradley, C. A. Sackett, and R. G. Hulet, Phys. Rev. Lett. **78**, 985 (1997); see also C. C. Bradley, C. A. Sackett, J. J. Tollett, and R. G. Hulet, Phys. Rev. Lett. **75**, 1687 (1995).
3. K. B. Davis, M.O. Mewes, N. Van Druten, D. Durfee, D. Kurn, and W. Ketterle, Phys. Rev. Lett. **75**, 3969 (1995).
4. D. Fried, T. Killian, L. Willmann, D. Landhuis, S. Moss, D. Kleppner, and T. Greytak, Phys. Rev. Lett. **81**, 3811 (1998).
5. C. Raman, M. Khl, R. Onofrio, D. S. Durfee, C. E. Kuklewicz, Z. Hadzibabic, and W. Ketterle, Phys. Rev. Lett. **83**, 2502 (1999); A.P. Chikkatur, A. Gorlitz, D.M. Stamper-Kurn, S. Inouye, S. Gupta, and W. Ketterle, cond-mat/0003387; R. Onofrio, C. Raman, J. M. Vogels, J. Abo-Shaeer, A. P. Chikkatur, and W. Ketterle, cond-mat/0006111.
6. D. Guéry-Odelin and S. Stringari, Phys. Rev. Lett. **83**, 4452 (1999).
7. O. M. Maragò, S. A. Hopkins, J. Arlt, E. Hodby, G. Hechenblaikner, and C. J. Foot, Phys. Rev. Lett. **84**, 2056 (2000).
8. M. R. Matthews, B. P. Anderson, P. C. Haljan, D. S. Hall, C. E. Wieman, and E. A. Cornell, Phys. Rev. Lett. **83**, 2498 (1999); B. P. Anderson, P.C. Haljan,C.E.Wieman, and E.A.Cornell, cond-mat/0005368
9. K. Madison, F. Chevy, W. Wohlleben, and J. Dalibard, Phys. Rev. Lett. **84**, 806 (2000).
10. K. Madison, F. Chevy, W. Wohlleben, and J. Dalibard, cond-mat/0004037.
11. F. Chevy, K. Madison, and J. Dalibard, cond-mat/ 0005221.
12. E.M. Lifshitz and L. P. Pitaevskii, *Statistical Physics, Part 2*, chap. III (Butterworth-Heinemann, 1980).
13. L. Onsager, Nuovo Cimento **6**, suppl. 2, 249 (1949).
14. R.P. Feynman, in "Progress in Low Temperature Physics", vol. 1, ed. by C. J. Gorter, North-Holland, Amsterdam.
15. R.J. Donnelly, *Quantized Vortices in Helium II*, (Cambridge, 1991).
16. Y. Castin and R. Dum, Phys. Rev. Lett. **77**, 5315 (1996).
17. E. Lundh, C. J. Pethick, and H. Smith, Phys. Rev. A **58**, 4816 (1998).
18. F. Dalfovo and M. Modugno, cond-mat/9907102.
19. F. Zambelli and S. Stringari, Phys. Rev. Lett. **81**, 1754 (1998).
20. A similar experiment has been performed at JILA by P. Haljan *et al*, QELS, May 7-12, postdeadline session.
21. W.F. Vinen, Nature **181**, 1524 (1958) and Proc. Roy. Soc. A **260**, 218 (1961).
22. G. Baym and C.J. Pethick, Phys. Rev. Lett. **76**, 6 (1996).
23. F. Dalfavo and S. Stringari, Phys. Rev. A **53**, 2477 (1996).
24. S. Sinha, Phys. Rev. A **55**, 4325 (1997).
25. E. Lundh, C. J. Pethick, and H. Smith, Phys. Rev. A **55**, 2126 (1997).
26. A. Fetter, J. Low. Temp. Phys. **113**, 189 (1998).
27. D. L. Feder, C. W. Clark, and B. I. Schneider, Phys. Rev. Lett. **82**, 4956 (1999).

28. Y. Castin and R. Dum, Eur. Phys. J. D. **7**, 399 (1999).
29. F. Dalfovo, S. Giorgini, M. Guilleumas, L. Pitaevskii, and S. Stringari, Phys. Rev. A **56**, 3840 (1997) and communication at "Rotating BEC", Trento, June 2000.
30. T. Isoshima, K. Machida, Phys. Rev. A **60**, 3313 (1999).
31. Y. Castin, private communication.
32. D. L. Feder, C. W. Clark, and B. I. Schneider, Phys. Rev. A **61**, 011601(R) (1999) and communication at "Rotating BEC", Trento, June 2000.
33. A. Svidzinsky and A. Fetter, cond-mat/0007139 and communication at "Rotating BEC", Trento, June 2000.
34. P. Fedichev and G. Shlyapnikov, Phys. Rev. A **60**, R1779 (1999).
35. P. O. Fedichev and A. E. Muryshev, con-mat/0004264.
36. U. Leonhardt and P. Piwnicki, Phys. Rev. Lett. **84**, 822 (2000).

Dynamics of two interacting Bose condensates in a magnetostatic trap

M. Modugno[1], F. Dalfovo[2], C. Fort[1], M. Inguscio[1], P. Maddaloni[1*], F. Minardi[1]

[1] INFM – LENS – Dipartimento di Fisica, Università di Firenze
L.go E. Fermi 2, I-50125 Firenze, Italy
[2] Dipartimento di Matematica e Fisica, Università Cattolica,
and INFM, Unità di Brescia,
Via Musei 41, I-25121 Brescia, Italy

Abstract. We experimentally and theoretically investigate the dynamics of two trapped condensates in a magnetostatic trap. The two condensates, being in different internal Zeeman state ($|F = 2, m = 2\rangle$ and $|F = 2, m = 1\rangle$), experience external parabolic potentials that differ by both curvature and equilibrium position.

The dynamics of the system, exhibiting large amplitude oscillations of the $|F = 2, m = 1\rangle$ condensate, is affected in a profound manner by the periodical inter-condensates collisions.

In particular, we show that both an amplitude damping and a frequency shift occur. Measuring the former allows to study the elastic scattering, whose suppression is a manifestation of the Bose condensates superfluidity, while the latter gives access to the cross scattering length a_{21}.

I INTRODUCTION

Since the realization of BEC in trapped alkali vapours [1], two condensate systems have appealed both on the experimental and theoretical side. A double condensate system is, for instance, the prototype for any kind of interference experiment and, to address the issue of the condensate phase, this is the simplest system we must deal with.

Two condensates may be distinguished either by their external or by their internal degrees of freedom. At MIT, two condensates prepared in adjacent potential wells formed a spatial interference pattern in the overlap region of the two clouds during ballistic expansion [2]: this was the first demonstration of interference with alkali Bose condensates. At JILA, two condensates almost overlapping in the trap were distinguishable thank to their hyperfine state and a time-domain, Ramsey-type, interference was observed [3]. Interference was also obtained with multiple

CP551, *Atomic Physics 17*, edited by E. Arimondo, P. DeNatale, and M. Inguscio
© 2001 American Institute of Physics 1-56396-982-3/01/$18.00

condensates, either sitting in an array of potential wells [4] or produced by Bragg scattering with different momenta [5].

As a preliminary step to the study of condensate phase and interference, we have investigated the dynamics of two ^{87}Rb condensates in the $|F = 2, m = 2\rangle$ and $|F = 2, m = 1\rangle$ states (hereafter denoted with $|2\rangle$ and $|1\rangle$ respectively) within a magneto-static trap. Due to the different magnetic moments, the overall potential, magnetic plus gravitational, is not the same for the two states. In particular, the equilibrium positions are displaced. As a result, the $|1\rangle$ center-of-mass undergoes large amplitude oscillations.

We find that interesting properties of the condensates already emerge from the analysis of this simple system. We will see that besides a frequency shift caused by interactions, the damping allows to study the condensates beyond mean-field approximation as it unveils the influence of "elastic scattering".

II   TWO CONDENSATES SYSTEM

We start with a single $|2\rangle$ condensate containing $1.5 \cdot 10^5$ ^{87}Rb atoms at a temperature below 130 nK, confined in an axially symmetric harmonic potential, as already described in [6]. The harmonic oscillator axial frequency is held fixed to $\nu_z = 12.6$ Hz, while the radial frequency can be varied by changing the minimum magnetic field B_0 according to the relation $\nu_\perp = 221$ Hz$/B_0[\text{G}]^{1/2}$. The density distribution is an inverted parabola typical of the Thomas-Fermi (TF) regime [7], with radii $R_\perp \simeq 44\,\mu$m, $R_z \simeq 3.4\,\mu$m and chemical potential $\mu/h \simeq 1.3$ kHz.

To put a fraction of atoms into the $|1\rangle$ state, we induce the magnetic dipole $|2\rangle \to |1\rangle$ by means of a rf oscillating magnetic field. We apply a short ($\sim$10 μs long) pulse, the amplitude and duration of which are adjusted to couple 13 % of the atoms into the second condensate. In the frequency domain, the pulse width exceeds by almost one order of magnitude the transition broadening due to the inhomogeneous magnetic field across the condensate. As a result, the coupling strength does not depend on position and $|1\rangle$ is produced with the same spatial density profile as $|2\rangle$. Furthermore, since the characteristic time scale of the collective dynamics is of the order of 1 ms, no evolution occurs for the wavefunctions during the pulse, but, immediately after, excitations start in both the condensates.

After the rf pulse, we hold the atoms in the magnetic trap for a variable time interval and then switch off the magnetic fields. We let the clouds drop and expand for 29 ms, before taking an absorption image on a CCD array. A couple of lenses ($f = 6$ cm and $f' = 50$ cm) image the shadow cast by the atoms on a probe beam of light resonant with the $5^2S_{1/2}, F = 2 \to 5^2P_{3/2}, F' = 3$ transition. The probe beam, having an intensity close to the saturation intensity $I_S = 1.6$ mW/cm^2 is shone for $\Delta t = 150\,\mu$s. Therefore, each atom scatters a number of photons N_{sc} of the order of $1.4 \cdot 10^3$, acquires an average velocity $v_r \cdot N_{sc} = 8$ m/s, at the end of the probing illumination, with a displacement of 0.6 mm ($v_r = 5.8$ mm/s is the single-photon recoil velocity). Due to the stochastic direction of spontaneously emitted photons, also a broadening of the condensate size takes place: roughly, this amounts to $v_r \Delta t \sqrt{N_{sc}/3}$, i.e. 19 μm in our experiment. For long expansion times, as those we adopted, the condensate size before the probing illumination is always at least 3 times larger than this spatial broadening, so that we can neglect convolving.

From the center-of-mass positions and the radii of both condensates, we study the dynamics of the two condensates in the trap before expansion. As the process of switching off the magnetic field lasts about 1 ms, of the same order of $\omega_{2\perp}^{-1}$, it is not really sudden. In our simplified model we mimic the effect of the dynamics during the turn-off (1) by adjusting the switch-off time in a 1 ms long window and (2) by imposing the atoms the acquired velocities experimentally observed, 1.4 cm/s and 0.7 cm/s, respectively for $|2\rangle$ and $|1\rangle$.

III THEORETICAL MODEL

At very low temperatures, where the thermal cloud can be ignored, we describe the two interacting Bose-Einstein condensates by two coupled Gross-Pitaevskii (GP) equations ($i = 1, 2$)

$$i\hbar \frac{\partial \Psi_1}{\partial t} = \left[-\frac{\hbar^2 \nabla^2}{2m} + V_1 + \frac{4\pi\hbar^2}{m} \left(a_{11} |\Psi_1|^2 + a_{21} |\Psi_2|^2 \right) \right] \Psi_1 \tag{1}$$

$$i\hbar \frac{\partial \Psi_2}{\partial t} = \left[-\frac{\hbar^2 \nabla^2}{2m} + V_2 + \frac{4\pi\hbar^2}{m} \left(a_{21} |\Psi_1|^2 + a_{22} |\Psi_2|^2 \right) \right] \Psi_2 \tag{2}$$

Here, Ψ_1 and Ψ_2 are the $|1\rangle$ and $|2\rangle$ condensates wavefunctions. The relevant scattering lengths for rubidium are known within 1 % [8]: $a_{22} = a_{12} = 98a_0$, $a_{11} = 94.8a_0$. Due to the different magnetic moments, the trapping potential V_1 and V_2 are different:

$$V_1(x, y, z) = \frac{m}{2} \omega_{\perp 1}^2 \left[(x^2 + (y + y_0)^2) + \lambda^2 z^2 \right] \tag{3}$$

$$V_2(x, y, z) = \frac{m}{2} \omega_{\perp 2}^2 \left[(x^2 + y^2) + \lambda^2 z^2 \right] \tag{4}$$

with $\omega_{\perp 1}^2 = \omega_{\perp 2}^2 / 2$. Gravity shifts the equilibrium positions for both states from the location of the minimum magnetic field ("sagging"), the relative displacement being $y_0 = g/\omega_{\perp 2}^2$. While the axial trapping frequency ω_{z2} was fixed to 12.6(2) Hz, we tuned the radial frequency $\omega_{\perp 2} \equiv \omega_{z2}/\lambda$, hence the asymmetry parameter λ, via the bias magnetic field B_0.

We simulate the system evolution after the rf pulse, with the boundary conditions at $t = 0^+$:

$$\Psi_1(x, y, z; 0) \propto \Psi_2(x, y, z; 0) \tag{5}$$

$$N_1 = 0.13N \qquad N = 1.5 \cdot 10^5 \tag{6}$$

by numerically solving the GP equations (1,2) . The simulation can be reduced to a 2D problem thanks to the geometry of our trap [9]. Due to the large number of atoms, the wavefunction $\Psi_2(x, y, z; 0)$ has the inverted parabola shape characteristic of the TF regime.

The ballistic expansion, following the trap switch-off, is treated by rescaling both the wavefunctions with the appropriate time-dependent factors introduced in [10]. In addition, we attach two different initial velocities to the condensates center-of-mass. These velocities, $v_{1y} = 0.7 \pm 0.1$ cm/s and $v_{2y} = 1.4 \pm 0.1$ cm/s, have been experimentally observed and are attributed to transient magnetic field gradients occurring during the trap switch-off. Due to this difference, the two condensates fall at different speed and they eventually overlap for some time in the fall. However, we ignore the mutual mean-field interaction during this process: such an approximation is justified if the overlap does not occur too early, as the density drops quickly.

The theoretical analysis reveals the profound effect of inter-condensate interactions: by playing with the cross scattering length a_{21}, we continuously pass from the non-interacting unphysical case $a_{21} = 0$ to the real experiment $a_{21} = 98\,a_0$.

IV   INTERNAL DYNAMICS

The most obvious consequence of transferring atoms out of the $|2\rangle$ condensate is the onset of its internal collective excitations. Indeed, this was the initial motivation of our work.

The reduced number of atoms in $|2\rangle$ imposes smaller equilibrium radii, the consequent implosion triggers the low frequency modes of the collective excitation spectrum. The cylindrical symmetry being preserved, only modes with $m = 0$ can be excited in this way. The frequencies of the lowest $m = 0$ modes have been analytically derived in the the TF regime, within the linear limit (i.e. for small amplitudes) [11].

For an asymmetry parameter $\lambda \ll 1$, to first order in λ we have:

$$\omega_+ \simeq 2\,\omega_{12}\,; \qquad \omega_- \simeq \sqrt{\frac{5}{2}}\,\lambda\,\omega_{12}\,. \tag{7}$$

These predictions are in very good agreement with the values measured by different groups [12,13,6].

To detect the onset of these collective excitations, we have measured the evolution of $|2\rangle$ and $|1\rangle$ aspect ratios (i.e. the ratios of axial to radial sizes) within the trap, by varying the permanence time before the magnetic field switch-off.

For $|2\rangle$, the collective excitations are triggered by the sudden change in the number of atoms N_2. However, as in the TF regime the radii R_i depend rather weakly on the number of atoms ($R_i \propto N_i^{1/5}$), a 13 % reduction of N_2 leaves a $|2\rangle$ condensate that is only 3% larger than its equilibrium size. The resulting amplitude of the monopole and quadrupole modes would be smaller than it actually is, the enhancement being due to the $|2\rangle - |1\rangle$collisions, occurring about every 8 ms (see Fig. 1.

On the other hand, as $|1\rangle$ is produced with the same density distribution as $|2\rangle$ ($R_{2r} = 3.4\,\mu$m and $R_{2z} = 44\,\mu$m), its radii are 32 % larger than their TF equilibrium values. Correspondingly, collective excitations have a larger amplitude. In this regime, we expect a strong coupling between different modes due to the non-linear interaction terms [14]. The numerical simulations also show that the density distribution of $|1\rangle$ undergoes a considerable fragmentation, as collisions occur. In this case, the ballistic expansion is certainly more complicated and we will perhaps need to replace the analysis based on the model of Ref. [10] with a fully numerical simulation. As a matter of fact, our GP model cannot account satisfactorily for the observed evolution of the $|1\rangle$ aspect ratio.

V   CENTER-OF-MASS MOTION

The motion of the center-of-mass of $|1\rangle$ displays interesting features that can be understood on a quantitative ground by comparison with theoretical simulation. Namely, the sloshing oscillation of the small fraction of atoms in $|1\rangle$ is both fre-quency shifted and damped by the presence of the remaining atoms in $|2\rangle$. The frequency shift is related to the mean-field repulsion between $|1\rangle$ and $|2\rangle$, while

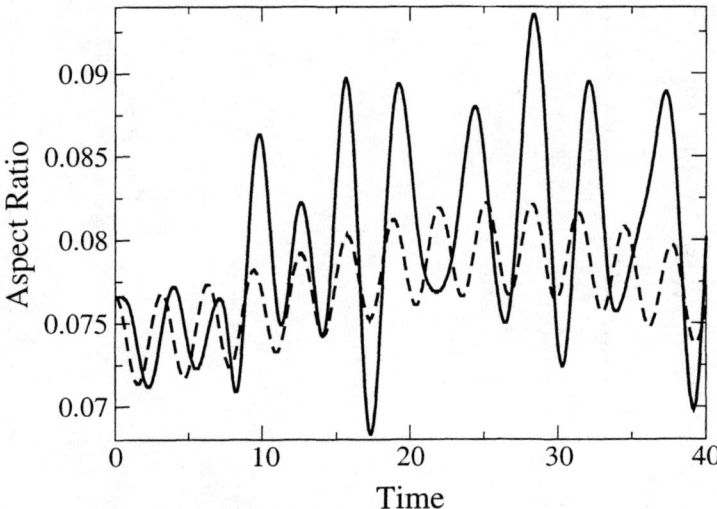

FIGURE 1. Evolution of the $|2\rangle$ aspect ratio ($R_{2\perp}/R_{2z}$) versus the permanence time in the trap (in units of $\omega_{2\perp}^{-1}$), between the rf pulse and the field switch off. Solid line: with condensates interaction, $a_{21} = 98a_0$; dashed line: without interaction $a_{21} = 0$

the damping is due to the energy transfer, occurring at each collision, toward the internal collective excitations of both condensates.

A   Frequency shift

The inter-condensates repulsion acts as an anharmonicity in the confining potential of the $|1\rangle$ condensate. Thus, while $|1\rangle$ oscillates with the expected frequency $\omega_{2\perp}/\sqrt{2}$ when no atoms are left in $|2\rangle$, its period decreases due to the presence of $|2\rangle$.

Our GP simulation reproduces this result, predicting a frequency shift of $+5.4\%$, slightly less than the observed 6.4%. By repeating the simulation, we find that the shift depends only weakly with the total number of atoms N: any reasonable systematic uncertainty on N cannot account for the observed discrepancy. Possibly, this is caused by the fragmentation of the $|1\rangle$ condensate: if the density distribution develops low density wings, we might be unable to detect them and this may have an influence on the measured center-of-mass.

On the experimental side, we find that the shift is generally increasing with the radial trapping frequency, see Fig. 2.

A useful insight into the $|1\rangle$ center-of-mass motion is provided by classical mechanics. The number of atoms in $|2\rangle$ being much larger than in $|1\rangle$, we can assume

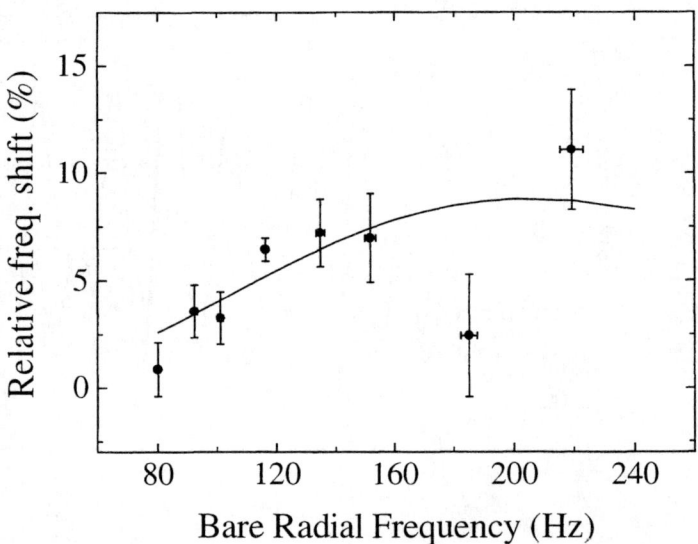

FIGURE 2. Relative shift of the $|1\rangle$ oscillation frequency versus its "bare" trapping frequency, i.e. in absence of interactions with $|2\rangle$: experimental data and prediction of the classical model (solid line).

that the former does not move at all upon collisions. Within this approximation, we consider the latter as a point particle moving in the 1-D external potential

$$V_{1,\text{cl}}(y) = \frac{m}{2}\omega_{\perp 1}^2 y^2 + g_{21}n_0 \max\{0, 1 - \frac{(y + y_0)^2}{R_{\perp 2}^2}\} \tag{8}$$

where $g_{21} = 4\pi\hbar^2 a_{21}/m$, $R_{\perp 2}$ and n_0 are the radius and the peak density of the $|2\rangle$ Thomas-Fermi distribution, respectively. We are assuming that $|1\rangle$ moves along the vertical axis y ($x = z = 0$), so that it crosses the point of $|2\rangle$ maximum density.

In particular, we obtain the oscillation period by calculating the integral:

$$T = \sqrt{2m} \int_{-y_0}^{y_i} (E - V_{1,\text{cl}}(y))^{-1/2} \, dy \tag{9}$$

where E and y_i indicate the classical energy and inversion point, such as $E = \frac{m}{2}\omega_{\perp 1}^2 y_i^2 = V_{1,\text{cl}}(-y_0)$. We find

$$T = \frac{2}{\omega_{\perp 1}}(\frac{\pi}{2} + \arcsin(\frac{1 - R_{\perp 2}/y_0}{\sqrt{1 + 2R_{\perp 2}^2/y_0^2}}) + \text{arcosh}(1 + \frac{R_{\perp 2}}{y_0})). \tag{10}$$

Despite the crudeness of the approximations involved, this result is not far from that of the GP simulation, as shown in Fig. 3, and it has the advantage to put in evidence the dependence of the frequency shift on the system parameters $N_2, a_{22}, a_{21}, \omega_{\perp 1}$.

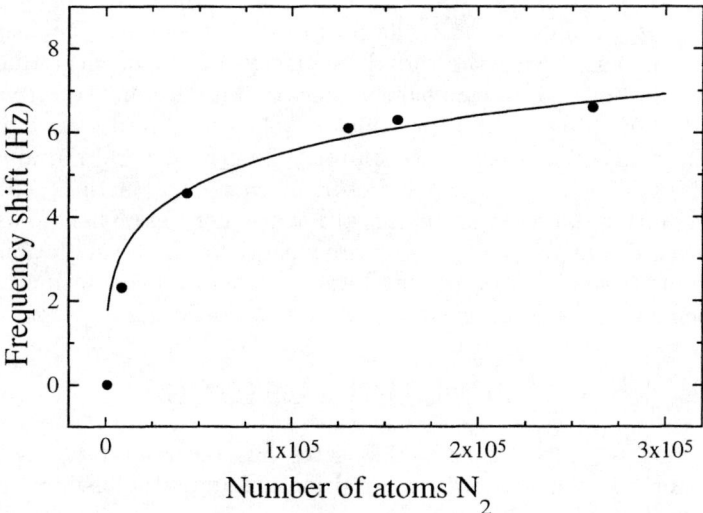

FIGURE 3. Comparison of the frequency shifts versus number of atoms in $|2\rangle$: classical model (solid line) and GP model (points).

B Damping

At each oscillation of $|1\rangle$, the two condensate overlap and interact via their mutual scattering length a_{21}: as a consequence energy is transferred to internal oscillations of both condensates and the $|1\rangle$ center-of-mass sloshing damps out.

The experimental evidence of this damping is clear if we compare the $|1\rangle$ sloshing with and without atoms left in $|2\rangle$ [15]. The GP model predicts the existence of such a damping, confirming our interpretation, but overestimates the damping time by a factor 2, i.e. the observed damping is stronger than calculated.

To sort out the reason of this discrepancy we have to consider that the two condensates collide at a velocity that is larger than the (maximum) speed of sound at the top of $|2\rangle$, $c_0 = \hbar\sqrt{4\pi n_2 a_{22}}/m = 2.5$ mm/s. As a consequence, we expect the GP mean-field approximation to break down and "elastic scattering" to be taken into account, as already done in the analysis of matter 4-waves mixing [16] and of the impurities motion inside a Bose condensate [17]. Thus, a more complete theoretical analysis of our experiment has been carried out by the group at NIST, in Gaithersbourg (USA): including elastic scattering greatly improves the agreement with measured damping time [18].

For the classical particle introduced above, the position-dependent velocity $u(\tilde{y})$, where $\tilde{y} = y + y_0$, reads

$$u^2(\tilde{y}) = g\tilde{y} + 2c_0^2\tilde{y}^2/R^2 \tag{11}$$

which is greater than the local sound velocity $c(\tilde{y}) = c_0\sqrt{1 - \tilde{y}^2/R_{21}^2}$ for $\tilde{y} > \bar{y} \equiv c_0^2/g$.

The ratio $\bar{y}/R_{21}$ scales as $\omega_{21}^{7/5}$: the portion of the parent condensate in which locally $u < c$ increases with the radial frequency, i.e. we expect that the elastic scattering should be less important for higher trapping frequencies, the density and sound velocity being higher and the collision velocity being lower.

As a check, we have measured the damping times for several radial trap frequencies: in Fig. 4, we plot the "quality factor" $Q = \tau \cdot \omega_{1\perp}$ of the oscillator against $\omega_{1\perp}$. The oscillation damping is smaller at low frequencies. This seems opposite to what we expected, but we need to compare with theoretical simulations in order to subtract the other sources of damping. At this stage, no conclusion can be drawn about the behavior of elastic scattering versus the ratio u/c.

VI CONCLUSIONS

A simple system of two Bose condensates has been used to investigate their mutual interactions. The mean-field repulsion is found responsible for a frequency shift of the oscillations within the trap. In principle, one could use this technique to determine the cross scattering length, a_{21}, were it known with poor accuracy.

The other interesting feature of the dynamics is that the damping seems to be linked to the elastic scattering, that lies beyond the mean-field theory. Since the

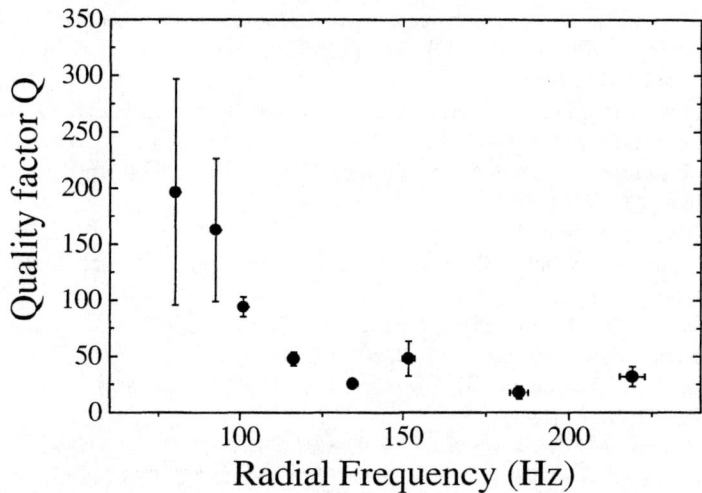

FIGURE 4. Measured "quality factor" for the |1⟩ oscillation versus its "bare" frequency.

suppression of elastic scattering below a critical collision velocity (of the order of the speed of sound), first investigated in the MIT experiment [17], is a manifestation of the superfluid character of Bose condensates, the damping of the |1⟩ sloshing motion may allow to unveil the Bose condensate superfluidity.

ACKNOWLEDGMENTS

We are grateful to Y. Band, J. Burke and P. Julienne for communicating their results prior to publication. We acknowledge also useful discussions with S.Stringari.

REFERENCES

* Dipartimento di Fisica, Università di Padova,
  via F. Marzolo 8, I-35131 Padova, Italy.
1. M. H. Anderson, J. R. Ensher, M. R. Matthews, C. E. Wieman, and E. A. Cornell, Science **269**, 198 (1995); K. B. Davis, M.-O. Mewes, M. R. Andrews, N. J. van Druten, D. S. Durfee, D. M. Kurn and W. Ketterle, Phys. Rev. Lett. **75**, 3969 (1995); C. C. Bradley, C. A. Sackett, J. J. Tollett and R. G. Hulet, Phys. Rev. Lett. **75**, 1687 (1995); **78**, 985 (1997).
2. M. R. Andrews *et al.*, Science **275**, 637 (1997).
3. D. S. Hall, M. R. Matthews, J. R. Ensher, C. E. Wieman, and E. A. Cornell, Phys. Rev. Lett. **81**, 1539 (1998).
4. B. P. Anderson and M. A. Kasevich, Science **282**, 1686 (1998).

5. E. W. Hagley, L. Deng, M. Kozuma, M. Trippenbach, Y. B. Band, M. Edwards, M. Doery, P. S. Julienne, K. Helmerson, S. L. Rolston, and W. Phillips, Phys. Rev. Lett. **83**, 3112 (1999).

6. C. Fort, M. Prevedelli, F. Minardi, F. S. Cataliotti, L. Ricci, G. M. Tino, and M. Inguscio, Eur. Phys. Lett. **48**, 8 (2000).

7. See, for instance: F. Dalfovo, S. Giorgini, L. P. Pitaevskii and S. Stringari, Rev. Mod. Phys. **71**, 463 (1999).

8. C. Williams, private comunication.

9. M. Modugno, F. Dalfovo, C. Fort, P. Maddaloni, and F. Minardi, Phys. Rev. A to be published.

10. Y. Castin and R. Dumm, Phys. Rev. Lett. **77**, 5315 (1996).

11. S. Stringari, Phys. Rev. Lett. **77**, 2360 (1996).

12. D. S. Jin, J. R. Ensher, M. R. Matthews, C. E. Wieman, and E. A. Cornell, Phys. Rev. Lett. **77**, 420 (1996).

13. M.-O. Mewes, M. R. Andrews, N. J. van Druten, D. M. Kurn, D. S. Durfee, C. G. Townsend, and W. Ketterle, Phys. Rev. Lett. **77**, 988 (1996).

14. Yu. Kagan, E.L. Surkov and G.V. Shlyapnikov, Phys. Rev. A **55**, R18 (1997).

15. P. Maddaloni, M. Modugno, C. Fort, F. Minardi and, M. Inguscio, Phys. Rev. Lett. **85**, 2413 (2000).

16. M. Trippenbach, Y. B. Band, and P. S. Julienne, Phys. Rev. A **62** 023608 (2000).

17. A. P. Chikkatur, A. Görlitz, D. M. Stamper-Kurn, S. Inouye, S. Gupta, and W. Ketterle, Phys. Rev. Lett. **85**, 483 (2000).

18. J. P. Burke Jr., Y. B. Band, and P. S. Julienne, private communication.

QUANTUM CONTROL AND COLLISIONAL DYNAMICS

Ultrafast Quantum Control in Atoms and Molecules

Philip H. Bucksbaum

Physics Department, University of Michigan
Ann Arbor, MI 48109-1120

Abstract. This paper reviews recent progress in experiments to control quantum dynamics in condensed phase and gas phase systems, using shaped ultrafast radiation. Many of the same techniques that have led to laser pulses in the 10 *fs* - 100 *ps* range can also be applied to the control of quantum systems with similar dynamical time scales. Systems under study include electron wave packets in Rydberg atoms, chemical dynamics in molecular liquids and lattice dynamics in crystalline solids. The feature common to each of these systems is ultrafast response. Shaped intense ultrafast radiation initiates the dynamics, which can then be studied using several new techniques.

INTRODUCTION

Quantum control aims to go beyond observations of atomic and molecular properties, to manipulate the dynamics of quantum systems using strong external fields. This paper summarizes recent techniques for controlling motion on the quantum scale, and also describes some recent applications in this new field. One of our goals has been the construction of a "learning machine," that is, an automated experimental apparatus that can use feedback signals from a quantum system to help optimize control strategies for a particular application. The learning machine consists of a learning algorithm, together with programmable experimental inputs and readouts. Learning machines could be particularly useful for controlling systems where the Hamiltonian is not known completely.

On a different level, wave packets are forms of information stored in the quantum system. We have studied some simple examples where information is stored as quantum phase in a Rydberg wave packet, and then efficiently extracted by converting the phase information to quantum amplitudes.

This paper will describe some basic aspects wave packet shaping and control, and then summarize our work on learning algorithms and quantum information storage.

CP551, *Atomic Physics 17,* edited by E. Arimondo, P. DeNatale, and M. Inguscio

SHAPING ULTRAFAST OPTICAL PULSES

Our optical pulses are formed in a Kerr-Lens mode-locked titanium sapphire oscillator[1]. Such lasers are capable of producing hundreds of nanometers of coherent bandwidth, although we typically only use about 10 nm. The output is amplified in a 10 Hz regenerative chirped-pulsed amplifier[2]. The output pulse is approximately 100 fsec long, with a central wavelength of about 790 nm.

Ultrafast laser pulses contain approximately two to 50 optical cycles, and last only a few femtoseconds. This is much faster than any current electronics and therefore shaping with fast time gates is very difficult. On the other hand, optical pulse bandwidths range from several tens of percent to only a few percent of the optical spectrum. Such large bandwidths are relatively easy to measure and to filter, and there are several techniques to shape the spectrum, and thereby shape the temporal pulse. In our pulse shaper, individual frequency components in the pulse are controlled by passing the light through two back-to-back spectrometers, which are configured to introduce zero net temporal dispersion: that is, all colors pass through the spectrometers in the same amount of time. The first spectrometer spreads the unshaped pulse spectrum along a line in the image plane, where the light intercepts a spatial amplitude and phase mask. The mask output then forms the entrance to a second spectrometer, which recombines the colors into a single shaped pulse.

The heart of the pulse shaper is the programmable mask that forms the Fourier filter. This mask must be capable of either attenuating the individual colors or shifting their phase. Two different electronically programmable masks that are capable of controlling both amplitude and phase have been demonstrated: a liquid crystal display (LCD) and an acousto-optic modulator (AOM).

We use an AOM design originally developed at Princeton.[3] The AOM consists of an anti-reflection coated Tellurium Dioxide *(TeO2)* crystal with a piezo-electric transducer glued onto one end. The central frequency of the acoustic wave is $\omega_c / 2\pi = 200 MHz$. The acoustic wave creates a transient transmission grating which diffracts the optical wave. at the Bragg angle. The acoustic velocity v_s in the crystal is 4.2km/s and the light pulse spends less than 10ps in the crystal, so the acoustic wave moves less than 0.002 $\lambda_{acoustic}$ during the transit of the light field through the crystal. Since the acoustic wave is essentially frozen as the optical pulse travels through the crystal, the complex amplitude of the acoustic wave traveling through the crystal in the y direction, $A(t)\cos\omega_c t = A(y/v_s)\cos\omega_c t$, is mapped onto the optical field $E(\omega)$ as it passes through the AOM. If some of the dispersed optical field encounters a weak acoustic wave, that frequency is attenuated; if the acoustic wave carrier is shifted by a phase angle ϕ, that phase shift is imposed on the optical field. Our pulse shaper has a total efficiency of about 10%, including the diffraction efficiency of the AOM and the diffraction efficiency of the gratings. We use the diffracted light and we discard the undiffracted "zero order" beam. This allows full modulation of both amplitude and phase in the shaped beam. The shaped beam then has the form

$$E_{shaped}(\omega) = E_{input}(\omega) \times a(\omega) \times e^{i\phi(\omega)t} \tag{1}$$

where $a(\omega)e^{i\phi(\omega)} = A[y(\omega)/v_s]$.

The damage threshold of the AOM is approximately $4J/cm^2$. The pulse shaper follows the first stage regenerative amplifier, and it precedes the final amplifier, in order to limit the input pulse energy to $100\mu J$. The shaped pulses are amplified further in an unsaturated multipass final amplifier. The resolution of an AOM-based shaper can be quoted in terms of "effective pixels". Our shaper was designed to have about 200 effective pixels over its complete length. However, we are currently using a shorter modulator, limiting our number of pixels to 170.

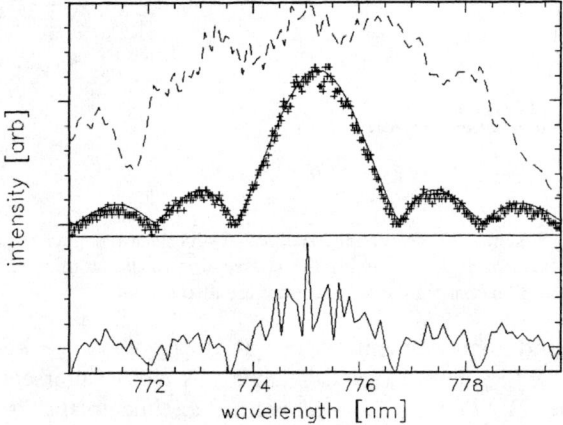

Figure 1. Sinc shaped spectrum with the model spectrum, unshaped spectrum and array sent to the waveform generator boards. The top curve is the unshaped spectrum. Below it is the measured spectrum with the overlaid solid curve showing the model spectrum. The bottom curve shows the array sent to the waveform generator for the measured spectrum.

The shaped pulses are measured using spectral interferometry. In this technique, the shaped laser pulse is merged with an unshaped reference pulse on a beam splitter, and then the combined pulses are analyzed in a spectrometer. The signal is a spectrally resolved interference. If the reference pulse is known to have flat spectral phase, then the amplitude beating of the output beam is a direct measure of the spectral phase function.

ATOMIC WAVE PACKETS

Sculpting technique

Rydberg atoms are a testing ground for quantum classical correspondence, and for quantum state preparation and measurement.[4] We can use Rydberg atoms to explore ideas in strong field quantum control and quantum information problems. This section describes our development of wave packet "sculpting" and the quantum interference techniques used to view these sculptures.[5].

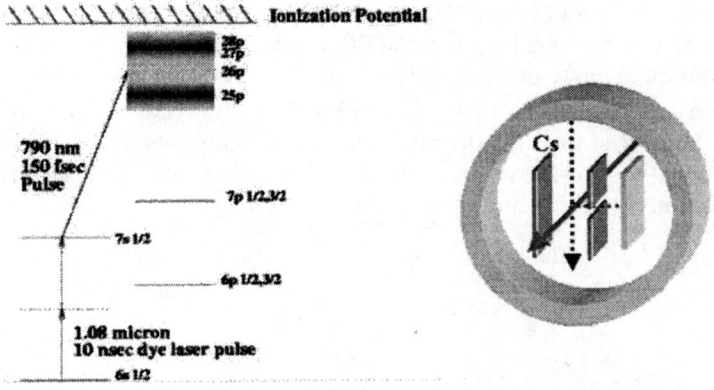

Figure 2. Excitation scheme for producing sculpted Rydberg wave packets in atomic Cs. Right: schematic of the interactions region, showing the intersection of the atomic beam and the laser beams. Field ionization plates and a charged particle detector are also shown.

The creation and measurement of our Rydberg wave packets occurs inside a vacuum chamber, where and effusive beam of Cs atoms intersect one or more laser beams (see Figure 2). Ground state atoms are excited to the 7s state using a two-photon transition at 1.08 μm, produced by Raman shifting the output of a pulsed dye laser. Sculpted laser pulses then excite the atoms to Rydberg states. This produces coherent superpositions of $p_{1/2}$ and $p_{3/2}$ states with principal quantum number between 24 and 35. In most applications we are only interested in tracking the shape over the first few picoseconds. The natural time scale for the spin-orbit interaction that splits the $p_{1/2}$ and $p_{3/2}$ states is much longer than this (on the order of 100 ps for the states considered here) so the wave packet is described quite well without electron spin, in the nlm basis. In that case, only $l=1$ states are excited. The value of the magnetic quantum number m then depends on the relative orientation of the quantization axis and the laser polarization that induces transitions from the 7s state. In most cases they are perpendicular, so we excite equal amplitudes of $m = \pm 1$ states.

The shape of a wave packet $\Psi(\mathbf{x},t)$ is determined by the amplitude and phase of the constituent eigenstates:

$$\Psi(\mathbf{x},t) = \sum_{np} a_n e^{i\Phi_n} u_{npm}(\mathbf{x}) e^{-i\omega_n t} \tag{2}$$

In the limit of weak excitation, the complex amplitude is simply given by Fermi's Golden Rule, so it is proportional to the dipole matrix element times the resonant electric field:

$$a_n e^{i\Phi_n} \propto E(\omega_{np} - \omega_{7s})\langle npm|z|7s\rangle \propto E(\omega_{np} - \omega_{7s})/n^3 \tag{3}$$

The quantum state amplitude and phase are therefore directly related to the amplitude and phase of the sculpted optical field.

Amplitude Measurement

The quantum state amplitudes are measured using state-selective field ionization[6]. A uniform electric field is slowly applied to the Rydberg system to induce field ionization. The field produces a saddle point in the coulomb potential. A hydrogenic Rydberg state with $m=0$ and binding energy E ionizes rapidly when the saddle point in the potential reaches the value $F=(4/9)E^{-2}$ in atomic units ($2000V/cm\times[20/n^4]$).

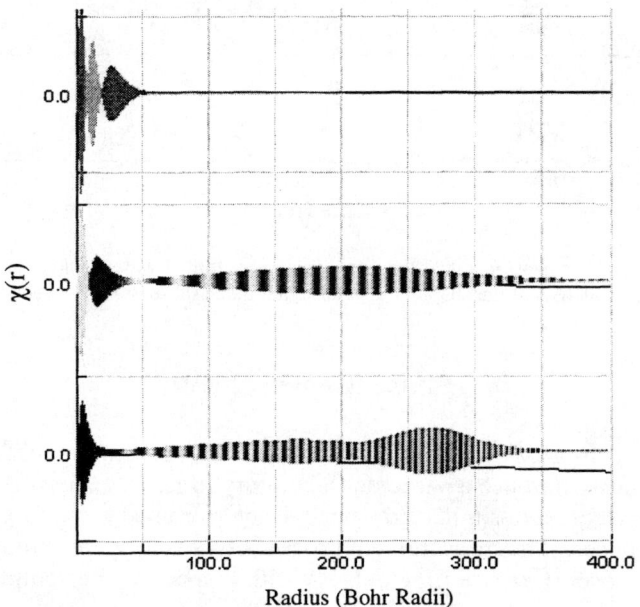

Figure 3. Calculation of a simple p-state radial wave packet undergoing field ionization. The different shades of gray represent the quantum phase of the wave packet. Top: Bound state wave packet consisting of the 2p, 3p, and 4p states of hydrogen with equal amplitude. Middle: Above the critical field for ionizing 4p, the wave packet splits into an ionized portion on the right, and a bound wave packet consisting of the 2p and 3p states. Bottom: For a larger field, only the 2p state remains bound.

The field must increase slowly compared to the characteristic dynamical time scale for the Rydberg state, which is the corresponding Kepler orbital period, $2\pi n^3$. The situation is similar for higher m and for atoms with a non-hydrogenic ion core: the critical ionization field is proportional to E^{-2}.

When a wave packet is subjected to the slow ramp, it does not necessarily ionize the moment that one of its n-states reaches the critical electric field. Instead, the ionization occurs with probability proportional to the square of the amplitude for that state. An ensemble of identical states will ionize with a distribution of critical fields that maps the squares of the individual state amplitudes, as shown in the calculation in figure 3, and the data in figure 4.

Ramped field measurements only reveal the amplitude of the states that make up a wave packet. The shape of the quantum sculpture also depends on the relative phase of these states, which can only be discovered through an interference measurement.

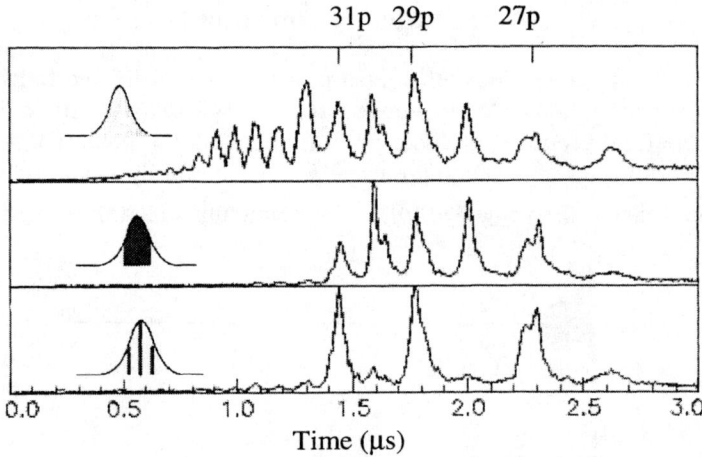

Figure 4. Ramped field ionization signal for three different sculpted wave packets in cesium. The insets show the amplitude function of the shaped light pulse that made these wave packets.

Phase Measurement

The relative phases of the quantum states evolve according to $e^{i\Delta\omega t}$, so any phase measurement must reference a particular laboratory time. We measure the phase using a holographic technique similar to spectral interferometry mentioned in the first section of this paper. The measurement proceeds by exciting the atoms with a second, *reference* wave packet with a time delay τ with respect to the sculpted packet. The reference excitation has all of its orbitals relatively real:

$$\Psi_{ref}(\mathbf{x},t) = e^{i\omega_{gs}\tau} \sum_{np} b_n u_{npm}(\mathbf{x}) e^{-i\omega_n t} \tag{4}$$

The optical field that excites such a wave packet is a transform-limited pulse arriving at time τ. The probability for state i following the excitation is

$$P_i = |a_i|^2 + |b_i|^2 + 2|a_i||b_i|\cos[(\omega_i - \omega_{gs})\tau - \phi_i] \tag{5}$$

In principle, ϕ_i can be extracted from this directly. In practice, though, the laser system is not stable with respect to frequencies on the scale of ω_i-ω_{gs}, which is an optical frequency. Instead, then, the phases are extracted by constructing a correlation function

$$r_{ij} = \frac{\langle P_i P_j \rangle - \langle P_i \rangle \langle P_j \rangle}{\langle \Delta P_i \rangle \langle \Delta P_j \rangle} = \cos[(\omega_i - \omega_j) - (\phi_i - \phi_j)] \tag{6}$$

where $\langle \Delta P_i \rangle$ is the root-mean-squared value of P_i averaged over many laser pulses.

An example of a wave packet sculpture reconstructed in this way is shown in figure 5.

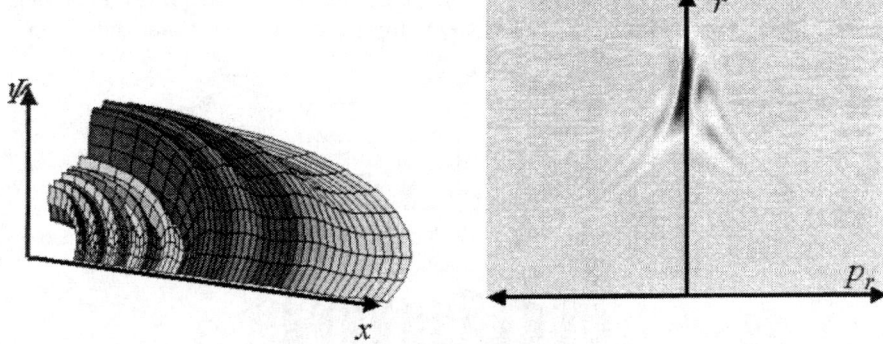

Figure 5. Two views of a sculpted wave packet. Left: Schrodinger wave function. The amplitude is shown as a function of x and z for a *p*-state wave packet oriented along x. Gray scale represents phase. Right: Wigner representation of the same wave packet in the *(r,p_r)* phase space plane. Here, gray scale represents amplitude.

QUANTUM INFORMATION APPLICATIONS

Quantum wave packets can also be used to store, retrieve, and manipulate information. This has been discussed recently in connection with the prospect of solving very large or difficult problems with a "quantum computer." While a Rydberg atom may be far too small to solve large problems, it contains the basic elements necessary for quantum computation. These are long coherence times, a large state space, and the entanglement of several degrees of freedom.

Consider, for example, the storage of a simple binary number, such as 000001000. A wave packet could encode this number in various ways. For example, one could load an *n*-state quantum wave packet with an *n*-bit number according to the prescription that at a specified time t, the phase is real and positive for binary 0, but real and negative for binary 1. Since every state in the wave packet has equal amplitude, ordinary spectroscopic techniques such as state-selective field ionization cannot reveal which state stores the binary 1. This bit is hidden from view. If the same data were stored in a classical binary register with n locations, one would have to search each location to find the marked bit. The search would take, on average, $n/2$ steps; however, the rules of quantum state manipulation provide some simple methods for revealing the marked bit.

If the quantum data register is formed by excitation from the atomic ground state in the weak field limit, then one may take advantage of the probability amplitude that remains in the ground state to find the marked bit by quantum holography. Figure 6 describes the method. A search, or "decoder" wave packet is prepared by exciting the atom with a simple pulse where all of the quantum phases are real and negative. This is, in fact, the very same "reference" pulse that was used to measure the phases in the sculpted wave in the previous section. In this case, the superposition results in destructive interference of any states where a binary 0 was stored, but constructive addition to any states with a binary 1. The combined wave packet then has the same information as before, but now it is encoded into quantum amplitudes rather than phases. These can then be read out easily using state-selective field ionization.[7]

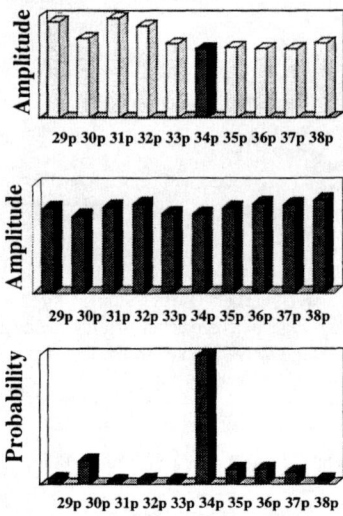

Figure 6. (a) Bar graph representing a Rydberg data register for the binary number 0000010000. The binary bit is encoded into the phase of the n-state, so that binary 0 states are real and positive, while binary 1 states are real and negative. (b) The decoding wave packet has equal amplitudes, and all negative real phases. (c) The superposition of (a) and (b) amplifies the binary 1 bits, while destructive interference destroys the binary 0 bits.

This is a simple example of a general class of search algorithms that make use of the properties of superposition and quantum interference, introduced by L.K. Grover.[8] Our particular implementation of a Grover-style search algorithm has some unresolved difficulties if the register is too large. The simple form of the decoding pulse only works in the perturbation theory limit, where almost all of the probability amplitude in the wave packet resides in the "launching" state ($7s$ for our work in Cs.) If there are too many states in the Rydberg wave packet, the launch state will become depleted. The data retrieval is still possible, but now the unitary transformation that amplifies the "1" bits and suppresses the "0" bits must depend on the total pulse energy, and possible other factors. In other words, we move into *the strong field regime*.

The connection between strong field dynamics and quantum information processing seems inescapable. To study this in Rydberg systems, we need to have methods to transfer large amounts of probability amplitude between Rydberg states. We have begun to study this with half cycle pulses of broadband terahertz radiation, which may be able to redistribute population, and amplify marked states in Rydberg systems.

LEARNING ALGORITHMS

Feedback and Learning Control

The techniques described in the previous sections are "feed-forward," since the construction and measurement processes make no use make no use of the measured results. The "feedback" loop is closed by the experimenters themselves, who may adjust the pulse shaper in response to information gathered by wave packet holography. This particular type of feedback is more properly called "learning," since the readjustment of the apparatus does not occur during a measurement, but rather follows the data acquisition, so that corrections can be made to future experiments.[9] We have implemented an automated form of learning control on wave packets in Rydberg states, and have found that an arbitrary wave packet can be made to converge on a "target" sculpture shape in only one or two iterations.[10]

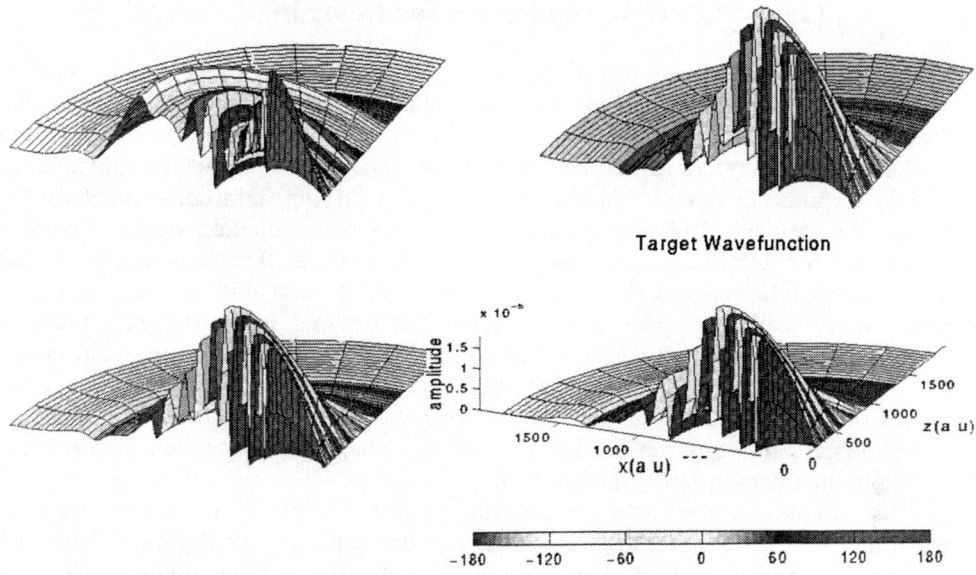

Figure 7. Learning feedback control of a Rydberg wave packet. The desired target wave packet is shown on the lower right. The initial guess produced by the pulse shaper is on the upper left. After two iterations, shown on the upper right and lower left, the experiment converges to the target shape.

The amplitude and relative phase of each eigenstates in the wave packet was determined by quantum holography and state-selective field ionization. Then, the system computer calculated the difference between each of these amplitudes and phases, and those of the desired target state. These phase and amplitude differences were then added to the phase and amplitude of each corresponding excitation wavelength in the Bragg modulator. This simple linear feedback algorithm is effective for two reasons: First, the system requires full access to the reconstruction information. The Rydberg system accomplishes this via quantum state holography.

Second, there must be a simple relationship function connecting the error signals and the correction information. This second requirement is particularly easy for the Bragg modulator, where each eigenstates uses a different wavelength, which traverses a different part of the device. In fact, convergence is very rapid, at least in principle, since the correction phase is the precise phase change for the corresponding excitation wavelength. A second iteration is only needed because of imprecise knowledge of the Bragg modulator response function. Nonlinearities in the Bragg detector response and other technical control problems are automatically corrected in the feedback loop.

One important consequence of automatic learning control is that the system is capable of adapting to changing initial conditions and maintaining a fixed output (the target wave function) with a changing input. We have demonstrated this by subjecting the pulse shaper to pulse shape changes or additional delays. The learning feedback control has no difficulty maintaining the target wave function, so long as the time delay between the shaped wave packet and the reference pulse is short enough to guarantee coherence

Genetic Search Algorithms for Learning Control

The linear feedback described in the preceding section is only effective for systems with a known Hamiltonian and eigenstates spectrum, and for laser interactions in the weak-field limit. The more general situation is more complicated, but also more useful. Nonlinear optical interactions are used to generate new coherent light sources, and also to probe dynamics. In principle, a strong driving field could be shaped to enhance the desired nonlinear interaction, so that pulse shaping could become a general tool for nonlinear dynamics. One obstacle is that the Hamiltonian for any complex molecular system is not known well enough, and with out it one cannot predict likely excitation pathways or derive the optimal pulse shapes. Learning algorithms may offer a solution to these problems. In this approach, the experiment runs itself by means of an intelligent feedback loop. It tries various pulse shapes, assesses their success in achieving the desired target excitation, and uses the knowledge gained in this way to improve the pulse shapes on subsequent experiments, all without the intervention of the researcher.

In its simplest form, learning control simply optimizes the pulse shape by controlling dispersion in the laser amplifier and the nonlinear medium, or by filtering the spectrum. The more general problem of pulse shaping is more daunting: Consider that the pulse shaper can control the amplitude and phase of well over 100 different frequency components with 8 bits of resolution (i.e. 256 different amplitudes) per frequency component. Therefore there are over 256^{100} different pulse shapes to try. Furthermore, each frequency component is not necessarily independent of the phase and amplitude of the others, so they cannot each be optimized independently. Additional requirements are that the algorithm be robust in the face of experimental noise and be capable of escaping local maxima in an enormous and rough potential energy landscape.

We have been studying learning control in several different nonlinear systems, using a search strategy known as the genetic algorithm (GA).[11] "Genetic" in this case means that the algorithm creates new pulses through a non-local approach based on splicing together traits of successful "parent" laser pulses, rather than by following a fitness gradient function, as in other evolutionary methods

In our GA implementation, each individual corresponds to a pulse shape, which is encoded as a string of floating point numbers (the individual genome) specifying the phase and amplitude at the various frequency components of the laser pulse. In the first generation, the population consists of sixty individual pulse shapes, chosen at random. We have studied quantum dynamics where the target measurement depended on the shape of the driving laser field. The experiment is performed for each pulse shape in turn, and the results assigned a numerical fitness value based on the result. This fitness value determines the chances that a particular pulse shape is selected to reproduce. For example, "roulette wheel" selection, an individual's reproduction probability is proportional to its fitness.

Adaptive Operators

Reproduction involves modifying and combining elements of previous individuals to create new individuals. Operators that act on the genomes of the pulse shapes carry this out. Operators that we use include *multi-point crossover, mutation, averaging, creep, smoothing, choice of genome basis, and polynomial phase mutation.*
Here is a brief description of how the different operators work:

- *Crossover* exchanges one or more sections of the genome from each of two or more parents. The resulting two gene strings are the two new children pulse shapes.
- *Mutation* randomly alters individual genes in the genome.
- *Averaging* produces children by averaging the gene values of two or more parents.
- *Creep* is mutation where the final gene value is constrained to fall close to the parent value.
- *Smoothing* averages nearby gene values in the genome.
- *Choice of basis* changes the basis vectors of the genetic code. For example, the genome representing frequency and phase might be replaced with its inverse Fourier transform genome, where the pulse shape is recorded as amplitude vs. time.
- *Polynomial-phase mutation* produces children by replacing a portion of a gene string with a polynomial fit.

A well-chosen set of operators can greatly enhance the performance of the GA and lend additional physical insight. But the proper choice is usually far from obvious, so we allow the algorithm to adapt itself by letting it choose how often to use a given operator to produce new children. The use of adaptive operators helps speed up convergence, and, perhaps more importantly, it helps shed light on the control mechanism at work. The reproductive operators compete to produce new pulse shapes. In other words, better operators produce more offspring.

A lower bound ensures that every operator always has some chance of being used during reproduction, a necessity if the operator is ever going to be able to increase its fitness. Some operators also involve an internal parameter, such as the mutation rate.

For example, we have found that the algorithm's performance is improved if two-point crossover operator has a large initial weighting. Crossover is more effective in the beginning of the algorithm when there is maximal genetic diversity, since it does a good job of mixing up the gene strings between parents. It becomes less effective as the GA converges to the best solutions, since at this point there is much less genetic diversity in the population, so there is not longer a need to drastically change the gene strings.

Control of Nonlinear Ionization of Na$_2$

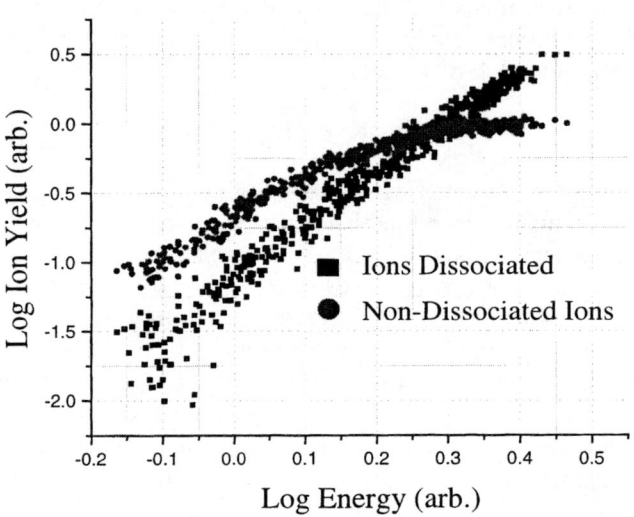

Figure 8. Ion yield vs. intensity for multiphoton ionization of Na$_2$.

The photoionization of diatomic sodium with 790 *nm* radiation provides an interesting test of the adaptive GA. There are two different final states for this process, which are reached by different nonlinear absorption pathways: *dissociative ionization* (Na$_2$ + 5γ → Na + Na$^+$ + e$^-$) and *non-dissociative ionization* (Na$_2$ + 4γ → Na$_2^+$ + e$^-$). A simple fitness function may be constructed from the ratio of [Na$^+$]/[Na$_2^+$] observed in the reaction chamber, which in our case is a supersonic molecular beam with an ion mass spectrometer. Conventional wisdom of high order perturbation theory predicts that the lower order non-dissociative process dominates at low intensities, but that the higher order dissociative rate increases faster as the intensity of the laser pulse increases. We also expect this could be modified by intermediate states, which could Stark shift into or out of lower order resonance with intensity. Indeed, a simple intensity vs. yield curve bears this out, and shows clearly how the non-dissociative process deviates as we enter the strong field regime (figure 8), in a way that favors the dissociative process even more strongly.

Recent work on resonant photoabsorption in atoms has shown that the *shape* of the laser pulse can also exert a strong influence on the absorption cross section.[12] This is because the shape of the pulse can alter the *nonlinear spectrum* (i.e. the Fourier transform of $E^n(t)$, even though it cannot change the *linear spectrum* (F.T. of $E(t)$).[13]

Could this effect also help to optimize the yield control over ionization of molecular sodium? We asked the GA to find out. Figure 9 shows the results for two different solutions, optimizing either the dissociative or non-dissociative yields.

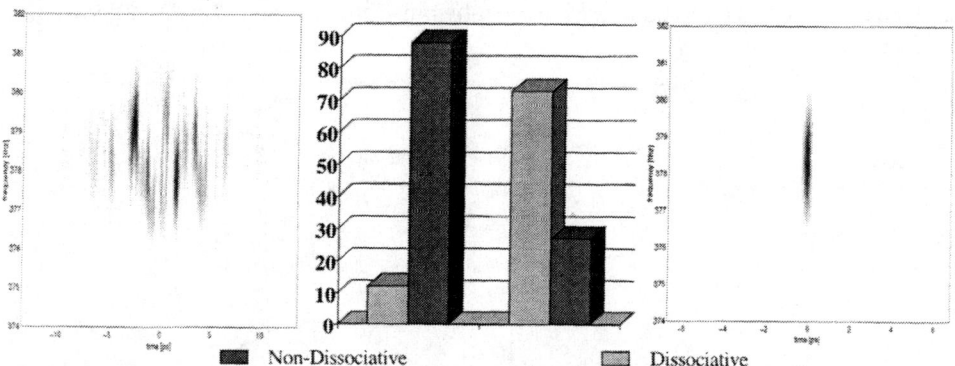

Figure 9. Optical spectrograms and yields for two solutions to multiphoton ionization of Na_2. The spectrograms give a qualitative picture of the pulse shapes, which show the laser intensity vs. time (horizontal) and frequency (vertical). Nondissociative ionization (spectrogram on the left) is favored when the pulse energy is spread over as long a time as possible, while dissociative ionization (spectrogram on the right) is favored when the light is concentrated into a very short time.

We found that pulse length, rather than pulse symmetry was the dominant mechanism for control in this case.

Selective Excitation of Quantum Modes in Molecular Liquids

Our most successful use of the GA in quantum systems thus far has been in selective excitation of C-H and C-D stretch modes in organic molecules in liquid phase.[14] Control has been demonstrated in methanol, benzene, and in mixtures of regular and deuterated benzene (C_6D_6). Liquids pose a special challenge to strong field physics, because of rapid relaxation, and also because many nonlinear processes influence the light and make it difficult to define an observable that is related to the desired fitness.

The results of control experiments in methanol illustrate some of the features of the control search algorithms. The laser pulses in these experiments were intense enough to produce self phase modulation in the $1 cm$ path length through the liquid. Stimulated Raman scattering into the C-H stretch modes created additional spectral structure on the output pulse. The Stokes shifts for the asymmetric and symmetric modes are 2942 cm^{-1} and 2847 cm^{-1} respectively, which therefore place them far outside the laser bandwidth, but within the white light continuum generated by self phase modulation. By Fourier's theorem the laser pulse duration is therefore long compared to the vibrational period of either stretch mode. In this non-impulsive regime, selective Raman modes cannot be seeded directly with the laser light. Nonetheless, we found that the GA could teach the laser to excite the symmetric and asymmetric stretch modes either together or individually by changing the shape of the driving pulse.

The fitness criterion was the integrated spectral intensity of stimulated Raman scattering into one or more Stokes peaks, which indicates excitation of the desired vibrational modes. Figure 10 shows evidence for selective excitation of the symmetric and asymmetric C-H stretch modes excited by pulse shapes found by the algorithm.

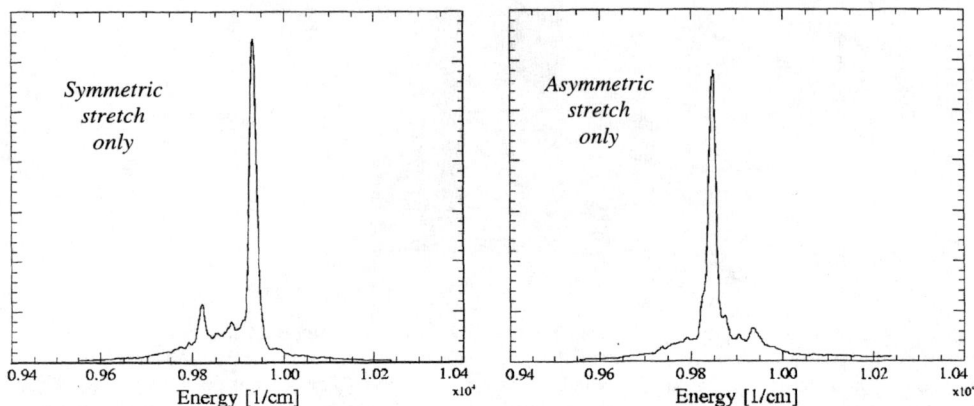

Figure 10. Selective Stokes peaks in the forward scattered light with a methanol sample. The driving laser pulse had a central frequency of 12,700 cm[-1]

Further insight into the selective excitation mechanism can be gained through examination of the shaped pulses that effected the control. Figure 11 shows the time and frequency map of the electromagnetic field pulses optimized by the GA for exciting the two C-H stretch modes in methanol.

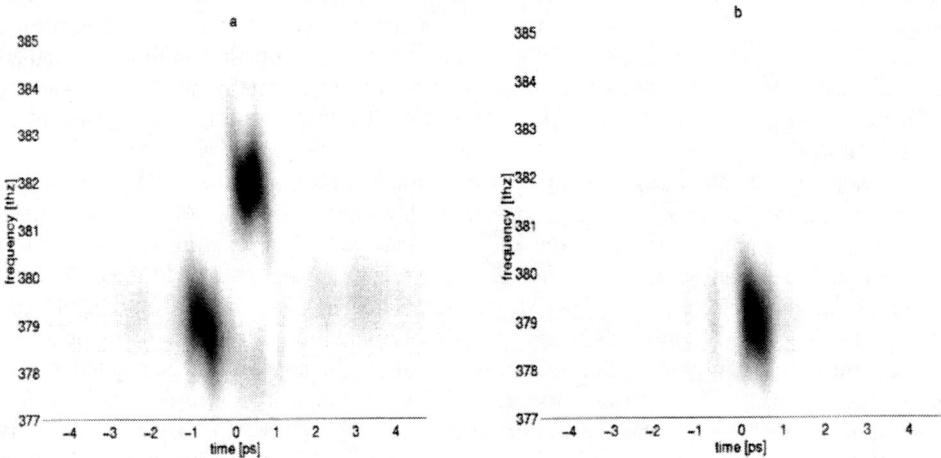

Figure 11. Spectral power distributions for shaped pulses optimized for driving the symmetric (on the left) and asymmetric (on the right) stretch modes in methanol.

476

ACKNOWLEDGMENTS

The experiments described in this paper were carried out by graduate students Thomas Weinacht, Jaewook Ahn, and Brett Pearson, and undergraduate student James White. We gratefully acknowledge support from the National Science Foundation, grant 9414335, and the Army Research Office.

REFERENCES

1 C. Spielman, P. F. Curley, T. Brabec, and F. Krausz, IEEE J. Quant Electron. QE-30 1100 (1994), and references therein.
2 D Strickland and G Mourou.. Opt. Commun., **55,** 447 (1985).
3 J. X. Tull, M. A. Dugan and W. S. Warren, Adv. Opt. Mag. Resonance 20, 1 (1990); A. M. Weiner D. E. Leird J. S. Patel, and J. R. Wullert , J. of Quantum Electronics **28** (1992).
4 T. Gallagher, Rydberg Atoms. Cambridge, Massachusetts: Cambridge Press, 1995.
5 T.C. Weinacht, J. Ahn, and P.H. Bucksbaum, Phys. Rev. Letters **80**, 5508 (1998).
6 T. Gallagher, *op. cit* 4.
7 J. Ahn, T.C. Weinacht, and P.H. Bucksbaum, Science, **287**, 463 (2000)
8 L.K. Grover, Phys. Rev. Lett. **79**, 325 (1997); Phys. Rev. Lett. **79**, 4709 (1997).
9 H. Rabitz and R. Judson, Phys. Rev. Lett. **68**, 1500 (1992).
10 T.C. Weinacht, J. Ahn, P.H. Bucksbaum, Nature **397**, 233 (1999).
11 Genetic algorithm: Davis, L. Ed. *Handbook of Genetic Algorithms*, Van Norstrand Reinhold: New York, (1991).
12 D. Meshulach, Y. Silberberg, Nature **398**, 298 (1998).
13 P.H. Bucksbaum, Nature **396**, 217 (1998)
14 T.C. Weinacht, J.L. White, P.H. Bucksbaum, Jour. of Phys. Chem. A, **103**, 10166 (1999).

van der Waals – induced level coupling in metastable atom on surface collisions

M. Boustimi, B. Viaris de Lesegno, F. Perales, J. Reinhardt,
J. Baudon, J. Robert and M. Ducloy

Laboratoire de Physique des Lasers (UMR 7538, CNRS), Université Paris 13
99 Avenue J.B. Clément, 93430 – Villetaneuse, France

Abstract. The symmetry of atomic wave functions is broken by the quadrupolar component of the van der Waals interaction between the atom and a planar surface. This results into a coupling able to mix levels of same parity such as $^3P_0$ and $^3P_2$ metastable levels of heavy rare gas atoms. The $^3P_0$-$^3P_2$ transition in Ar and Kr has been evidenced in a time-of-flight experiment using as a surface the edge of a copper slit.

INTRODUCTION

The atom-surface van der Waals interaction originates from the interaction between the fluctuating atomic dipole and the polarisation charges induced in the solid, i.e., in the non-retarded regime, the instantaneous image of the atom [1]. It depends on the atom-surface distance z as z^{-3}. Because of the cylindrical symmetry around the normal to the surface, it is anisotropic and contains a quadrupolar part proportional to $D_z^2 - D^2/3$, where D is the atomic dipole operator [2]. This off-diagonal term, *via* virtual couplings to higher levels, can mix atomic states of same parity, such as metastable states $^3P_0$, $^3P_2$ of rare gas atoms heavier than He. In the thermal energy range ($E_0 \sim 60$ meV), the $^3P_2$-$^3P_0$ transition corresponds to a closed channel : the only exoenergic reversed transition is allowed to occur. For Argon and Krypton atoms, the energy gain ($\Delta E = 174$ meV and 650 meV, resp.) is large compared to the initial kinetic energy E_0. Actually this energy is gained by the external motion in a direction normal to the surface, which gives rise to a large deflection angle. For an atom of velocity V_0 impinging the surface at a small angle θ_0 (see fig.1),the final velocity is :

$$V_f = (V_0^2 + 2\Delta E / m)^{1/2} \qquad (1)$$

CP551, *Atomic Physics 17*, edited by E. Arimondo, P. DeNatale, and M. Inguscio
© 2001 American Institute of Physics 1-56396-982-3/01/$18.00

where m is the atomic mass. The final velocity makes with the surface an angle θ_f given by the momentum conservation rule :

$$V_0 \cos \theta_0 = V_f \cos \theta_f \qquad (2)$$

The deflection angle is well defined as soon as the initial velocity V_0 is selected. These specific features of the kinematics have allowed us to observe the $^3P_0$ - $^3P_2$ transition in spite of a very low counting rate [3].

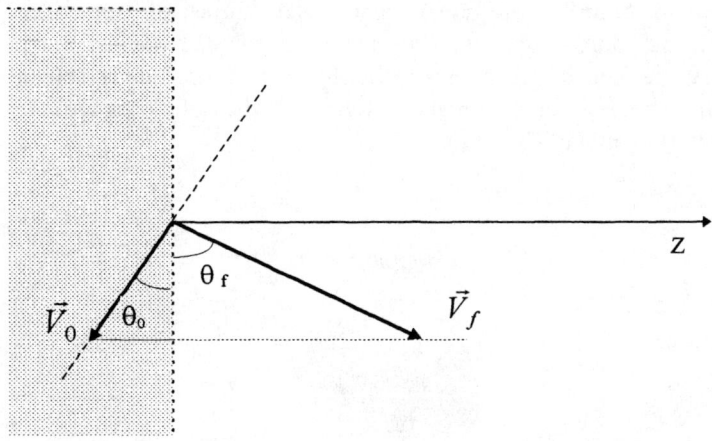

FIGURE 1. Velocity diagram of the collision $^3P_0 \rightarrow {}^3P_2 + \Delta E$; the initial velocity is $\vec{V}_0$, the final one is $\vec{V}_f$.

EXPERIMENT

The principle of this experiment is quite simple (see fig 2) [4]. A metastable atom beam is past through a copper slit (width :100 μm , thickness :50 μm), one edge of which plays the role of a surface. Metastable atoms (Ar*, Kr*) are produced by electron bombardment of an effusive beam. The resulting velocity distribution is wide (it is of a Maxwellian type). In view of performing time of flight measurements and reducing the angular range of the inelastic process, a velocity selection is made by using a pulsed bombardment (duration :30-40 μs, period : 2 ms) and a slotted disk synchronized with the source. A relative velocity spread of about 10% is achieved by this double chopping technique. It should be noticed that the incident beam consists of a mixture (1:5) of states $^3P_0$ and $^3P_2$. However, as mentioned before, atoms in state $^3P_2$ cannot participate to any inelastic process. Because of the weakness of the expected signal (3-5 10^{-3} cnt/s), the detector is a very important element of the experiment. This detector consists of a polarised secondary-emission Aluminium plate followed by a channeltron. It can be rotated around the slit axis from -30° up to +75°. The efficiency of the detector is estimated to be about 30%. The noise of the detector itself, measured with no gas, no source filament heating nor voltage, is extremely low (about 10^{-4} cnt/s). Unfortunately this noise is not the only background contribution since at such low counting rates any detectable particle directly or indirectly reaching the detector (i.e. the secondary-emission plate or the channeltron itself) contributes to the signal. In order to get an estimate of this background contribution under normal conditions of operation, the delay time between the source pulse and the opening of the disk slit has been chosen in such a way that neither UV photons from the source nor metastable atoms could pass through. The counting rate measured over the whole range of a time of flight spectrum (0-1 ms) is 10^{-3} cnt/s.

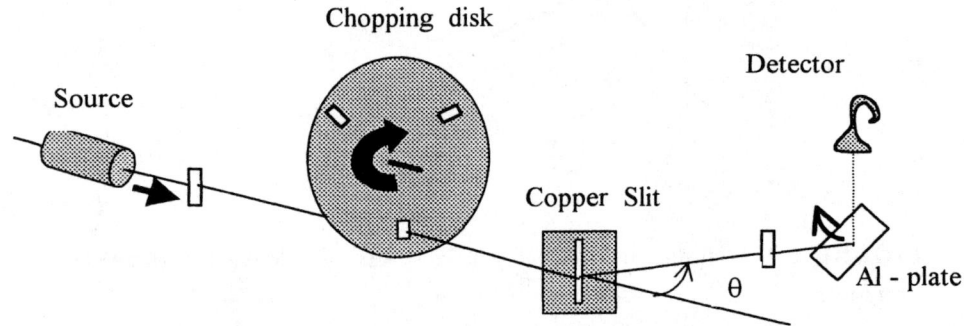

FIGURE 2. Experimental set up : the source of metastable atoms consists of a pulsed electron bombardment of an effusive beam ; the slotted disk is synchronised with the source pulse (velocity selection); the edge of the copper slit is used as a surface ; the rotatable detector consists of a secondary-emission plate (Al plate) followed by a channel electron multiplier. The distance from the source to the slit is 121 mm, the distance from the slit to the Al plate is 164 mm.

Once chosen a selected incident velocity (V_0 = 535 and 356 m/s for Ar*, 436 m/s for Kr*), time-of-flight spectra have been accumulated (1-2 days each) for a series of deflection angles ranging from 40° to 75°. Figure 3 shows examples of such spectra obtained with Kr* atoms (V_0 = 436 m/s), at scattering angles 60°, 73°. It is seen that a

" super elastic " peak, at time of flight $t_{inel} = L_0/V_0 + L_1/V_f$ (where $L_0 = 121$ mm is the distance from the source to the slit and $L_1 = 164$ mm is the distance from the slit to the detector), only appears at the predicted deflection angle $\theta_f \approx 73°$. The width of this peak is mainly determined by the incident velocity spread.

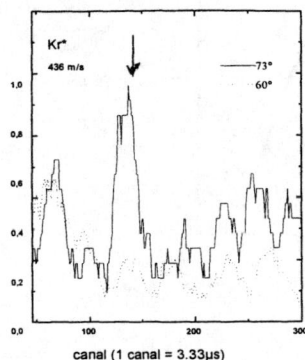

FIGURE 3. Time-of-flight spectra with Kr* metastable atoms ($V_0 = 436$ m/s) at angles $\theta = 60°$ (dotted line) and 73° (full line). The vertical arrow shows the time-of-flight calculated for the inelastic process.

The acquisition times needed to get a reasonable signal-to-noise ratio are very long (about 2 weeks for one V_0 value). Therefore because of unavoidable experimental drifts a direct comparison of the spectra would be meaningless. To make a relative calibration we have used the relatively intense peak (0.1 and 0.8 cnt/s for Ar* and Kr*) of UV photons reaching the detector at time zero even when the source-disk delay is adjusted to let pass the atoms. Time-of-flight spectra have been accumulated at the same angles as before, over much shorter times (1000-1500 s). Once calibrated the areas of the inelastic peaks have been corrected from the background and also from the variation of the angular acceptance as a function of the deflection angle θ. Indeed the opposite edge of the slit eliminates atoms leaving the surface at a too large angle. Finally one gets the inelastic intensity I_{in} as a function of θ, i.e. the differential cross section (DCS) of the process. Figure 4a shows these DCS-s for Ar* atoms at $V_0 = 535$ and 356 m/s and figure 4b the DCS for Kr* atoms at 436 m/s. Both the location of the inelastic peak and the sharp peaking of the DCS-s are signatures of the $^3P_0$-$^3P_2$ transition.

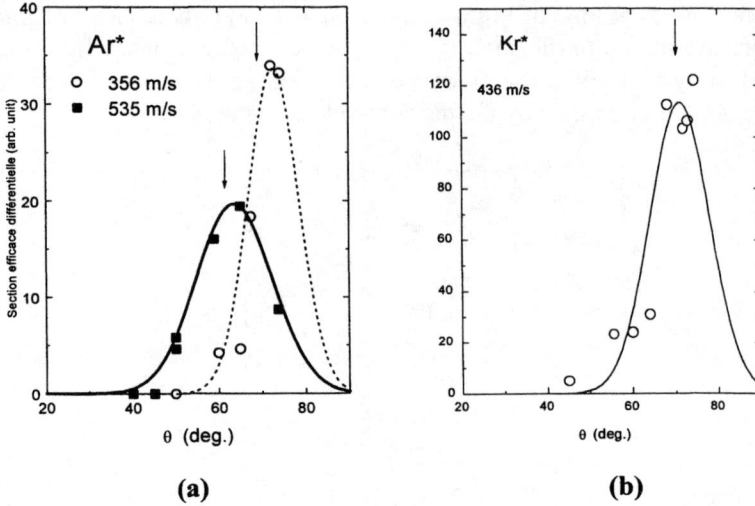

FIGURE 4. Inelastic differential cross sections in relative values as functions of the scattering angle θ: (a) Ar* atoms at 535 m/s (full squares) and 356 m/s (open circles) ; (b) Kr* atoms at 436 m/s. The vertical arrows indicate the predicted angle (θ_f).

It is worth noticing that the angular width of the DCS-s is larger than that of the detector (4°) added to the angular spread $\delta\theta_f$ originating from the velocity spread δV_0 of the incident beam :

$$\delta\theta_f = sin(2\theta_f)\frac{\delta V_0}{2V_0} \tag{3}$$

$\delta\theta_f = 2.5°$ and 2° for Ar* at 535 and 356 m/s, and 1.3° for Kr* at 436 m/s. The excess of angular spread in the DCS-s is probably due to the surface corrugation.

THEORY

In the present case the only coordinate along which the external motion is not free is z normal to the surface. For this coordinate the interaction includes a scalar potential $-C/z^3$ in the incoming channel ($^3P_0$), a scalar potential of the same type in the outgoing channel ($^3P_2$) and a coupling (assumed to be a real quantity), Γ/z^3. In the following we shall assume a common constant C for both scalar potentials. Obviously these forms of interaction are no longer valid when the distance to the surface becomes too small ($z <$ 2-3 au) since the metastable state is quenched with a probability very close to unity. This

482

feature can be rather easily taken into account in a semi-classical treatment involving atomic trajectories. In a time-independent treatment, asymptotic behaviours at large z must be imposed as follows : (i) in the $^3P_0$ channel an incident plane wave propagating towards the surface (the reflection factor is 0 and the surface is a perfect absorber), (ii) in the $^3P_2$ channel, an outgoing plane wave along $z > 0$. In the present paper we shall present only the semi-classical approach. The time-independent treatment will be presented in a forthcoming paper.

Semi-classical approach

Such an approach is justified because of the relatively large energies involved in our problem. However a difficulty usually encountered in the treatment of inelastic atomic collisions occurs, namely the absence of a common trajectory for both channels. The classical trajectory in the entrance channel for an impact parameter ρ is easily found to be :

$$\sqrt{\frac{2C}{m}}\, t = \rho^{5/2}\, F(z/\rho) \quad ; \quad x = V_0\, t \tag{4}$$

where :

$$F(u) = \int_u^1 dy \left(y^{-3} - 1\right)^{-1/2}, \, u \le 1 \tag{5}$$

This trajectory ends on the surface at time $t_o = F(0)\left(\dfrac{m}{2C}\right)^{1/2} \rho^{5/2}$. If the transition occurs at a time $\tau(0 \le \tau \le t_o(\rho))$, then an amount ΔE is abruptly added to the kinetic energy and the atom follows with a higher velocity a new trajectory asymptotically leading to the deflection angle θ_f. We shall assume that once the atom follows this new trajectory, it has almost no chance to experience the reversed transition. Obviously this assumption eliminates the problem of the absence of a common trajectory. To mimic the fast elimination of the atom once it is in state $|b\rangle$, a width γ is attributed to this state, $1/\gamma$ being comparable to a typical exit time, i.e. a few 10^{-4} au. The sensitivity of the results to the value of γ will be examined further. Under such conditions the evolution of the internal state of the atom is governed by a time-dependent effective Hamiltonian which is modelised in the basis set $\{|a\rangle, |b\rangle\}$ by the non hermitic matrix (in atomic units) :

$$H = \begin{pmatrix} \dfrac{\delta}{2} & \varsigma(t) \\ \varsigma(t) & -\dfrac{\delta}{2} - i\gamma \end{pmatrix} \qquad (6)$$

States a and b are respectively $^3P_0$ and $^3P_2$. The coupling term is $\zeta = \Gamma/z^3$ and δ is simply ΔE in atomic units ($\delta = 0.0064$ au for Ar*). Let $\alpha(t)$, $\beta(t)$ be the components of the state on $|a\rangle, |b\rangle$. These amplitudes obey the coupled equations :

$$i\,\dot\alpha = \frac{\delta}{2}\alpha + \varsigma\,\beta$$

$$i\,\dot\beta = -\left(\frac{\delta}{2} + i\gamma\right)\beta + \varsigma\,\alpha \qquad (7)$$

These equations are easily decoupled to give for instance :

$$\ddot\alpha + \left(\gamma - \frac{\dot\varsigma}{\varsigma}\right)\dot\alpha + \left(\varsigma^2 + \frac{\delta^2}{4} + i\frac{\delta}{2}\left(\gamma - \frac{\dot\varsigma}{\varsigma}\right)\right)\alpha = 0 \qquad (8)$$

where the only time dependent coefficients are ς and $\dot\varsigma$, through $z(t)$.

Owing to the rapidly oscillating character of the solution, a slow-envelop or WKB approximation can be used. Let us assume a solution of the type :

$$A(t)\ exp\left[i\int^t \omega(t')\,dt'\right] \qquad (9)$$

where $A(t)$ is a slow envelope (compared to the exponential) and ω is a root of :

$$\omega^2 - i\left(\gamma - \frac{\dot\varsigma}{\varsigma}\right)\omega - \left(\varsigma^2 + \frac{\delta^2}{4} + i\frac{\delta}{2}\left(\gamma - \frac{\dot\varsigma}{\varsigma}\right)\right) = 0 \qquad (10)$$

Neglecting $\ddot A$ in (8), one gets :

$$A(t) = C\ exp\left[-\int^t \frac{\dot\omega}{2\omega - i(\gamma - \dot\varsigma/\varsigma)}\,dt'\right] \qquad (11)$$

It turns out that the quantity $\dot\varsigma / \varsigma$ is generally small compared to γ. Under such conditions, $A(t)$ simplifies into a familiar WKB amplitude :

$$A(t) = C \left[4\varsigma^2 + (\delta - i\gamma)^2 \right]^{-1/4} \tag{12}$$

Now one is able to explicitly calculate $\alpha(t)$ and $\beta(t)$, subjected to the initial conditions $\alpha(0) = 1$, $\beta(0) = 0$. As we have mentioned before these solutions rapidly oscillate with time, especially when z is small, i.e. within the range where the coupling is the most efficient. Apart from a rather small oscillating part, the squared norm of the atomic state : $N^2 = |\alpha|^2 + |\beta|^2$ monotonously decreases from $t = 0$ to $t = t_i$, t_i being the time at which the atom reaches the distance z_i beyond which it is quenched. Therefore the quantity $P(\rho) = 1 - N(t_i)^2$ is the probability that the transition occurs along the trajectory of impact parameter ρ.

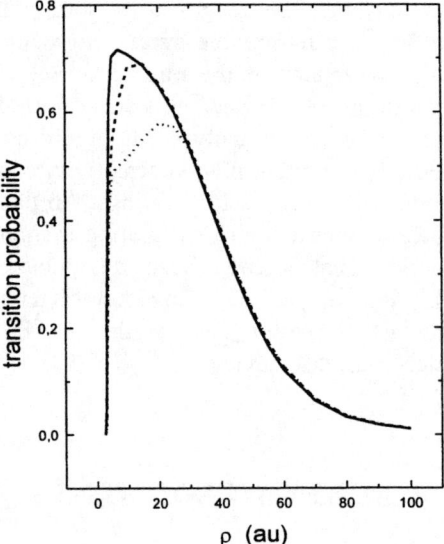

FIGURE 5. Transition probability P as a function of the impact parameter ρ. Full line : $\gamma = 10^{-3}$ au ; dashed line : $\gamma = 10^{-4}$ au ; dotted line : $\gamma = 10^{-5}$ au.

$P(\rho)$ has been calculated using the following sets of parameters (in atomic units) : $C = 4$; $\Gamma = 0.3$; $z_i = 3$; $\gamma = 10^{-5}$, 10^{-4} and 10^{-3} au. The results are shown in fig.5. They show to be relatively insensitive to the value of γ. The integral $\int_{z_i}^{\infty} P(\rho)\, d\rho$ represents a mean impact parameter ρ_m for the transition, and the ratio ρ_m / w , where w is the width of the slit, is equal to the ratio I_{inel} / I_0 , where I_{inel} is the total outgoing flux in the inelastic channel and I_0 the incident flux. In the present case the calculation gives a value for ρ_m of about 28 ± 5 au which gives a ratio $I_{inel} / I_0 \approx 1.46\ (\pm\ 0.15)\ 10^{-5}$, a value in good agreement with the experiment ($1.5 \pm 0.5\ 10^{-5}$) which warrants the validity of the model and the choice of the parameters in our calculation.

CONCLUSION

We have experimentally demonstrated transitions between metastable levels of Argon and Krypton atoms induced by a off-diagonal term of the van der Waals interaction with a Copper surface. The results of a semi-classical treatment of the inelastic collision problem are in a reasonable agreement with experiment. It is worth noticing that because of the symmetry of the interaction, $^{3}P_2$ atoms produced by the transition are fully polarised in the M = 0 Zeeman state (referred to the z axis normal to the surface). This effect has not been yet analysed. It should be rather easily evidenced by means of an optical method. The experimental sensitivity should be highly improved by the use of metallic or dielectric transmission gratings. An experiment on the inelastic diffraction of Ar* atoms by a dielectric micro-slit grating is in progress. The importance of van der Waals interactions has already been recognised in ground state atom interferometry [5]. Similar but larger and somewhat different coherent effects are expected in the case of metastable atoms, such as the $^{3}P_0$-$^{3}P_2$ transition studied here. They could lead to new interferometric devices.

ACKNOWLEDGMENTS

The authors thank P. Berman for enlightening discussions about theoretical aspects. M. Boustimi thanks the Association Louis de Broglie d'Aide à la Recherche for providing him with a Nicolas-Claude Fabri de Peiresc grant.

REFERENCES

1. Lennard-Jones J.E., *Trans. Faraday Soc.*, **28**, 333 (1932)
2. Ducloy M., *Nanoscale Science and Technology*, edited by N. Garcia *et al.*, Kluwer, the Netherlands, 1998, p. 235
3. Surface-induced inelastic rotational transfers have already been studied. See for instance : Whaley K.B. et al., *J. Chem. Phys.*, **83**, 4235 (1985)
4. A preliminary report on this work has been presented at the 6th *French-Israeli Symposium on Nonlinear and Quantum Optics* (*FRISNO6*), Les Houches, France (2000)
5. Grisenti R.E. *et al.*, *Phys. Rev. Lett.*, 83, 1775 (1999)

AUTHOR INDEX

S

Sackett, C. A., 173
Salomon, C., 23
Santarelli, G., 23
Savalli, V., 397
Schmidt-Kaler, F., 130
Scully, M. O., 204
Sengstock, K., 426
Silver, J., 282
Sokolov, A. V., 189
Sortais, Y., 23
Stamper-Kurn, D. M., 103
Steane, A. M., 158
Stevens, D., 397

T

Taichenachev, A. V., 204
Tamarat, P., 121
Thompson, J. K., 73
Turchette, Q. A., 173

U

Udem, T., 58

V

Velichansky, V. L., 204
Vernooy, D. W., 103
Viaris de Lesegno, B., 478

Vuletić, V., 356

W

Walker, D. R., 189
Westbrook, C. I., 397
Westbrook, N., 397
Wicht, A., 43
Wieman, C. E., 325
Wineland, D. J., 173
Wohlleben, W., 442

Y

Yavuz, D. D., 189
Ye, J., 103
Yin, G. Y., 189
Young, B. C., 43
Young, L., 367
Yudin, V. I., 204

Z

Zhang, S., 23
Zibrov, A. S., 204